# Transformative Design in Healthcare

Ali Akbar Firoozi · Ali Asghar Firoozi

# Transformative Design in Healthcare

## Infection Control and Patient Well-Being Through Architectural Innovation

Springer

Ali Akbar Firoozi
Department of Civil Engineering
Faculty of Engineering and Technology
University of Botswana
Gaborone, Botswana

Strategic Research Institute (SRI)
Asia Pacific University of Technology
and Innovation (APU)
Kuala Lumpur, Malaysia

Ali Asghar Firoozi
Department of Civil Engineering
Faculty of Engineering
National University of Malaysia (UKM)
Selangor, Malaysia

ISBN 978-3-032-08018-9 ISBN 978-3-032-08019-6 (eBook)
https://doi.org/10.1007/978-3-032-08019-6



This Springer imprint is published by the registered company Springer Nature Switzerland AG
The registered company address is: Gewerbestrasse 11, 6330 Cham, Switzerland

# Preface

In the ever-evolving field of healthcare, the role of architecture in shaping the outcomes of medical care and patient experiences cannot be overstated. *Transformative Design in Healthcare: Infection Control and Patient Well-Being Through Architectural Innovation* emerges at a critical juncture in the global healthcare landscape, where the built environment's intrinsic value in promoting health, preventing disease, and enhancing patient and staff well-being is increasingly recognized. This book aims to delineate the pivotal intersection of architectural innovation, healthcare delivery, and infection control, highlighting how strategic design elements can significantly influence the efficacy of healthcare environments.

The impetus for this comprehensive exploration is derived from the recent global challenges posed by infectious diseases, notably the COVID-19 pandemic, which has underscored the necessity for healthcare facilities that not only respond to the immediate demands of patient care but also anticipate and mitigate the risks associated with healthcare-associated infections. The lessons learned from such global health crises have catalyzed a reevaluation of traditional healthcare design paradigms, pushing towards environments prioritizing infection control without compromising the therapeutic and human-centric aspects of care settings.

This book is co-authored by Dr. Ali Akbar Firoozi and Dr. Ali Asghar Firoozi, who bring together a wealth of expertise from their extensive backgrounds in civil engineering with a focus on sustainable and resilient infrastructure. Dr. Ali Akbar Firoozi, based at the University of Botswana, and Dr. Ali Asghar Firoozi, stationed at the National University of Malaysia (UKM), have collaborated to bridge their insights from diverse geographic and socio-economic contexts, offering a unique perspective on how architectural design can be leveraged to enhance health outcomes globally.

Our discussions extend beyond conventional architectural solutions, venturing into innovative strategies that harness the latest advancements in materials science, ventilation systems, and space utilization to foster inherently safer environments and more conducive to healing. Each chapter of this book is meticulously crafted to introduce readers to a spectrum of design considerations—from the micro-environmental controls necessary for infection mitigation to the macro-scale planning of healthcare facilities that integrate seamlessly into community health networks.

Furthermore, *Transformative Design in Healthcare* delves into architectural elements' psychological and emotional impacts on all facility users, recognizing that the well-being of patients and healthcare staff is symbiotically linked to their physical surroundings. Through a blend of case studies, theoretical analyses, and evidence-based design principles, the book provides healthcare administrators, architects, and planners with actionable insights aimed at creating more resilient and adaptive healthcare environments.

As co-authors, we extend our deepest gratitude to our academic and professional colleagues, whose invaluable insights and critiques have significantly shaped the narrative and depth of this book. We also acknowledge the contributions of healthcare professionals who, through their day-to-day experiences in various healthcare settings, have highlighted critical areas for improvement that have informed much of our discussion.

This book is intended for a broad audience, including architects, healthcare facility managers, policymakers, and students of healthcare management, architectural design, and civil engineering. It serves as both a foundational text for those new to the field and a strategic guide for professionals seeking to enhance their knowledge of the role of design in advancing healthcare outcomes. Through this work, we aspire to inspire ongoing dialogue and innovation in healthcare facility design, to foster environments that are not only safe and functional but also uplifting and healing.

We are confident that *Transformative Design in Healthcare* will serve as a pivotal resource for all stakeholders involved in healthcare infrastructure development, urging a forward-thinking approach that places patient and staff well-being at the core of architectural and operational strategies. Our collective journey toward transformative healthcare environments continues, and we hope that this book contributes meaningfully to that endeavor.

Gaborone, Botswana
Selangor, Malaysia

Dr. Ali Akbar Firoozi
Dr. Ali Asghar Firoozi

**Acknowledgments** We express our profound gratitude to everyone who contributed to the realization of this book, *Transformative Design in Healthcare: Infection Control and Patient Well-Being Through Architectural Innovation.* This work represents not only our professional insights but also the collective wisdom of numerous individuals who have generously shared their knowledge and experiences.

First and foremost, we extend our deepest appreciation to our academic peers and colleagues at the University of Botswana and the National University of Malaysia (UKM). Their rigorous reviews and thoughtful critiques have been invaluable, pushing us to refine our arguments and enhance the quality of our work. We are also immensely grateful to the healthcare professionals and facility managers who provided real-world insights into the operational challenges and successes of healthcare environments. Their contributions have ensured that our discussions are not only theoretically sound but also practically applicable.

Special thanks are due to our editorial team and the staff at the publishing house, whose diligence and patience through countless drafts have helped shape this manuscript into its final form. Their expertise in navigating the complexities of publication has been instrumental in bringing this work to print.

We must also acknowledge the role of our educational institutions, which have supported our research endeavors and provided us with the resources necessary to undertake this significant project. Their commitment to academic excellence and innovation continues to inspire us.

On a more personal note, we are eternally grateful to our parents, whose unwavering support and encouragement have guided us throughout our lives, both personally and professionally. They have instilled in us the values of hard work and perseverance and have always believed in our potential to contribute meaningfully to our fields. This book is a testament to their love and dedication, and we hope it makes them proud.

Lastly, our thanks goes to you, the reader, for your interest in advancing healthcare through architectural innovation. We hope this book inspires you to consider how thoughtful design can make a tangible difference in the health and well-being of patients and healthcare workers alike.

Dr. Ali Akbar Firoozi
Dr. Ali Asghar Firoozi

**Conflict of Interest** The authors declare no conflict of interest regarding the publication of this book, *Transformative Design in Healthcare: Infection Control and Patient Well-Being Through Architectural Innovation.* The research and writing were conducted without any commercial or financial relationships that could be construed as potential conflicts of interest. The views expressed in this book are solely those of the authors, Ali Akbar Firoozi and Ali Asghar Firoozi, and do not necessarily represent those of their affiliated institutions or any other party. The content of this book has been written independently and with full academic freedom.

# Contents

# Chapter 1
# Architectural Therapeutics: Advancing Public Health Through Hospital Design Innovation

In this chapter, we delve into the pivotal role of modern hospital design at the intersection of architecture, health science, and technology, highlighting its profound implications for public health, patient care, and environmental sustainability. The design of healthcare facilities has evolved significantly, particularly underscored by recent global health crises like the COVID-19 pandemic, which emphasized the need for adaptable, resilient, and patient-centric environments. This chapter explores how strategic design elements such as natural lighting, air quality control, and the integration of green building practices not only enhance patient recovery rates and reduce infection risks but also contribute to operational efficiency and staff productivity. By combining a comprehensive analysis of current trends with global case studies, this monograph aims to equip healthcare professionals, architects, and policymakers with the insights needed to develop healthcare environments that promote holistic well-being and meet the complex demands of modern medical care. Through a structured exploration of evolution, current practices, and future directions in hospital design, the chapter sets the stage for understanding how thoughtful architecture can transform healthcare delivery and outcomes.

## 1.1 Scope and Significance of Modern Hospital Design

The design of modern hospital facilities represents a crucial intersection of architecture, health science, and technology, with profound implications for public health, patient care, and environmental sustainability. In an era marked by rapid technological advancements and increasing health challenges, the architecture of healthcare facilities has transcended its traditional boundaries. It now plays a pivotal role in infection control, patient recovery rates, and overall health system efficiency. The significance of hospital design has been further underscored by global health crises,



A. A. Firoozi and A. A. Firoozi, *Transformative Design in Healthcare*,
https://doi.org/10.1007/978-3-032-08019-6_1

such as the COVID-19 pandemic, which highlighted the need for adaptable, resilient, and patient-centric healthcare environments.

Modern hospital design integrates a wide array of considerations, from aesthetic and therapeutic environments to the functional requirements of advanced medical technologies. It also addresses the psychological needs of patients and the operational needs of healthcare staff, creating spaces that facilitate optimal care delivery and improve patient outcomes. The scope of modern hospital design thus encompasses not only the physical structure but also the comprehensive planning of interior spaces to promote health, safety, and well-being.

Strategic design elements such as natural lighting, noise control, and air quality have been shown to have significant impacts on patient recovery times, infection rates, and staff productivity. Additionally, the incorporation of green building practices reflects an increasing commitment to sustainability within the healthcare sector. These practices not only reduce the environmental footprint of healthcare facilities but also contribute to a healthier indoor environment conducive to healing and recovery. Thus, the design of hospitals is a dynamic field that continuously adapts to new scientific insights and societal needs, aiming to balance the complexities of medical functions with the simplicity required for universal accessibility.

Table 1.1 compares design features systematically between traditional and modern hospital architectures. The table highlights the evolution in hospital design, emphasizing elements such as natural lighting and green practices. It outlines the contrast in structural and aesthetic elements, material use, patient-focused considerations, and environmental sustainability practices. This comparison shows the advancements in architectural design and reflects the shift toward creating more patient-centric and environmentally conscious healthcare facilities. This detailed evaluation is a crucial reference for medical doctors, architects, and health administrators aiming to understand the dynamics of hospital design evolution and its implications on healthcare delivery and patient well-being.

Figure 1.1 illustrates the stark contrast between traditional and modern hospital architectures, highlighting the transformative trends in healthcare facility design. On the left, traditional hospital design is characterized by concrete and steel constructions, artificial lighting, and closed spaces, reflecting an era when functionality dominated architectural considerations. On the right, modern hospital design embraces sustainable materials, integrated technology, natural lighting, and open, adaptable layouts. These contemporary designs prioritize patient comfort, environmental sustainability, and adaptability to technological advancements, showcasing a significant evolution in the infrastructure of healthcare facilities. This comparison not only underscores the shift in architectural priorities but also indicates a broader movement toward creating health-promoting environments in medical facilities.

**Table 1.1** Comparison of features in traditional versus modern hospital design

| Feature | Traditional hospital design | Modern hospital design |
|---|---|---|
| Structural elements | Predominantly closed spaces with defined and rigid structures | Open, flexible spaces that can be easily adapted to changing needs |
| Aesthetic elements | Minimal aesthetic consideration, functional focus | Aesthetic appeal integrated with functionality, use of art, and soothing color schemes |
| Material use | Common use of concrete and steel, less emphasis on sustainability | Increased use of sustainable materials, eco-friendly construction practices |
| Natural lighting | Limited, with reliance on artificial lighting | Extensive use of natural lighting to enhance mood and energy efficiency |
| Green practices | Rare or minimal inclusion of green spaces | Incorporation of green roofs, healing gardens, and other biophilic designs |
| Patient comfort | Functional designs with limited customization for patient comfort | Patient-centric designs, and customizable features to enhance the patient experience |
| Technological integration | Basic, primarily for operational functions | Advanced technology integrated for patient care, monitoring, and facility management |

## 1.2 Objectives and Structure of the Monograph

This monograph aims to provide a comprehensive analysis of the transformative role of architectural innovation in healthcare settings, with a focus on advancing infection control and enhancing patient well-being. The objectives are twofold: to critically examine the current trends and innovations in hospital design and to offer insights into the integration of these innovations into future healthcare facilities. By doing so, this work seeks to equip architects, facility managers, healthcare professionals, and policymakers with the knowledge and tools necessary to develop healthcare environments that meet clinical needs and advance the well-being of patients and staff.

The structure of the monograph is designed to facilitate a deep understanding of the complex factors involved in hospital design. It begins with this introduction to establish the scope and significance of modern hospital architecture, followed by an exploration of the historical and current advancements in the field. Subsequent chapters delve into specific design strategies for infection control, patient-centric environments, and operational efficiencies, each illustrated with global case studies to provide practical insights and real-world applications. The monograph concludes with a discussion on regulatory and policy considerations and offers a forward-looking perspective on the challenges and future directions in hospital design. This

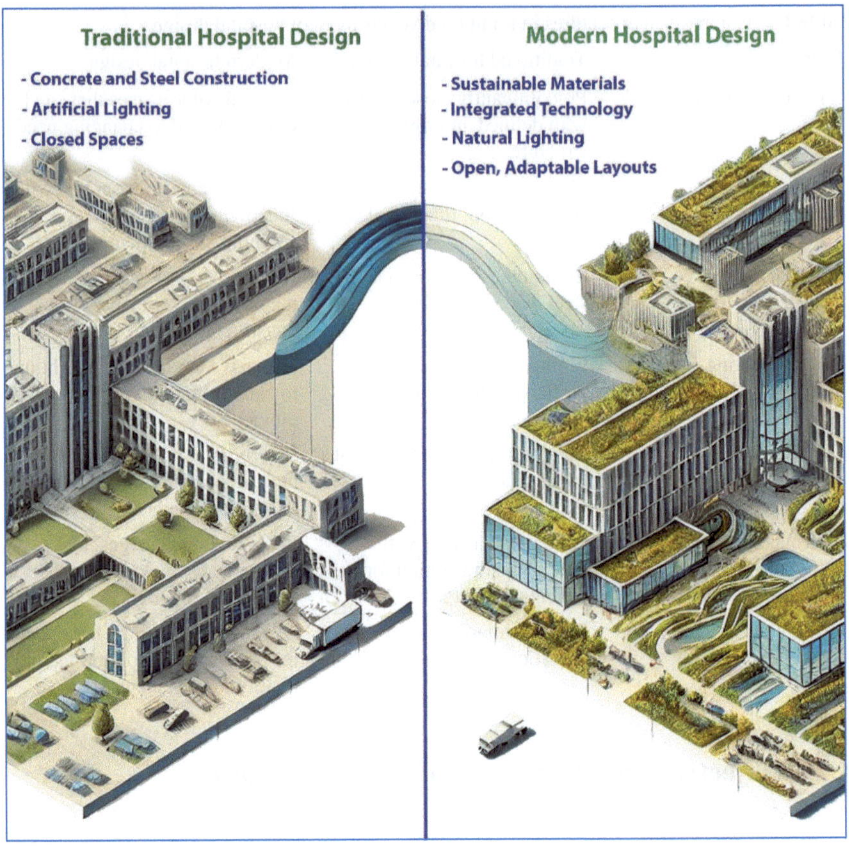


**Fig. 1.1** Comparative evolution of hospital architecture: traditional versus modern design

structured approach ensures comprehensive coverage of the topic, enabling readers to grasp the multifaceted nature of designing modern healthcare facilities.

Figure 1.2 provides a structured visual representation of the monograph, detailing the progression from a foundational introduction to detailed explorations and forward-looking analyses. The introduction sets the stage by outlining the significance of modern hospital architecture, followed by a review of historical and current advancements that have shaped hospital design. Subsequent sections delve into specific design strategies, such as infection control, patient-centric environments, and operational efficiencies, each bolstered by global case studies that offer practical insights and real-world applicability. The discussion then transitions into regulatory and policy considerations, addressing the legal frameworks and policy impacts that influence architectural practices. The monograph concludes with a discussion on future directions, anticipating upcoming challenges and innovations in hospital design. This sequential flow ensures a comprehensive understanding, equipping architects, healthcare professionals, and policymakers with the knowledge to innovate and improve healthcare environments in line with contemporary demands and prospects.

## 1.3 Evolution of Healthcare Facilities

The evolution of healthcare facilities over the centuries reflects broader changes in medicine, society, and technology. From the rudimentary infirmaries of the medieval period to the advanced, digitally integrated hospitals of the twenty-first century, the development of healthcare facilities has been inextricably linked to the evolution of medical practices and societal needs. Historically, hospitals were often places of last resort, where the sick were isolated from society. However, with advancements in medical science and public health policy, hospitals have been transformed into centers of healing, research, and education.

The twentieth century witnessed significant transformations with the introduction of antibiotics and surgical innovations, prompting changes in hospital designs to accommodate these medical advancements. The concept of specialized wards evolved, and facilities became more complex, incorporating a range of services from emergency care to long-term treatment in a single institution. The late twentieth and early twenty-first centuries saw the rise of patient-centered design, emphasizing the importance of creating environments that facilitate healing rather than merely housing the sick.

Today, the design of healthcare facilities is guided by a holistic understanding of health that includes physical, psychological, and social well-being. Modern hospitals are designed to be patient-friendly, with considerations for privacy, comfort, and access to nature, all of which are recognized for their therapeutic benefits. The use of technology has also been pivotal, with digital health records, telemedicine, and robotic surgery becoming integral to hospital operations. The ongoing evolution of healthcare facilities now focuses on flexible designs that can quickly adapt to emerging health threats, embodying the principles of resilience and adaptability that are crucial for future healthcare challenges.

## 1.4 Current Trends in Healthcare Design

In recent years, the field of healthcare design has witnessed transformative shifts that are shaping the future of hospitals and medical facilities around the globe. A deeper understanding of these contemporary trends is essential as they significantly influence how healthcare environments are planned, built, and managed. One of the most pivotal trends is the growing emphasis on patient-centered care, which focuses on creating environments that enhance patient comfort, privacy, and emotional well-being. This approach has led to designs that include more natural lighting, quieter

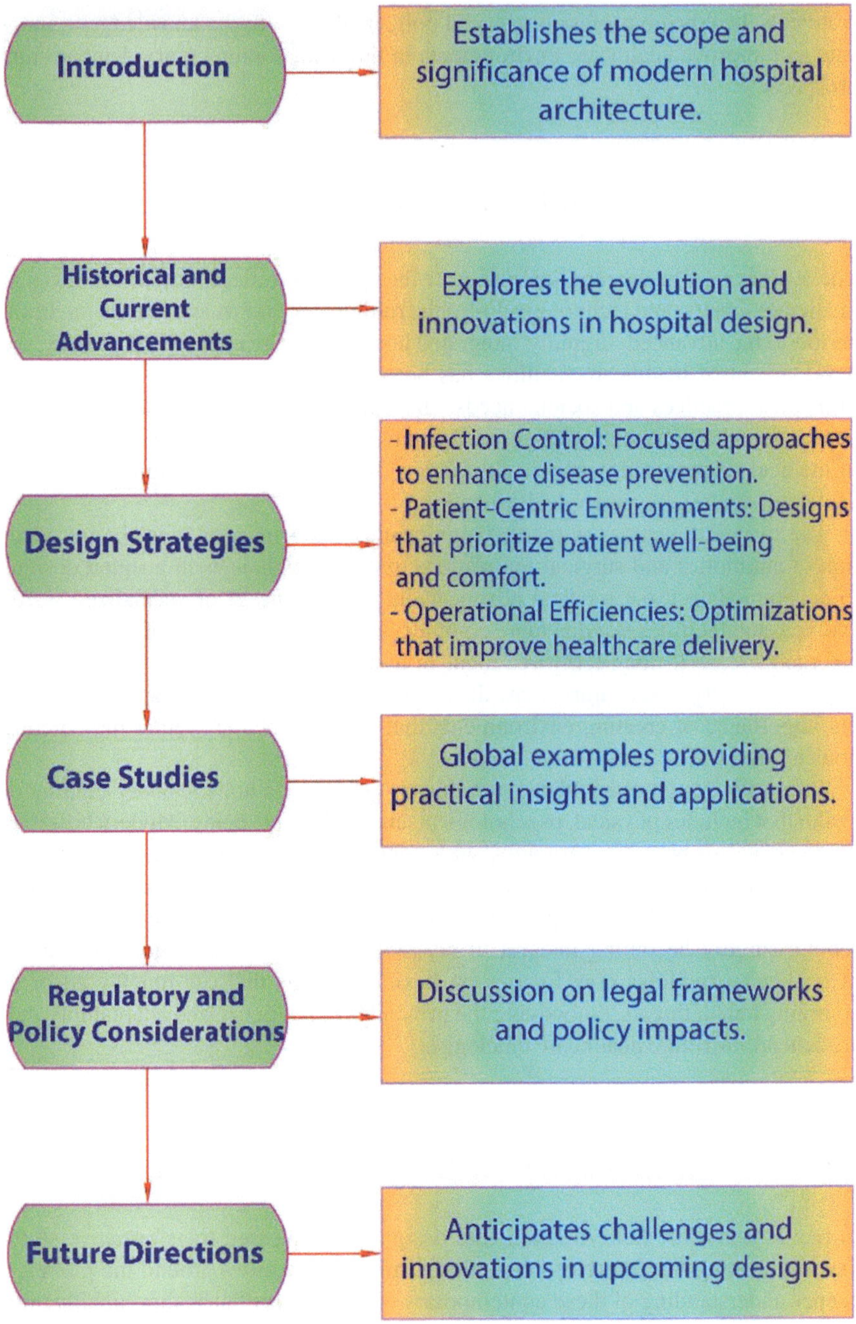


**Fig. 1.2** Structural layout of the monograph on hospital design innovations

wards, private rooms, and spaces that accommodate family members, recognizing that a supportive environment can accelerate healing and improve outcomes.

Parallel to the shift toward patient-centered designs is the rapid integration of technology into healthcare settings. From telemedicine facilities that connect patients with specialists across the globe to advanced patient monitoring systems that gather real-time data, technology is becoming an integral part of the healthcare infrastructure. These technological advances are not only improving the quality of care but are also making healthcare more accessible and efficient. As a result, architects and planners are increasingly tasked with creating spaces that seamlessly incorporate a range of technologies, from digital kiosks for patient check-in to integrated systems for electronic health records.

Another significant trend is the increasing focus on sustainability in healthcare design. This involves more than just the use of eco-friendly materials or the inclusion of green spaces within hospital grounds. Sustainable design in healthcare encompasses energy-efficient buildings that reduce the carbon footprint, water-saving fixtures that conserve resources, and waste-reduction strategies that promote recycling and minimize environmental impact. Moreover, sustainable hospitals are designed to have a positive impact on the health of their occupants by using non-toxic materials and improving indoor air quality. This holistic approach to sustainability not only aligns with global environmental goals but also supports the core mission of healthcare facilities to promote health and well-being.

Each of these trends—patient-centered care, technology integration, and sustainability—reflects a broader understanding of health that transcends traditional medical treatments. They highlight a shift toward environments that support comprehensive health outcomes, including physical, mental, and environmental well-being. As these trends continue to evolve, they will shape the fundamental principles of healthcare design, influencing everything from the architectural blueprint to the daily operations of healthcare facilities. This evolving landscape requires continuous adaptation and innovative thinking from those involved in the design and operation of healthcare environments, ensuring that facilities not only meet the current standards of care but are also prepared for future challenges and advancements.

Table 1.2 delineates various contemporary trends in healthcare design and their direct impact on healthcare outcomes. This table details key design trends including patient-centered care, technology integration, and sustainability, and examines how each influences patient recovery times, staff efficiency, and overall facility effectiveness. The insights provided in this table are essential for understanding how modern design strategies contribute to enhanced patient care, reduced environmental impact, and improved operational efficiency. By offering a comprehensive overview of each trend paired with its benefits, this table serves as a vital tool for professionals in healthcare and design sectors, aiding in the decision-making process for future healthcare projects.

**Table 1.2** Current trends in healthcare design and their impacts on outcomes

| Design trend | Description | Impact on healthcare outcomes |
|---|---|---|
| Patient-centered care | Design approaches that prioritize patient comfort, privacy, and empowerment | Improved patient satisfaction, reduced stress, and better overall patient health outcomes |
| Technology integration | Incorporation of cutting-edge technologies for diagnosis, treatment, and facility management | Enhanced diagnostic accuracy, streamlined operations, and more personalized patient care |
| Sustainability | Implementing eco-friendly materials and practices to minimize environmental impact | Reduced operational costs, better air quality, and increased community and staff satisfaction |
| Flexible design | Spaces designed to adapt to changing medical practices and patient needs | Ability to quickly respond to pandemics or changes in medical care demand, minimizing disruptions |
| Safety enhancements | Enhanced safety measures to prevent infections and accidents | Decreased infection rates, improved patient and staff safety, and compliance with health regulations |
| Biophilic design | Integration of natural elements into the building design to promote connection with nature | Decreased patient and staff stress levels, improved mental health, and shorter recovery times |
| Smart automation | Use of automated systems for clinical tasks, monitoring, and patient management | Increased efficiency, reduced human error, and better resource management |

## 1.5 Impact of Design on Healthcare Outcomes

The design of healthcare facilities plays a crucial role in determining healthcare outcomes, affecting everything from patient recovery times to staff efficiency and overall operational effectiveness. Thoughtful design not only enhances the aesthetic appeal of a facility but more importantly, contributes to the health and well-being of its users—patients, staff, and visitors alike.

A well-designed hospital environment can significantly reduce patient recovery times. Studies have shown that patients with access to natural light and views of nature tend to require less pain medication and experience shorter hospital stays. For instance, surgical recovery rooms with views of gardens or natural landscapes have been linked with improved patient attitudes, less anxiety, and fewer post-surgical complications. Similarly, incorporating elements such as enhanced natural lighting, noise control, and private rooms can dramatically improve a patient's comfort and privacy, thereby facilitating a quicker and more comfortable recovery.

The impact of hospital design extends beyond patient care to influence staff efficiency and satisfaction. Efficiently designed workspaces can reduce the physical strain on healthcare workers and decrease the time they spend moving between different areas of the facility. For example, centralized nursing stations that offer clear sightlines to patient rooms allow staff to monitor patients effectively without the need for constant physical movement. Additionally, providing dedicated spaces for staff to rest and recharge, away from the clinical environment, can help reduce burnout and improve overall job satisfaction.

Operational effectiveness is another critical aspect influenced by hospital design. The layout of a healthcare facility affects how services are delivered—from emergency responses to day-to-day patient care. Optimized layouts that minimize travel distances between frequently used spaces can enhance the speed and quality of care delivery. Moreover, designing flexible spaces that can adapt to changing needs, such as expandable operating rooms or multipurpose patient areas, ensures that the facility can respond effectively to varying healthcare demands without the need for constant physical expansions or redesigning.

Furthermore, the incorporation of sustainable design principles in hospitals not only supports environmental health but also promotes better patient and staff outcomes. The use of non-toxic materials improves indoor air quality, and energy-efficient systems contribute to creating a safer and healthier environment for everyone within the facility. This holistic approach to healthcare facility design demonstrates a commitment to health at every level, emphasizing that the physical environment is as crucial to healthcare delivery as the medical treatment itself.

In sum, the design of healthcare facilities is a critical determinant of healthcare outcomes. By aligning architectural strategies with the needs of patients and healthcare providers, facilities can create environments that are not only functional and beautiful but also inherently therapeutic, making them vital instruments in the delivery of effective healthcare.

Equation 1.1 offers a structured model for quantifying the influence of hospital design elements on patient recovery times, underscoring the critical role that architectural decisions play in enhancing healthcare outcomes. The equation is positioned within the broader discussion of how specific design variables, such as natural light, spatial layouts, and material choices, directly correlate with the speed and effectiveness of patient recovery. This quantitative analysis serves as a crucial tool for architects and healthcare professionals aiming to optimize hospital environments for better health outcomes. The correlation coefficient ($R$) derived from this equation helps in determining the strength and direction of the relationship between design elements and recovery times, providing evidence-based insights that can guide future hospital design decisions. This model not only underscores the importance of patient-centered design in healthcare settings but also provides a mathematical basis for evaluating how different design elements contribute to recovery processes. The inclusion of such precise modeling is essential for substantiating design decisions with empirical data, thereby aligning architectural practices with medical objectives to foster environments that actively support patient health and recovery.

$$R = \beta_0 + \beta_1 X_1 + \beta_2 X_2 + \ldots + \beta_n X_n \tag{1.1}$$

In this equation, $R$ represents the recovery time of patients, which is the dependent variable. Each $X$ represents different design elements within the hospital environment (e.g., amount of natural light, room size, noise levels), and $\beta$ values are coefficients that quantify the influence of these variables on recovery time. $\beta_0$ is the intercept, indicating the baseline recovery time independent of the measured variables. This model allows for a detailed understanding and analysis of how specific changes in hospital design can directly impact patient health outcomes, thereby providing a rigorous framework to support design enhancements in healthcare facilities.

## 1.6 Challenges Facing Modern Healthcare Facilities

Modern healthcare facilities are confronted with a complex array of challenges that necessitate continuous adaptation and innovation to deliver high-quality care effectively and sustainably. These challenges range from the swift pace of technological advancements to the escalating demands of healthcare services and the urgent need to address global health threats.

One of the primary hurdles is adapting to rapid technological changes. The healthcare sector is continuously transformed by advancements such as robotic surgery, telemedicine, advanced diagnostics, and electronic health records. To integrate these technologies effectively, healthcare facilities must be designed with flexibility, allowing for the incorporation of new technologies as they become available. This requires a forward-thinking architectural approach that anticipates future technological needs and incorporates adaptable infrastructure systems, ensuring that facilities can evolve without needing extensive overhauls.

Another significant challenge is managing increasing healthcare demands driven by aging populations and the rising prevalence of chronic diseases. As patient numbers grow, so do expectations for high-quality and personalized care. Healthcare facilities must therefore expand and adapt their environments to accommodate more patients while also enhancing efficiency and quality of care. This includes designing multipurpose spaces, creating scalable areas that can adjust to fluctuating patient volumes, and optimizing patient flow to minimize wait times and prevent overcrowding.

In addition to these operational challenges, modern healthcare facilities must be equipped to address global health threats, such as pandemics and antibiotic-resistant infections. This necessitates designs that prioritize infection control, facilitate effective quarantine procedures, and support large-scale vaccination efforts. Architectural solutions might include advanced air filtration systems, antimicrobial surfaces, and layouts that can be swiftly reconfigured in response to health emergencies. Moreover, facilities must be well-integrated into broader public health networks to ensure seamless coordination with other institutions and agencies during global health crises.

Environmental sustainability poses yet another challenge. With growing pressure to reduce environmental impacts, healthcare facilities are increasingly expected to adopt sustainable practices. This involves utilizing green building materials, harnessing renewable energy sources, and designing buildings that leverage natural light and airflow to minimize reliance on mechanical heating and cooling systems. Sustainability efforts also extend to retrofitting existing facilities with energy-efficient systems and technologies to meet contemporary environmental standards.

Lastly, modern healthcare facilities must navigate the cultural and social complexities of the communities they serve. This requires designing spaces that are culturally sensitive and accessible to diverse populations. Facilities need to consider linguistic diversity, cultural norms related to health and sickness, and the specific needs of various demographic groups to ensure inclusivity and accessibility in healthcare delivery.

By tackling these challenges through innovative design and strategic planning, healthcare facilities can significantly enhance their capacity to provide effective and responsive care. This proactive approach not only improves patient outcomes but also ensures that facilities remain viable and relevant in the face of a rapidly evolving healthcare landscape.

Figure 1.3 illustrates the diverse challenges encountered by modern healthcare facilities and the strategic responses designed to address them effectively. It highlights the necessity for flexible designs that can accommodate rapid technological advancements, ensuring facilities remain at the forefront of medical innovation. The figure also emphasizes the importance of expandable spaces that can adapt to the increasing demands of healthcare delivery, particularly considering aging populations and the prevalence of chronic diseases. Furthermore, it underscores the critical need for configurable layouts equipped with advanced infection control systems to manage global health threats such as pandemics. The commitment to environmental sustainability is depicted through the adoption of green building practices, which reduce the ecological footprint of healthcare facilities. Lastly, it portrays the integration of inclusive designs that ensure accessibility and cater to the cultural and social nuances of diverse communities. This comprehensive approach not only enhances the efficiency and effectiveness of healthcare delivery but also ensures that facilities can adapt to future challenges while providing high-quality care.

## 1.7 The Role of Regulatory Frameworks in Shaping Design

Regulatory frameworks play a pivotal role in shaping the design and construction of healthcare facilities, setting standards for safety, accessibility, and quality of care. These regulations ensure that healthcare environments not only support the well-being and recovery of patients but also provide safe, functional, and efficient spaces for healthcare providers and staff. Understanding these frameworks is essential for architects and planners as they navigate the complex landscape of healthcare design, ensuring that facilities meet both national and international standards.

Healthcare facilities are subject to a variety of regulatory standards that influence every aspect of their design, from structural requirements to environmental controls. For instance, building codes dictate specifications for fire safety, structural integrity,

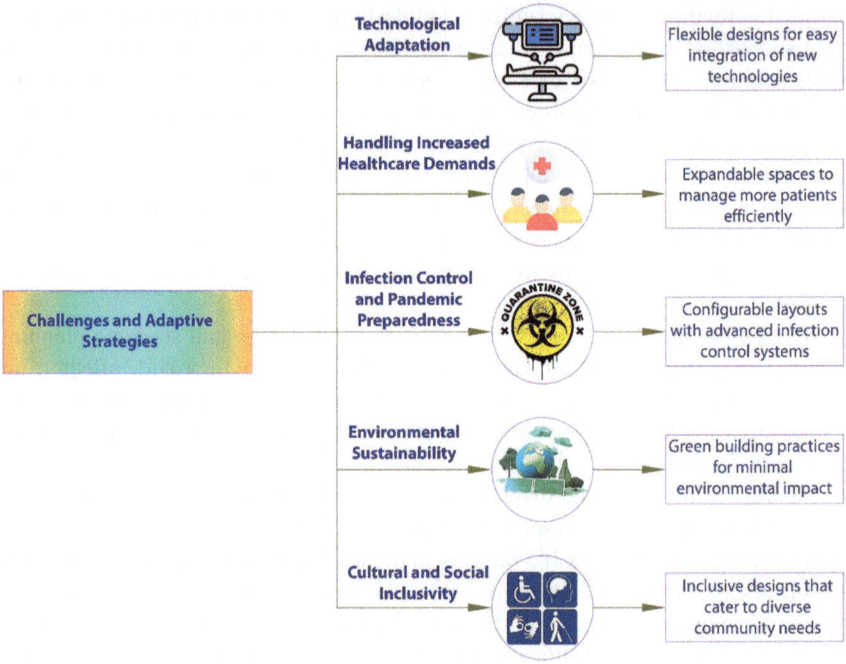


**Fig. 1.3** Strategic responses to modern healthcare challenges

and emergency egress, ensuring that facilities can protect occupants in the event of disasters. Similarly, health codes govern aspects of the operational environment within healthcare facilities, including air quality, infection control measures, and the management of medical waste, which are critical for preventing hospital-acquired infections and ensuring a safe environment for patient care.

Accessibility regulations, such as those outlined in the Americans with Disabilities Act (ADA) in the United States or similar standards globally, also significantly impact hospital design. These regulations ensure that all individuals, regardless of disability, can access and use healthcare facilities without barriers. This includes features like wheelchair-accessible entrances, corridors, and patient rooms; visually accessible signage; and auditory systems for the hearing impaired.

Furthermore, environmental regulations are becoming increasingly important as the healthcare sector seeks to reduce its carbon footprint and promote sustainability. These might include requirements for energy efficiency, sustainable building materials, and systems designed to minimize water and energy consumption. Compliance with these standards not only supports environmental sustainability but also often results in operational cost savings and improved patient and staff comfort.

In addition to complying with mandatory regulations, healthcare facilities often strive to meet voluntary certification standards that promote best practices and excellence in healthcare design. Certifications such as Leadership in Energy and Environmental Design (LEED) or the WELL Building Standard focus on various aspects of environmental and human health, offering frameworks through which facilities can demonstrate their commitment to exceeding the minimum requirements.

Navigating these regulatory frameworks requires a collaborative approach among architects, healthcare administrators, and regulatory experts. Effective communication and planning can help align the design process with regulatory requirements, ensuring that healthcare facilities not only comply with the law but also advance the goal of providing exceptional patient care. Moreover, by proactively incorporating regulatory requirements into the design process, healthcare facilities can avoid costly retrofits and redesigns, ensuring that they are well-positioned to serve their communities both now and in the future.

Table 1.3 provides a detailed comparison and summary of key regulatory frameworks from around the world and their implications for the design of healthcare facilities. This table explores various international standards, including those from the United States, the European Union, and other regions, highlighting how these regulations influence aspects such as building safety, accessibility, environmental sustainability, and patient privacy. Understanding these regulatory impacts is crucial for healthcare administrators, architects, and planners to ensure that new or renovated hospital designs comply with the required standards and best support patient care while ensuring safety and operational efficiency.

## 1.8 Global Perspectives on Hospital Design

The design of healthcare facilities is deeply influenced by cultural, environmental, and socio-economic factors that vary widely across the globe. This diversity in design principles and practices reflects not only the unique challenges and resources of each region but also their distinct cultural attitudes toward health and community. Understanding these global perspectives is crucial for appreciating the complexity and creativity inherent in hospital design and for fostering a more comprehensive approach to healthcare architecture worldwide.

In many Western countries, hospital design has increasingly emphasized patient comfort and privacy, integrating features such as single-patient rooms, family-friendly spaces, and nature-infused environments that support healing and reduce stress. These designs often incorporate advanced technology systems for health monitoring and patient care, reflecting a high-tech approach to medical treatment that is prevalent in these regions.

Conversely, in parts of Asia and Africa, community and family play a more central role in the healing process. Hospitals in these regions may include larger communal areas and facilities designed to accommodate extended family visits. The integration of traditional medicine and practices alongside Western medical technology can

**Table 1.3** Summary of key regulatory frameworks and their implications for hospital

| Country/ region | Regulatory body | Key standards | Implications for design |
|---|---|---|---|
| United States | The Joint Commission (TJC) | Safety and health regulations, patient privacy standards | Mandatory safety features, accessibility for all patients, emphasis on patient confidentiality areas |
| European Union | European Committee for Standardization (CEN) | EN 15224 (Healthcare Quality), EN 13485 (Medical Devices) | High standards for building quality, focus on infection control and patient safety |
| Canada | Canadian Standards Association (CSA) | CSA Z8000 (Canadian Healthcare Facilities) | Specifics on room sizes, air quality control, and sustainable building practices |
| United Kingdom | National Health Service (NHS) | Health Building Notes, Health Technical Memoranda | Detailed guidelines on design and operation, focus on patient dignity and staff efficiency |
| Australia | Australian Commission on Safety and Quality in Healthcare | National Safety and Quality Health Service Standards | Requirements for patient-centered care and building safety, strong focus on environmental design |
| Asia (General) | Various national standards | Varying by country, often influenced by WHO guidelines | Adaptation to local needs, emphasis on accessibility, and disaster resilience |
| Middle East | Health Authority—Abu Dhabi (HAAD), Saudi FDA | Facility guidelines and safety standards | Integration of cultural considerations, high standards for patient environment, and safety |

also be a significant feature, reflecting the cultural synthesis that characterizes the healthcare systems in these areas.

In regions with limited resources, hospital design often focuses on sustainability and efficiency. For instance, in some developing countries, facilities are designed to maximize natural ventilation and lighting to reduce energy costs and ensure comfort in the absence of advanced climate control systems. These designs might also incorporate local materials and building techniques, which not only reduces costs but also helps the facility blend with local architectural styles and community expectations.

The impact of environmental factors on hospital design is also a critical consideration. For example, in earthquake-prone areas like Japan, hospitals are designed with advanced structural reinforcements that ensure they can continue to operate in the aftermath of a disaster. In contrast, in northern Europe, where sunlight is limited for much of the year, hospital designs often focus on maximizing light exposure to combat the effects of seasonal affective disorder and improve patient and staff well-being.

This global approach to hospital design not only enhances functionality but also enriches the aesthetic and cultural relevance of healthcare facilities. By examining and integrating diverse architectural practices, designers and planners can create more adaptive and inclusive healthcare environments that reflect the global landscape of health needs and practices.

Exploring these varied approaches sets the stage for later discussions on global case studies, offering readers a richer understanding of how hospitals can serve as cultural bridges, connecting diverse practices with universal health goals. This perspective not only broadens the reader's view of healthcare architecture but also highlights the innovative ways designers respond to different cultural expectations and environmental challenges.

## Further Reading

Abdelaal, M. S., & Soebarto, V. (2019). Biophilia and Salutogenesis as restorative design approaches in healthcare architecture. *Architectural Science Review*, 62(3), 195–205. https://doi.org/10.1080/00038628.2019.1604313

Barcelo, D. (2020). An environmental and health perspective for COVID-19 outbreak: meteorology and air quality influence, sewage epidemiology indicator, hospitals disinfection, drug therapies and recommendations. *Journal of Environmental Chemical Engineering*, 8(4), 104006. https://doi.org/10.1016/j.jece.2020.104006

Demir, E., Southern, D., Rashid, S., & Lebcir, R. (2018). A discrete event simulation model to evaluate the treatment pathways of patients with cataract in the United Kingdom. *BMC Health Services Research*, 18, 1–15. https://doi.org/10.1186/s12913-018-3741-2

Fadda, J. (2019). Quality of healthcare: A review of the impact of the hospital physical environment on improving quality of care. In Sustainable Building for a Cleaner Environment: Selected Papers from the World Renewable Energy Network's Med Green Forum 2017 (pp. 217–253). *Springer International Publishing*. https://doi.org/10.1007/978-3-319-94595-8_20

Ghaffari, F., Shabak, M., Norouzi, N., & Nayyeri Fallah, S. (2023). Hospital salutogenic public spaces: a conceptual framework of effective perceptional environment quality components on patients' satisfaction. *International Journal of Building Pathology and Adaptation*, 41(5), 965–987. https://doi.org/10.1108/IJBPA-05-2021-0071

Golbazi, M., & Aktas, C. B. (2020). LEED certification and patient wellbeing in green healthcare facilities. *Journal of Green Building*, 15(4), 3–18. https://doi.org/10.3992/jgb.15.4.3

Hosseini, M. R., Martek, I., Zavadskas, E. K., Aibinu, A. A., Arashpour, M., & Chileshe, N. (2018). Critical evaluation of off-site construction research: A Scientometric analysis. *Automation in Construction*, 87, 235–247. https://doi.org/10.1016/j.autcon.2017.12.002

Jiang, B. (2019). A topological representation for taking cities as a coherent whole. *The Mathematics of Urban Morphology*, 335–352. https://doi.org/10.1007/978-3-030-12381-9_15

Jovanović, N. (2022). Hospital architecture in times of crisis. *International Review of Psychiatry*, 34(7–8), 861–867. https://doi.org/10.1080/09540261.2022.2154642

Leising, E., Quist, J., & Bocken, N. (2018). Circular Economy in the building sector: Three cases and a collaboration tool. *Journal of Cleaner Production*, 176, 976–989. https://doi.org/10.1016/j.jclepro.2017.12.010

Li, S. A., Jeffs, L., Barwick, M., & Stevens, B. (2018). Organizational contextual features that influence the implementation of evidence-based practices across healthcare settings: a systematic integrative review. *Systematic Reviews*, 7, 1–19. https://doi.org/10.1186/s13643-018-0734-5

Petrowski, K., Bührer, S., Albus, C., & Schmalbach, B. (2021). Increase in cortisol concentration due to standardized bright and blue light exposure on saliva cortisol in the morning following sleep laboratory. *Stress*, 24(3), 331–337. https://doi.org/10.1080/10253890.2020.1803265

Slama, A., Śliwczyński, A., Woźnica, J., Zdrolik, M., Wiśnicki, B., Kubajek, J., ... & Franek, E. (2019). Impact of air pollution on hospital admissions with a focus on respiratory diseases: a time-series multi-city analysis. *Environmental Science and Pollution Research*, 26, 16998–17009. https://doi.org/10.1007/s11356-019-04781-3

Twohig-Bennett, C., & Jones, A. (2018). The health benefits of the great outdoors: A systematic review and meta-analysis of greenspace exposure and health outcomes. *Environmental Research*, 166, 628–637. https://doi.org/10.1016/j.envres.2018.06.030

Ulrich, R. S., Bogren, L., Gardiner, S. K., & Lundin, S. (2018). Psychiatric ward design can reduce aggressive behavior. *Journal of Environmental Psychology*, 57, 53–66. https://doi.org/10.1016/j.jenvp.2018.05.002

Zhang, Y., Tzortzopoulos, P., & Kagioglou, M. (2019). Healing built-environment effects on health outcomes: Environment–occupant–health framework. *Building Research & Information*, 47(6), 747–766. https://doi.org/10.1080/09613218.2017.1411130

# Chapter 2
# Architectural Chronicles in Healthcare: From Ancient Sanatoriums to Modern Medical Facilities

This chapter traces the evolution of healthcare architecture from ancient healing spaces to modern, high-tech hospitals, illustrating how the physical environments of healthcare have mirrored advances in medical science and societal changes. Starting with early sanatoriums and asylums to today's sophisticated healthcare facilities, architecture has evolved to meet the increasing complexity of medical treatments and patient care. We explore the shift from hospitals as places of basic care to centers of advanced treatment and research, highlighting pivotal developments like the introduction of pavilion-style hospitals that emphasized sanitation, fresh air, and natural light principles that significantly improved patient outcomes. The chapter then transitions to modern advancements, where hospital design now incorporates patient-centered approaches, technological integrations, and sustainability. These contemporary designs prioritize patient privacy, comfort, and environmental sustainability, integrating green building practices and advanced technology to enhance operational efficiency and patient care. Through a historical lens, we provide a comprehensive overview of how architectural innovation in healthcare settings has adapted to the changing needs of medical science and patient care, setting the stage for a detailed exploration of specific design strategies and their impacts on patient safety and hospital functionality in subsequent chapters. This narrative not only enriches our understanding of hospital design but also underscores the importance of architectural evolution in enhancing healthcare delivery and patient outcomes.

## 2.1 Historical Developments in Healthcare Architecture

The history of healthcare architecture provides a fascinating lens through which to view the evolution of medical science and its physical manifestations in healthcare facilities. From the early sanatoriums and asylums to modern, high-tech hospitals,



A. A. Firoozi and A. A. Firoozi, *Transformative Design in Healthcare*,
https://doi.org/10.1007/978-3-032-08019-6_2

the architecture of healthcare facilities has evolved significantly, reflecting broader societal changes and advances in medical knowledge.

In ancient times, healing spaces were often located in temples and other religious sites, integrating spiritual care with physical health treatments. The design of such facilities was primarily focused on spiritual and natural healing, utilizing the benefits of sunlight, fresh air, and the serene environment typical of sacred sites. As medical practices advanced during the Greek and Roman periods, the concept of hospitals began to take a more structured form, although still heavily influenced by religious institutions.

The Middle Ages saw the rise of hospitals as charitable institutions run by religious orders, providing care primarily for the destitute. These early hospitals were simple in design, focusing on basic shelter and care rather than specialized medical treatment. The Renaissance period brought about a renewed interest in science and anatomy, leading to more sophisticated designs that included separate wards for different diseases and improved ventilation systems, which were believed to help prevent the spread of illness.

The nineteenth century marked a significant turning point with the advent of the Industrial Revolution, which introduced new materials and technologies, reshaping hospital architecture. The pavilion-style hospital, pioneered by Florence Nightingale, emphasized sanitation, fresh air, and natural light, dramatically improving patient recovery rates. This design principle dominated hospital architecture well into the twentieth century.

The twentieth century brought radical changes with the incorporation of technological advances in medical science. Hospitals have been transformed from passive places of care to active research and treatment centers equipped with laboratories, imaging technologies, and surgical facilities. The design of hospitals became increasingly complex, incorporating specialized facilities for diagnostics, treatment, and long-term care, all requiring sophisticated architectural solutions to support the diverse functionalities of modern medicine.

Today, the historical evolution of healthcare architecture continues to influence current design practices, reminding us of the importance of adapting design strategies to meet the changing needs of medical science and patient care. This historical perspective not only enriches our understanding of the field but also provides valuable lessons that inform contemporary architectural practices in healthcare. As we delve deeper into the modern advancements in the next section, we carry forward the wisdom gleaned from centuries of architectural development in healthcare.

Table 2.1 illustrates the evolution of architectural innovations from the Middle Ages through the nineteenth century and their impact on patient care and hospital functionality. This comparative analysis highlights pivotal changes in hospital architecture, such as the introduction of sanitary measures, centralized heating, and the design of isolated wards. Each period's architectural advancements are linked to significant improvements in patient outcomes, infection control, and overall healthcare delivery efficiency. This table serves as an insightful resource for historians, architects, and medical professionals interested in understanding how historical shifts in design philosophy have contributed to modern healthcare practices.

**Table 2.1** Key architectural innovations and their impact on patient care

| Period | Architectural innovations | Impact on patient care |
|---|---|---|
| Middle ages | Introduction of hospices with communal halls | Provided basic shelter and care, focusing on spiritual healing rather than medical treatment |
| Renaissance | Emphasis on aesthetics and human-centric designs in hospital structures | Improved patient morale and care through more comfortable and visually appealing environments |
| Seventeenth century | Development of specialized facilities like sanatoriums for specific diseases | Enhanced treatment efficacy by isolating patients, reducing the spread of infectious diseases |
| Eighteenth century | Implementation of systematic hospital designs with better ventilation and lighting | Improved patient recovery rates through better environmental conditions and reduced cross-infection |
| Nineteenth century | Introduction of pavilion-style hospitals with separate wards | Significantly reduced infection rates through increased air circulation and patient isolation |

Figure 2.1 visualizes the evolution of healthcare architecture, tracking its development from ancient healing spaces to the high-tech hospitals of today. Starting in ancient times, where temples served as the primary healing environments emphasizing spiritual care, the timeline progresses through the Greek and Roman periods when early structured hospitals emerged under religious influence. During the Middle Ages, hospitals functioned as basic charitable shelters managed by religious orders. The Renaissance brought scientific advances, leading to the design of specialized disease wards. The nineteenth century introduced pavilion-style hospitals, which promoted improved sanitation and access to natural light, significantly impacting patient recovery. The twentieth century marked a significant transformation with technological integration into hospital design, reflecting the rapid advancements in medical technology. Today, we see hospitals as complex, multifunctional facilities designed to meet the diverse and sophisticated needs of advanced medical care. This timeline not only highlights key architectural milestones but also reflects the societal shifts and technological advancements that have shaped modern healthcare facilities.

## 2.2 Review of Modern Advancements in Hospital Design

Modern advancements in hospital design represent a significant departure from historical practices, shaped by a combination of technological innovation, evolving medical practices, and a deeper understanding of the human factors in healthcare environments. These advancements are not merely aesthetic or functional improvements but are strategic responses to the complex challenges of contemporary healthcare delivery. They aim to enhance the efficiency of medical operations, improve patient

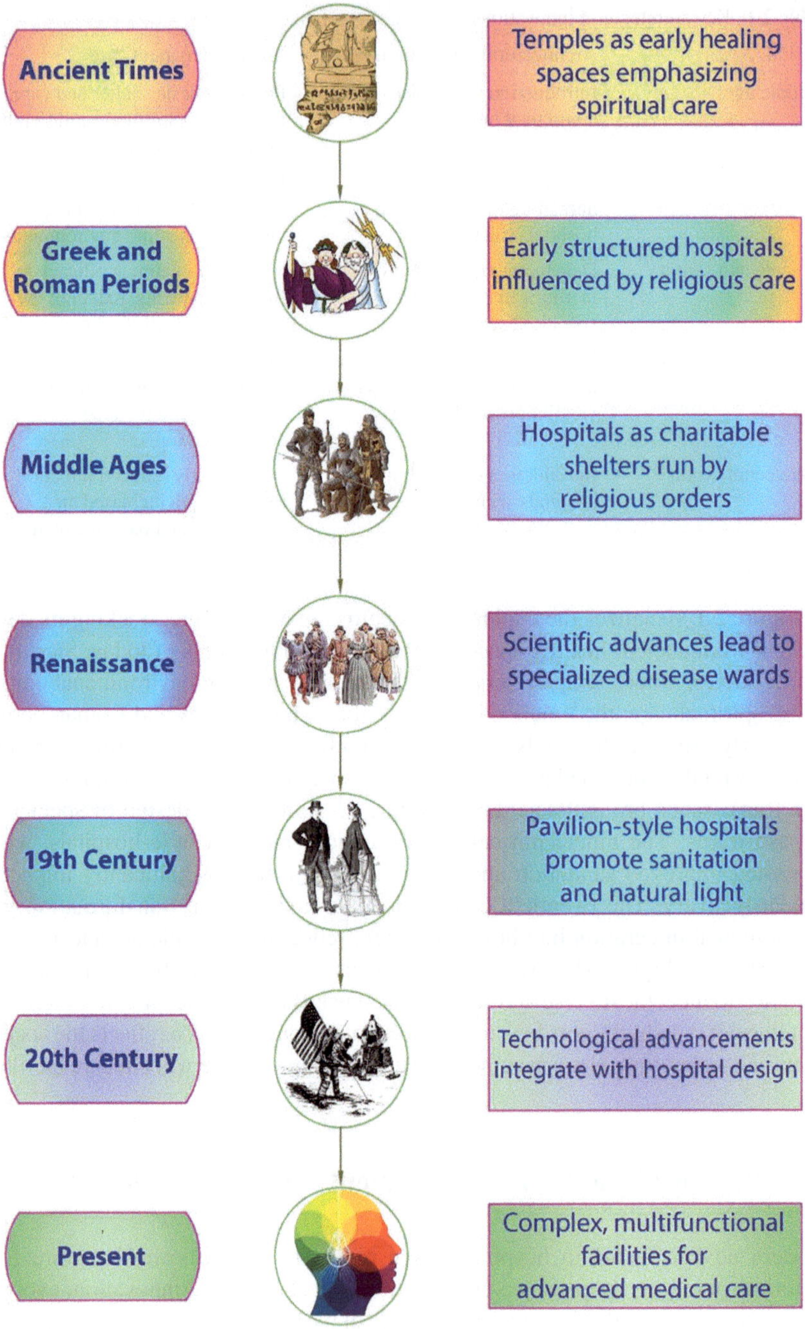


**Fig. 2.1** Timeline of healthcare architecture development

outcomes, and foster environments that support the well-being of both patients and healthcare providers.

One of the most significant advancements in recent decades has been the shift toward patient-centered design. This approach focuses on creating spaces that enhance patient privacy, comfort, and control, which are factors known to contribute positively to health outcomes. Hospital rooms are now commonly designed as single occupancy with private bathrooms, providing a quiet, personal space for patients to recover with dignity. This layout not only reduces the spread of infections but also allows for personalized care, significantly improving patient satisfaction and emotional well-being.

Technological integration into hospital design has also seen substantial growth, particularly with the advent of smart hospital concepts. These facilities utilize advanced technologies like IoT (Internet of Things) sensors, AI (Artificial Intelligence), and automated systems to enhance various aspects of healthcare delivery, from patient monitoring to facility management. For example, real-time location systems (RTLS) are used to track equipment, staff, and patients, improving workflow and safety. Similarly, automated patient handling technologies reduce physical strain on healthcare workers and increase the efficiency of patient care.

Sustainability has become another crucial focus in hospital design, driven by an awareness of the environmental impact of healthcare facilities and the need for cost-effective operations. Modern hospitals are increasingly incorporating green building practices, such as energy-efficient systems, green roofs, and sustainable materials, to reduce their carbon footprint and operating costs. These practices not only contribute to environmental sustainability but also create healthier indoor environments for patients and staff.

Additionally, the design of modern hospitals now often includes flexible spaces that can adapt to changing needs and technologies. These adaptable designs allow hospitals to expand or modify their facilities with minimal disruption, accommodating advances in medical technology and changes in healthcare practices. For instance, modular operating rooms can be quickly reconfigured to incorporate new equipment or adapt to different surgical procedures, ensuring that facilities remain at the forefront of medical technology.

The ongoing COVID-19 pandemic has further accelerated certain trends, particularly the design of infection control measures into the architectural fabric of healthcare facilities. Air filtration systems, materials that minimize the spread of pathogens, and layouts that facilitate social distancing have become standard considerations in hospital design.

These modern advancements reflect a holistic approach to hospital design, integrating architectural innovation with technological and operational strategies to create healthcare environments that are not only functional but also humane and sustainable. As we continue to explore specific architectural innovations and their impact on patient safety in the next section, these themes of integration and holistic design will be further elaborated.

Equation 2.1 articulates a mathematical model that captures the relationship between the integration of advanced technologies in hospital settings and the resulting improvements in operational efficiency and patient care quality. This model is central to the discussions in the section dedicated to technological integration within healthcare facilities, providing a quantitative foundation for evaluating the impact of technology on healthcare delivery systems. The equation establishes a direct link between the level of technological adoption (*T*) and key performance indicators such as reduced operational costs (*C*), decreased patient waiting times (*W*), and enhanced patient outcomes (*O*). By employing this model, hospital administrators and healthcare planners can assess the tangible benefits derived from investments in technology, facilitating data-driven decisions to optimize hospital operations and patient experiences.

$$E = \alpha_0 + \alpha_1 T_C + \alpha_2 T_W + \alpha_3 T_O \tag{2.1}$$

In this formula, *E* represents the overall efficiency of hospital operations, which is the dependent variable influenced by different technological integrations. The variables $T_C$, $T_W$, and $T_O$ denote the extent of technology adoption specifically geared towards cost reduction, wait time reduction, and outcome improvement, respectively. The coefficients $\alpha_1$, $\alpha_2$, and $\alpha_3$ measure the impact of each technological aspect on the overall efficiency, while $\alpha_0$ serves as the baseline efficiency in the absence of these technological integrations. This model not only highlights the critical role of technology in modern healthcare but also provides a robust framework to quantify its benefits, supporting the strategic enhancement of hospital operations through technological advancements. This rigorous approach ensures that the deployment of new technologies is aligned with the overarching goal of improving healthcare delivery and patient satisfaction.

Figure 2.2 highlights the significant differences between traditional and modern hospital rooms, emphasizing the evolution toward patient-centered design. Traditional rooms, characterized by multiple occupancy and limited privacy, reflect an older model of care with minimal technological integration and basic infection control measures. In contrast, modern rooms feature single occupancy with en-suite bathrooms, integrating advanced smart technologies for personalized care and superior infection control with safe materials. Additionally, the modern design focuses on patient comfort through ergonomic and comfort-driven amenities. This shift not only enhances patient privacy and satisfaction but also significantly improves the overall efficiency of healthcare delivery and patient outcomes. By adopting these modern design principles, healthcare facilities can better meet contemporary healthcare demands and provide a more humane and effective care environment.

## 2.3 Architectural Innovations and Patient Safety

Architectural innovations in hospital design have increasingly focused on enhancing patient safety, a critical concern in healthcare settings. The intersection of architecture and patient safety is multifaceted, involving the careful consideration of spatial

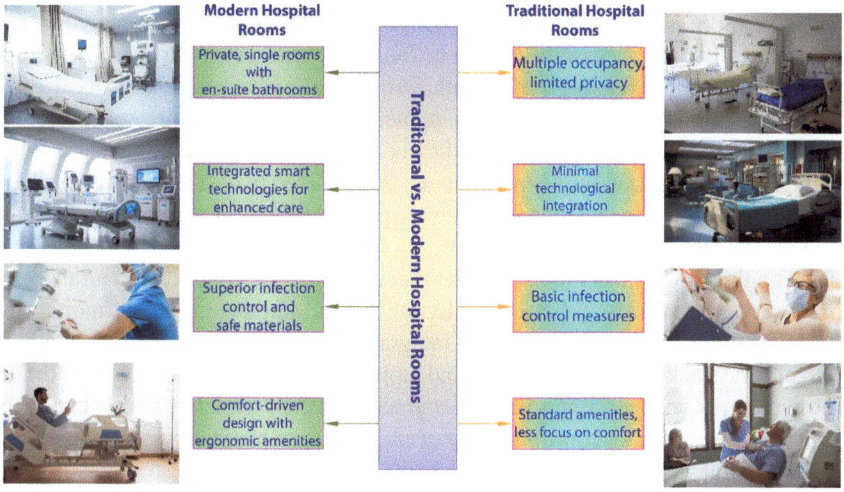


**Fig. 2.2** Comparison of traditional versus modern hospital rooms

configurations, material choices, and technological integration to minimize risks and improve health outcomes. These innovations are designed to address a range of safety concerns, from infection control to the prevention of medical errors and the facilitation of emergency responses.

One of the key areas where architectural innovation has significantly impacted patient safety is in the design of hospital layouts that optimize workflow and reduce the potential for errors. Corridor designs, for instance, are now strategically planned to separate the flow of patients, staff, and supplies, thereby minimizing cross-contamination and ensuring that critical areas such as operating rooms and intensive care units are kept isolated from high-traffic zones. This separation not only supports stringent infection control protocols but also enhances operational efficiency by reducing congestion and improving the accessibility of essential services.

Material innovations also play a pivotal role in patient safety. Non-porous materials and antimicrobial coatings are increasingly used in high-contact areas to prevent the growth and spread of pathogens. Flooring and wall surfaces are selected for their ability to be cleaned easily and withstand rigorous sanitization protocols. Additionally, the use of shatterproof glass and slip-resistant flooring helps prevent accidents, ensuring that the physical environment itself does not pose risks to patients or staff.

The integration of natural elements through architectural design has also been shown to improve patient safety and recovery. Access to natural light, views of nature, and incorporation of indoor and outdoor healing gardens are known to reduce stress and anxiety, which can significantly impact patient recovery rates and safety. These elements not only create a more pleasant and calming hospital environment but have been scientifically linked to reduced post-operative recovery times and decreased reliance on pain medications.

Furthermore, the design of patient rooms has evolved to enhance safety through better visibility and accessibility. Patient rooms are now often designed with large windows that allow nursing staff to maintain visual contact with patients without constantly entering the room, thereby reducing disturbances while still providing constant monitoring. The layout of the room, including the placement of beds, medical equipment, and nurse call systems, is optimized to ensure that patients have easy access to assistance and that healthcare providers can perform their duties with minimal obstacles.

Technological advancements have also been incorporated into architectural design to enhance patient safety. For example, the use of automated patient monitoring systems and electronic health records integrated directly into patient rooms allows for real-time tracking of patient status and rapid response to any changes in condition. These systems not only improve the accuracy of patient care but also enhance the security of patient data, contributing to overall safety in healthcare environments.

As hospital architecture continues to evolve, the focus remains steadfast on innovations that enhance patient safety. Each architectural decision is made with the intention of creating a safe, supportive, and healing environment, reflecting the core mission of healthcare to not harm. Moving forward, the next section will delve into the sustainable practices that are being integrated into healthcare facilities, ensuring that patient safety and environmental responsibility go hand in hand with modern hospital design.

Table 2.2 details the materials commonly used in modern hospital construction, emphasizing their properties and contributions to safety and cleanliness. This table explores various innovative materials such as antibacterial coatings, non-porous surfaces, and advanced composite materials. Each material is assessed for its effectiveness in promoting a sterile environment, reducing the spread of infections, and enhancing overall safety in hospital settings. This comprehensive overview aids architects, facility managers, and healthcare professionals in making informed decisions about material selection to optimize the health and safety standards of healthcare facilities.

Figure 2.3 outlines the critical architectural strategies employed in hospital design to ensure infection control and enhance patient safety. It details how strategic zoning minimizes cross-contamination by effectively separating the flow of patients, staff, and supplies. The use of non-porous materials and antimicrobial coatings in high-contact areas helps prevent the spread of pathogens, ensuring a sterile environment. Natural elements like light and gardens are integrated to promote well-being and speed up recovery, demonstrating a commitment to creating therapeutic environments. Enhanced visibility and accessibility in inpatient room design allow for better monitoring and reduced disturbances, improving both safety and patient comfort. Lastly, the integration of advanced technological systems into these spaces supports real-time health tracking, enabling timely medical responses and continual patient care. Collectively, these strategies illustrate a comprehensive approach to modern hospital architecture that prioritizes patient health, safety, and comfort.

**Table 2.2** Material innovations and their safety implications in modern hospitals

| Material | Properties | Safety implications |
|---|---|---|
| Antibacterial coatings | Contains antimicrobial agents that prevent the growth of bacteria and other pathogens | Reduces surface contamination, and lowers infection rates in clinical settings |
| Non-porous surfaces (e.g., medical-grade stainless steel, solid surface materials) | Easy to clean, does not allow bacteria to penetrate or accumulate | Enhance cleanliness, critical in areas like operating rooms and intensive care units |
| Advanced composites (e.g., fiber-reinforced plastics) | Durable, lightweight, resistant to a wide range of chemicals and pollutants | Used in areas requiring high durability and hygiene, contributes to longer-lasting, safer building materials |
| Smart glass | Can change its transparency based on light intensity or electrical control, offers privacy without curtains or blinds | Reduces the need for fabric that can harbor pathogens, increases patient privacy and control |
| Eco-friendly paints | Low VOC (Volatile Organic Compounds) and antimicrobial properties | Improves air quality, reduces potential for mold growth and spread of airborne pathogens |
| Recycled and natural materials | Often sourced sustainably; includes natural stone, recycled glass, and wood with protective coatings | Supports environmental sustainability without compromising sterility, especially when treated for hospital use |

## 2.4 Sustainable Practices in Healthcare Facilities

Sustainable practices in healthcare facilities have become increasingly important as the global community becomes more aware of environmental issues and seeks to mitigate the impact of human activities on the planet. Healthcare facilities, known for their high energy consumption and waste production, are in a unique position to lead by example in the implementation of sustainable and environmentally friendly practices. The integration of sustainability into hospital design not only supports environmental conservation but also promotes healthier indoor environments and can lead to significant cost savings in the long run.

Energy efficiency is a cornerstone of sustainable hospital design. Modern healthcare facilities are incorporating state-of-the-art technologies to reduce energy consumption, such as LED lighting, high-efficiency HVAC systems, and advanced building insulation techniques. Solar panels and other renewable energy sources are also increasingly common, providing a clean and sustainable energy supply that reduces reliance on fossil fuels and lowers greenhouse gas emissions. The design often includes large windows and strategically placed skylights to maximize natural light, reducing the need for artificial lighting and contributing to energy conservation.

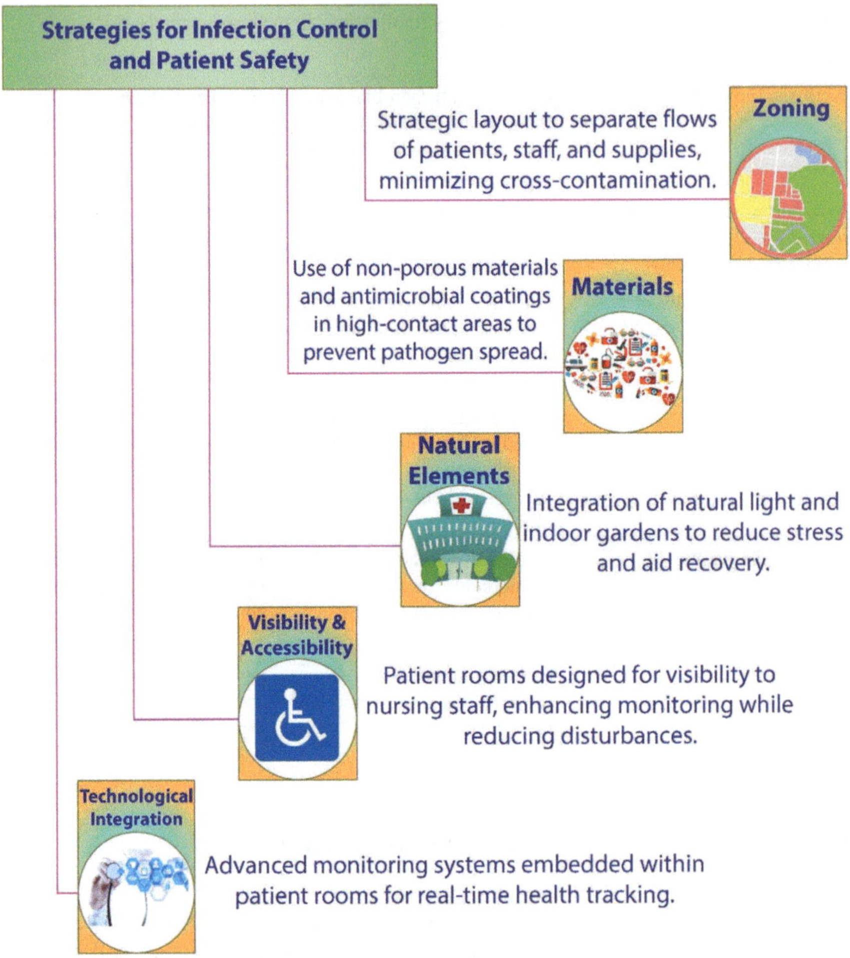


**Fig. 2.3** Design strategies for infection control and patient safety

Water conservation is another critical aspect of sustainable healthcare architecture. Hospitals consume large quantities of water daily, and reducing this consumption through efficient fixtures and smart water management systems is a priority. Low-flow toilets and faucets, efficient laundry and sterilization equipment, and rainwater harvesting systems are examples of technologies and practices that help reduce water usage. Some facilities also treat and reuse wastewater significantly reducing their overall water footprint.

The choice of materials also plays a vital role in the sustainability of healthcare facilities. Sustainable hospital designs often utilize recycled, renewable, or sustainably sourced materials, reducing the environmental impact associated with construction and maintenance. These materials include recycled steel, low-VOC (volatile organic compounds) paints and finishes, and sustainably harvested wood, all of which contribute to a healthier indoor environment by reducing pollutants and improving air quality.

Waste management is a complex challenge in healthcare settings due to the variety of waste generated, including hazardous and bio-medical waste. Sustainable hospital designs incorporate comprehensive waste management systems that promote recycling and minimize landfill use. Strategies include the segregation of waste at the source, the use of reusable medical equipment where possible, and the safe disposal of hazardous materials. Education and training for staff on sustainable waste practices are also integral to the success of these programs.

Moreover, the overall architectural design of the building itself can contribute to sustainability. Green roofs, for example, not only provide insulation and reduce runoff but also create peaceful green spaces for patients and staff to enjoy. These spaces can improve mental health and patient outcomes while also promoting biodiversity and reducing the urban heat island effect.

In sum, sustainable practices in healthcare facilities are multifaceted and involve a comprehensive approach to design and operations. By adopting these practices, healthcare facilities can significantly reduce their environmental impact while also providing a healthier and more pleasant environment for patients and staff. As we continue to explore how technology is integrated into hospital design in the next section, we will see how sustainability and technological advancements often go hand in hand to create the hospitals of the future.

Table 2.3 offers a quantitative comparison of energy consumption in hospitals before and after the implementation of green initiatives. It outlines the significant reductions in energy use resulting from various energy-efficient technologies such as LED lighting, HVAC improvements, and solar panel installations. This table is designed to demonstrate the tangible benefits of sustainable practices in healthcare facilities, providing clear data on energy savings and environmental impact reductions. It serves as a valuable resource for hospital administrators and environmental engineers who are considering or evaluating the effectiveness of green technologies in their facilities.

**Table 2.3** Comparison of energy consumption before and after green initiatives in hospitals

| Energy-efficient technology | Energy consumption before (kWh) | Energy consumption after (kWh) | Percentage reduction (%) |
|---|---|---|---|
| LED lighting | 1,000,000 | 700,000 | 30 |
| Improved HVAC systems | 2,500,000 | 1,750,000 | 30 |
| Solar panel installations | 2,000,000 | 1,000,000 | 50 |
| High-efficiency boilers | 1,200,000 | 840,000 | 30 |
| Automated energy management | 1,800,000 | 1,260,000 | 30 |

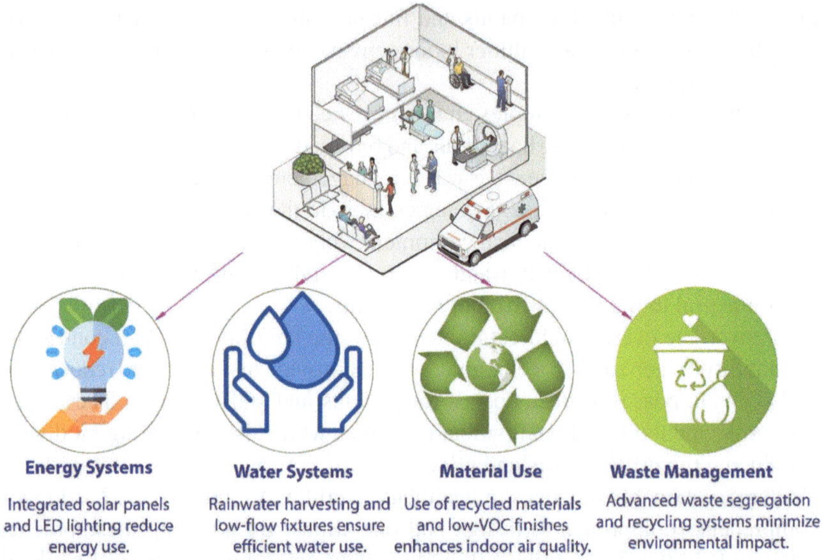


**Fig. 2.4** Schematic of sustainable systems in modern hospitals

In addition, Fig. 2.4 visually illustrates the integration of sustainable systems within modern hospital architecture, focusing on four key areas: energy systems, water systems, material use, and waste management. Energy systems in the hospital include solar panels and LED lighting, which help reduce reliance on non-renewable energy sources and decrease overall energy consumption. Water conservation is achieved through rainwater harvesting and the use of low-flow fixtures, ensuring efficient use of water resources. Sustainable material use is highlighted by the adoption of recycled materials and low-VOC finishes, which support environmental health and enhance indoor air quality. Lastly, advanced waste management practices are shown, involving rigorous waste segregation and recycling protocols that minimize environmental impact and support sustainability in healthcare operations. This schematic demonstrates how modern hospitals incorporate comprehensive sustainable practices to not only provide health care but also promote environmental stewardship and resource conservation, aligning healthcare facilities with broader global sustainability goals.

## 2.5 Technological Integration: From Theory to Practice

The integration of technology into hospital design represents one of the most significant trends shaping modern healthcare facilities. This fusion of technology and architecture is not just about incorporating new gadgets or systems; it's about rethinking the entire approach to healthcare delivery to make it safer, more efficient, and more patient centered. Technological advancements have the power to transform every aspect of hospital operations, from patient diagnostics and treatment to facility management and staff workflows.

One of the key areas where technology has made a profound impact is patient care. Digital health technologies, including telemedicine, wearable devices, and remote monitoring systems, allow for continuous patient care without the need for constant physical presence. This not only maximizes the efficiency of healthcare providers but also enhances patient comfort by allowing them to remain in more familiar, less intimidating environments whenever possible. Integrated patient care systems enable a seamless flow of information across departments, ensuring that every member of the healthcare team has access to up-to-date patient data, thus reducing errors and improving outcomes.

In terms of facility management, smart building technologies are revolutionizing how healthcare facilities operate. Building automation systems control lighting, heating, ventilation, and air conditioning (HVAC) systems more efficiently, adjusting in real-time to external conditions and internal requirements. These systems help reduce energy costs and improve the environmental comfort of both patients and staff. Furthermore, IoT sensors can monitor everything from the structural health of the building to the cleanliness of surfaces, ensuring that the facility is always safe and well-maintained.

Another critical area of technological integration is in the design of surgical and diagnostic areas. Advanced imaging technologies, robotic surgery systems, and high-tech diagnostic tools require specialized spaces designed to accommodate their specific needs. These areas must be flexible enough to adapt to rapid technological changes without requiring complete redesigns. Modular design elements and reconfigurable spaces are becoming increasingly common in hospital architecture to address this need.

Technological integration also extends to emergency and disaster readiness. Hospitals are now equipped with advanced communication systems that can operate under various scenarios, including power outages and natural disasters. These systems ensure that hospitals can maintain operations and communicate effectively both internally and with external emergency services during critical times.

The practical application of these technologies in hospital design requires a multidisciplinary approach that combines insights from architecture, engineering, medical staff, and IT professionals. This collaborative approach ensures that technological solutions are not only state-of-the-art but also tailored to the specific needs and workflows of each healthcare facility.

As hospitals continue to embrace these advanced technologies, the design principles guiding their architecture must also evolve to ensure that technological integration is practical, effective, and enhances the overall patient experience. The next section will further explore the design principles that impact patient outcomes, demonstrating how strategic design decisions can contribute to more effective healthcare environments.

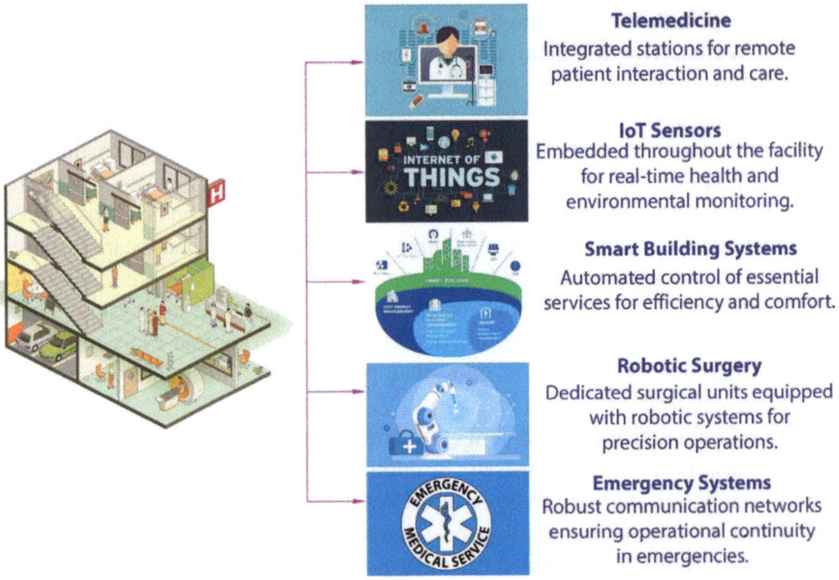


**Fig. 2.5** Integration of advanced technologies in hospital design

Figure 2.5 provides a comprehensive schematic of the advanced technologies integrated into modern hospital design, highlighting how each system enhances healthcare delivery. Telemedicine stations facilitate remote patient interaction, expanding care beyond traditional hospital boundaries. IoT sensors distributed throughout the facility enable real-time monitoring of health metrics and environmental conditions, ensuring optimal operational efficiency and patient safety. Smart building systems automate the control of lighting, heating, ventilation, and air conditioning, significantly reducing energy usage while maintaining comfort. Robotic surgery units offer precise surgical interventions, minimizing human error and improving patient outcomes. Emergency communication systems ensure the hospital remains operational and connected even during critical situations, enhancing disaster readiness. This figure illustrates not only the individual contributions of these technologies but also their combined impact on creating a hospital environment that is safe, efficient, and patient-centered.

## 2.6 Design Principles Impacting Patient Outcomes

The design of healthcare facilities significantly impacts patient outcomes by influencing factors ranging from patient recovery times to their overall experience during hospital stays. Understanding and implementing effective design principles is crucial for creating environments that promote healing, reduce stress, and improve efficiency of healthcare delivery. These principles encompass various aspects of architectural design, from spatial configurations and material selections to acoustic management and wayfinding systems.

### *2.6.1 Spatial Configuration and Healing Environments*

The architectural layout of healthcare facilities significantly influences the therapeutic journey of patients. This realization has led to a paradigm shift in hospital design from purely functional spaces to environments that embrace patient well-being as a fundamental aspect. Research robustly supports the idea that recovery is not only about medical treatments but also the environment where healing occurs. Natural light plays a pivotal role. Studies have consistently demonstrated that exposure to daylight can help regulate patients' sleep cycles, improve mood, and reduce depression and pain, thereby diminishing the length of hospital stays and dependency on pain-relieving medications.

Moreover, the integration of quiet spaces and natural vistas into hospital design has evolved from aesthetic choices to evidence-based practices. For example, incorporating elements such as indoor gardens or water features can significantly mitigate stress and anxiety among both patients and staff. Such features not only provide a sensory respite from the clinical environment but also facilitate informal interactions that are vital for mental health. The presence of green spaces has been linked to lower levels of stress hormones and improved mental well-being, which are crucial in a high-stress environment like a hospital.

The spatial arrangement within hospitals is meticulously strategized to enhance efficiency while minimizing stress-inducing factors such as noise, which can be detrimental to patient recovery. Patient rooms are increasingly designed to maximize privacy and tranquility, often positioned away from high-traffic areas to reduce noise pollution. Similarly, corridors and common areas are thoughtfully laid out to ensure a seamless flow that prevents crowding and reduces the overall noise levels. The strategic placement of nursing stations and the use of sound-absorbing materials further contribute to a calm atmosphere, enabling a more healing-centric environment.

The design philosophy that underpins modern hospital layouts is centered around creating spaces that are not only medically efficient but also holistically healing. The incorporation of art, music, and access to communal spaces are examples of how the healing environment is tailored to address the psychological and emotional needs of patients, facilitating a comprehensive approach to recovery that transcends traditional medical treatments. Each design decision, from the choice of color palettes to the type of furniture used, is made to enhance patient comfort and well-being and reflect a deep understanding of the complex interplay between space and health.

In conclusion, the spatial configuration of healthcare facilities is a crucial component of modern medical care. It reflects a broader, more inclusive understanding of health that considers the physical, psychological, and emotional dimensions of healing. By designing spaces that foster recovery, reduce stress, and promote well-being, healthcare architecture not only supports medical outcomes but also elevates the overall quality of care. This integrative approach is fundamental in crafting environments that truly heal, demonstrating the vital role of architecture in health and recovery.

Figure 2.6 showcases the essential design elements that contribute to healing environments within hospital settings. It highlights the strategic integration of natural light through large windows and skylights, enhancing patient recovery by providing optimal daylight exposure, which is known to improve mood and reduce recovery times. Quiet spaces are thoughtfully designed away from noise and high-traffic areas, offering patients a tranquil space for rest and recuperation. Indoor gardens and water features are incorporated to reduce stress and promote relaxation, creating a serene and aesthetically pleasing atmosphere conducive to healing. Additionally, the use of sound-absorbing materials in walls and ceilings helps to minimize noise pollution, further enhancing the calming and restorative environment. These design elements collectively work to support patient recovery by addressing both the psychological and physiological needs of patients, ultimately fostering a healthcare setting that prioritizes well-being and reduces stress.

### *2.6.2 Material Selection and Infection Control*

The selection of materials in the construction and outfitting of healthcare facilities is a foundational element in the architectural and operational strategy to prevent and control infections. Materials used within hospitals must adhere to rigorous health and safety standards to ensure they contribute effectively to infection control. These materials are selected not only for their functional and aesthetic qualities but also for their ability to meet the complex demands of a sanitary environment.

Central to these considerations is the requirement for surfaces to be non-porous and easily cleanable. Non-porous materials, such as certain polymers and metals, prevent the absorption and retention of pathogens, making them ideal for areas that require frequent sanitization. The ability of these materials to resist the growth of bacteria, viruses, and fungi is crucial in minimizing the risk of hospital-acquired infections (HAIs), which remain a significant challenge in healthcare settings.

Recent advances in material science have further enhanced the capability of these environments to control infections through the development of antimicrobial surfaces. These innovative materials are imbued with antimicrobial agents—such as metallic ions like copper and silver—which have natural properties that disrupt the vital processes of microbes, effectively reducing the load of pathogens on surfaces. The application of such technology extends from high-touch areas like doorknobs and handrails to more extensive surfaces such as countertops and bedside tables.

Moreover, the integration of these advanced materials must be thoughtfully planned to ensure that they do not inadvertently contribute to microbial resistance, a growing concern in modern medicine. The selection process, therefore, involves

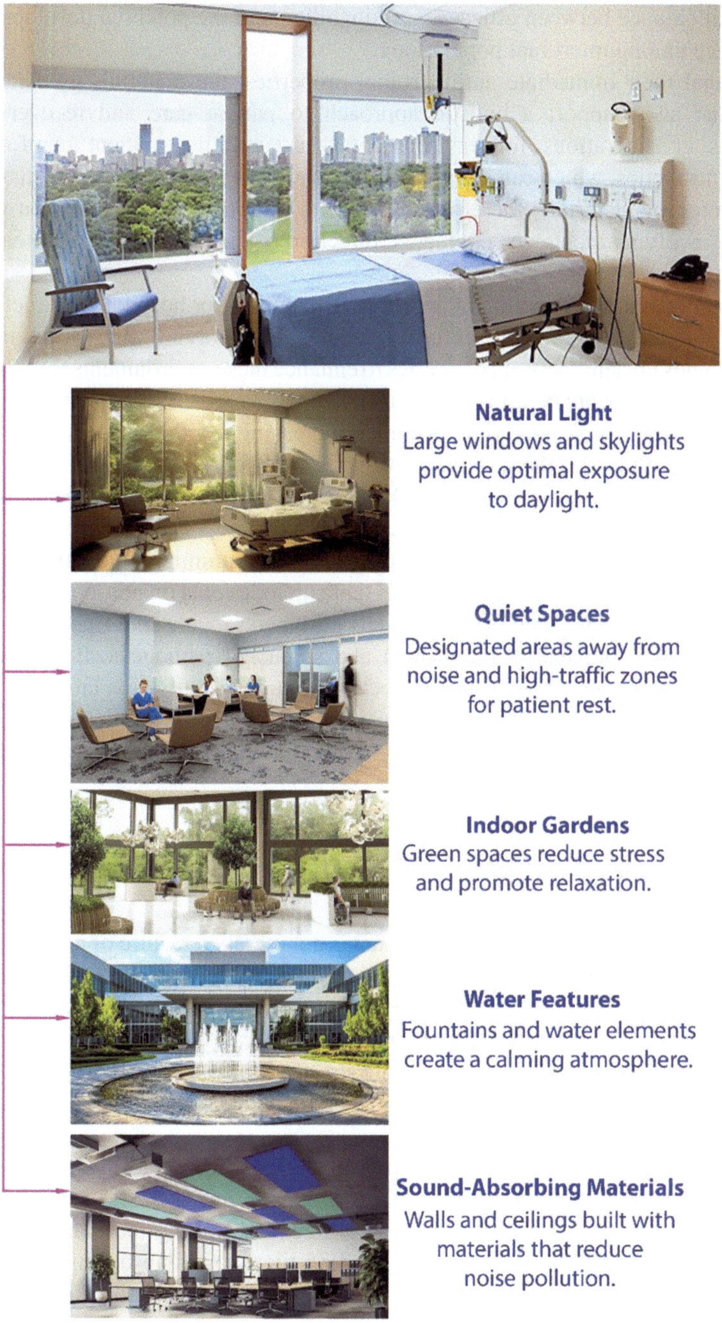


**Fig. 2.6** Healing environment design elements

a careful balance between efficacy, sustainability, and the potential development of resistance among microbial populations.

Beyond their immediate antimicrobial properties, the materials used in hospitals must also support a holistic approach to patient care and recovery. This includes considerations for environmental sustainability, patient comfort, and overall aesthetics, which can significantly impact patient outcomes and satisfaction. The choice of materials, therefore, becomes a multidisciplinary decision-making process involving architects, engineers, infection control specialists, and healthcare providers.

In conclusion, the selection of appropriate materials for hospital construction is a critical component of infection control strategies. The evolving landscape of material technologies presents new opportunities to enhance these environments for safety and efficacy. As healthcare facilities continue to face challenges in infection management, the role of innovative materials becomes increasingly vital in supporting health, safety, and wellness in these critical settings.

Table 2.4 provides an exhaustive list of materials commonly used in hospital settings, detailing their properties, typical uses, and effectiveness in infection control. It highlights the importance of material selection in minimizing infection risks and enhancing overall hospital hygiene. This table includes data on non-porous surfaces, antimicrobial coatings, and other materials designed to withstand rigorous cleaning protocols while preventing the growth and spread of pathogens. By showcasing various materials and their specific infection control benefits, this table serves as a crucial guide for architects, facility managers, and healthcare professionals tasked with designing or renovating hospital spaces to maximize cleanliness and safety.

### 2.6.3 *Acoustic Comfort*

Acoustic comfort in hospital environments is not merely a feature of modern healthcare architecture; it is a fundamental aspect that impacts patient well-being, staff efficiency, and overall healthcare quality. Noise in hospitals, often characterized by a mix of voices, equipment, and operational activities, can significantly impede patient recovery by disrupting sleep and elevating stress and anxiety levels. The significance of managing these noise levels extends beyond patient comfort, influencing clinical outcomes and patient satisfaction.

To address this critical issue, healthcare facilities incorporate advanced architectural and design strategies aimed at noise reduction and acoustic control. One of the primary approaches includes the use of sound-absorbing materials throughout the facility. Materials such as acoustic ceiling tiles, sound-absorbing wall panels, and floors that reduce sound transmission are strategically installed to mitigate the propagation of noise. These materials are carefully chosen not only for their noise-canceling properties but also for their safety, durability, and ease of maintenance, ensuring they align with the rigorous demands of a healthcare environment.

**Table 2.4** Materials used in hospitals and their infection control properties

| Material | Properties | Uses in hospitals | Infection control effectiveness |
|---|---|---|---|
| Stainless steel | Non-porous, easy to clean, durable | Sinks, countertops, surgical instruments | High; does not harbor pathogens easily |
| Copper and alloys | Natural antimicrobial properties | Door handles, bed rails, and other high-touch surfaces | Very high; kills bacteria and viruses on contact |
| Medical-grade plastics | Non-porous can be infused with antimicrobials | Trays, cabinets, patient room furniture | High; prevents growth and easy to sanitize |
| Antimicrobial coatings | Contains agents that kill or inhibit microbes | Applied to walls, floors, furniture | Very high; actively reduces the microbial load on surfaces |
| Laminar airflow systems | Ensures unidirectional airflow, filters pathogens | Operating rooms, intensive care units | Very high; minimizes airborne pathogens |
| Glass | Non-porous, easy to clean | Windows, interior partitions | Moderate; easy to clean but does not actively kill pathogens |
| Epoxy flooring | Seamless, impervious to water, contains antimicrobial additives | Hallways, operating rooms | Very high; durable and inhibits microbial growth |

Moreover, the design and layout of hospitals play a crucial role in controlling sound. Architectural planning often involves creating dedicated pathways for staff and service equipment that are isolated from patient recovery areas. This segregation helps minimize the intrusion of operational noises into spaces designed for rest and recuperation. Additionally, the layout of patient rooms, nursing stations, and treatment areas is configured to prevent the direct transmission of sounds, employing buffering zones such as storerooms or utility areas that absorb and block noise.

Another aspect of enhancing acoustic comfort is the technological integration of sound-masking systems. These systems emit subtle, controlled background sounds, often akin to airflow, which help mask disruptive noises, making the environment more serene without being intrusively quiet or unnaturally silent. Such systems are especially beneficial in areas where total silence is not achievable, providing a consistent auditory backdrop that helps reduce the perceptibility of intermittent noises.

Beyond these measures, hospitals also consider the psychological impact of sound by integrating natural soundscapes, such as recordings of natural sounds, which have been shown to reduce stress and improve patient and staff well-being. This approach not only aids in acoustic comfort but also enhances the overall sensory environment, contributing to a more holistic healing atmosphere.

In conclusion, acoustic comfort in hospitals is a multidimensional aspect of healthcare design that involves a combination of materials, technology, and strategic planning. By effectively managing noise levels, healthcare facilities can create a conducive healing environment that significantly improves patient outcomes and enhances the working conditions for medical staff. This commitment to acoustic excellence reflects a profound understanding of the intricate relationship between the physical environment and health, essential for modern medical practices.

Equation 2.2 is meticulously designed to calculate the Noise Reduction Coefficient (NRC), a critical measure in assessing the effectiveness of various sound-absorbing materials used in hospital settings. This equation is strategically placed within the discourse on acoustic comfort, emphasizing the pivotal role of ambient noise control in creating conducive environments for patient recovery and staff operations. The NRC calculation is fundamental for healthcare architects and facility managers to determine how different materials can mitigate noise levels, enhancing the overall acoustic quality within medical facilities. By quantifying the sound absorption capabilities of materials, this equation aids in the selection process, ensuring that choices are made based on empirical data to optimize acoustic comfort in patient care areas.

$$\mathrm{NRC} = \frac{S_1 a_1 + S_2 a_2 + \cdots + S_n a_n}{S_{\mathrm{total}}} \tag{2.2}$$

In this formula, NRC represents the Noise Reduction Coefficient, which is a dimensionless index ranging from 0 (no sound absorption) to 1 (complete sound absorption), reflecting the average sound absorption performance of a material at specific frequencies. The variables $S_i$ and $a_i$ represent the surface area and the sound absorption coefficient of each material used within the hospital, respectively, where $i$ corresponds to different materials. $S_{\mathrm{total}}$ is the total surface area covered by all the sound-absorbing materials. This equation allows for the precise calculation of the overall effectiveness of the acoustic design in reducing environmental noise, which is crucial for maintaining a quiet and serene hospital atmosphere conducive to healing and efficient healthcare service delivery. By integrating such detailed calculations into the design process, healthcare facilities can significantly improve the acoustic environment, directly contributing to patient comfort and recovery rates.

### *2.6.4 Wayfinding Systems*

Effective wayfinding systems in hospitals are crucial not only for functional navigation but also for alleviating the inherent stress and anxiety associated with medical environments. Patients and their families, often under considerable emotional distress, benefit significantly from clear and intuitive guidance. Wayfinding in healthcare settings extends beyond mere signage; it encompasses a comprehensive approach involving architectural design, strategic signage, and accessible information systems that together create a navigable and reassuring environment.

The architectural layout of a hospital plays a pivotal role in intuitive navigation. Corridors, lobbies, and transition spaces are designed with clear visual cues that naturally guide individuals through space. This could include distinct color schemes for different wings or floors, architectural cues like illuminated pathways, or even flooring patterns that subtly direct traffic flow. These elements are integrated into the building's design from the outset, promoting a sense of orientation without reliance on active navigation efforts.

Signage forms the backbone of any effective wayfinding system. In healthcare facilities, this means more than just functional signs; it involves the strategic placement of directional cues that are easy to read and understand briefly. Signs need to accommodate a diverse range of users, including those with visual impairments or cognitive challenges. To this end, hospitals often employ large, clear typefaces, high-contrast colors, and international symbols that transcend language barriers. Additionally, the consistency of these signs throughout the facility prevents confusion and ensures that once a visitor has learned the system, they can navigate the entire hospital more confidently.

Moreover, digital wayfinding solutions are increasingly becoming integral to modern hospital designs. Interactive kiosks, mobile apps, and digital directories offer dynamic navigation aids that can adapt to the user's needs, providing step-by-step directions that are easy to follow. These systems often include features such as estimated walking times, which can be particularly helpful for those with mobility issues. By integrating technology into the wayfinding system, hospitals can offer up-to-date information about room changes or closures, enhancing the visitor's ability to navigate complex pathways efficiently.

Accessibility is another critical aspect of wayfinding, ensuring that all patients and visitors, regardless of physical ability or health condition, can use navigation aids effectively. This includes auditory systems for the visually impaired, tactile paths for those who require physical navigation aids, and signage at appropriate heights for wheelchair users.

In conclusion, wayfinding systems in healthcare facilities are a critical component of the overall patient and visitor experience. By reducing navigational stress, these systems contribute significantly to a supportive healing environment. Effective wayfinding is achieved through a blend of thoughtful architectural design, consistent and clear signage, and the strategic use of technology, all tailored to meet the diverse needs of healthcare facility users. This holistic approach not only aids navigation but also enhances the safety, efficiency, and overall satisfaction of patients and their families within the hospital setting.

Figure 2.7 illustrates the integration of various wayfinding systems within a hospital environment, showcasing how architecture and technology come together to facilitate easy navigation and enhance patient and visitor experiences. Strategic signage with large, clear typefaces and high contrast ensures readability and accessibility, guiding visitors smoothly through different departments. Color coding is used to differentiate various hospital wings, providing quick orientation and reducing confusion. Digital directories positioned at key points offer interactive assistance, enabling visitors to navigate complex routes with real-time updates. Architectural cues, such as illuminated pathways and unique flooring patterns, intuitively direct movement within the hospital, ensuring that visitors and patients can find their way with minimal stress. This cohesive approach to wayfinding not only aids in navigation but also significantly alleviates the anxiety often associated with hospital visits, creating a more welcoming and patient-friendly environment.

### 2.6.5 Patient Privacy and Dignity

Respecting patient privacy and dignity is not merely a courtesy but a fundamental ethical and legal requirement in healthcare settings. The design of hospital spaces plays a crucial role in ensuring that these values are upheld, affecting not only compliance with privacy regulations but also significantly impacting patients' well-being and satisfaction. Effective design strategies must therefore go beyond the provision of private rooms, addressing the broader issues of sightlines, acoustics, and the overall patient environment.

Architectural design can profoundly influence how privacy is maintained within hospital settings. Careful planning of the hospital layout is essential to control sightlines and prevent direct views into patient areas from public spaces or high-traffic corridors. For instance, the use of frosted glass panels, strategic placement of walls, and partially enclosed spaces can safeguard patient privacy while still allowing natural light and visibility necessary for medical monitoring and patient safety.

Moreover, the orientation of patient beds, the design of doorways and windows, and even the layout of furniture are considered to minimize exposure during medical examinations or treatments. Ensuring that functional spaces like nursing stations or consultation areas are strategically positioned away from the public gaze also contributes to a sense of security and dignity for patients.

Acoustic privacy is another critical aspect of hospital design. Soundproofing materials and technologies are used to prevent the travel of voices and noises, protecting sensitive conversations from being overheard and preserving the confidentiality of personal and medical information. This is particularly important in open-plan wards, emergency rooms, and consultation areas where conversations are frequent, and privacy needs are high.

In addition to physical modifications, the incorporation of technology can further enhance privacy protections. For example, electronic access controls can restrict entry to sensitive areas, and digital systems can replace paper records to minimize the risk of information breaches. The thoughtful integration of technology not only improves security but also enhances the overall patient experience by facilitating smoother operations and reducing stress.

Patient dignity is closely tied to these privacy concerns, impacting how individuals perceive their treatment and care. Design elements that promote a sense of control and autonomy, such as adjustable privacy screens, personal storage spaces, and accessible personal controls for lighting and temperature, empower patients in their care environment. Providing spaces where patients can spend time away from their beds, such as private gardens or lounge areas, can also help maintain their

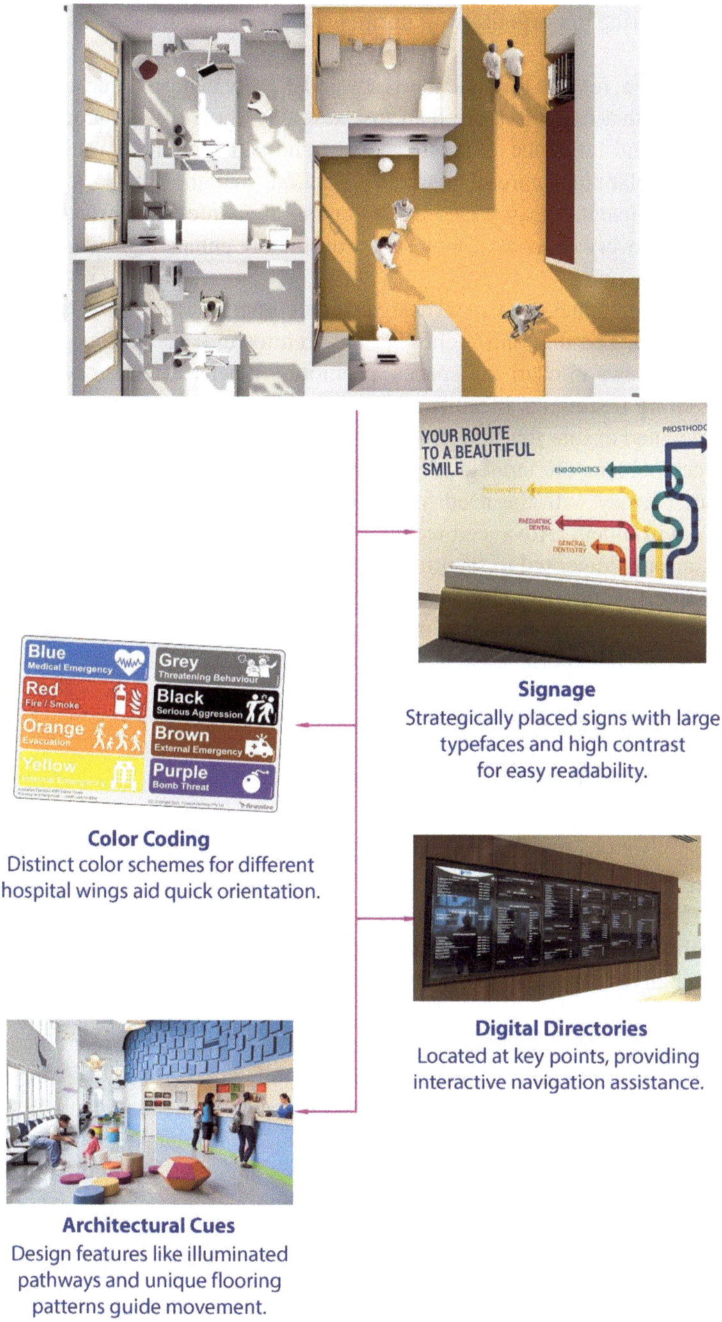


**Fig. 2.7** Wayfinding system integration in hospital layouts

dignity by offering opportunities for normalcy and relaxation away from the clinical setting.

In conclusion, respect for patient privacy and dignity in hospital design is a critical consideration that impacts more than just aesthetic choices—it shapes the healing environment and influences patient recovery and satisfaction. Through thoughtful architectural planning, careful material selection, and the integration of modern technologies, hospitals can create environments that safeguard privacy and honor the dignity of every individual they serve, ultimately enhancing the quality of care and patient outcomes.

Figure 2.8 illustrates the critical design elements employed in hospitals to enhance patient privacy. The diagram highlights the architectural layout designed to minimize public views into private spaces, promoting a sense of security for patients. Soundproofing treatments in walls and ceilings are essential to maintain conversation privacy and ensure that sensitive discussions remain confidential. Frosted glass and privacy screens are strategically used in patient room windows and partitions to balance privacy with the need for natural light, thus creating a comforting and discreet environment. Advanced technological solutions, such as electronic access systems, are implemented to secure sensitive areas, preventing unauthorized access and enhancing overall patient safety. Together, these design strategies form a comprehensive approach to preserving patient dignity and confidentiality, thereby improving the overall patient experience in healthcare settings.

### *2.6.6 Adaptable and Flexible Spaces*

The principle of adaptability in hospital design is a critical consideration that ensures healthcare facilities remain functional and relevant over time, responsive to evolving medical technologies, changing patient needs, and emerging healthcare practices. This flexibility is particularly vital in areas of the hospital that experience the most dynamic changes and demands, such as emergency rooms, intensive care units, and outpatient facilities.

Adaptable spaces in healthcare settings are designed with modular elements that can be easily reconfigured or updated without significant disruptions to the facility's operations. This might involve structural choices such as movable walls, adjustable lighting systems, and modular furniture that can be reorganized to suit different medical functions or to accommodate fluctuations in patient volume. The design of utility systems—such as power, data, and plumbing—is also crucial, incorporating redundancy and flexibility to allow for technological upgrades and changes in medical equipment needs without extensive renovations.

Furthermore, the concept of adaptability extends to the very infrastructure of the hospital. For instance, floor systems designed to accommodate heavy loads, and multiple configurations can support a variety of medical equipment and can be adapted for different medical specialties as needs evolve. Ceiling systems that can

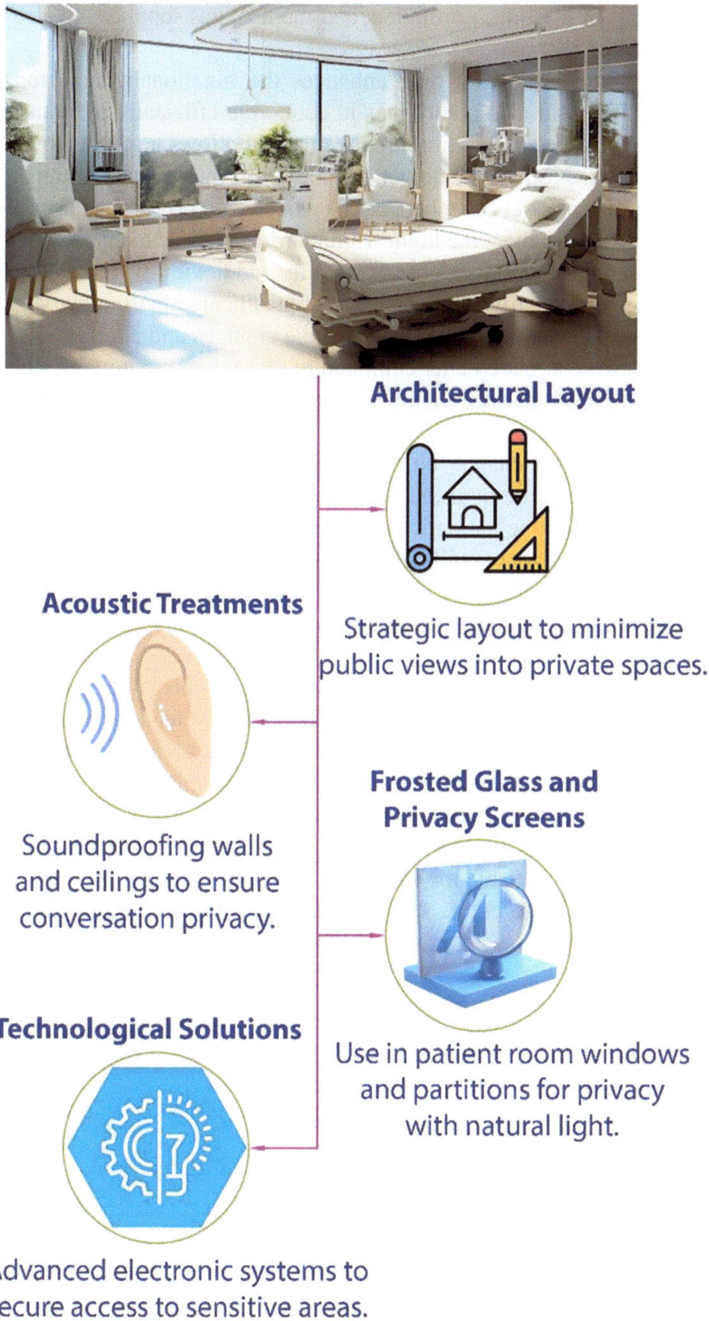


**Fig. 2.8** Design strategies for enhancing patient privacy

integrate a range of lighting, ventilation, and medical gas supplies are also essential for spaces that must serve multiple uses over time.

This design philosophy not only enhances the functionality and longevity of healthcare facilities but also contributes to economic efficiency by minimizing the need for costly and disruptive future renovations. It allows hospitals to expand and contract service areas such as emergency rooms and intensive care units rapidly in response to surges in patient numbers or changes in care protocols, ensuring that the facility can continue to meet the highest standards of care delivery.

By embracing these principles of adaptability and flexibility, architects and planners contribute significantly to the creation of healthcare environments that are not just physically enduring but also resilient to the changes and challenges of modern healthcare. These spaces are better equipped to serve their communities by aligning the physical environment with the operational and therapeutic needs of healthcare delivery.

As we conclude this chapter on the foundational principles of hospital design, we pave the way for more nuanced discussions on specific strategies that enhance infection control, promote patient-centered care, and optimize operational efficiencies. The upcoming chapters will delve deeper into how these design strategies are implemented and their profound impact on health outcomes and hospital functionality, underscoring the indispensable role of thoughtful architectural design in the advancement of healthcare delivery.

Table 2.5 explores the elements of flexible design in hospitals and their specific applications across various departments. It details how features such as movable walls, modular furniture, and reconfigurable spaces can enhance the adaptability and functionality of healthcare environments. By examining these elements, the table underscores the importance of designing facilities that can quickly adjust to changing medical needs, patient volumes, and technological advancements. This insight is particularly valuable for hospital administrators, designers, and healthcare professionals who aim to create spaces that can evolve with emerging healthcare trends and requirements.

**Table 2.5** Flexible design elements and their applications in hospital settings

| Flexible design element | Description | Applications in hospital departments |
|---|---|---|
| Movable walls | Walls that can be easily repositioned to change room sizes and layouts | Emergency rooms, outpatient clinics, and multi-use areas to accommodate varying patient loads |
| Modular furniture | Furniture designed to be reconfigured or moved as needed | Patient rooms, waiting areas, and staff lounges to provide adaptable seating and usage configurations |
| Reconfigurable spaces | Areas designed to serve multiple purposes through quick adjustments | Conference rooms that convert into additional treatment areas during peak times |
| Retractable partitions | Temporary partitions that can be extended or retracted to divide or open spaces | Intensive care units and recovery rooms, allowing privacy or open monitoring as required |
| Telescopic seating systems | Seating that can be expanded or contracted based on need | Lecture halls and assembly areas within the hospital for educational sessions or large meetings |
| Demountable systems | Temporary structures that can be assembled or disassembled as needed | Temporary screening areas or expanded administrative spaces during health crises |

# Further Reading

Abdelaal, M. S., & Soebarto, V. (2019). Biophilia and Salutogenesis as restorative design approaches in healthcare architecture. *Architectural science review*, 62(3), 195–205. https://doi.org/10.1080/00038628.2019.1604313

Alhmoud, S. H., & Arcan, E. F. (2020). Improving interior environmental quality using sustainable design in Jordanian hospital bedrooms. *European Journal of Sustainable Development*, 9(3), 443–443. https://doi.org/10.14207/ejsd.2020.v9n3p443

Alzunitan, M. A., Edmond, M. B., Alsuhaibani, M. A., Samuelson, R. J., Schweizer, M. L., & Marra, A. R. (2022). Positive deviance in infection prevention and control: a systematic literature review. *Infection Control & Hospital Epidemiology*, 43(3), 358–365

Arakawa Martins, B., Barrie, H., Visvanathan, R., Daniel, L., Martins, L. A., Ranasinghe, D., ... & Soebarto, V. (2021). A multidisciplinary exploratory approach for investigating the experience of older adults attending hospital services. *HERD: Health Environments Research & Design Journal*, 14(1), 141–163. https://doi.org/10.1177/1937586720920858

Baron, J. (2018). A brief history of evidence-based policy. *The ANNALS of the American Academy of Political and Social Science*, 678(1), 40–50. https://doi.org/10.1177/0002716218763128

Bate, P., & Robert, G. (2023). Bringing user experience to healthcare improvement: the concepts, methods and practices of experience-based design. *CRC Press*

Brambilla, A., Rebecchi, A., & Capolongo, S. (2019). Evidence based hospital design. A literature review of the recent publications about the EBD impact of built environment on hospital occupants' and organizational outcomes. *Annali di Igiene Medicina Preventiva e di Comunita*, 31(2), 165–180. https://hdl.handle.net/11311/1074776

Buse, C., Martin, D., & Nettleton, S. (2018). Conceptualising 'materialities of care': making visible mundane material culture in health and social care contexts. *Materialities of care: Encountering health and illness through artefacts and architecture*, 1–13. https://doi.org/10.1002/9781119499749.ch1

Capolongo, S., Gola, M., Brambilla, A., Morganti, A., Mosca, E. I., & Barach, P. (2020). COVID-19 and healthcare facilities: a decalogue of design strategies for resilient hospitals. *Acta Bio Medica: Atenei Parmensis*, 91(9-S), 50. https://doi.org/10.23750/abm.v91i9-S.10117

Carayon, P., Wooldridge, A., Hoonakker, P., Hundt, A. S., & Kelly, M. M. (2020). SEIPS 3.0: Human-centered design of the patient journey for patient safety. *Applied ergonomics*, 84, 103033. https://doi.org/10.1016/j.apergo.2019.103033

Cole, L. B. (2019). Green building literacy: a framework for advancing green building education. *International Journal of STEM Education*, 6, 1–13. https://doi.org/10.1186/s40594-019-0171-6

DuBose, J., MacAllister, L., Hadi, K., & Sakallaris, B. (2018). Exploring the concept of healing spaces. *HERD: Health Environments Research & Design Journal*, 11(1), 43–56. https://doi.org/10.1177/1937586716680567

Dushkova, D., & Ignatieva, M. (2020). New trends in urban environmental health research: from geography of diseases to therapeutic landscapes and healing gardens. Geography, Environment, *Sustainability*, 13(1), 159–171. https://doi.org/10.24057/2071-9388-2019-99

Eklund, J. H., Holmström, I. K., Kumlin, T., Kaminsky, E., Skoglund, K., Höglander, J., ... & Meranius, M. S. (2019). "Same same or different?" A review of reviews of person-centered and patient-centered care. *Patient education and counseling*, 102(1), 3–11. https://doi.org/10.1016/j.pec.2018.08.029

Fagerström, L., Kinnunen, M., & Saarela, J. (2018). Nursing workload, patient safety incidents and mortality: an observational study from Finland. *BMJ open*, 8(4), e016367. https://doi.org/10.1136/bmjopen-2017-016367

Gaminiesfahani, H., Lozanovska, M., & Tucker, R. (2020). A scoping review of the impact on children of the built environment design characteristics of healing spaces. *HERD: Health Environments Research & Design Journal*, 13(4), 98–114. https://doi.org/10.1177/1937586720903845

Ghaffari, F., Shabak, M., Norouzi, N., & Nayyeri Fallah, S. (2023). Hospital salutogenic public spaces: a conceptual framework of effective perceptional environment quality components on patients' satisfaction. *International Journal of Building Pathology and Adaptation*, 41(5), 965–987. https://doi.org/10.1108/IJBPA-05-2021-0071

Ghaffarianhoseini, A., AlWaer, H., Omrany, H., Ghaffarianhoseini, A., Alalouch, C., Clements-Croome, D., & Tookey, J. (2018). Sick building syndrome: are we doing enough?. *Architectural Science Review*, 61(3), 99–121. https://doi.org/10.1080/00038628.2018.1461060

Gilbert, T., Neuburger, J., Kraindler, J., Keeble, E., Smith, P., Ariti, C., ... & Conroy, S. (2018). Development and validation of a Hospital Frailty Risk Score focusing on older people in acute care settings using electronic hospital records: an observational study. *The Lancet*, 391(10132), 1775–1782

Gharaveis, A., Kirk Hamilton, D., Shepley, M., Pati, D., & Rodiek, S. (2019). Design suggestions for greater teamwork, communication and security in hospital emergency departments. *Indoor and Built Environment*, 28(8), 1126–1139. https://doi.org/10.1177/1420326X19836209

Grover, S., Fitzpatrick, A., Azim, F. T., Ariza-Vega, P., Bellwood, P., Burns, J., ... & Ashe, M. C. (2022). Defining and implementing patient-centered care: An umbrella review. *Patient education and counseling*, 105(7), 1679–1688. https://doi.org/10.1016/j.pec.2021.11.004

Gusmano, M. K., Maschke, K. J., & Solomon, M. Z. (2019). Patient-centered care, yes; patients as consumers, no. *Health affairs*, 38(3), 368–373. https://doi.org/10.1377/hlthaff.2018.05019

Halawa, F., Madathil, S. C., Gittler, A., & Khasawneh, M. T. (2020). Advancing evidence-based healthcare facility design: a systematic literature review. *Health Care Management Science*, 23, 453–480. https://doi.org/10.1007/s10729-020-09506-4

Hammouni, Z. (2020). Innovation and Technology in One New Hospital in Montreal: A Lived Experience of Healthcare Professionals. In Human Interaction, Emerging Technologies and Future Applications II: Proceedings of the 2nd International Conference on Human Interaction and Emerging Technologies: Future Applications (IHIET–AI 2020), April 23–25, 2020, Lausanne, Switzerland (pp. 389–395). *Springer International Publishing.* https://doi.org/10.1007/978-3-030-44267-5_58

Jewkes, Y. (2018). Just design: Healthy prisons and the architecture of hope. *Australian & New Zealand Journal of Criminology*, 51(3), 319–338. https://doi.org/10.1177/0004865818766768

Laurent, J. G. C., Williams, A., Oulhote, Y., Zanobetti, A., Allen, J. G., & Spengler, J. D. (2018). Reduced cognitive function during a heat wave among residents of non-air-conditioned buildings: An observational study of young adults in the summer of 2016. *PLoS medicine*, 15(7), e1002605. https://doi.org/10.1371/journal.pmed.1002605

Lee, S. M., & Lee, D. (2021). Opportunities and challenges for contactless healthcare services in the post-COVID-19 Era. *Technological Forecasting and Social Change*, 167, 120712. https://doi.org/10.1016/j.techfore.2021.120712

Maina, J., Ouma, P. O., Macharia, P. M., Alegana, V. A., Mitto, B., Fall, I. S., ... & Okiro, E. A. (2019). A spatial database of health facilities managed by the public health sector in sub-Saharan Africa. *Scientific data*, 6(1), 134. https://doi.org/10.1038/s41597-019-0142-2

Marcus, C. C. (2018). Therapeutic landscapes. In Environmental psychology and human well-being (pp. 387–413). Academic Press. https://doi.org/10.1016/B978-0-12-811481-0.00015-9

Matarneh, S. T., Danso-Amoako, M., Al-Bizri, S., Gaterell, M., & Matarneh, R. (2019). Building information modeling for facilities management: A literature review and future research directions. *Journal of Building Engineering*, 24, 100755. https://doi.org/10.1016/j.jobe.2019.100755

McPhearson, T., M. Raymond, C., Gulsrud, N., Albert, C., Coles, N., Fagerholm, N., ... & Vierikko, K. (2021). Radical changes are needed for transformations to a good Anthropocene. *Npj urban sustainability*, 1(1), 5. https://doi.org/10.1038/s42949-021-00017-x

Morag, I., & Pintelon, L. (2021). Digital wayfinding systems in hospitals: A qualitative evaluation based on managerial perceptions and considerations before and after implementation. *Applied Ergonomics*, 90, 103260. https://doi.org/10.1016/j.apergo.2020.103260

Murai, I. H., Fernandes, A. L., Sales, L. P., Pinto, A. J., Goessler, K. F., Duran, C. S., ... & Pereira, R. M. (2021). Effect of a single high dose of vitamin D3 on hospital length of stay in patients with moderate to severe COVID-19: a randomized clinical trial. *Jama*, 325(11), 1053–1060. https://doi.org/10.1001/jama.2020.26848

Nightingale, F. (2022). Notes of hospitals. BoD–Books on Demand.

Plunkett, A., & Plunkett, E. (2022). Positive approaches to safety: learning from what we do well. *Pediatric Anesthesia*, 32(11), 1223–1229. https://doi.org/10.1111/pan.14509

Reis, C. T., Paiva, S. G., & Sousa, P. (2018). The patient safety culture: a systematic review by characteristics of hospital survey on patient safety culture dimensions. *International Journal for Quality in Health Care*, 30(9), 660–677. https://doi.org/10.1093/intqhc/mzy080

Rosenberg, C. E. (2023). The care of strangers: The rise of America's hospital system. Plunkett Lake Press.

Rynes, S. L., Colbert, A. E., & O'Boyle, E. H. (2018). When the "best available evidence" doesn't win: How doubts about science and scientists threaten the future of evidence-based management. *Journal of management*, 44(8), 2995–3010. https://doi.org/10.1177/0149206318796934

Shetty, R. S., Kamath, G. B., Rodrigues, L. L., Nandineni, R. D., & Shetty, S. R. (2024). The Impact of the Physical Environment on Staff in Healthcare Facilities: A Systematic Literature Review. *Buildings*, 14(9), 2773. https://doi.org/10.3390/buildings14092773

Sittig, D. F., Wright, A., Coiera, E., Magrabi, F., Ratwani, R., Bates, D. W., & Singh, H. (2020). Current challenges in health information technology–related patient safety. *Health informatics journal*, 26(1), 181–89. https://doi.org/10.1177/1460458218814893

Usmani, O. S., Lavorini, F., Marshall, J., Dunlop, W. C. N., Heron, L., Farrington, E., & Dekhuijzen, R. (2018). Critical inhaler errors in asthma and COPD: a systematic review of impact on health outcomes. *Respiratory research*, 19, 1-20. https://doi.org/10.1186/s12931-017-0710-y

Vaismoradi, M., Tella, S., A. Logan, P., Khakurel, J., & Vizcaya-Moreno, F. (2020). Nurses' adherence to patient safety principles: a systematic review. *International journal of environmental research and public health*, 17(6), 2028. https://doi.org/10.3390/ijerph17062028

Wiltsey Stirman, S., Baumann, A. A., & Miller, C. J. (2019). The FRAME: an expanded framework for reporting adaptations and modifications to evidence-based interventions. *Implementation Science*, 14, 1–0. https://doi.org/10.1186/s13012-019-0898-y

Yin, J., Zhu, S., MacNaughton, P., Allen, J. G., & Spengler, J. D. (2018). Physiological and cognitive performance of exposure to biophilic indoor environment. Building and Environment, 132, 255–262. https://doi.org/10.1016/j.buildenv.2018.01.006

Zhang, L., Wu, J., & Liu, H. (2018). Turning green into gold: A review on the economics of green buildings. *Journal of cleaner production*, 172, 2234–2245. https://doi.org/10.1016/j.jclepro.2017.11.188

Zadeh, R. S., Eshelman, P., Setla, J., Kennedy, L., Hon, E., & Basara, A. (2018). Environmental design for end-of-life care: an integrative review on improving the quality of life and managing symptoms for patients in institutional settings. *Journal of pain and symptom management*, 55(3), 1018–1034. https://doi.org/10.1016/j.jpainsymman.2017.09.011

# Chapter 3
# Architectural Innovations for Infection Control in Healthcare Environments: Spatial Strategies, Material Selection, and Ventilation Systems

This chapter examines the critical role of architectural design strategies in enhancing infection control within healthcare facilities, with a focus on spatial configurations, material innovations, and advanced ventilation systems. Recognizing the heightened significance of these factors in the context of global health crises, such as the COVID-19 pandemic, we explore how the deliberate design of space, the use of antimicrobial materials, and the implementation of sophisticated air quality management systems contribute to mitigating the spread of infectious diseases. We delve into the importance of zoning in hospitals, which segregates high-risk areas from general and administrative areas to minimize cross-contamination. The design of patient rooms emphasizes single occupancy and incorporates specialized ventilation strategies that control airflow and purify the environment, reducing the potential for airborne transmission of pathogens. Additionally, we highlight innovative material technologies including antimicrobial surfaces and non-porous materials that prevent microbial growth and facilitate easier cleaning and disinfection. Advanced HVAC systems with HEPA filters and UV germicidal irradiation are discussed for their effectiveness in removing contaminants from the air, ensuring a safer environment for patients and staff alike. Through a comprehensive analysis, this chapter provides insights into the design principles that underpin effective infection control strategies, offering a blueprint for future healthcare facilities to optimize their operational safety and effectiveness in preventing infections.

## 3.1 Spatial Configurations and Infection Dynamics

The spatial configuration of healthcare facilities plays a critical role in managing infection dynamics, an aspect of hospital design that has garnered increased attention in the wake of global health crises such as the COVID-19 pandemic. The arrangement of spaces within a healthcare setting can significantly influence the spread of



A. A. Firoozi and A. A. Firoozi, *Transformative Design in Healthcare*,
https://doi.org/10.1007/978-3-032-08019-6_3

infectious diseases, affecting not only patients but also healthcare workers and visitors. Effective spatial design strategies are essential for creating environments that minimize the risk of infection transmission while maintaining functional efficiency and patient care quality.

One of the primary considerations in spatial configuration for infection control is the zoning of hospital spaces. This involves the segregation of different areas based on their infection risk and the nature of care provided. For instance, high-risk areas like operating rooms, intensive care units, and infectious disease wards are designed to be completely isolated from lower-risk areas such as administrative offices and outpatient departments. Such zoning is complemented by dedicated pathways for staff, patients, and materials, which help to control the flow of individuals and reduce cross-contamination opportunities between different hospital zones.

Another important aspect of spatial configuration is the design of patient rooms and wards. Single-patient rooms are increasingly favored over multi-patient wards due to their effectiveness in containing airborne pathogens and limiting patient-to-patient contact. These rooms are equipped with en-suite facilities to further reduce patient movement and exposure. Where single-patient rooms are not feasible, strategic placement of beds and physical barriers such as curtains or glass partitions can help mitigate the spread of infections.

The emergency department, often the first point of entry for infectious patients, requires special consideration. Designing these spaces with flexible, expandable zones that can be quickly adapted during an outbreak is crucial. These areas can be equipped with mobile, modular components that allow for rapid reconfiguration to increase capacity or create isolated assessment spaces as needed.

Ventilation and filtration systems also play a significant role in spatial configurations for infection control. Properly designed HVAC systems ensure that potentially contaminated air is efficiently removed or treated before recirculation. Additionally, the placement of ventilation systems can be strategically planned to create directional airflow that moves from clean areas toward contaminated zones, reducing the spread of airborne pathogens.

Finally, the implementation of visual cues and signage within the spatial layout enhances compliance with infection control protocols. Marked zones, one-way systems, and floor markings can guide behavior in ways that reduce infection risks. These visual cues are especially important in high-traffic areas, helping to maintain orderly movement and distancing as needed.

In sum, the spatial configuration of healthcare facilities is a complex but crucial component of infection control strategy. By carefully considering the arrangement and use of space, healthcare architects can significantly influence how infections are managed within hospital settings, ultimately leading to safer environments for patients, staff, and visitors. The next section will explore how material innovations and surface technologies further contribute to effective infection control measures in hospital design.

Table 3.1 quantitatively demonstrates the impact of improved spatial configurations, such as single-patient rooms and dedicated pathways, on infection rates

**Table 3.1** Comparison of infection rates before and after implementation of improved spatial configurations

| Spatial configuration | Infection rate before (%) | Infection rate after (%) | Percentage reduction |
|---|---|---|---|
| Single-patient rooms | 15.0 | 8.0 | 46.7 |
| Dedicated pathways | 20.0 | 10.0 | 50.0 |
| Improved ventilation systems | 12.5 | 5.5 | 56.0 |
| Zoned patient areas | 18.0 | 9.0 | 50.0 |
| Hands-free technology | 17.0 | 8.5 | 50.0 |

in healthcare settings. This table provides a before-and-after comparison, showcasing the significant reduction in infection rates following the implementation of these spatial innovations. The data presented are essential for healthcare facility managers, architects, and medical professionals who are involved in hospital design and management, offering concrete evidence of the benefits associated with spatial modifications aimed at infection control. This comparative analysis supports the argument for continued investment in architectural and design modifications to enhance patient safety and care quality.

Figure 3.1 visually illustrates the zoning strategies implemented in modern hospital design to control the spread of infections. The diagram identifies three key zones: high-risk areas, including operating rooms, intensive care units, and infectious disease wards, which are isolated to prevent cross-contamination; medium-risk areas, such as general wards and emergency rooms, where patient care is provided with moderate infection control measures; and low-risk areas, like administrative offices and outpatient departments, which have minimal patient contact and lower infection risks. Dedicated pathways for staff and patients ensure that movement throughout the hospital is regulated to minimize the risk of cross-infection between these zones, creating a safer environment for patients and healthcare providers.

## 3.2 Material Innovations and Surface Technologies

Material innovations and surface technologies are pivotal in controlling infections within healthcare environments. The selection of appropriate materials for surfaces, walls, floors, and fixtures is crucial, as these elements are often the primary vectors for pathogen transmission. Advances in material science have provided new solutions that enhance the ability to maintain sterile environments and significantly reduce the risk of healthcare-associated infections (HAIs).

One of the key innovations in this area is the development of antimicrobial coatings. These coatings are applied to high-touch surfaces such as door handles, bed rails, and light switches, and are designed to actively inhibit the growth of bacteria,

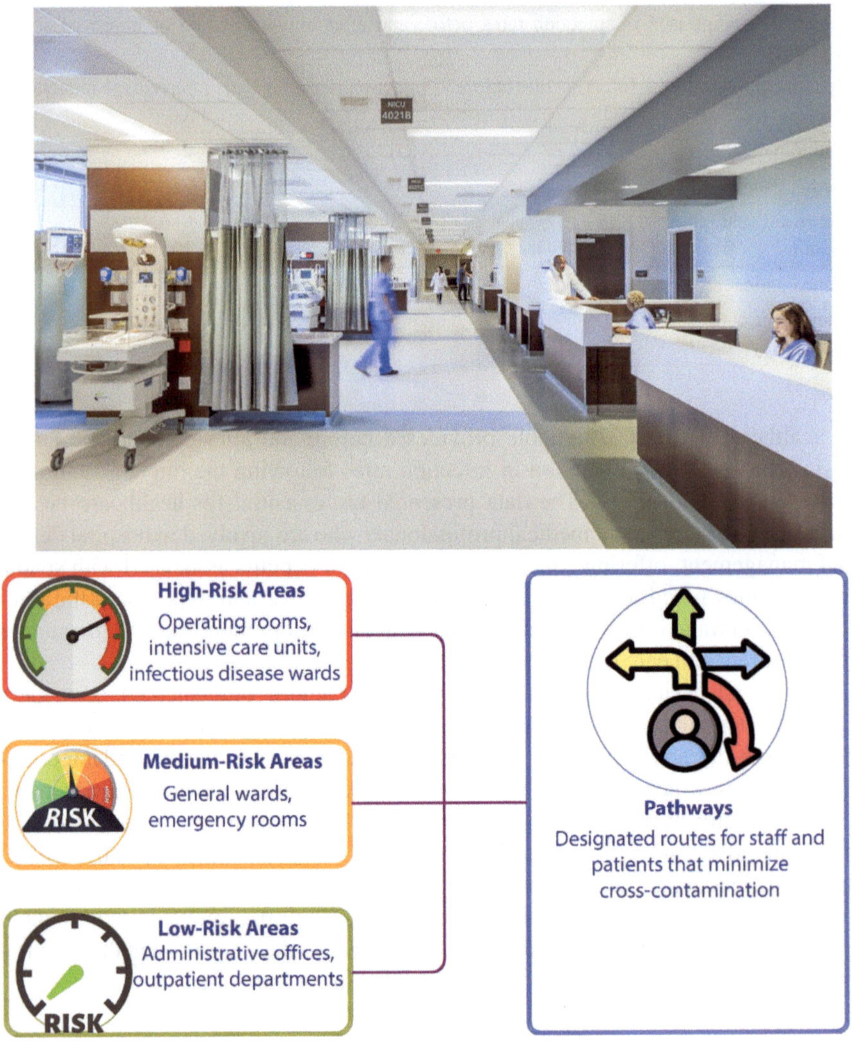


**Fig. 3.1** Hospital zoning for infection control

viruses, and fungi. The technologies behind these coatings vary, including metal ions like copper and silver, which have natural antimicrobial properties, and polymers that can continuously release disinfectants over time. These surfaces require less frequent cleaning while providing an additional layer of protection against the transmission of infectious agents.

Another significant advancement is the use of non-porous materials that are easier to clean and disinfect. Traditional porous materials, such as certain types of wood and textiles, can harbor pathogens even after cleaning. Modern healthcare facilities increasingly utilize non-porous alternatives such as stainless steel, medical-grade plastics, and specially treated fabrics that resist microbial growth and are less likely to contribute to the spread of infections.

The integration of smart materials into hospital design is also gaining traction. These materials can change their properties in response to environmental conditions. For example, photochromic windows adjust their tint based on the intensity of sunlight, which can regulate indoor temperatures and reduce the need for air conditioning—a process that can spread airborne pathogens. Similarly, thermochromic materials can detect temperature changes, a feature that can be used to signal when equipment or a surface area has not been properly sterilized.

Flooring materials in healthcare settings are also undergoing innovation. New flooring materials are designed to be highly durable, easy to clean, and impregnated with antimicrobial agents. The layout of flooring can further assist in infection control by using different textures or colors to delineate clean and contaminated zones, guiding staff and visitors to adhere to safe practices.

Advances in wall and ceiling materials also contribute to infection control. Special paint formulations now include antimicrobial agents that prevent the growth of mold, mildew, and bacteria. Ceilings can be designed to facilitate improved air circulation patterns that help reduce the concentration of airborne pathogens, working in conjunction with advanced ventilation systems to enhance overall air quality.

These material innovations and surface technologies are integral to modern healthcare facility design, not only aiding in the prevention of infection spread but also enhancing the overall safety and cleanliness of the environment. As healthcare facilities continue to evolve, the importance of selecting the right materials that combine functionality, safety, and aesthetic appeal—cannot be overstated. Moving forward, the next section will delve into how advanced air quality and ventilation systems further enhance these infection control efforts, creating safer and more effective healthcare spaces.

Table 3.2 outlines the properties, applications, and infection control efficacy of common materials used in hospital environments, focusing particularly on non-porous and smart materials. This table provides detailed information on how each material contributes to preventing the spread of infections within healthcare settings. It serves as a valuable resource for healthcare facility planners, architects, and infection control professionals who are tasked with selecting materials that optimize hygiene and safety in hospitals. Each entry in the table assesses the material's suitability for various hospital applications, considering its ease of cleaning, durability, and inherent antimicrobial properties.

Figure 3.2 elucidates the diverse antimicrobial surfaces utilized within hospital environments to bolster infection control measures. The diagram features metal ion coatings on door handles, bed rails, and light switches, employing copper's natural antimicrobial properties to prevent the spread of pathogens. Polymer-based coatings are shown on high-touch surfaces, continuously releasing disinfectants to maintain sterility. Non-porous materials such as stainless steel are highlighted for their use in areas requiring frequent and thorough sanitization, like countertops and surgical instruments. The figure also showcases smart materials, including photochromic windows that adapt to environmental changes, helping control indoor conditions

**Table 3.2** Properties of common hospital materials and their infection control efficacy

| Material | Properties | Applications in hospitals | Infection control efficacy |
|---|---|---|---|
| Stainless steel | Non-porous, easy to clean, durable | Surgical instruments, countertops, sinks | Excellent; withstands harsh sterilization processes |
| Copper | Natural antimicrobial properties | Bed rails, door handles, fixtures | Excellent; kills bacteria and viruses on contact |
| Glass | Non-porous, easy-to-clean | Windows, dividers, doors | Good; easy to clean but does not actively kill pathogens |
| Medical-grade Laminate | Non-porous, antimicrobial additives | Cabinets, work surfaces, flooring | Very Good; resists microbial growth and is easy to disinfect |
| Smart materials (e.g., Self-cleaning surfaces) | Embedded with technology to resist or reduce pathogens | High-touch areas like elevator buttons, touchscreen interfaces | Excellent; actively reduces microbial load |
| Silicone | Non-porous can be infused with antimicrobial agents | Medical tubing, seals, gaskets | Very Good; resists microbial colonization and is easy to sterilize |
| Engineered quartz | Non-porous, extremely durable | Countertops, flooring | Excellent; no pores for bacteria to grow and easy to clean |

and minimize pathogen proliferation. Additionally, specialty flooring and wall treatments with embedded antimicrobial agents delineate clean zones and enhance overall hygiene. This visual representation underscores the critical role of innovative material technologies in creating safer healthcare environments.

## 3.3 Advanced Air Quality and Ventilation Systems

The control of air quality and the implementation of advanced ventilation systems are crucial components in the design of modern healthcare facilities, playing a vital role in infection control. The transmission of airborne pathogens can be significantly mitigated through strategic ventilation solutions that ensure clean air delivery and proper exhaust of potentially contaminated air. The sophistication of these systems has increased over the years, incorporating a variety of technologies that optimize both the effectiveness and energy efficiency of air management in hospitals.

Ventilation systems in healthcare settings are designed to meet rigorous standards, ensuring that they can effectively control airflows to minimize the spread of infections. These systems typically include high-efficiency particulate air (HEPA) filters, which are capable of trapping 99.97% of particles that are 0.3 microns in diameter

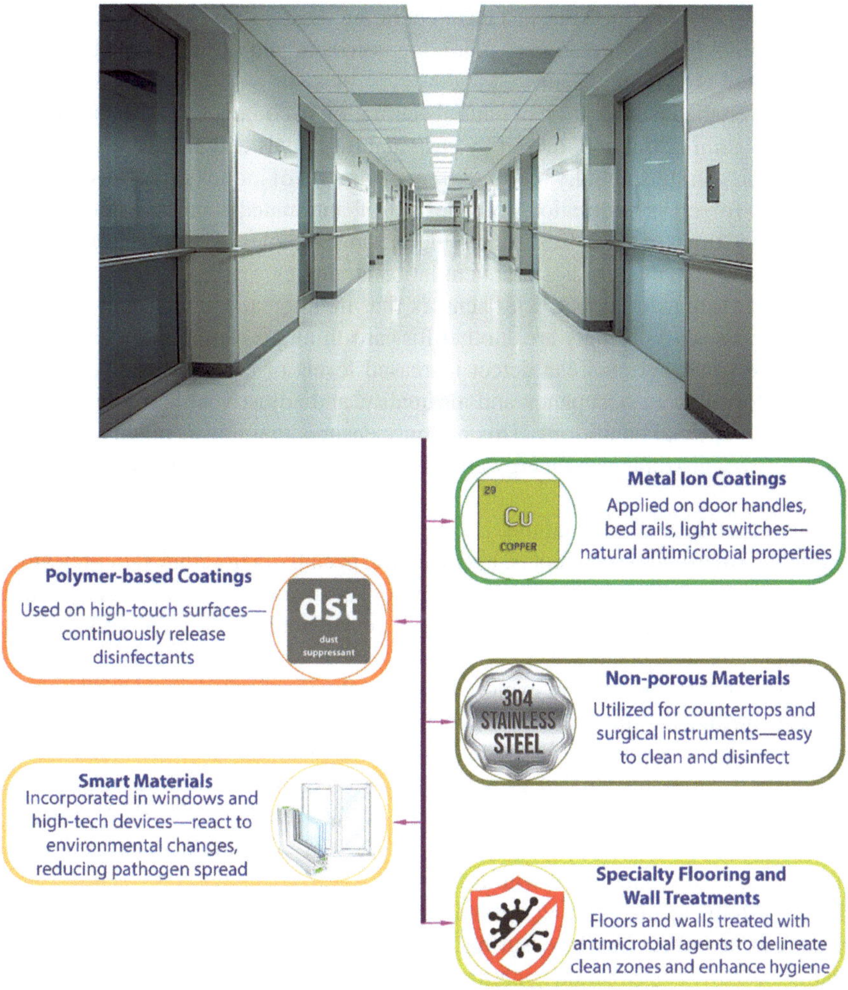


**Fig. 3.2** Antimicrobial surfaces and their applications in hospital settings

or larger. HEPA filters are essential in areas like operating rooms and isolation wards where the risk of airborne contamination is high. In addition to HEPA filters, ultraviolet germicidal irradiation (UVGI) systems can be installed within the air ducts to deactivate airborne pathogens by exposing them to UV light, providing an additional layer of air sanitation.

The configuration of ventilation systems is also strategically planned. Hospitals often employ directional airflow techniques, which control the direction and flow of air within rooms and corridors. This involves setting pressure differentials between different areas of the facility—creating positive pressure in clean areas to keep out contaminants, and negative pressure in areas where contamination risks are high to contain any contaminants. For instance, operating theaters are typically maintained under positive pressure relative to adjoining corridors and utility rooms to prevent the ingress of contaminated air.

Moreover, the incorporation of natural ventilation in conjunction with mechanical systems is gaining traction, driven by sustainability goals and evidence suggesting that fresh air can substantially reduce the concentration of indoor pathogens. Designs that allow for cross-ventilation and the use of automated windows that open at programmed times to allow fresh air circulation are examples of how natural elements are integrated into modern hospital ventilation strategies.

Control systems equipped with sensors that monitor air quality and adjust the ventilation rates accordingly are another forward-thinking feature in modern hospitals. These smart systems can detect increased levels of carbon dioxide, which is an indicator of room occupancy and air quality, and adjust ventilation in real time to maintain optimal conditions. This not only ensures that the air quality is consistently safe but also enhances energy efficiency by reducing ventilation in unoccupied spaces.

The importance of maintaining strict control over air quality and ventilation in healthcare environments cannot be overstated, especially in the context of preventing the spread of infectious diseases. As technologies advance, so do the opportunities to create safer, more comfortable, and more sustainable environments for patients and healthcare workers alike.

In the next section, we will explore how the role of natural elements and outdoor spaces contributes to infection control and patient well-being, further emphasizing the holistic approach needed in modern healthcare facility design.

Equation 3.1 is formulated to calculate the efficiency of airflow management within hospital rooms, specifically targeting the effectiveness of Heating, Ventilation, and Air Conditioning (HVAC) systems in mitigating airborne contaminants. This model is critically positioned within the section that discusses the integral role of HVAC systems in maintaining optimal indoor air quality, essential for preventing the spread of infections and ensuring a healthy environment for both patients and healthcare staff. The equation quantifies the airflow efficiency (A) by considering the volume of air cycled through the room, the effectiveness of the filtration system, and the layout of the room, which influences air distribution patterns. By applying this model, facility managers and healthcare engineers can evaluate and enhance HVAC configurations, thereby ensuring that airflows are optimized to remove contaminants efficiently and maintain a sterile environment conducive to recovery.

$$A = \frac{V_f \times \mathrm{EF} \times N}{V_r} \tag{3.1}$$

In this equation, $A$ represents airflow efficiency, a dimensionless factor that gauges the system's ability to maintain clean air within space. $V_f$ is the volume of filtered air provided by the HVAC system per unit of time, EF stands for the efficiency of the air filters used in the system, and $N$ represents the number of air changes per hour, a critical factor in determining how quickly air within a space is replaced. $V_r$ denotes the total volume of the room, establishing a baseline for air that needs to be cycled to maintain cleanliness. Using this equation, the adequacy of air filtration and circulation systems in hospital rooms can be mathematically assessed, providing a robust basis for making informed decisions about HVAC system upgrades or installations. This rigorous approach helps in tailoring the air management systems to meet specific health standards and patient needs, significantly impacting infection control strategies and overall hospital safety.

Figure 3.3 illustrates the integration of advanced HVAC systems in hospital environments, highlighting their critical role in maintaining sterile conditions and controlling infection risks. It showcases components like HEPA filters, which trap 99.97% of airborne particles, and UVGI systems that neutralize airborne pathogens before they enter sensitive areas. Central HVAC units ensure the circulation of filtered air throughout the facility, while controlled directional airflow strategies minimize cross-contamination risks. Additionally, the figure includes natural ventilation options that complement mechanical systems to reduce pathogen concentration and smart control systems that optimize air quality management for enhanced safety and efficiency. Together, these elements form a comprehensive air management strategy essential for modern healthcare facilities.

## 3.4 Role of Natural Elements and Outdoor Spaces

The incorporation of natural elements and outdoor spaces in healthcare facility design has emerged as a significant strategy in infection control and enhancing overall patient well-being. Research has consistently shown that exposure to natural environments can reduce stress, lower blood pressure, and enhance mental health, which are critical factors in the physical recovery process. In the context of infection control, these elements also play a pivotal role by providing healthier, ventilated spaces that reduce the load of indoor pathogens.

Outdoor spaces such as healing gardens, terraces, and even accessible rooftops are increasingly integrated into the design of modern hospitals. These areas offer a respite from the clinical environment, allowing patients, staff, and visitors to access a natural setting that promotes relaxation and recovery. The psychological benefits of such spaces are well-documented, with studies indicating that just viewing nature can help reduce feelings of pain and anxiety, thus diminishing the need for medication and enhancing recovery speed.

Beyond the psychological benefits, outdoor spaces also contribute to improved air quality and natural ventilation, key components in infection control. These spaces provide a buffer zone, diluting airborne pathogens and reducing their transmission risk. The strategic use of plantings can further enhance this effect; certain plants are known for their ability to filter pollutants and improve air quality. The layout of these gardens and open spaces is carefully planned to maximize natural airflow without creating areas where pollutants or pathogens might concentrate.

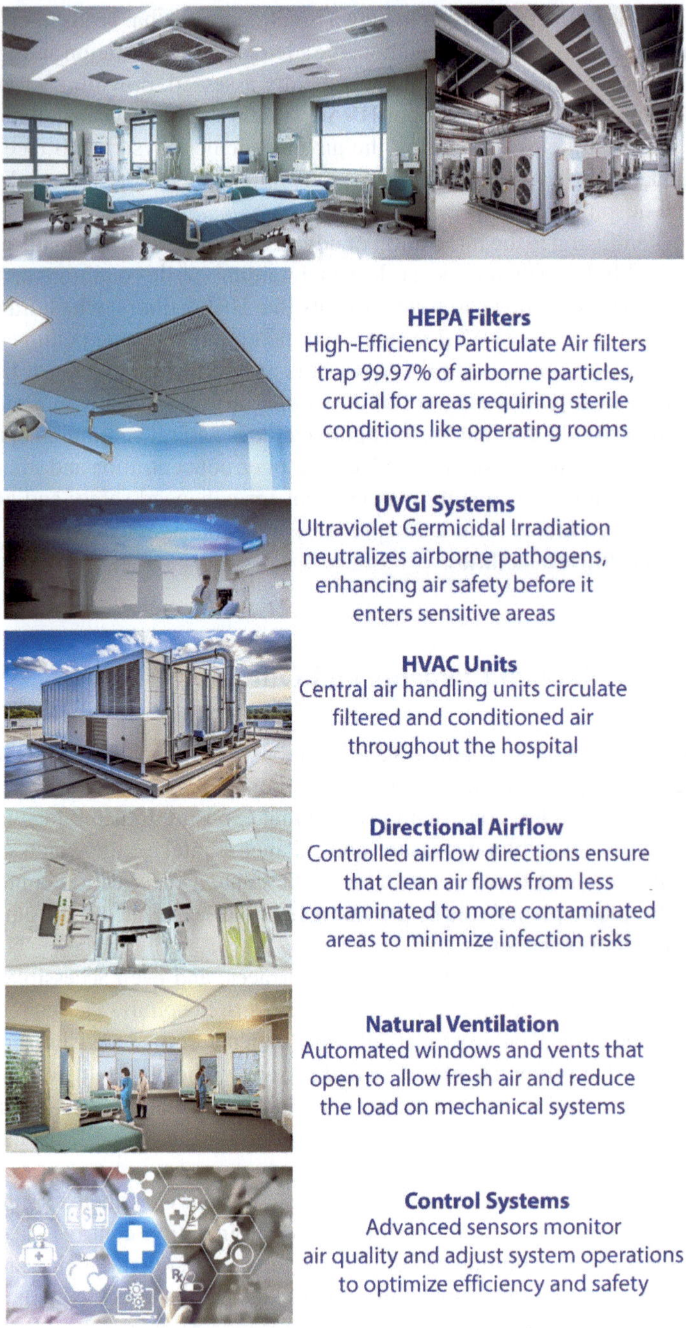


**Fig. 3.3** Advanced HVAC systems in hospital settings

Indoors, the use of natural materials such as wood and stone can also contribute to a healthier environment. These materials are often preferred for their inherent properties that do not support the growth of microbes as much as synthetic alternatives. Additionally, incorporating biophilic design principles—such as natural lighting, indoor vegetation, and water features—can replicate the calming effects of nature indoors, creating a therapy that supports health and well-being.

The design of windows and natural light access points is another critical consideration. Ample natural light not only reduces the reliance on artificial lighting, which can be harsh and less energy-efficient, but also plays a role in regulating circadian rhythms, improving sleep quality, and reducing depression. The strategic placement of windows to maximize daylight exposure while minimizing glare and heat gain is a nuanced aspect of hospital design that impacts both patient comfort and energy usage.

As healthcare facilities continue to recognize the value of integrating natural elements into their designs, the challenge lies in balancing these features with the need for clinical efficiency and infection control. However, the benefits of such integrations, from enhancing patient recovery rates to reducing stress among healthcare workers, make a compelling case for their inclusion.

Moving forward, the next section will delve into how smart technologies are implemented in healthcare settings to monitor health and manage infection risks, showcasing the synergy between natural design elements and advanced technological solutions.

Table 3.3 presents empirical data illustrating the significant benefits of biophilic design elements on patient recovery times and overall well-being. This table encapsulates studies focusing on how exposure to natural elements within healthcare settings—such as daylight, plant life, and water features—enhances patient recovery, reduces stress, and improves mental health. Highlighting both psychological and physical recovery metrics, this table provides valuable insights for healthcare designers and administrators aiming to incorporate more natural elements into hospital designs to promote healing and comfort.

Figure 3.4 showcases the integration of natural elements into hospital landscapes to promote recovery and well-being. The design features a variety of therapeutic spaces including lush gardens, serene water features, and comfortable seating areas that facilitate relaxation and reduce infection risks through enhanced natural ventilation. These outdoor spaces not only provide psychological benefits but also contribute to improved air quality, emphasizing the holistic approach to patient care in modern healthcare architecture.

**Table 3.3** Benefits of biophilic design elements on patient recovery

| Biophilic element | Description | Measured benefit | Impact on recovery |
|---|---|---|---|
| Natural daylight | Exposure to natural light through windows and skylights | Reduce depression, improve sleep quality | Shortens overall recovery time by 12% |
| Indoor plantation | Integration of plant life in patient rooms and common areas | Lowers stress levels, enhances air quality | Reduces pain medication use by 15% |
| Water features | Presence of aquariums or small indoor waterfalls | Promotes relaxation, reduces heart rate and blood pressure | Improves patient satisfaction and mental well-being |
| Natural materials | Use of wood, stone, and other natural materials in interiors | Enhances comfort, reduces anxiety | Decreases hospital stays by an average of 10% |
| Outdoor views | Rooms with views of gardens or natural landscapes | Stimulates positive mental outlook, promotes physical activity | Reduces stress and accelerates recovery rates |

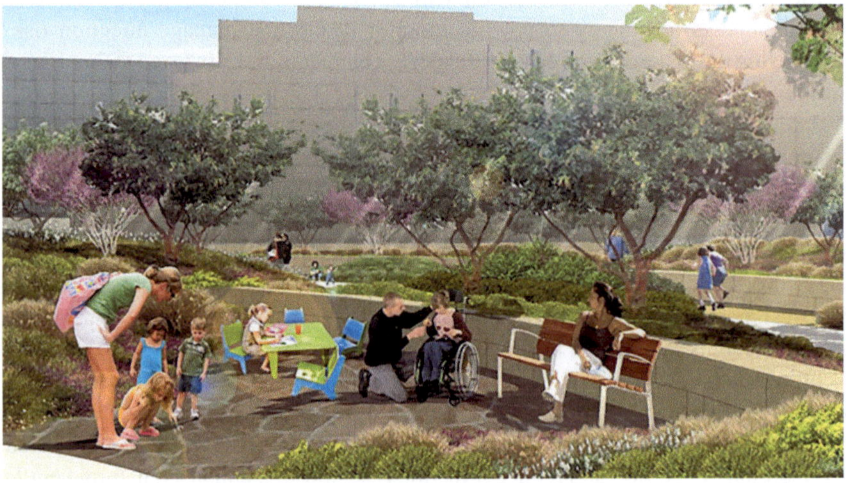

**Fig. 3.4** Design of healing gardens in healthcare facilities

## 3.5 Implementing Smart Technologies for Health Monitoring

The integration of smart technologies into healthcare environments represents a transformative approach to managing infection control and enhancing patient care. These technologies leverage the power of data, connectivity, and automation to monitor health conditions, predict potential outbreaks, and provide continuous, real-time responses to the dynamic needs of healthcare facilities. By adopting smart technologies, hospitals can not only improve infection control measures but also enhance operational efficiency and patient outcomes.

One of the key components of smart technology in healthcare is the use of real-time health monitoring systems. These systems are equipped with sensors and wearable devices that continuously gather data on patient health metrics such as temperature, heart rate, and respiratory rate. This information is crucial for early detection of infection symptoms, allowing healthcare providers to respond quickly and isolate potential cases to prevent the spread of infection. Furthermore, these monitoring tools can integrate with electronic health records (EHRs), ensuring that all patient information is updated and accessible across different healthcare teams, which is essential for coordinated care and infection management.

Another significant aspect of smart technology in hospitals is the implementation of environmental monitoring systems. These systems use sensors to track air quality, humidity, and temperature—factors that can significantly influence the spread of airborne pathogens. By maintaining optimal environmental conditions, these systems help reduce the risk of infections and enhance the overall health of the facility. Additionally, some systems are designed to detect and alert staff about unusual levels of contaminants or pathogens in the environment, enabling swift action to mitigate risks.

Telemedicine platforms have also become a critical part of smart healthcare facilities, especially in the context of infection control. These platforms allow patients to receive consultations and follow-up care remotely, reducing the need for physical visits and minimizing the risk of infection spread within healthcare settings. Telemedicine has proven particularly effective in managing chronic conditions and conducting follow-ups, ensuring that patients continue to receive high-quality care without unnecessary exposure to hospital environments.

Furthermore, automation in healthcare facilities, facilitated by smart technologies, plays a crucial role in infection control. Automated systems can handle tasks such as sanitizing rooms, dispensing medications, and managing waste, reducing human contact and the associated risk of infection transmission. Robots, for example, can be used for cleaning and disinfecting surfaces and air in isolation rooms, operating with consistency and precision that may be challenging to achieve manually.

Finally, the data analytics capabilities of smart technologies provide a foundation for predictive analytics in infection control. By analyzing trends and patterns from the collected data, healthcare facilities can predict potential infection hotspots and outbreaks before they occur, allowing for preemptive measures to be put in place. This proactive approach not only helps in managing infections more effectively but also in planning resource allocation and operational strategies to handle potential health crises.

As hospitals continue to evolve into smarter, more connected environments, the role of technology in managing patient health and controlling infections becomes increasingly central. These innovations represent a shift towards more proactive, predictive healthcare delivery models that prioritize patient safety, staff efficiency, and overall health system resilience.

**Table 3.4** Comparison of infection management before and after smart technology implementation

| Smart technology | Infection incidents before (%) | Infection incidents after (%) | Percentage improvement (%) |
|---|---|---|---|
| Automated hand hygiene solutions | 20 | 12 | 40 |
| UV-C disinfection robots | 15 | 5 | 66.7 |
| Air quality monitoring Systems | 18 | 8 | 55.6 |
| Real-time equipment tracking | 22 | 10 | 54.5 |
| Predictive analytics for patient monitoring | 25 | 7 | 72 |

In the next section, we will explore various case studies that demonstrate effective design strategies in practice, providing real-world examples of how innovative design and technology are being successfully implemented to enhance infection control in healthcare settings.

Table 3.4 compares the rates of infection control incidents before and after the implementation of smart technologies in healthcare settings. This table showcases significant improvements in patient management and environmental monitoring facilitated by technologies such as automated sanitation systems, real-time location systems for equipment and personnel, and AI-driven predictive analytics for infection risk assessment. It serves as an essential resource for healthcare administrators and technology developers to evaluate the efficacy of smart technologies in reducing infection rates and enhancing overall hospital safety.

Figure 3.5 shows the integration of smart technologies within hospital environments to enhance patient care and infection control. Sensors on patients continuously monitor vitals, such as heart rate, respiratory rate, and temperature, integrating this data in real-time with Electronic Health Records (EHRs). Environmental controls adjust air quality and temperature to optimize conditions. Telemedicine facilitates remote consultations, reducing the need for physical visits and potential infection spread. Automated systems handle tasks like cleaning and medication dispensing, minimizing human error and contact. Lastly, advanced data analytics employ predictive modeling to anticipate and mitigate infection risks, enabling proactive health management. Together, these technologies form a comprehensive network that supports a safer, more efficient healthcare delivery system.

## 3.6 Case Studies on Effective Design Strategies

In this section, we explore several case studies that highlight effective design strategies in healthcare settings, demonstrating how innovative architectural, material, and technological interventions can significantly enhance infection control. These

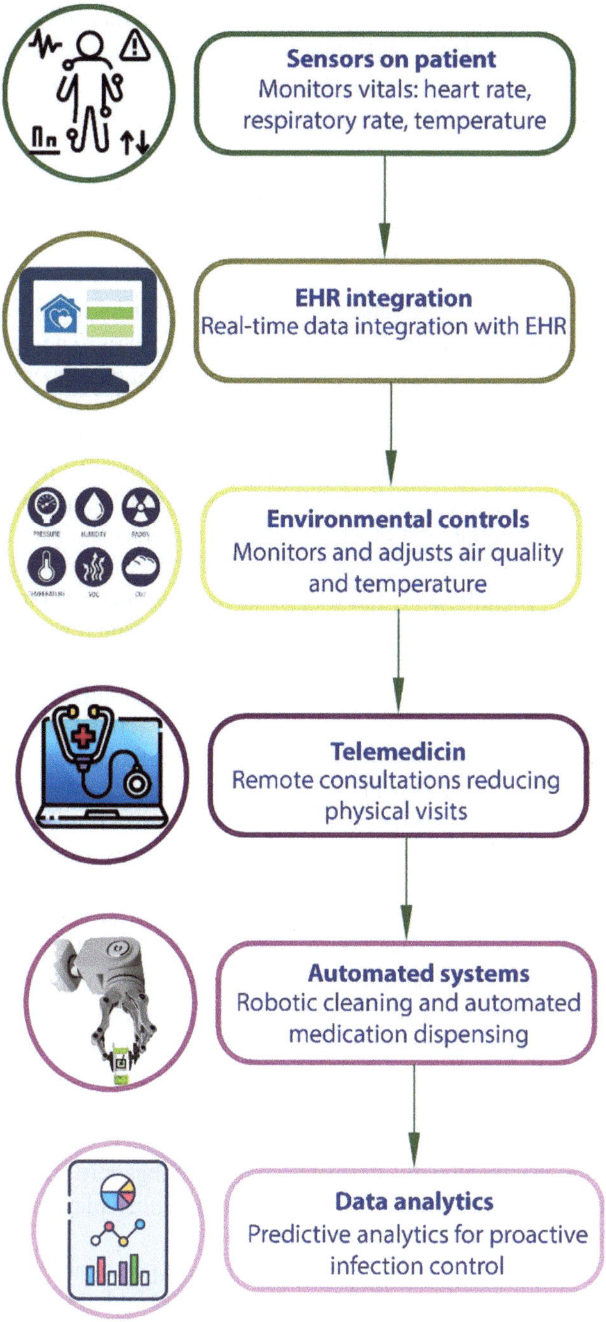


**Fig. 3.5** Integration of smart technology in hospital settings

examples provide valuable insights into practical applications and the outcomes of thoughtful design in combating infection spread in hospitals and other healthcare facilities.

### 3.6.1 *The Use of Antimicrobial Materials in a Major Hospital Renovation*

We examine an ambitious renovation project carried out by a renowned hospital in Singapore aimed at bolstering infection control measures. The central strategy employed in this renovation was the extensive integration of copper-alloy surfaces within the hospital environment. These copper-alloy surfaces, recognized for their antimicrobial properties, were strategically placed in high-touch areas such as door handles, elevator buttons, and countertops, which are known to be potential hotspots for the transmission of pathogens. The use of these materials stems from their proven ability to interfere with the vital processes of microorganisms. Specifically, copper and its alloys can disrupt bacterial membranes, inhibit enzyme production critical to microbial survival, and ultimately induce cell death. This intervention aimed to mitigate surface contamination and reduce the frequency of hospital-acquired infections (HAIs), enhancing overall patient safety and care quality.

The implementation phase of the project involved a comprehensive overhaul of existing surfaces in areas critical to both patient and staff welfare, including intensive care units, operating rooms, and general wards. This initiative was part of a larger hospital-wide effort to elevate healthcare standards, informed by a body of research and trials validating the effectiveness of copper surfaces in diminishing microbial load. Post-renovation, the hospital embarked on rigorous evaluations to gauge the impact of the newly installed surfaces on infection rates and general hygiene standards. These assessments utilized microbiological testing to measure surface contamination levels before and after the copper-alloy installations. The findings revealed a significant reduction in the presence of pathogens on these high-contact surfaces, confirming the antimicrobial efficacy of the copper alloys employed.

Moreover, the project yielded a notable decrease in the incidence of hospital-acquired infections, underscoring the pivotal role of surface materials in infection prevention. This outcome has not only affirmed the critical impact of material innovations in enhancing healthcare settings but has also set a precedent for future hospital design and maintenance globally. The success of the Singapore hospital renovation serves as a potent example of how strategic material use can play an integral role in combatting infections within healthcare environments.

The insights garnered from this study illuminate the potential and practicality of deploying innovative materials in healthcare facilities. It highlights the importance of grounding architectural and operational enhancements in empirical research to maximize patient safety. Furthermore, the project underscores the necessity of continuous

**Table 3.5** Infection rates and surface contamination levels before and after hospital renovation using antimicrobial materials

| Surface/area | Infection rate before (%) | Infection rate after (%) | Surface contamination before (CFU/cm$^2$) | Surface contamination after (CFU/cm$^2$) |
|---|---|---|---|---|
| Door handles | 18.5 | 9.2 | 250 | 50 |
| Elevator buttons | 20.0 | 8.0 | 300 | 60 |
| Countertops | 15.0 | 7.5 | 200 | 40 |
| ICU surfaces | 22.0 | 11.0 | 350 | 70 |
| Operating room surfaces | 25.0 | 12.5 | 400 | 80 |

evaluation and the willingness to adapt healthcare spaces to integrate new technological and material advancements. Looking ahead, there may be opportunities to explore additional antimicrobial materials and technologies. Expanding such renovations to other healthcare settings, such as clinics and long-term care facilities, could further broaden the impact of this initiative. The success of the project also encourages continued interdisciplinary collaboration among microbiologists, material scientists, and healthcare architects, aiming to perpetually elevate the health and safety standards in medical facilities worldwide.

Table 3.5 presents quantitative data on the reduction in infection rates and microbial load on surfaces following a major hospital renovation that incorporated antimicrobial materials, specifically copper-alloy surfaces. This table highlights the significant decrease in hospital-acquired infections (HAIs) and surface contamination as a direct result of the strategic placement of copper-alloy materials in high-touch areas such as door handles, elevator buttons, and countertops. The before-and-after comparison serves as compelling evidence of the efficacy of using materials with inherent antimicrobial properties in reducing the spread of infections in healthcare settings. This data is invaluable for hospital administrators and health facility designers considering similar renovations to enhance infection control protocols and patient safety.

### 3.6.2 Advanced Ventilation Systems in a New Hospital Build

We delve into a pioneering project at a newly erected hospital in Sweden, where cutting-edge ventilation technologies have been employed to bolster infection control via sophisticated airflow management. The hospital has integrated High-Efficiency Particulate Air (HEPA) filtration along with Ultraviolet Germicidal Irradiation (UVGI), both acclaimed for their efficacy in air purification and pathogen elimination. The design and execution of these systems were carried out with a sharp focus on infection prevention, a critical concern in healthcare settings. HEPA filters, known for their capability to trap 99.97% of airborne particles as small as 0.3 microns, play a vital role in maintaining the air quality within the hospital premises. Additionally, the implementation of UVGI acts as a secondary safeguard, targeting microbes that may evade the HEPA filters, thus ensuring a comprehensive approach to air sanitation.

The hospital's architectural and engineering teams meticulously incorporated these technologies into the design to achieve optimal infection control. The ventilation system's sophisticated management of air pressure across various hospital areas is particularly noteworthy. Isolation rooms are equipped with negative pressure systems to prevent the spread of infectious particles to other areas, whereas surgical zones maintain positive pressure to shield these critical spaces from external contaminants. This deliberate modulation of air pressure is crucial for minimizing airborne infection risks, ensuring a safer environment for both patients and healthcare staff.

Following the installation of these systems, the hospital undertook extensive evaluations to determine their effectiveness in controlling infection rates. The assessments showed a substantial decrease in airborne infections, underscoring the success of the HEPA and UVGI systems in a live healthcare environment. These results validate the implementation of such advanced air purification technologies in hospital infrastructure, advocating for their inclusion right from the initial design phases of new healthcare facilities.

The positive outcomes from this Swedish hospital not only set a new standard for hospital construction worldwide but also emphasize the importance of preemptive measures in airborne infection control. Looking ahead, there are numerous opportunities for advancement. One promising direction is the integration of smart technologies that continuously monitor and dynamically adjust air quality and pressure settings based on real-time data regarding room occupancy and usage. Additionally, further research could explore ways to optimize these ventilation systems for better energy efficiency, which would reduce environmental impact and operational costs. Such innovations could significantly enhance the efficacy and sustainability of infection control measures in healthcare settings, ensuring safer environments for patients and staff alike.

Table 3.6 provides measurable improvements in air quality and the corresponding reduction in airborne infections following the installation of advanced ventilation systems in a newly constructed hospital in Sweden. This table outlines the effectiveness of High-Efficiency Particulate Air (HEPA) filtration combined with Ultraviolet Germicidal Irradiation (UVGI) in maintaining optimal air quality and controlling airborne pathogens. The data showcase significant decreases in airborne infection rates and enhancements in air purity, validating the strategic integration of these technologies into hospital design. This evidence is crucial for hospital administrators and healthcare facility designers considering similar systems to ensure a clean, safe, and infection-controlled environment.

**Table 3.6** Air quality and infection control metrics post-installation of advanced ventilation systems

| Area of installation | Air quality before (particulates/m$^3$) | Air quality after (particulates/m$^3$) | Airborne infection rates before (%) | Airborne infection rates after (%) |
|---|---|---|---|---|
| General wards | 150 | 30 | 2.5 | 0.8 |
| ICU | 100 | 20 | 3.0 | 0.5 |
| Surgical areas | 50 | 5 | 1.5 | 0.2 |
| Outpatient areas | 200 | 40 | 3.5 | 1.0 |
| Isolation rooms | 120 | 15 | 4.0 | 0.7 |

### *3.6.3 Smart Hospital Technology for Remote Patient Monitoring*

In Case Study 3, we explore a progressive initiative by a healthcare network in the United States that has effectively harnessed Internet of Things (IoT) technology to enable continuous remote monitoring of patient vitals. This technological leap has significantly streamlined patient care processes and bolstered infection control measures within the facility. Central to this initiative was the extensive deployment of IoT devices throughout the hospital network, which were intricately integrated into patient care systems. These devices continually track critical health metrics such as heart rate, blood pressure, and body temperature, automating the data collection process. This automation crucially reduces the need for physical interaction during routine checks, thereby lowering the risk of infectious transmissions between patients and healthcare staff.

Further enhancing this system, the integration included environmental monitoring sensors, designed to detect changes in air quality and alert the hospital staff accordingly. This capability is particularly vital for managing infection risks, enabling swift hospital responses to potential airborne threats through adjustments in ventilation systems or the initiation of targeted cleaning protocols. The impact of implementing IoT-based remote patient monitoring has been profoundly positive, marked by significant enhancements in patient care and operational efficiencies within the hospital. A standout benefit has been the system's ability to enable early detection of infection symptoms, a crucial factor in preventing potential outbreaks. Real-time data provided by the IoT system allows medical personnel to rapidly identify any changes in a patient's condition, enabling prompt and precise medical interventions.

Additionally, the environmental sensors have ensured consistently safe hospital environments by maintaining strict air quality standards. These sensors are particularly instrumental during high-risk scenarios for airborne infections, allowing preemptive actions that prevent wider infection spread. The success of this smart

**Table 3.7** Comparison of patient monitoring efficiency before and after IoT implementation

| Metric | Before IoT implementation | After IoT implementation | Percentage improvement (%) |
|---|---|---|---|
| Average response time (minutes) | 10 | 4 | 60 |
| Rate of routine check efficiency (%) | 65 | 95 | 46.15 |
| Patient satisfaction rating | 80 | 95 | 18.75 |
| Infection control compliance (%) | 85 | 98 | 15.29 |
| Environmental monitoring alerts (per day) | 5 | 1 | 80 |

hospital initiative sheds light on the substantial advantages of integrating sophisticated technologies into healthcare settings. It underscores the potential of IoT devices not only in enhancing patient monitoring and environmental safety but also in elevating the overall quality of patient care and infection control.

Looking to the future, the application of IoT technology in healthcare appears ripe for expansion. Prospective advancements may include the development of more sophisticated predictive analytics that use the extensive data collected by IoT devices to foresee potential health deteriorations before they manifest. Moreover, the enhancement of environmental monitoring could evolve to manage hospital microclimates more precisely, optimizing conditions to further support infection control, patient comfort, and recovery processes. Such technological integrations promise to drive forward the capabilities of healthcare systems, ensuring higher standards of care and operational efficiency.

Table 3.7 provides statistical evidence of the improvements in patient monitoring, response times, and overall patient care efficiency following the integration of Internet of Things (IoT) devices within a healthcare network in the United States. This table quantifies the enhancements brought about by the IoT in terms of more timely and accurate patient care, demonstrating significant reductions in the time required for detecting and responding to patient needs. It also highlights the reduced necessity for direct physical interactions, which contributes to infection control. These results underscore the transformative impact of IoT on healthcare delivery, offering crucial data for healthcare administrators and technology developers aiming to further harness IoT capabilities in patient care processes.

### *3.6.4 Healing Gardens for Patient Recovery and Infection Control*

In Case Study 4, we explore an innovative approach implemented by a hospital in Australia, which incorporated healing gardens and outdoor therapy areas into its infrastructure. This strategic integration was designed to enhance patient recovery and serve as an effective component of the hospital's infection control strategy. Recognizing the profound therapeutic benefits of natural environments, the hospital embarked on creating healing gardens that encompass a diversity of plants, serene water features, and meandering walking paths. These elements collectively foster a tranquil and restorative atmosphere conducive to patient healing. The gardens and therapy areas were thoughtfully situated to ensure easy accessibility from patient wards while adhering to the stringent safety and hygiene standards required in medical environments. Additionally, the outdoor therapy areas were outfitted with comfortable, shaded seating to accommodate use under various weather conditions.

The design of these spaces was meticulously planned to cater to patients with different mobility levels and health conditions. The pathways were constructed to be wheelchair accessible, and the seating arrangements were designed to offer options for privacy or social interaction, tailored to individual patient preferences and medical requirements. The impact of introducing these healing gardens and outdoor therapy areas was markedly positive, significantly aiding the hospital's infection control efforts and enhancing patient recovery outcomes. By providing appealing and safe alternatives to indoor settings, these outdoor spaces helped minimize patient congestion in indoor common areas, reducing the opportunities for airborne infection transmission.

Moreover, the initiative led to notable improvements in patient mental health and overall satisfaction. Reports from patients highlighted reduced stress levels and enhanced relaxation, which are beneficial for more effective and swifter recovery processes. The psychological advantages of engaging with natural surroundings, such as diminished perceptions of pain and expedited recovery times, were reflected in patient feedback and clinical outcomes. The success of this project in Australia provides valuable insights into the integration of natural elements into healthcare settings globally. It underscores those therapeutic landscapes, beyond their aesthetic appeal, can significantly contribute to infection control and patient care strategies.

These environments advocate a holistic healthcare approach, where patient well-being is considered a synergistic outcome of medical treatments and environmental conditions. Looking ahead, there is potential for expanding the application of therapeutic natural spaces in various medical contexts, such as long-term care facilities and mental health centers, to leverage these benefits on a broader scale. Further research into the specific natural elements that most effectively support patient recovery and contribute to infection control could refine the design and implementation of these therapeutic landscapes, optimizing their impact on patient health and hospital environments.

Table 3.8 provides a comparative analysis of recovery times and patient satisfaction rates between patients who utilized the healing gardens and those who did not at a hospital in Australia. This table highlights the health benefits associated with access to outdoor spaces, illustrating the therapeutic impact of healing gardens on patient recovery and overall satisfaction. It quantifies improvements in psychological well-being and physical health outcomes, underscoring the significance of integrating

**Table 3.8** Patient recovery statistics: indoor versus outdoor recovery

| Recovery setting | Average recovery time (days) | Patient satisfaction rate (%) |
|---|---|---|
| Indoor only | 10 | 75 |
| Access to healing gardens | 7 | 92 |

natural environments into healthcare settings. This data serves as a powerful argument for the inclusion of therapeutic landscapes in hospital design, emphasizing their role not only in enhancing patient recovery but also in contributing to effective infection control measures.

### *3.6.5 Telemedicine Integration in Response to a Pandemic*

In Case Study 5, we analyze the strategic enhancement of telemedicine services by a hospital in Italy during the COVID-19 pandemic. This adjustment was crucial for managing patient care amidst the challenges posed by the global health crisis. The rapid deployment of digital health services became indispensable in ensuring the continuity of care while simultaneously minimizing the risk of virus transmission within the healthcare setting. As the pandemic intensified, the hospital confronted the dual task of treating patients with COVID-19 and providing ongoing care for other medical conditions without increasing the exposure risk for both patients and staff. The response involved a rapid expansion of the hospital's telemedicine infrastructure. This expansion was multifaceted, including boosting server capacities to accommodate a higher volume of virtual consultations, swiftly training staff in telehealth tools, and establishing detailed protocols for various types of virtual interactions.

The telemedicine services implemented encompassed virtual consultations, follow-up sessions, and remote monitoring of chronic conditions. A critical component was the introduction of a telemedicine-based triage system, designed to assess the necessity of in-person visits, significantly reducing unnecessary physical contact and potential exposure. The impact of integrating telemedicine during the pandemic was profound. It drastically curtailed the number of non-critical patients physically entering the hospital, thereby playing a pivotal role in controlling the virus's spread. By facilitating remote care, the hospital not only protected the health of patients and staff but also optimized the allocation of inpatient resources to critically ill COVID-19 patients.

Moreover, the shift towards telemedicine proved essential during lockdown periods, enabling uninterrupted medical care when patient movement was heavily restricted. The high patient satisfaction rates reflected the value placed on the safety, convenience, and effectiveness of receiving healthcare remotely. The insights from this case study are invaluable for understanding the scalability and adaptability of healthcare systems during crises. The rapid deployment of telemedicine services by this Italian hospital highlights the critical need for healthcare infrastructures that are

**Table 3.9** Telemedicine service utilization and patient outcomes during the pandemic

| Telemedicine service | Pre-pandemic usage (%) | Pandemic usage (%) | Patient outcome improvements (%) | Infection control effectiveness (%) |
|---|---|---|---|---|
| Virtual consultations | 10 | 70 | 30 | 50 |
| Remote patient monitoring | 15 | 80 | 40 | 60 |
| Virtual follow-ups | 20 | 85 | 35 | 55 |
| Triage via telemedicine | 5 | 60 | 25 | 45 |

flexible and resilient, capable of adapting swiftly to sudden shifts in care delivery demands.

This case underscores the importance of incorporating robust telemedicine frameworks into the standard operational practices of healthcare facilities, not merely as contingency measures but as integral components of healthcare provision. This example serves as a model for hospitals worldwide, advocating for enhanced infection control measures and care strategies to better prepare for any future healthcare disruptions like the COVID-19 pandemic.

Table 3.9 details the significant increase in telemedicine usage and the associated improvements in patient outcomes and infection control during the COVID-19 pandemic, as experienced by a hospital in Italy. This table quantifies the expansion of telehealth services and illustrates how this technological shift contributed to effective patient management and virus spread mitigation under challenging conditions. The data highlights the role of telemedicine in reducing hospital visits, thus minimizing exposure risks, and maintaining continuous care for non-COVID-19 conditions, showcasing its effectiveness in ensuring patient safety and satisfaction during widespread health crises.

## 3.7 Ethical Considerations in Hospital Design

Ethical considerations in hospital design are fundamental to ensuring that healthcare environments uphold the dignity, rights, and well-being of all patients while facilitating effective infection control measures. This subsection explores the ethical dimensions that must be considered when designing spaces that balance patient safety with privacy, autonomy, and equitable access to care.

### 3.7.1 Respect for Patient Privacy and Dignity

In healthcare environments, particularly those that necessitate strict infection control protocols such as isolation units, the importance of upholding patient privacy and dignity cannot be overstated. Isolation measures are essential for controlling the spread of infectious diseases, but they can inadvertently compromise patient privacy and dignity by confining patients to small, often stark clinical spaces where their interactions with others are highly restricted. To address these challenges, healthcare architects and designers are turning to innovative design solutions that help preserve the privacy and dignity of patients while still adhering to necessary health protocols.

One key strategy for maintaining privacy in isolated environments involves the use of architectural elements like frosted glass. Frosted glass allows natural light to fill the room, which can mitigate the sterile atmosphere typical of isolation units, while also obscuring the patient's space from an external view. This ensures that patients are not subject to the scrutiny of passersby or other hospital residents, providing a sense of security and personal space that is crucial for psychological well-being in vulnerable times. Moreover, the incorporation of soundproofing measures is critical in these settings. Isolation rooms often become venues for sensitive discussions regarding a patient's health status and treatment options. To protect this sensitive information, the application of acoustic panels, double-glazed windows, and specialized ceiling tiles are employed to prevent sound from traveling outside the room, thereby safeguarding patient confidentiality.

Additionally, while continuous monitoring is critical in isolation wards to ensure patient safety and facilitate timely medical interventions, it is imperative to balance this need with the patient's right to privacy. To achieve this balance, hospitals are increasingly utilizing advanced, less intrusive monitoring technologies. Sensors and other remote monitoring tools can alert medical staff to changes in a patient's condition without the need for constant visual monitoring, thus reducing the feeling of being under surveillance and helping maintain a dignified environment.

Beyond structural and technological solutions, the overall design and atmosphere of patient care areas are also crafted with dignity in mind. Providing patients with elements of control over their environment—such as adjustable lighting, access to entertainment, and facilities to communicate with family and healthcare providers—can significantly enhance their comfort and autonomy. Personal touches, such as the ability to personalize the space, even minimally, can help dispel the institutional feel of isolation rooms, fostering a more homelike and dignified patient experience.

In conclusion, while the primary objective in isolation units is to prevent the spread of infections, this goal should not be pursued at the expense of patient privacy and dignity. Thoughtful integration of design elements that promote privacy and dignity is crucial. These measures not only enhance patient satisfaction but are integral to delivering holistic healthcare that supports the physical and emotional well-being of patients, ultimately contributing to more positive health outcomes. By prioritizing both infection control and respect for individual dignity, healthcare facilities can provide care that is truly comprehensive and compassionate.

### 3.7.2 Equitable Access to Care

Ensuring equitable access to healthcare is a fundamental principle that must be meticulously maintained, especially in environments where infection control measures are stringent. This requires an unwavering commitment to design and operational strategies that guarantee all patients, irrespective of their age, disability status, or other vulnerable characteristics, receive high-quality medical care without discrimination or diminished access. The design of healthcare facilities, particularly isolation rooms, plays a critical role in fulfilling this commitment.

Inclusive design in isolation rooms is essential to ensure that they are accessible to patients with special needs. This includes equipping rooms with adjustable beds that cater to various body sizes and physical conditions, ensuring doorways are sufficiently wide to accommodate wheelchair access, and designing bathrooms with safety features such as grab bars and non-slip flooring. The overall layout of these rooms must facilitate easy maneuverability for individuals with physical disabilities, allowing them the same ease of access and use as other patients. Such thoughtful design considerations are crucial in promoting an equitable healthcare environment.

Furthermore, the accessibility of hygiene facilities such as hand sanitation stations is vital in controlling infections and must be designed inclusively. Placing these stations at multiple heights accommodates users of all statues and those in wheelchairs, and employing mechanisms that can be operated via elbow or automatic sensors ensures that everyone, regardless of physical capability, can utilize them independently. This inclusive approach extends to all aspects of the healthcare facility's design, promoting autonomy and dignity for all patients.

Effective communication is another pillar of equitable healthcare access. Healthcare facilities must ensure that infection control protocols are communicated to all patients and visitors. This involves using signage that includes multiple languages relevant to the community, as well as incorporating Braille and tactile symbols for those with visual impairments. Such measures not only aid in the effective dissemination of critical health information but also contribute to creating an environment that is welcoming and inclusive of diverse patient needs.

Beyond the physical infrastructure, the training of healthcare staff in inclusive care practices is paramount. This training should encompass effective communication and assistance strategies for interacting with patients across a spectrum of ages, abilities, and cultural backgrounds. Staff should be proficient in addressing the needs of patients with cognitive impairments, language barriers, or physical disabilities, ensuring that all individuals receive compassionate and effective care.

In conclusion, achieving equitable access to healthcare requires a comprehensive approach that integrates thoughtful design, effective communication, and specialized staff training. By implementing these strategies, healthcare facilities can ensure that their infection control measures do not inadvertently exclude or marginalize vulnerable populations. Instead, they uphold the dignity and rights of all patients, fostering an environment that not only meets diverse needs but also enhances patient satisfaction and health outcomes. Through these efforts, healthcare providers can truly deliver care that is equitable and inclusive, reflecting the fundamental values of the medical profession.

### 3.7.3 Balancing Safety with Autonomy

In healthcare settings, particularly those that incorporate stringent infection control measures like isolation rooms, the challenge of balancing safety protocols with patient autonomy is significant. While ensuring patient safety is paramount, facilitating a degree of control over their immediate environment can greatly enhance a patient's well-being and willingness to comply with necessary health interventions. This dual focus on safety and autonomy not only improves patient outcomes but also respects their dignity and independence during vulnerable times.

Design features that enhance autonomy in patient rooms are crucial for fostering a sense of normalcy and control. Adjustable lighting systems, which patients can modify not only in terms of brightness but also color temperature, allow individuals to tailor their environment to their comfort and mood. Such features can be particularly comforting in a hospital setting, where personal choice is often limited. Additionally, providing patients with the ability to adjust window blinds offers control over natural light and privacy, further enhancing their sense of personal space and autonomy. These adjustments help mitigate the stress and alienation that can come from long-term confinement in a clinical setting.

Technological advancements also play a pivotal role in promoting autonomy within isolation units. Interactive systems that enable patients to communicate effectively with healthcare providers, manage room settings, or control entertainment devices are essential. These technologies not only reduce feelings of isolation but also empower patients by giving them active roles in their care process. This sense of involvement can significantly alleviate the psychological burdens of treatment in restrictive environments.

Offering patients choices regarding their treatment spaces where feasible is another strategy that enhances autonomy. Allowing patients to select room decor from predefined themes or choose their room within certain parameters can make a substantial difference in their comfort and satisfaction. While medical and safety considerations may limit these choices, the availability of even minimal options can greatly influence a patient's experience and their adherence to treatment protocols.

However, these autonomy-enhancing measures must be carefully balanced with safety requirements. Controls provided to patients need to be designed in such a way that they cannot be misused in a manner that might compromise their health or safety. Moreover, all patient-controlled features must comply with strict hygiene and sanitation protocols, particularly in infection-control settings. This balance ensures that while autonomy is supported, it does not detract from the essential safety protocols that protect both patients and healthcare staff.

In conclusion, integrating patient autonomy into the design and operation of healthcare facilities, particularly those involving infection control, is crucial.

By allowing patients some measure of control over their environment and care choices, healthcare providers can boost patient compliance, satisfaction, and overall health outcomes. This approach does not merely serve medical objectives but also profoundly respects and upholds the dignity of those in care. It is a testament to the evolving nature of healthcare, where patient-centered approaches are increasingly recognized as integral to successful medical outcomes.

### 3.7.4 Transparency and Communication

In healthcare settings, particularly those requiring strict infection control measures like isolation wards, the architectural design of the facility is pivotal in promoting effective and transparent communication. Transparent communication is essential not only for delivering high-quality care but also for maintaining patient trust and satisfaction, especially in scenarios involving isolation or quarantine.

One innovative design strategy to enhance communication while maintaining necessary safety protocols is the incorporation of transparent barriers. Using materials such as glass or clear acrylic panels facilitates visual connectivity between isolated patients and their caregivers or visiting family members. This approach is particularly valuable in intensive care units or quarantine areas, where physical isolation is mandatory. Transparent barriers help reduce the psychological burden of isolation by alleviating feelings of loneliness and anxiety, as they allow patients to remain visually connected to their surroundings and the people caring for them without compromising the stringent infection control required.

Furthermore, the design of healthcare facilities must also consider spaces that enable confidential yet accessible consultations. Creating consultation areas that offer both privacy and comfort is crucial for fostering open communication between patients and healthcare professionals. These areas can be enhanced with soundproofing materials to ensure that conversations are kept private, which is vital for building trust and encouraging patients to adhere to medical advice. For discussions involving particularly sensitive health information, rooms should be specifically designed with acoustics in mind to prevent eavesdropping. This might include technologies that mask sound or architectural layouts that minimize sound leakage, ensuring that patients feel secure enough to share personal health details critical to their care.

Technology also plays a crucial role in facilitating communication, particularly under conditions where traditional face-to-face interactions may be limited. Digital tools, such as patient portals, offer secure platforms for messaging and information exchange between patients and healthcare providers. These tools are especially beneficial for individuals in isolation, as they provide a continuous line of communication with their medical team, allowing for updates on health status and care plans without the need for direct physical contact. This use of technology ensures that despite the physical barriers imposed by infection control protocols, communication remains as open and clear as possible.

In conclusion, an effective healthcare design that prioritizes transparent communication and privacy not only supports clinical outcomes but also significantly enhances the patient's experience. By reducing anxiety and promoting a sense of dignity, well-thought-out design and technology integration play crucial roles in developing healthcare facilities that are safe, supportive, and patient-centered. As healthcare environments continue to evolve, the implementation of design strategies that facilitate clear and compassionate communication will be essential in creating settings that meet the complex needs of patients and healthcare providers alike.

### 3.7.5 Ethical Use of Resources

In the realm of healthcare facility design and operation, the ethical use of resources stands as a critical consideration, especially during periods of crisis or high demand. Ethical resource management requires making informed decisions that not only enhance patient care and bolster infection control efforts but also consider the broader impacts on community health and environmental sustainability.

Strategic resource allocation is vital in ensuring that investments in healthcare infrastructure deliver maximum benefits. This often involves prioritizing resources in areas most critical to infection control, such as investing in advanced air filtration systems, utilizing durable and easily cleanable surface materials, and designing adaptable patient care areas. These priorities should reflect a commitment to serving the greatest number of people effectively, particularly in parts of the facility that are at increased risk of infection transmission. It is crucial, however, that enhancing certain areas does not compromise the functionality or resource availability in other parts of the hospital. Efficient design and planning can ensure that resources are utilized optimally, promoting versatility in facility use and avoiding the need for redundant investments.

Sustainable practices are also a cornerstone of ethical resources used in hospital design. Choosing sustainable materials that are responsibly sourced and have minimal environmental impact is essential for reducing the ecological footprint of healthcare facilities. Similarly, incorporating energy-efficient designs can significantly decrease operational costs and the overall environmental impact of the hospital. Features like green roofs, energy-efficient HVAC systems, and LED lighting not only contribute to reduced energy consumption but also support broader environmental goals. Additionally, integrating natural elements into the hospital design can aid in patient recovery—as demonstrated in studies on healing gardens—and enhance biodiversity, thereby enriching the surrounding ecosystem.

Promoting and adhering to ethical standards in resource use within healthcare facilities not only supports clinical and safety objectives but also respects and promotes humanistic aspects of healthcare delivery. This ethical approach fosters community trust and cooperation, as patients and their families recognize the facility's commitment to sustainability and ethical practices. Moreover, healthcare institutions that lead by example in ethical resource management set benchmarks for industry, encouraging other facilities to adopt similar sustainable and responsible practices.

In conclusion, by embedding ethical considerations into their resource allocation strategies and facility designs, healthcare institutions can provide superior care, advance sustainability, and fulfill their societal obligations. These practices ensure that healthcare facilities are not merely centers for immediate medical needs but also contributors to the long-term health and well-being of the community and the environment. Such a comprehensive approach to healthcare delivery fosters a legacy of trust and ethical integrity that extends beyond the walls of the hospital, influencing broader industry standards and societal expectations.

## Further Reading


Abdul-Aziz, M. H., Alffenaar, J. W. C., Bassetti, M., Bracht, H., Dimopoulos, G., Marriott, D., ... & Infections in the ICU and Sepsis Working Group of International Society of Antimicrobial Chemotherapy (ISAC). (2020). Antimicrobial therapeutic drug monitoring in critically ill adult patients: a Position Paper#. *Intensive care medicine*, 46, 1127–1153. https://doi.org/10.1007/s00134-020-06050-1

Abou-Nassar, E. M., Iliyasu, A. M., El-Kafrawy, P. M., Song, O. Y., Bashir, A. K., & Abd El-Latif, A. A. (2020). DITrust chain: towards blockchain-based trust models for sustainable healthcare IoT systems. *IEEE access*, 8, 111223–111238. https://doi.org/10.1109/ACCESS.2020.2999468

Ahmad, S., & Alfouzan, W. (2021). Candida auris: epidemiology, diagnosis, pathogenesis, antifungal susceptibility, and infection control measures to combat the spread of infections in healthcare facilities. *Microorganisms*, 9(4), 807. https://doi.org/10.3390/microorganisms9040807

Ainsworth, B., Miller, S., Denison-Day, J., Stuart, B., Groot, J., Rice, C., ... & Yardley, L. (2021). Infection control behavior at home during the COVID-19 pandemic: observational study of a web-based behavioral intervention (germ defence). *Journal of Medical Internet Research*, 23(2), e22197. https://doi.org/10.2196/22197

Alazab, M., Khan, L. U., Koppu, S., Ramu, S. P., Iyapparaja, M., Boobalan, P., ... & Aljuhani, A. (2022). Digital twins for healthcare 4.0—Recent advances, architecture, and open challenges. *IEEE Consumer Electronics Magazine*, 12(6), 29–37. https://doi.org/10.1109/MCE.2022.3208986

Allen, J. G., & Ibrahim, A. M. (2021). Indoor air changes and potential implications for SARS-CoV-2 transmission. *Jama*, 325(20), 2112–2113. https://doi.org/10.1001/jama.2021.5053

Alhumaid, S., Al Mutair, A., Al Alawi, Z., Alsuliman, M., Ahmed, G. Y., Rabaan, A. A., ... & Al-Omari, A. (2021). Knowledge of infection prevention and control among healthcare workers and factors influencing compliance: a systematic review. *Antimicrobial Resistance & Infection Control*, 10(1), 86. https://doi.org/10.1186/s13756-021-00957-0

Ang, H., & Sun, X. (2018). Risk factors for multidrug-resistant Gram-negative bacteria infection in intensive care units: a meta-analysis. *International journal of nursing practice*, 24(4), e12644. https://doi.org/10.1111/ijn.12644

Awada, M., Becerik-Gerber, B., Hoque, S., O'Neill, Z., Pedrielli, G., Wen, J., & Wu, T. (2021). Ten questions concerning occupant health in buildings during normal operations and extreme events including the COVID-19 pandemic. *Building and Environment*, 188, 107480. https://doi.org/10.1016/j.buildenv.2020.107480

Azuma, K., Yanagi, U., Kagi, N., Kim, H., Ogata, M., & Hayashi, M. (2020). Environmental factors involved in SARS-CoV-2 transmission: effect and role of indoor environmental quality in the strategy for COVID-19 infection control. *Environmental health and preventive medicine*, 25, 1–16. https://doi.org/10.1186/s12199-020-00904-2

Battisto, D., Li, X., Dong, J., Hall, L., & Blouin, J. (2023). Research methods used in evidence-based design: An analysis of five years of research articles from the HERD journal. *HERD: Health Environments Research & Design Journal*, 16(1), 56–82. https://doi.org/10.1177/19375867221125940

Bharathi, R., Abirami, T., Dhanasekaran, S., Gupta, D., Khanna, A., Elhoseny, M., & Shankar, K. (2020). Energy efficient clustering with disease diagnosis model for IoT based sustainable healthcare systems. *Sustainable Computing: Informatics and Systems*, 28, 100453. https://doi.org/10.1016/j.suscom.2020.100453

Blot, S., Ruppé, E., Harbarth, S., Asehnoune, K., Poulakou, G., Luyt, C. E., ... & Zahar, J. R. (2022). Healthcare-associated infections in adult intensive care unit patients: Changes in epidemiology, diagnosis, prevention and contributions of new technologies. *Intensive and Critical Care Nursing*, 70, 103227. https://doi.org/10.1016/j.iccn.2022.103227

Boatman, J. D., Stevens, M. P., & Markley, J. D. (2022). Infection Control in the Outpatient Setting. In *Infection Prevention: New Perspectives and Controversies* (pp. 47–71). Cham: Springer International Publishing. https://doi.org/10.1007/978-3-030-98427-4_6

Boswihi, S. S., & Udo, E. E. (2018). Methicillin-resistant Staphylococcus aureus: An update on the epidemiology, treatment options and infection control. *Current Medicine Research and Practice*, 8(1), 18–24. https://doi.org/10.1016/j.cmrp.2018.01.001

Brambilla, A., Rebecchi, A., & Capolongo, S. (2019). Evidence based hospital design. A literature review of the recent publications about the EBD impact of built environment on hospital occupants' and organizational outcomes. *Annali di Igiene Medicina Preventiva e di Comunita*, 31(2), 165–180. https://re.public.polimi.it/handle/11311/960586

Brosseau, L. M., Gold, D., Materna, B., McPhaul, K., Rosen, M., Seminario, P., & Thomason, J. (2024). Preventing Aerosol-Transmissible Diseases in Healthcare Settings: The Need for Protective Guidelines and Standards—Workshop Report. *NEW SOLUTIONS: A Journal of Environmental and Occupational Health Policy*, 33(4), 236–247. https://doi.org/10.1177/10482911231215498

Bueno de Mesquita, P. J., Delp, W. W., Chan, W. R., Bahnfleth, W. P., & Singer, B. C. (2022). Control of airborne infectious disease in buildings: Evidence and research priorities. *Indoor Air*, 32(1), e12965. https://doi.org/10.1111/ina.12965

Capolongo, S., Gola, M., Brambilla, A., Morganti, A., Mosca, E. I., & Barach, P. (2020). COVID-19 and healthcare facilities: a decalogue of design strategies for resilient hospitals. *Acta Bio Medica: Atenei Parmensis*, 91(9-S), 50. https://doi.org/10.23750/abm.v91i9-S.10117

Carayon, P., Wust, K., Hose, B. Z., & Salwei, M. E. (2021). Human factors and ergonomics in health care. Handbook of human factors and ergonomics, 1417–1437. https://doi.org/10.1002/9781119636113.ch53

Cheng, V. C., Wong, S. C., Chen, J. H., Yip, C. C., Chuang, V. W., Tsang, O. T., ... & Yuen, K. Y. (2020). Escalating infection control response to the rapidly evolving epidemiology of the coronavirus disease 2019 (COVID-19) due to SARS-CoV-2 in Hong Kong. *Infection Control & Hospital Epidemiology*, 41(5), 493–498. https://doi.org/10.1017/ice.2020.58

Cruz-López, F., Martínez-Meléndez, A., & Garza-González, E. (2023). How does hospital microbiota contribute to healthcare-associated infections?. *Microorganisms*, 11(1), 192. https://doi.org/10.3390/microorganisms11010192

Despotovic, A., Milosevic, B., Milosevic, I., Mitrovic, N., Cirkovic, A., Jovanovic, S., & Stevanovic, G. (2020). Hospital-acquired infections in the adult intensive care unit—Epidemiology, antimicrobial resistance patterns, and risk factors for acquisition and mortality. *American journal of infection control*, 48(10), 1211–1215. https://doi.org/10.1016/j.ajic.2020.01.009

Dexter, F., Parra, M. C., Brown, J. R., & Loftus, R. W. (2020). Perioperative COVID-19 defense: an evidence-based approach for optimization of infection control and operating room management. *Anesthesia & Analgesia*, 131(1), 37–42. https://doi.org/10.1213/ANE.0000000000004829

Difficile Update, C. Guidelines for Environmental Infection Control in Health-Care Facilities.

Eijkelenboom, A., & Bluyssen, P. M. (2022). Comfort and health of patients and staff, related to the physical environment of different departments in hospitals: a literature review. *Intelligent Buildings International*, 14(1), 95–113. https://doi.org/10.1080/17508975.2019.1613218

Emmanuel, U., Osondu, E. D., & Kalu, K. C. (2020). Architectural design strategies for infection prevention and control (IPC) in health-care facilities: towards curbing the spread of Covid-19. *Journal of environmental health science and engineering*, 18, 1699–1707. https://doi.org/10.1007/s40201-020-00580-y

Escombe, A. R., Ticona, E., Chávez-Pérez, V., Espinoza, M., & Moore, D. A. (2019). Improving natural ventilation in hospital waiting and consulting rooms to reduce nosocomial tuberculosis transmission risk in a low resource setting. *BMC infectious diseases*, 19, 1-7. https://doi.org/10.1186/s12879-019-3717-9

Gan, W. H., Lim, J. W., & Koh, D. (2020). Preventing intra-hospital infection and transmission of coronavirus disease 2019 in health-care workers. *Safety and health at work*, 11(2), 241–243. https://doi.org/10.1016/j.shaw.2020.03.001

Gandhi, T. K., Kaplan, G. S., Leape, L., Berwick, D. M., Edgman-Levitan, S., Edmondson, A., ... & Wachter, R. (2018). Transforming concepts in patient safety: a progress report. *BMJ Quality & Safety*, 27(12), 1019–1026. https://doi.org/10.1136/bmjqs-2017-007756

Gordon, D., Ward, J., Yao, C. J., & Lee, J. (2021). Built environment airborne infection control strategies in pandemic alternative care sites. *HERD: Health Environments Research & Design Journal*, 14(2), 38–48. https://doi.org/10.1177/1937586720979832

Gu, G. Y., Chen, M., Pan, J. C., & Xiong, X. L. (2023). Risk of multi-drug-resistant organism acquisition from prior bed occupants in the intensive care unit: a meta-analysis. *Journal of Hospital Infection*, 139, 44–55. https://doi.org/10.1016/j.jhin.2023.06.020

Guo, M., Xu, P., Xiao, T., He, R., Dai, M., & Miller, S. L. (2021). Review and comparison of HVAC operation guidelines in different countries during the COVID-19 pandemic. *Building and Environment*, 187, 107368. https://doi.org/10.1016/j.buildenv.2020.107368

Halawa, F., Madathil, S. C., Gittler, A., & Khasawneh, M. T. (2020). Advancing evidence-based healthcare facility design: a systematic literature review. *Health Care Management Science*, 23, 453–480. https://doi.org/10.1007/s10729-020-09506-4

Hignett, S., Lu, J. L., & Fray, M. (2009). ICU design and human factors: The role of the physical environment in patient outcomes. *Critical Care Nursing Quarterly*, 32(3), 171–179.

Islam, M. S., Rahman, K. M., Sun, Y., Qureshi, M. O., Abdi, I., Chughtai, A. A., & Seale, H. (2020). Current knowledge of COVID-19 and infection prevention and control strategies in healthcare settings: A global analysis. *Infection Control & Hospital Epidemiology*, 41(10), 1196–1206. https://doi.org/10.1017/ice.2020.237

Izadyar, N., & Miller, W. (2022). Ventilation strategies and design impacts on indoor airborne transmission: A review. *Building and Environment*, 218, 109158. https://doi.org/10.1016/j.buildenv.2022.109158

Kirk Hamilton, D. (2020). Evidence-based practice: Four levels revisited. *HERD: Health Environments Research & Design Journal*, 13(3), 26–29. https://doi.org/10.1177/1937586720931064

Lipinski, T., Ahmad, D., Serey, N., & Jouhara, H. (2020). Review of ventilation strategies to reduce the risk of disease transmission in high occupancy buildings. *International Journal of Thermofluids*, 7, 100045. https://doi.org/10.1016/j.ijft.2020.100045

Munoz-Price, L. S., Bowdle, A., Johnston, B. L., Bearman, G., Camins, B. C., Dellinger, E. P., ... & Birnbach, D. J. (2019). Infection prevention in the operating room anesthesia work area. *Infection Control & Hospital Epidemiology*, 40(1), 1–17. https://doi.org/10.1017/ice.2018.303

Nadi, Z. B., Raisali, F., Jafari, N., & Bayramzadeh, S. (2024). The influence of physical environment on health care–associated infections: A literature review. *American Journal of Infection Control*, 52(2), 229–242. https://doi.org/10.1016/j.ajic.2023.06.010

Nardell, E. A. (2021). Air disinfection for airborne infection control with a focus on COVID-19: why germicidal UV is essential. *Photochemistry and Photobiology*, 97(3), 493–497. https://doi.org/10.1111/php.13421

Navaratnam, S., Nguyen, K., Selvaranjan, K., Zhang, G., Mendis, P., & Aye, L. (2022). Designing post COVID-19 buildings: Approaches for achieving healthy buildings. *Buildings*, 12(1), 74. https://doi.org/10.3390/buildings12010074

O'Neill, L., Park, S. H., & Rosinia, F. (2018). The role of the built environment and private rooms for reducing central line-associated bloodstream infections. *Plos one*, 13(7), e0201002. https://doi.org/10.1371/journal.pone.0201002

Pinheiro, M. D., & Luís, N. C. (2020). COVID-19 could leverage a sustainable built environment. *Sustainability*, 12(14), 5863. https://doi.org/10.3390/su12145863

Plachouras Diamantis, Kärki Tommi, Hansen Sonja, Hopkins Susan, Lyytikäinen Outi, Moro Maria Luisa, Reilly Jacqui, Zarb Peter, Zingg Walter, Kinross Pete, Weist Klaus, Monnet Dominique L., Suetens Carl, the Point Prevalence Survey Study Group. Antimicrobial use in European acute care hospitals: results from the second point prevalence survey (PPS) of healthcare-associated infections and antimicrobial use, 2016 to 2017. *Euro Surveill.* 2018;23(46)=1800393. https://doi.org/10.2807/1560-7917.ES.23.46.1800393

Salman, A., Sattineni, A., Azhar, S., & Leousis, K. (2022). A systematic review of building systems and technologies to mitigate the spread of airborne viruses. *Journal of Facilities Management*, 20(3), 369–384. https://doi.org/10.1108/JFM-01-2021-0015

Santos, A. F., Gaspar, P. D., Hamandosh, A., Aguiar, E. B. D., Guerra Filho, A. C., & Souza, H. J. L. D. (2020). Best practices on HVAC design to minimize the risk of COVID-19 infection within indoor environments. *Brazilian Archives of Biology and Technology*, 63, e20200335. https://doi.org/10.1590/1678-4324-2020200335

Shajahan, A., Culp, C. H., & Williamson, B. (2019). Effects of indoor environmental parameters related to building heating, ventilation, and air conditioning systems on patients' medical outcomes: A review of scientific research on hospital buildings. *Indoor Air*, 29(2), 161–176. https://doi.org/10.1111/ina.12531

Suleyman, G., Alangaden, G., & Bardossy, A. C. (2018). The role of environmental contamination in the transmission of nosocomial pathogens and healthcare-associated infections. *Current infectious disease reports*, 20, 1–11. https://doi.org/10.1007/s11908-018-0620-2

Tang, J. W., Bahnfleth, W. P., Bluyssen, P. M., Buonanno, G., Jimenez, J. L., Kurnitski, J., ... & Dancer, S. J. (2021). Dismantling myths on the airborne transmission of severe acute respiratory syndrome coronavirus-2 (SARS-CoV-2). *Journal of Hospital Infection*, 110, 89–96. https://doi.org/10.1016/j.jhin.2020.12.022

Timsit, J. F., Bassetti, M., Cremer, O., Daikos, G., De Waele, J., Kallil, A., ... & Barbier, F. (2019). Rationalizing antimicrobial therapy in the ICU: a narrative review. *Intensive care medicine*, 45, 172–189. https://doi.org/10.1007/s00134-019-05520-5

Tsang, T. W., Mui, K. W., & Wong, L. T. (2023). Computational Fluid Dynamics (CFD) studies on airborne transmission in hospitals: A review on the research approaches and the challenges. *Journal of Building Engineering*, 63, 105533. https://doi.org/10.1016/j.jobe.2022.105533

Ward, M. E., De Brún, A., Beirne, D., Conway, C., Cunningham, U., English, A., ... & McAuliffe, E. (2018). Using co-design to develop a collective leadership intervention for healthcare teams to improve safety culture. *International journal of environmental research and public health*, 15(6), 1182. https://doi.org/10.3390/ijerph15061182

Xu, C., & Liu, L. (2018). Personalized ventilation: one possible solution for airborne infection control in highly occupied space?. *Indoor and Built Environment*, 27(7), 873–876. https://doi.org/10.1177/1420326X18777383

Xu, C., Wei, X., Liu, L., Su, L., Liu, W., Wang, Y., & Nielsen, P. V. (2020). Effects of personalized ventilation interventions on airborne infection risk and transmission between occupants. *Building and environment*, 180, 107008. https://doi.org/10.1016/j.buildenv.2020.107008

# Chapter 4
# Enhancing Quality of Care Through Patient-Centric Design: Architectural and Operational Strategies for Improved Patient Satisfaction

This chapter delves into the transformative impact of patient-centric hospital environments on enhancing the quality of care and optimizing patient satisfaction. It systematically explores how hospitals can integrate both architectural and operational strategies to prioritize patient comfort and well-being at every stage of medical care. Focused on a deep understanding of patient needs, this approach goes beyond basic medical treatments to address crucial emotional and psychological aspects through strategic design choices. The discussion extends to the development of spaces that effectively cater to diverse patient populations, emphasizing inclusivity and accessibility. It also highlights the importance of incorporating therapeutic natural elements, sophisticated art, and ergonomic design, which collectively foster a healing atmosphere conducive to recovery. Additionally, the chapter discusses the critical role of flexible spaces and advanced technologies in adapting to the evolving needs of healthcare delivery and enhancing patient engagement. This holistic approach to hospital design not only aims to improve the efficiency of healthcare delivery but also enriches the patient experience by creating environments that promote recovery, comfort, and overall well-being. The insights presented are particularly valuable for healthcare administrators, architects, and designers dedicated to shaping environments that truly resonate with and support the needs of patients.

## 4.1 Understanding Patient Needs and Requirements

The concept of patient-centric design in healthcare facilities prioritizes the needs and experiences of patients at every stage of medical care. This approach is rooted in a deep understanding of what patients require not only medically but also emotionally and psychologically during their healthcare journey. Recognizing these needs and incorporating them into the architectural and operational fabric of healthcare facilities is fundamental to creating environments that promote healing and comfort.



A. A. Firoozi and A. A. Firoozi, *Transformative Design in Healthcare*,
https://doi.org/10.1007/978-3-032-08019-6_4

To truly grasp patient needs and requirements, it is essential to consider the holistic experience of individuals who enter a healthcare facility. This involves acknowledging the vulnerabilities, anxieties, and expectations patients typically face. For instance, a patient's sense of control is often diminished upon entering a hospital due to the unfamiliar environment and the critical nature of their health condition. Addressing this involves creating spaces that are not only functionally apt but also reassuring and welcoming.

Medical needs are the primary concern, which includes easy accessibility to treatment areas, minimal travel distances within the facility, and the availability of medical information. This can be achieved through thoughtful architectural planning that reduces the physical strain on patients and their families, such as placing parking areas close to entrance points, minimizing navigation complexities, and ensuring that essential services are easily accessible.

Emotional and psychological needs are also paramount. The environment should foster a sense of calm and reassurance rather than stress or disorientation. This can be facilitated by incorporating elements of nature, using soothing colors and materials, and ensuring ample natural light, all of which are known to help reduce stress and anxiety. Quiet zones, private spaces, and areas for family members to stay close to their loved ones are also important, as they contribute significantly to the emotional comfort and overall well-being of patients.

Understanding patient needs also extends to cultural sensitivity, which involves recognizing and respecting the diverse backgrounds and values of patients. Healthcare facilities must cater to these varied needs by providing spaces for cultural practices, dietary requirements, and multilingual services to ensure that all patients feel respected and cared for.

Involving patients and their families in the design process through surveys, focus groups, and feedback mechanisms is another effective way to understand and implement patient-centric designs. This direct engagement not only helps in gathering real insights into patient needs but also makes the patients feel valued and involved in their own care processes.

In sum, understanding patient needs and requirements in hospital design is not just about architectural and operational efficiency; it's about creating a healing environment that respects and responds to the comprehensive needs of those it serves. As we delve deeper into the specific strategies for designing for diverse patient populations in the next section, the focus will remain on how these designs can be inclusively tailored to meet the wide-ranging needs of all patients.

Table 4.1 provides quantitative data derived from a patient needs assessment survey, highlighting patient priorities and satisfaction levels regarding various aspects of hospital care and facility design. This table is integral to understanding and addressing the specific needs and preferences of patients, thereby supporting the creation of healthcare spaces that enhance patient comfort, safety, and satisfaction. The results emphasize the importance of patient-centered design and highlight areas requiring attention to improve overall patient experience in healthcare settings.

**Table 4.1** Patient needs assessment survey results

| Patient need category | Percentage prioritizing need (%) | Average satisfaction level (%) |
|---|---|---|
| Privacy and confidentiality | 85 | 75 |
| Access to natural light | 65 | 60 |
| Quiet and noise control | 75 | 55 |
| Proximity to support Services (e.g., pharmacies) | 80 | 70 |
| Comfortable waiting areas | 70 | 65 |
| Accessibility for disabilities | 90 | 80 |
| Technology integration for health management | 60 | 50 |

## 4.2 Designing for Diverse Patient Populations

Designing healthcare facilities to accommodate diverse patient populations is a critical aspect of patient-centric hospital environments. This approach requires a broad understanding of the demographic, cultural, and individual differences that influence healthcare needs. By acknowledging and planning for diversity, healthcare facilities can ensure that they serve all patient groups effectively and compassionately.

Diverse patient populations include individuals of various ages, cultures, languages, physical abilities, and health conditions. Each of these groups may have unique needs that must be considered in hospital design to provide equitable and effective care. For example, elderly patients may require different amenities compared to younger adults or children, such as non-slip floors, enhanced lighting, and easily accessible emergency communication systems. Similarly, children's hospital areas should be designed with lower countertops, vibrant colors, and interactive elements that reduce anxiety and foster a friendly environment.

Cultural diversity is another crucial factor in designing patient-centric environments. Hospitals serve increasingly multicultural communities, and this diversity should be reflected in the design and operations of healthcare facilities. This includes multilingual signage, culturally appropriate decor, and spaces designated for various cultural practices and religious observances. Providing dietary options that cater to different cultural preferences in hospital cafeterias can also make a significant difference in patient comfort and satisfaction.

Accessibility is a fundamental design consideration that ensures inclusivity for patients with disabilities. This includes physical accessibility, such as wheelchair ramps, wide corridors, and adaptive restroom facilities, as well as sensory adaptations for those with visual or auditory impairments, such as braille signage and hearing loop systems. Advanced technology can also play a role here, with the use of voice-activated systems or other assistive technologies enhancing accessibility and independence for patients with various disabilities.

**Table 4.2** Comparison of patient satisfaction by demographic

| Demographic group | Satisfaction before (%) | Satisfaction after (%) | Improvement (%) |
|---|---|---|---|
| Pediatric patients | 70 | 85 | 21.4 |
| Elderly patients | 60 | 80 | 33.3 |
| Female patients | 65 | 80 | 23.1 |
| Male patients | 68 | 82 | 20.6 |
| Patients with disabilities | 55 | 76 | 36.4 |
| Chronic condition patients | 58 | 77 | 32.8 |

Gender inclusivity also plays a role in patient-centric design. Facilities should provide safe and accessible spaces for all gender identities, including gender-neutral bathrooms and changing rooms. This not only respects individual identity and privacy but also enhances comfort and security for transgender and non-binary patients, who may otherwise face significant stress in environments that do not acknowledge their gender identity.

Healthcare facilities must also address the needs of patients with chronic conditions or severe illnesses by designing spaces that reduce the risk of infection, provide easy access to frequent treatments, and include comfortable waiting areas for patients and their families who spend significant amounts of time in medical settings.

By integrating these diverse needs into the design process, healthcare facilities can create environments that are not only functional and efficient but also welcoming and supportive for every patient, regardless of their background or condition. The importance of such thoughtful design extends beyond mere accommodation; it impacts patient recovery, satisfaction, and overall health outcomes. Moving forward, the next section will explore the therapeutic aspects of hospital design, which further enhance the healing environment by integrating medical care with supportive physical and emotional conditions.

Table 4.2 presents data comparing patient satisfaction before and after the implementation of diverse design features tailored to different patient demographics. This table highlights how specific adjustments in hospital design can significantly enhance satisfaction levels across various groups, including age, gender, and specific health needs. The results underscore the effectiveness of personalized design strategies in improving patient experience and satisfaction, providing a clear metric for healthcare facilities aiming to become more inclusive and responsive to the needs of all patients.

## 4.3 Therapeutic Aspects of Hospital Design

The therapeutic aspects of hospital design encompass a range of elements that contribute to the healing environment beyond the direct provision of medical care. These design elements are intended to enhance the physical, psychological, and emotional well-being of patients, fostering an atmosphere that promotes healing and reduces the stress commonly associated with hospital stays.

### 4.3.1 Natural Healing Environments

The profound integration of natural elements within hospital environments signifies a transformative approach in modern healthcare design, aiming to bolster patient recovery and optimize health outcomes. This strategy is underpinned by extensive research which underscores the restorative effects of natural settings. Incorporating features such as gardens, water elements, and ample natural light within the hospital infrastructure plays a pivotal role in alleviating patient stress, anxiety, and pain—key obstacles in the healing process. The therapeutic advantages of natural elements are well-documented, with studies indicating significant health benefits such as reduced blood pressure, diminished stress hormone levels, and enhanced mental well-being. These benefits are particularly salient within the stressful milieu of hospital stays. For instance, patients who enjoy nature views from their rooms tend to experience quicker recovery rates, owing to decreased stress and heightened satisfaction with the care environment. Additionally, hospital gardens provide serene sanctuaries not only for patients and their families but also for hospital staff, creating a more positive and rejuvenating atmosphere for all involved.

When it comes to designing these natural healing environments, several critical considerations must be addressed to fully harness their therapeutic potential. Strategic planning in the placement of windows and the orientation of patient rooms is essential to afford a maximum number of patients direct views of natural vistas. The architectural use of large, unobstructed windows enhances this effect, facilitating deeper penetration of natural light into the building and fostering an ambiance that is both open and invigorating. Beyond visual exposure, engaging the tactile and olfactory senses through the integration of textured natural materials and aromatic plants can significantly enrich the sensory experience. Additionally, the incorporation of auditory elements like water features—fountains or cascades—can help mask hospital noise and promote tranquility.

Aligning with sustainable design practices, the implementation of natural healing environments not only cuts down on artificial lighting, thereby reducing energy consumption but also improves indoor air quality, enhancing both environmental health and sustainability. These features are designed to be adaptable to various climatic conditions, ensuring their effectiveness and durability throughout the year. In conclusion, the deliberate incorporation of natural elements in hospital design transcends mere aesthetic enhancement. It is a strategic and evidence-based approach to healthcare that fosters healing and well-being, transforming patient experiences and establishing new benchmarks in healthcare environments. By integrating natural healing environments that are supported by empirical research and sustainable principles, hospitals can offer profound benefits not only to patients and staff but also to the broader community, embodying a holistic approach to health and wellness.

Equation 4.1 is a scientifically grounded formula designed to quantify the benefits of natural light exposure on patient recovery times, illustrating a direct correlation between daylight in hospital environments and accelerated healing processes. Positioned within the analysis of how architectural elements influence health outcomes, this equation underscores the therapeutic properties of natural light in medical settings. Providing a measurable framework, it enables healthcare designers and administrators to evaluate and enhance the amount of daylight incorporated into patient care areas, aligning hospital design with evidence-based health benefits.

$$\Delta R = -\gamma \log_{10}(1 + E) \tag{4.1}$$

In this formula, $\Delta R$ represents the change in recovery time, with negative values indicating a reduction in the duration of hospital stays. $E$ denotes the exposure to natural light, measured in lux hours, which is the cumulative amount of light a patient receives during their stay. $\gamma$ is a coefficient that quantifies the sensitivity of recovery time to changes in light exposure, derived from empirical studies. The logarithmic relationship captures the diminishing returns of increased light exposure, meaning while initial increases in natural light can significantly enhance recovery, the benefits taper off at higher levels of exposure.

This model is instrumental in emphasizing the role of natural lighting as more than just an aesthetic feature by showcasing its practical health implications. It provides hospital planners and architects with a powerful tool to advocate for designs that incorporate more natural elements, potentially leading to shorter hospital stays and improved patient satisfaction. Through such quantification, the adoption of natural light strategies can be more effectively justified and optimized, ensuring that hospital environments are not only functional but also healing spaces conducive to recovery.

Figure 4.1 illustrates an exemplary patient room designed to maximize exposure to natural light and outdoor views, enhancing the therapeutic environment within the hospital. The room features expansive windows that provide patients with calming and restorative views of nature, which are integral for psychological and emotional well-being. The thoughtful layout includes a patient bed positioned to face the windows, ensuring that patients can enjoy these views directly from their beds, thereby aiding in stress reduction and accelerating recovery. The room is also equipped with modern medical amenities and designed with a soft color palette that contributes to a serene and comfortable atmosphere. The inclusion of functional furniture, such as adjustable bedside tables and ample seating for visitors, ensures practicality while maintaining a focus on patient comfort and care.

### 4.3.2 Art and Aesthetics

The integration of art and aesthetic design within hospital settings transcends traditional decor, serving as a pivotal element in fostering a therapeutic environment

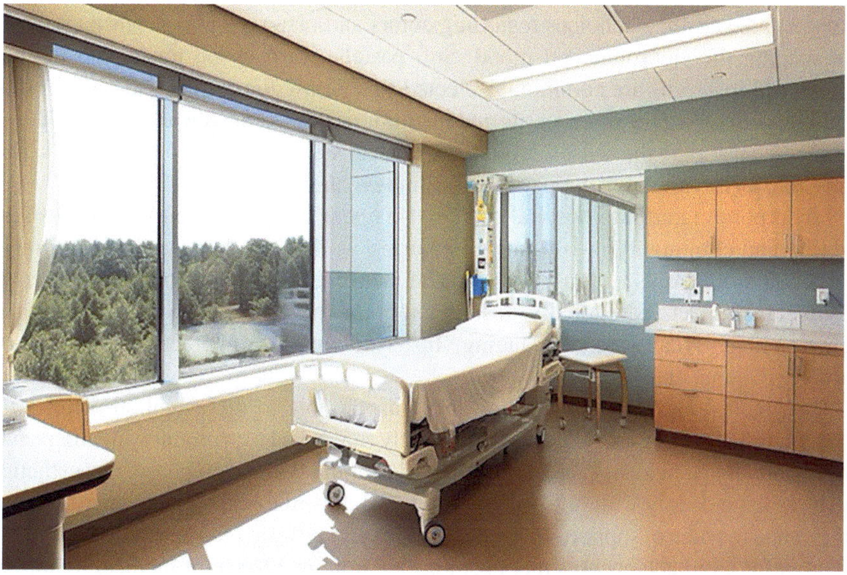

**Fig. 4.1** Patient room design optimized for natural light and scenic views

conducive to healing. Art installations and thoughtful aesthetic choices play a critical role in transforming the atmosphere of healthcare facilities, offering numerous benefits including stress reduction, elevation of spirits, and promotion of psychological well-being among patients, visitors, and staff. The therapeutic benefits of incorporating art in healthcare environments are substantial; art provides a beneficial distraction from the clinical aspects of medical treatment, allowing patients to focus on beautiful or intriguing visuals instead of their pain or anxiety. Research underscores that interacting with art can stimulate mental engagement and evoke positive emotional responses, which are vitally important during the recovery process. For example, soothing landscapes, abstract visuals, and nature-inspired artworks not only aesthetically pleasing but also contribute to lower blood pressure and reduced reliance on pain medications.

Moreover, art can act as a beacon of inspiration and hope. Artworks featuring uplifting imagery or motivational phrases can profoundly resonate with patients' experiences, offering them solace and encouragement amidst health challenges. Art that reflects cultural stories or values enhances a sense of connection and belonging, particularly comforting patients and families from diverse backgrounds who find familiarity and reassurance in these expressions.

Strategic integration of art and aesthetics into hospital design is crucial. It involves careful selection of artworks that appeal to a wide audience while steering clear of themes that might invoke distress. The placement of art also holds significant therapeutic potential; for instance, art pieces in waiting areas can alleviate anxiety, whereas those in patient rooms are best suited to foster relaxation and introspection. Beyond artwork, the aesthetic choices regarding colors and materials play a substantial role in shaping the environmental mood. Soft pastels, warm wood tones, and the use of natural materials are commonly selected to create a comforting and warm atmosphere, countering the often sterile and institutional feel of healthcare settings. These choices can soothe the senses, reduce the clinical starkness of hospital interiors, and promote an environment supportive of calmness and recovery.

Incorporating art into hospital design also opens avenues for embracing sustainability and community involvement. Featuring works from local artists or utilizing sustainable materials not only supports the local economy but also minimizes environmental impact. Additionally, hosting temporary art exhibits can keep the hospital environment dynamic and engaging, fostering ongoing interaction and interest among patients and visitors.

In conclusion, art and aesthetics are indispensable to the evolution of hospital design, transforming these spaces from merely functional to profoundly healing environments. By judiciously selecting art and designing with an emphasis on aesthetics that resonate deeply with both patients and staff, hospitals can significantly enhance the care experience, supporting more effective and holistic recovery processes. This thoughtful approach ensures that hospitals are not just places for medical treatment but sanctuaries of recovery and hope, essential for the comprehensive well-being of those they serve.

Table 4.3 provides statistical evidence of the positive impact of art installations on patient recovery rates and overall satisfaction within healthcare environments. This table quantifies the effects of integrating various forms of art—such as paintings, sculptures, and interactive installations—on improving patient outcomes and enhancing the aesthetic quality of hospital spaces. It demonstrates how art contributes not only to the beautification of medical facilities but also plays a crucial role in therapeutic interventions, aiding in faster recovery and increasing patient well-being.

Figure 4.2 illustrates the effective use of large-scale murals in pediatric hospital corridors, transforming them into visually engaging and comforting spaces for young patients. The artwork, rich in color and nature motifs, not only beautifies the space but also serves a therapeutic purpose by creating a more pleasant and less intimidating environment. Such installations demonstrate the thoughtful integration of art within healthcare settings, aiming to uplift spirits and aid in the healing processes of children. The choice of themes often reflects familiar and serene landscapes, contributing to a sense of tranquility and distraction from the clinical setting.

**Table 4.3** Art and patient recovery statistical analysis

| Type of art installation | Improvement in recovery time (%) | Increase in patient satisfaction (%) |
|---|---|---|
| Paintings | 15 | 20 |
| Sculptures | 10 | 18 |
| Interactive installations | 20 | 25 |
| Nature art | 18 | 22 |
| Digital art displays | 12 | 15 |

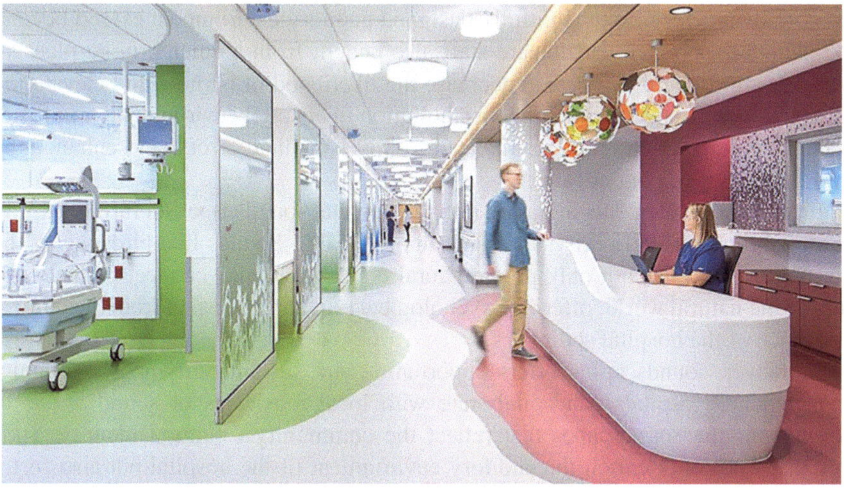

**Fig. 4.2** Nature-inspired art installations enhancing pediatric hospital corridors

### *4.3.3 Soundscaping*

Soundscaping in hospital design represents a critical innovation, acknowledging the auditory environment as an integral component of patient care. Traditionally, hospitals are replete with disruptive sounds—from the incessant beeping of medical equipment to the relentless activity in corridors—contributing to an atmosphere laden with stress and anxiety. A well-implemented soundscape strategy can significantly attenuate these negative auditory impacts, transforming the sounds within a hospital into a therapeutic tool that aids in patient recovery and enhances overall well-being.

Effective soundscaping involves a multi-layered approach that includes structural interventions and the deliberate introduction of soothing sounds. Key to this strategy is soundproofing, which incorporates architectural, and design choices aimed at minimizing noise transmission. This may involve the use of specialized acoustic panels, double-glazed windows, and wall and floor materials designed to absorb rather than reflect sound. Such measures are particularly vital in sensitive areas like intensive care units and patient rooms, where tranquility is paramount for recovery.

Beyond structural soundproofing, the ambient noise level in hospitals can be meticulously managed through the infusion of background music or natural sounds. Soft, instrumental music, or recordings of natural soundscapes—such as flowing water, gentle rain, or forest sounds—can be played through speaker systems to cultivate a calming atmosphere. These sounds serve not only to enhance the sensory environment but also to mask the jarring noises typical of hospital settings. The psychological impact of these auditory elements is profound; they can help lower blood pressure, reduce stress, and improve mental health, contributing to a more serene hospital environment.

The advantages of a thoughtfully designed auditory environment extend beyond the patient experience. Healthcare staff also benefit from a quieter, more pleasant working atmosphere, which can decrease the likelihood of burnout and elevate job satisfaction. Moreover, reducing noise pollution in hospital environments can foster better communication among staff, lead to fewer medical errors, and enhance the overall efficacy of patient care.

Incorporating elements of nature into the hospital soundscape can further exploit the biophilic links between nature and human health. Sounds like birdsong or the murmuring of water not only provide aural comfort but also connect patients and staff to the natural world, offering psychological relief and a respite from the clinical ambiance of the hospital setting.

Moreover, soundscaping offers opportunities for sustainability and community engagement. Hospitals can collaborate with local musicians or sound artists to develop unique soundscapes that reflect the community's cultural identity. Such initiatives not only enrich the auditory environment of the hospital but also foster stronger community connections and support local artistic talent.

In conclusion, prioritizing soundscaping in hospital design underscores the importance of the auditory environment in healthcare settings. By meticulously crafting the soundscape, hospitals can significantly boost patient recovery, enhance staff well-being, and forge a more humane and healing environment. This strategic focus on auditory design demonstrates a comprehensive approach to healthcare, emphasizing the need to consider all sensory experiences in fostering a holistic atmosphere conducive to health and wellness.

Figure 4.3 highlights key acoustic challenges identified in hospital settings, with data collected from survey respondents illustrating prevalent noise issues such as disturbances from medical equipment and high-traffic areas. The infographic underscores the critical need for effective soundscaping strategies to mitigate these issues, enhancing patient recovery and staff well-being. It visually represents how soundproofing measures, strategic placement of soothing sound sources, and architectural adjustments can transform a noisy hospital environment into a therapeutic and tranquil space. This alignment with soundscaping principles demonstrates a targeted approach to reducing stress and promoting healing through improved auditory conditions.

### *4.3.4 Ergonomics and Layout*

The ergonomic design and physical layout of hospital spaces play a crucial role in enhancing therapeutic care and ensuring the healthcare environment is safe, navigable, and patient-oriented. Employing ergonomic principles in the design of hospital interiors is aimed at optimizing human well-being and overall system performance, which are essential for facilitating recovery and minimizing stress among patients, visitors, and healthcare providers.

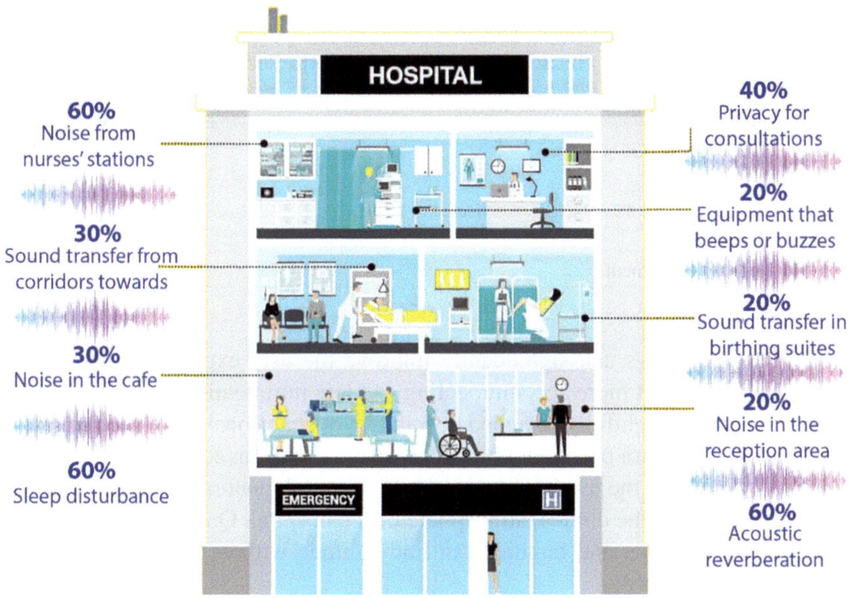


**Fig. 4.3** Hospital acoustic issues and soundscaping strategies

In patient rooms, ergonomic considerations significantly influence a patient's comfort and their ability to recover with dignity. For instance, adjustable beds that cater to various medical conditions and patient preferences are fundamental, as they ensure comfort and support during treatment and recovery. Additionally, bathrooms designed with accessibility in mind—equipped with safety features such as grab bars, non-slip floors, and appropriately height-adjusted toilets and sinks—are crucial for fostering patient independence and reducing fall risks.

The incorporation of user-friendly technology in patient rooms, such as remote controls for blinds, lighting, and temperature settings, allows patients to tailor their environment to their preferences without constantly needing to call for assistance. These features not only boost patient comfort but also promote a sense of autonomy and control, which have therapeutic benefits.

The overall layout of hospital corridors, wards, and service areas must be designed intuitively to facilitate easy navigation. This involves creating logical, straightforward pathways that naturally guide patients and visitors to their destinations, alleviating unnecessary stress or confusion. Clear, multilingual signage, sufficient lighting, and the elimination of physical barriers are essential for creating an environment that is easily navigable, especially crucial for individuals with limited mobility or cognitive challenges.

Incorporating color coding, pictograms, and tactile paths can further enhance navigation, particularly in larger facilities where the risk of disorientation is higher. Strategically placing essential services like pharmacies, cafeterias, and restrooms

**Table 4.4** Ergonomic design impact analysis

| Ergonomic feature | Improvement in patient comfort (%) | Reduction in staff stress-related incidents (%) |
|---|---|---|
| Adjustable beds | 25 | N/A |
| Ergonomic seating | 20 | 15 |
| Workstation layouts | N/A | 20 |
| Mobile equipment | 18 | 10 |
| Patient handling equipment | 22 | 25 |

along these pathways can significantly improve the user experience, making the hospital environment more accommodating and less intimidating.

Moreover, a thoughtfully designed hospital layout can considerably reduce stress and anxiety, which are prevalent in hospital settings and can adversely affect health outcomes. Ensuring the availability of rest areas, ample natural light, and views of nature can mitigate the clinical atmosphere of the hospital. Quiet zones and private spaces where patients and families can find solitude without leaving the hospital premises are invaluable.

In conclusion, integrating ergonomics and a well-considered layout in hospital design transcends aesthetic appeal or standard compliance; it is fundamentally about enhancing the care experience. By creating spaces that support both physical and psychological well-being, hospitals can provide a more therapeutic environment conducive to recovery. These design choices not only cater to the physical health of patients but also address their psychological needs, embodying a holistic approach to healthcare that respects and promotes patient dignity and well-being.

Table 4.4 details the improvements in patient and staff reports of physical comfort and reductions in stress-related incidents resulting from ergonomic enhancements in patient rooms and facilities. This table quantifies the positive effects of integrating ergonomic design elements such as adjustable beds, ergonomic seating, and thoughtfully designed workstations that facilitate ease of use and minimize physical strain. By emphasizing ergonomic considerations, healthcare facilities can significantly improve comfort and safety, reducing the incidence of stress and injury among patients and staff alike.

### *4.3.5 Flexible Spaces*

The concept of flexibility in hospital design is increasingly recognized as crucial for adapting to the diverse and changing needs of patients, offering an environment that is not only medically adept but also emotionally supportive. This approach underscores a comprehensive understanding of healing as a multifaceted process, where the physical environment can significantly impact recovery and well-being.

One of the primary ways hospitals incorporate flexibility is through the design of multi-purpose rooms. These spaces are ingeniously crafted to serve various functions, easily transforming to accommodate medical consultations, therapy sessions, private family gatherings, or social activities. This adaptability allows hospitals to address the psychological and social needs of patients alongside their medical care. For example, a room used for clinical examinations during the day can be reconfigured to host family celebrations or patient support groups in the evening, facilitating much-needed social interaction and mental stimulation.

The therapeutic benefits of social interactions in the recovery process are well-documented. Engaging with family, friends, and fellow patients can significantly enhance a patient's mood, diminish feelings of isolation, and foster a positive mental outlook—all vital components of effective recovery. Flexible spaces that promote such interactions are not merely auxiliary to medical care; they are integral to creating a therapeutic environment that supports holistic healing.

Designing flexibility requires meticulous planning concerning space usability, furniture selection, and accessibility. Furniture in these spaces should be lightweight, easily movable, and reconfigurable to allow quick alterations in room setup without significant effort or time. Additionally, incorporating elements such as movable partitions or curtains can help manage privacy and noise control, enabling these spaces to be swiftly adapted into serene environments suitable for therapy sessions or private consultations.

This approach to hospital design highlights the importance of holistic healthcare architecture, which seeks to create environments that nurture both the medical and emotional needs of patients. By crafting spaces that can dynamically adapt to the evolving requirements of patients, hospital architecture itself acts as a therapeutic tool, enhancing the efficacy of healthcare delivery and improving patient outcomes.

In conclusion, the integration of flexible spaces within hospitals is a pivotal design strategy that significantly contributes to the comprehensive well-being of patients. These spaces exemplify the transformative impact of thoughtful design in healthcare settings, turning them into sanctuaries of healing that cater not just to physical ailments but also to the psychological and social needs of patients. As we delve into the subsequent sections, we will further explore how elements such as privacy, comfort, and patient-centric design are seamlessly integrated into hospital environments to augment the therapeutic experience and reinforce the role of hospitals as true refuges for recovery and restoration.

Figure 4.4 illustrates an exemplary multi-purpose patient room designed to adapt seamlessly to various needs throughout the day. The room features a patient bed equipped with modern medical facilities, easily movable furniture, and spaces designed for family interactions and private consultations. The design incorporates ample natural light and views outdoors, enhancing the therapeutic ambiance. This flexibility not only aids in clinical efficiency but also supports the psychological and social well-being of patients, making it possible for the room to transform from a clinical setting to a more relaxed, family-friendly environment. This approach underpins the concept that hospital design can profoundly influence recovery and overall

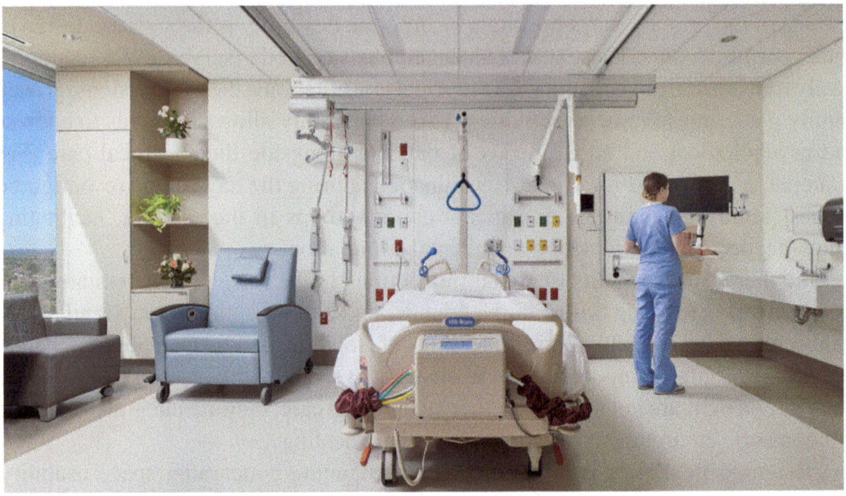

**Fig. 4.4** Multi-purpose patient room design

patient care by providing spaces that cater to both medical needs and the emotional comfort of patients and their families.

## 4.4 Privacy, Comfort, and Patient Well-Being

Privacy, comfort, and patient well-being are interlinked facets of hospital design that directly impact patient satisfaction and recovery outcomes. Ensuring that these aspects are adequately addressed is crucial in creating a patient-centric environment where individuals feel cared for, respected, and comfortable during their medical treatments and recovery.

### *4.4.1 Privacy*

Privacy in healthcare settings is a complex necessity, pivotal not only for adhering to legal standards but also for providing care that is compassionate and respectful. In hospital environments, the scope of privacy extends beyond simply providing private spaces; it encompasses auditory, visual, and informational dimensions that significantly influence a patient's sense of security and dignity.

Auditory and visual privacy are critical in designing patient-centric spaces. For auditory privacy, hospitals are progressively utilizing soundproofing materials in the construction of patient rooms. Soundproof walls and ceilings are essential to prevent the transmission of sounds such as personal conversations, medical discussions, and signs of distress. This ensures that sensitive information remains confidential, and that the environment maintains respect for patient privacy.

Visual privacy is addressed through thoughtful architectural and interior design choices. The use of frosted glass on windows and doors permits natural light while obstructing clear visual access from outside the patient's room. Strategically orienting patient beds away from direct sightlines from doorways or windows further safeguards visual privacy, allowing patients to rest and receive care without the discomfort of feeling exposed.

The design also incorporates private bathrooms within patient rooms, enhancing comfort and dignity, particularly vital during extended stays or periods of reduced mobility. In shared rooms, individual patient areas delineated by partitions or curtains provide a pragmatic privacy solution when single rooms are not available. These arrangements offer personal space while facilitating the efficient use of space within the facility.

Beyond physical privacy, confidentiality in communication is equally crucial. Hospitals are increasingly designing spaces that ensure private conversations between patients and healthcare providers remain confidential. This includes consultation areas equipped with acoustic sealing or sound-masking technologies, preventing discussions about patient care from being overheard by others.

In common areas such as waiting rooms, the thoughtful arrangement of seating, the strategic placement of privacy screens, and the orientation of conversation spaces are all designed to minimize the risk of inadvertent eavesdropping. Such measures not only protect the confidentiality of patient information but also enhance the overall sense of comfort and trust within the healthcare environment.

In conclusion, addressing privacy in hospital design is fundamental to reducing patient stress, preserving dignity, and enhancing satisfaction with the care experience. By implementing comprehensive design strategies that cater to auditory, visual, and informational privacy, hospitals can create environments that not only facilitate medical healing but also respect and protect the rights and comforts of patients. These efforts are crucial for nurturing an atmosphere of trust and confidentiality, which are foundational to delivering quality healthcare.

Equation 4.2 is meticulously designed to calculate the Sound Reduction Index (SRI), a key metric in evaluating the effectiveness of soundproofing materials used in hospital settings. This formula is crucially positioned within the discussion on soundproofing technologies, highlighting the importance of auditory privacy and noise control in medical environments. The Sound Reduction Index quantifies the ability of a barrier or material to obstruct sound transmission, thereby enhancing the auditory comfort of patients and staff. This metric is vital for architects, engineers, and healthcare facility managers in selecting appropriate materials that effectively reduce noise, a factor critical to patient recovery and privacy.

$$\mathrm{SRI} = 10 \log_{10}\left(\frac{I_i}{I_t}\right) \tag{4.2}$$

In this equation, SRI represents the Sound Reduction Index, measured in decibels (dB), which quantifies the sound insulation effectiveness of a material or assembly. $I_i$ is the intensity of the incident sound wave, typically measured in watts per square meter, while $I_t$ is the intensity of the transmitted sound wave that passes through the material or barrier. The logarithmic term $10\log_{10}(I_i/I_t)$ calculates the ratio of the incident sound power to the transmitted sound power, providing a measure of how much sound is attenuated by the barrier.

This model is fundamental in the strategic placement and selection of soundproofing materials within hospital designs, ensuring that sound levels are kept within therapeutic limits to facilitate a conducive healing environment. By accurately calculating the SRI, healthcare facilities can implement more effective sound insulation strategies, significantly improving the quality of care through enhanced patient comfort and confidentiality. The use of this equation supports evidence-based decisions in architectural design, promoting the development of quieter, more private hospital spaces essential for modern healthcare delivery.

Figure 4.5 details the systematic approach to maintaining patient information privacy from admission to checkout in hospital settings. Each step illustrates a crucial phase in safeguarding both auditory and visual privacy as well as the confidentiality of patient data. Patient Admission marks the initiation of privacy protocols, ensuring each patient's entry into the system is secure. Assign Privacy-Enhanced Room shows how patients are accommodated in environments designed to optimize privacy through soundproofing and visual barriers. Confidential Consultations highlight areas specially equipped to keep discussions between patients and healthcare providers private. Privacy Screens in Waiting Areas are strategically placed to prevent unintended information leakage. Secure Information Handling underscores the importance of protecting patient data through advanced security measures in both digital and physical formats. Finally, Patient Checkout ensures all personal information is appropriately managed as the patient departs, maintaining confidentiality until the end of their stay. This visual representation underscores the comprehensive steps taken by healthcare facilities to protect patient privacy, reflecting a commitment to high ethical standards and patient care quality.

### *4.4.2 Comfort*

Comfort in hospital environments is a critical element of patient care, impacting not only physical recovery but also encompassing emotional and psychological well-being. This holistic approach to healing necessitates thoughtful design and operational strategies within healthcare facilities to provide multifaceted comfort to patients, their families, and staff.

Ergonomic furniture is essential for physical comfort in patient rooms and communal areas. Adjustable hospital beds allow patients to find the most suitable position for their specific conditions, easing mobility and reducing the risk of complications such as bedsores. Additionally, ergonomically designed chairs in

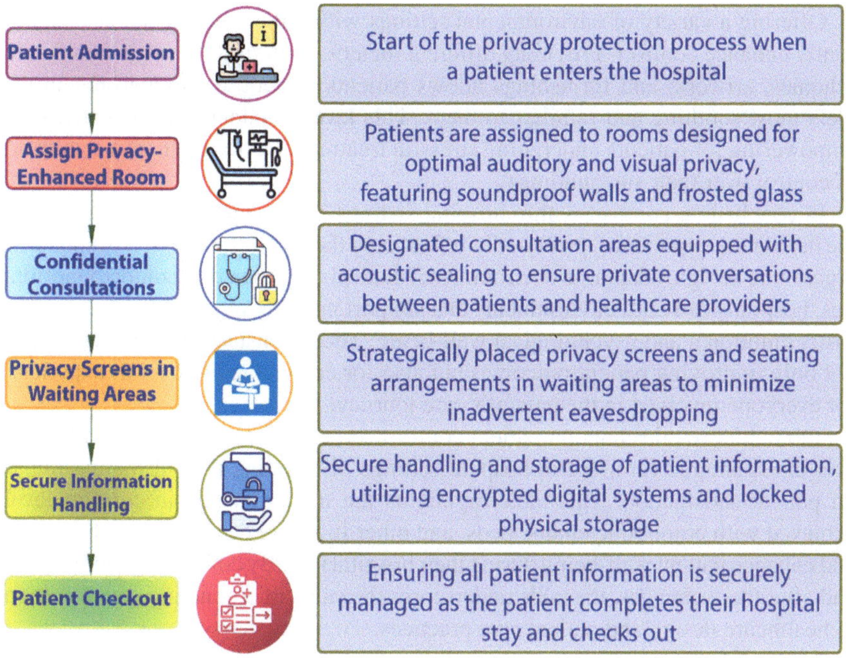


**Fig. 4.5** Steps to ensure patient information privacy in hospital settings

patient rooms and waiting areas provide necessary support during prolonged periods of sitting, enhancing comfort for both patients and visitors. It is also vital to select furniture that can be easily cleaned and maintained to uphold hygiene standards without compromising comfort.

Lighting significantly affects both physical and psychological comfort. Adjustable lighting systems that emulate natural light patterns help sustain the body's circadian rhythms, which is crucial for promoting good sleep quality and overall well-being. Allowing patients to adjust the brightness and color temperature in their rooms enables them to tailor the environment to their preferences, fostering a sense of autonomy and comfort.

Effective climate control is another cornerstone of comfort in healthcare settings. Advanced HVAC systems that efficiently filter air while maintaining consistent temperature and humidity levels are indispensable. Such systems ensure the environment is not only comfortable but also safe from allergens and pathogens, which is particularly important in a healthcare setting.

Personalization of space is profoundly beneficial for emotional and psychological comfort. Allowing patients to decorate their space with personal items such as photos, drawings, or familiar objects can make the hospital setting feel less intimidating and more homely. This is especially significant for patients facing long-term stays, as personal touches can greatly improve mental health and overall disposition.

Offering a variety of environmental settings within the hospital can also significantly enhance comfort. Providing different themes or room styles with varied color schemes, artwork, and furnishings allows patients to choose an environment that feels most soothing and familiar to them. This level of choice can be particularly empowering for patients undergoing stressful treatments, offering them a semblance of control over their surroundings.

In conclusion, comfort within hospital settings is a dynamic and vital aspect of the healthcare experience, profoundly influencing the healing process. By integrating ergonomic design, adaptable environmental controls, and personalization opportunities, hospitals can craft environments that support not only physical healing but also foster emotional and psychological well-being. These considerations are crucial for not only improving patient outcomes but also for enhancing the overall experience for everyone involved in the patient's care journey.

Table 4.5 provides data on patient comfort scores before and after the introduction of ergonomic furniture in hospital settings, illustrating the substantial impact on patient satisfaction. This table highlights the improvements in comfort levels achieved with ergonomic chairs, beds, and other furniture designed to reduce strain and enhance patient well-being during their hospital stay. By comparing scores before and after these changes, the table underscores the value of ergonomic considerations in healthcare design and patient care practices.

Figure 4.6 illustrates an ergonomically designed patient room optimized for both physical comfort and psychological well-being. The room features an adjustable hospital bed that supports different patient needs and postures, enhancing mobility and reducing potential complications such as bedsores. Ergonomic chairs provide essential back support, crucial for patients and visitors during prolonged periods of sitting. The room is designed with large windows that allow natural light, promoting a healthy circadian rhythm and improving overall patient mood and recovery. Ample space and a neutral color palette contribute to a calming environment, vital for patient recovery. Personal items can be placed around the room to add a sense of homeliness, crucial for long-term stays. This design approach underscores the significance of a holistic environment that supports comprehensive healing by integrating comfort, functionality, and aesthetics.

**Table 4.5** Ergonomic furniture and patient comfort scores

| Furniture type | Comfort score before | Comfort score after | Improvement (%) |
|---|---|---|---|
| Ergonomic beds | 70 | 85 | 21.4 |
| Ergonomic chairs | 65 | 80 | 23.1 |
| Overbed tables | 60 | 78 | 30.0 |
| Patient room sofas | 58 | 77 | 32.8 |
| Mobile workstations | 55 | 72 | 30.9 |

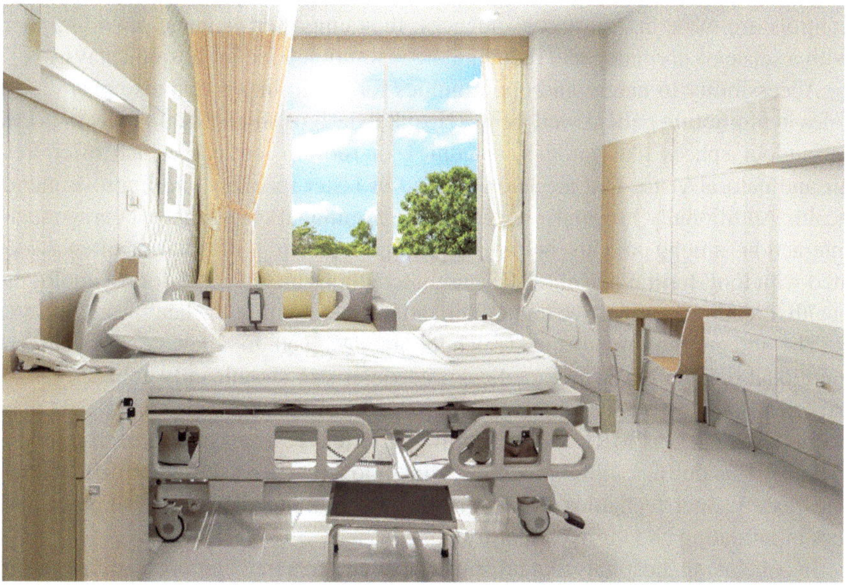

**Fig. 4.6** Ergonomic and comfortable patient room design

### *4.4.3 Patient Well-Being*

The concept of patient well-being within a hospital setting encompasses a broad spectrum of care that includes physical, emotional, and psychological dimensions. A hospital environment that excels in patient care goes beyond fulfilling clinical needs; it is designed to foster a holistic healing experience that considers the complete well-being of its occupants. This involves the strategic incorporation of both therapeutic and aesthetic elements to create a space that promotes recovery and tranquility.

Therapeutic environments within hospitals are integral to enhancing patient well-being. The use of art and aesthetic design in healthcare settings plays a critical role in uplifting patients' spirits and providing a distraction from the clinical atmosphere and treatment-related stress. Art installations, such as murals and sculptures, along with temporary art exhibitions, serve not only as visual stimulants but also as tools for psychological relief and emotional upliftment. The inclusion of natural elements like indoor plants, water features, and ample natural lighting is deliberately aimed at crafting a serene atmosphere that aids in stress reduction and anxiety alleviation. These features are not merely decorative additions but are foundational in creating an environment conducive to recovery.

Another crucial aspect of patient well-being is the ability to exercise control over one's personal space within the hospital. Allowing patients to adjust their immediate surroundings—such as lighting, window shades, and room temperature—gives them a sense of autonomy and comfort, especially during prolonged hospital stays. Such controls are more than just conveniences; they empower patients, providing them with a sense of normalcy and reducing the institutional feel of hospital stays.

Accessibility to nature and opportunities for social interaction also play pivotal roles in promoting patient well-being. Hospital designs should facilitate easy access to outdoor spaces like gardens, balconies, or rooftop terraces, which offer fresh air and natural vistas that are known to aid in faster recovery and improve mental health. Additionally, hospitals should include communal spaces that encourage social interactions among patients, helping to prevent the feelings of isolation often associated with long-term medical care. These interactions are essential for psychological health, creating a supportive community atmosphere within the healthcare setting.

The integration of advanced technology designed to alleviate physical and emotional discomfort during medical procedures can significantly enhance patient experiences. For instance, employing noise-reducing headphones during loud diagnostic tests, such as MRI scans, or utilizing virtual reality systems to distract and calm patients during painful procedures are innovative ways to reduce patient stress and anxiety. Such technological integrations are vital in making medical procedures less intimidating and more patient friendly.

In conclusion, hospitals that prioritize a patient-centered approach in their design focus on aspects such as privacy, comfort, aesthetic appeal, and personal control, transforming their spaces into sanctuaries of comprehensive care and well-being. Healthcare facilities must adopt design strategies that are not only functional and hygienic but also warm and psychologically supportive, ensuring that patients feel valued and holistically cared for. The upcoming discussion will delve into inclusive design strategies that ensure these benefits are accessible to all patients, creating an environment that is welcoming and accommodating to individuals from all walks of life, regardless of their specific needs or circumstances.

Equation 4.3 introduces a pivotal model known as the Patient Autonomy Impact Formula, which quantifies the significant effects of allowing patients personal control over their environmental settings—such as lighting, temperature, and sound—on their overall well-being and recovery rates. This equation is crucially positioned within discussions about the patient personalization of space, emphasizing the concept that greater patient autonomy in controlling their environment can lead to improved health outcomes and enhanced satisfaction with care received.

$$\mathrm{PAI} = k_0 + k_1 L + k_2 T + k_3 S \tag{4.3}$$

In this formula, PAI denotes the Patient Autonomy Index, which is a composite measure reflecting the overall impact of environmental control on patient recovery and well-being. $L$, $T$, and $S$ represent the levels of control a patient has over lighting, temperature, and sound, respectively. Each variable is scored based on the extent of control provided, with higher scores indicating greater autonomy. The coefficients $k_1$, $k_2$, and $k_3$ quantify the sensitivity of the patient's well-being to changes in control over these specific environmental factors, while $k_0$ serves as a baseline measure of well-being in the absence of such controls.

This model serves as a sophisticated tool for healthcare providers and facility designers, illustrating how the design of patient spaces can be optimized to support autonomy and, consequently, improve health outcomes. By applying Eq. 4.3, healthcare facilities can effectively assess and implement strategies that empower patients, providing them with the ability to personalize their surroundings to suit their preferences and needs. Such empowerment not only enhances comfort and satisfaction but also actively contributes to the speed and effectiveness of the recovery process, showcasing the profound intersection of environmental psychology and healthcare.

Table 4.6 compares patient recovery metrics, such as stress levels and recovery times, in environments with and without therapeutic design elements. This table highlights the measurable benefits of integrating therapeutic design features—like calming color schemes, natural lighting, and noise-reduction materials—into healthcare settings. By showcasing the differences in patient outcomes between enhanced and standard environments, this data underscores the significance of therapeutic design in promoting patient well-being and accelerating recovery.

Figure 4.7 captures the essence of holistic care within hospital settings, where every element is meticulously designed to promote patient well-being. The room features ergonomic furniture that offers both comfort and functionality, adjustable to meet the specific needs of patients. Natural lighting and a soothing color palette help maintain circadian rhythms, crucial for promoting rest and recovery. The inclusion of nature through window views and indoor plants adds a serene touch, aiding in stress reduction and emotional uplift. Personal touches are evident, allowing patients to feel more at home, and further alleviating the clinical feel of hospital stays. This environment exemplifies how thoughtful design can transform a patient room into a sanctuary of healing, focusing on all aspects of patient care—physical comfort, emotional tranquility, and psychological support.

**Table 4.6** Patient well-being metrics

| Design element | Stress level in standard environment | Stress level in enhanced environment | Recovery time in standard environment (days) | Recovery time in enhanced environment (days) |
|---|---|---|---|---|
| Calming colors | High | Low | 10 | 7 |
| Natural lighting | Moderate | Very low | 12 | 8 |
| Noise reduction | High | Moderate | 11 | 9 |
| Ergonomic furniture | Moderate | Low | 10 | 8 |
| Art installations | High | Very low | 14 | 10 |

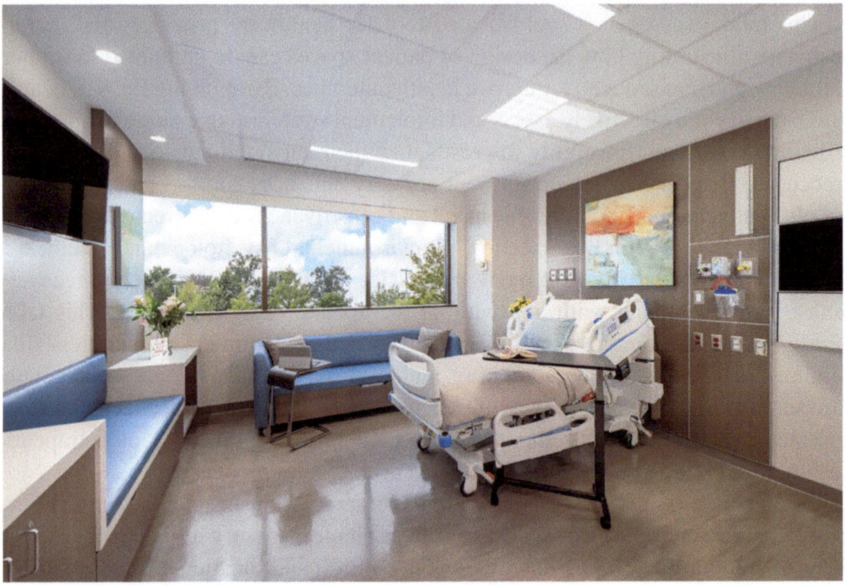

**Fig. 4.7** Enhancing patient well-being through holistic room design

## 4.5 Inclusive Design Approaches

Inclusive design in healthcare settings is essential for ensuring that all patients, regardless of their physical abilities, age, cultural background, or health conditions, receive equitable care in a setting that respects their dignity and individual needs. This approach to design not only enhances accessibility but also promotes a sense of belonging and respect, which are crucial for the psychological well-being of patients.

### *4.5.1 Universal Accessibility*

Universal accessibility in healthcare settings is a fundamental aspect of inclusive design, ensuring that all individuals, irrespective of their physical, sensory, or cognitive abilities, have unimpeded access to hospital facilities and services. This commitment is not only a matter of complying with legal and ethical mandates but is also a profound expression of the healthcare sector's dedication to compassionate and patient-centered care.

A primary consideration in creating an accessible healthcare environment is the incorporation of physical accessibility features from the initial stages of architectural planning. Healthcare facilities must be designed with features such as ramps and automatic doors at all entry points to cater to patients who use wheelchairs, walkers, or other mobility aids. For buildings with multiple floors, it is essential to include elevators that can accommodate several users with mobility aids simultaneously, ensuring there is enough space and that the controls are easily operable.

Within the hospital, the spatial layout requires meticulous planning to eliminate potential barriers. Corridors should be sufficiently wide to allow smooth passage for mobility devices and should remain free from obstructions that could create hazards. Additionally, selecting non-slip flooring materials is crucial for enhancing safety for all users, especially those facing mobility challenges, thereby preventing slips and falls.

Visual and cognitive accessibility is equally vital. Hospitals can significantly improve navigation for visually impaired patients through the implementation of tactile pathways that facilitate movement throughout the facility. Complementary to visual signage, Braille signs should be installed to provide crucial information regarding room numbers, directions, and emergency exits, thus fostering independence among visually impaired users.

Signage must also be designed to support cognitive accessibility. Employing clear, straightforward language and high-contrast colors helps in making the signs easy to read and understand for individuals with cognitive impairments or those who process visual information differently. The adoption of universal symbols in signage aids in bridging language gaps and cognitive variances, ensuring information is comprehensible to a diverse patient population.

Inclusive design transcends physical adjustments to encompass a comprehensive approach toward accessibility. This includes auditory accommodation such as public address systems that can deliver announcements in a clear, understandable manner. In environments like patient rooms and consultation areas, features such as adjustable lighting and the provision of quiet spaces are essential to cater to individuals with sensory sensitivities.

Additionally, the training of hospital staff plays a crucial role in ensuring that all personnel are prepared to assist and effectively communicate with patients having diverse needs. This training should include the proficient use of assistive technologies, the interpretation of non-verbal cues, and the development of skills for clear and empathetic communication.

In conclusion, universal accessibility within hospitals is about creating an inclusive environment where every patient feels welcomed, respected, and valued. By embedding comprehensive accessibility features into the design of healthcare facilities, these institutions can not only enhance patient satisfaction and outcomes but also bolster community trust in the healthcare system. The ongoing commitment to universal accessibility should manifest in every aspect of hospital design and operations, setting a high standard for inclusivity and ensuring that all patients receive the care they deserve in an environment that is tailored to their specific needs.

### *4.5.2 Cultural Sensitivity*

Cultural sensitivity in hospital design is an imperative that acknowledges the varied backgrounds of both patients and staff, aiming to foster an environment that respects and reflects a wide range of cultural and religious needs. This approach is essential not only to provide equitable care but also to ensure that all individuals feel comfortable and supported within the healthcare setting.

The creation of multicultural spaces is a core element of culturally sensitive design. For instance, hospitals should include dedicated prayer rooms that cater to various faiths. These rooms ought to be equipped with features accommodating the distinct practices of different religions, such as wash areas for ritual cleansing and non-denominational symbols. Additionally, directional signs indicating prayer orientations necessary in Islamic practices should be considered. Beyond facilities for structured religious practices, spaces for quiet meditation and reflection are equally vital, catering to those who seek moments of spiritual or emotional solace in a more private setting.

Art and decorations significantly contribute to creating a welcoming hospital environment. Incorporating artwork that represents a diverse array of cultural backgrounds can make patients from various ethnic groups feel seen and respected. This might involve displaying paintings, sculptures, or crafts that celebrate different cultural heritages within common areas such as lobbies, corridors, and waiting rooms. Such visuals do more than enhance the aesthetics of space; they foster a sense of inclusivity and promote mutual understanding among patients and staff.

Moreover, cultural sensitivity extends to communication strategies within the hospital. Ensuring that signage, informational brochures, and digital displays are available in multiple languages helps non-English speaking patients navigate the facility and comprehend their treatment plans without substantial barriers. The employment of multilingual staff or the availability of translation services can significantly improve communication, ensuring that all patients receive precise and comprehensible medical information.

Staff training is crucial in effectively implementing culturally sensitive practices. Healthcare workers should receive education regarding the cultural norms and religious beliefs of the communities they serve. Such training helps prevent cultural misunderstandings and ensures the delivery of care that respects patients' cultural or religious preferences. Additionally, awareness programs aimed at handling cultural diversity can enhance interactions between patients and staff, ultimately improving the quality of healthcare delivery.

In conclusion, integrating cultural sensitivity into hospital design transcends the mere addition of specific features; it involves nurturing an environment that genuinely respects and celebrates diversity. By developing spaces that accommodate various cultural and religious needs and incorporating artistic and linguistic elements reflective of the multicultural community, hospitals can transcend their traditional roles. They become sanctuaries where every individual feels acknowledged and valued. This dedication to cultural sensitivity not only enriches the patient's experience

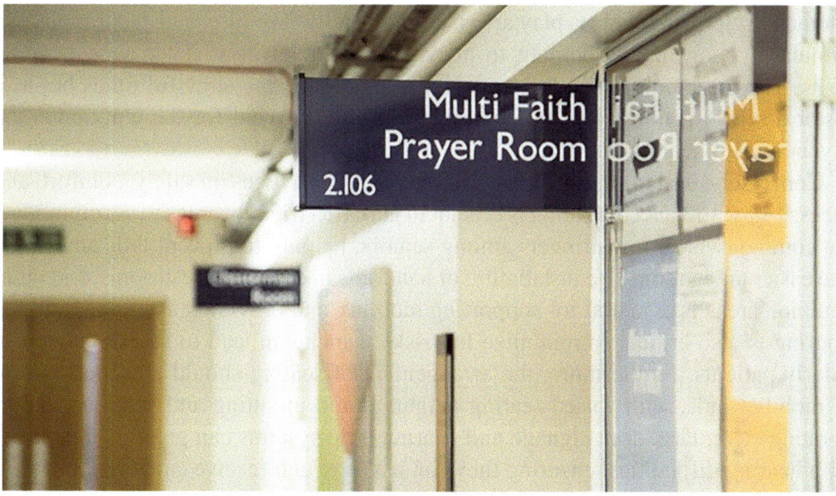


**Fig. 4.8** Integrating multicultural respect: a multi-faith prayer room in a hospital setting

but also solidifies the community's trust in the healthcare system, fostering greater understanding and acceptance among diverse populations.

Figure 4.8 depicts a hospital corridor leading to a "Multi-Faith Prayer Room," emblematic of the commitment to cultural sensitivity in hospital design. Including such spaces is essential for accommodating the spiritual and religious practices of diverse patient and staff populations, fostering an environment where all individuals feel respected and supported. Hospitals that provide designated areas for prayer and meditation respect individual religious practices and promote emotional and spiritual well-being. This approach enhances the overall healthcare experience by ensuring that cultural and spiritual needs are recognized and honored. Implementing such features demonstrates a profound understanding of the diverse backgrounds of those served by the facility, contributing to a more inclusive and comforting healthcare environment.

### 4.5.3 Age Considerations

Designing healthcare facilities with careful consideration of age-specific needs is crucial for creating inclusive, safe, and supportive environments for all patients. Different age groups have distinct needs and sensitivities that significantly influence their experience and interaction with healthcare services. For instance, pediatric environments present unique challenges as hospitals can appear intimidating to children. To counteract this, designing these areas to be bright, vibrant, and engaging is essential. Utilizing themes from nature, children's storybooks, or popular cartoons can help transform intimidating clinical spaces into lively and inviting settings. Including interactive elements such as play areas, child-friendly furniture, and decorations can stimulate young minds, helping to alleviate the anxiety associated with hospital visits. These features not only serve as distractions during stressful times but also foster a sense of normalcy and enjoyment, which are vital for the emotional and psychological health of children.

Conversely, designing for elderly patients requires a focus on safety, comfort, and accessibility. Enhanced lighting is crucial in geriatric-friendly areas to accommodate the common visual impairments among seniors, helping to prevent falls and facilitate easier navigation. The installation of handrails in corridors, restrooms, and other common areas is essential for supporting mobility and providing stability. Non-slip flooring is also critical to minimize fall risks, which can lead to severe injuries in elderly patients. Furthermore, the arrangement of seating should consider the use of mobility aids, with varied seating heights to aid in sitting and standing. Incorporating clear, large-font signage and hearing loop systems can greatly assist those with hearing difficulties, ensuring they can navigate and receive care effectively.

While certain features are specifically tailored to either pediatric or geriatric users, some universal design principles should be implemented throughout healthcare facilities to accommodate all ages. These include designing wide corridors that support both the playful activities of children, and the mobility needs of elderly individuals, quiet zones that offer spaces for rest and recovery, and flexible spaces that can be adjusted according to the daily needs or demographic profiles of the patients being served.

Incorporating these age-specific considerations into hospital design not only meets the clinical requirements of patients but also significantly enhances their comfort, safety, and emotional well-being. By adopting a sensitive and inclusive approach to the unique needs of different age groups, healthcare facilities can create supportive environments that promote healing and recovery across the lifespan. Thoughtfully designed spaces improve patient outcomes and enrich the overall experience for patients, families, and caregivers, making the healthcare environment a cornerstone of community wellness.

Figure 4.9 illustrates an exemplary healthcare waiting area designed with age considerations in mind. The space seamlessly integrates elements suitable for both children and adults, ensuring comfort and engagement across generations. Children are provided with interactive and engaging elements such as toys and child-friendly furniture, promoting a positive and calming atmosphere. Simultaneously, adult seating areas offer a serene environment for conversations or contemplation. The strategic use of vibrant art and natural light enhances the overall ambiance, contributing to a supportive and healing environment. This design approach not only reduces anxiety but also fosters a sense of community and belonging among patients of all ages, illustrating the profound impact of thoughtful, inclusive design in healthcare settings.

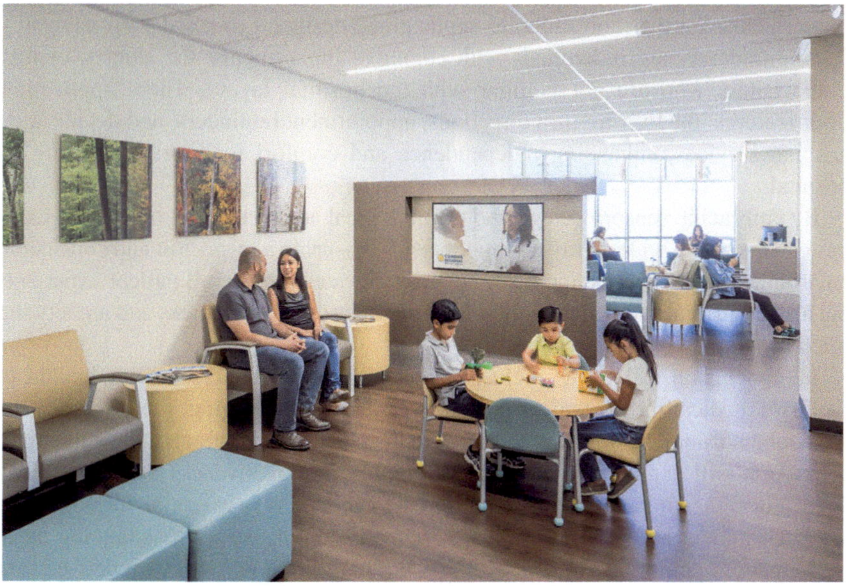

**Fig. 4.9** Inclusive healthcare environment for all ages

### *4.5.4 Sensory Design*

Sensory design in healthcare settings plays a vital role in creating an accommodating environment for patients with sensory processing disorders and those who may find traditional hospital atmospheres overwhelming. This design approach focuses on constructing spaces that reduce sensory overload, which is prevalent in the bustling and often hectic hospital environment.

The integration of low-stimulation areas within hospitals serves as an effective strategy to support individuals who benefit from reduced sensory input. These spaces are defined by their use of subdued lighting, avoiding harsh fluorescent lights in favor of soft, natural lighting or adjustable lighting systems that can be tailored to individual preferences. Noise reduction is another critical aspect, achieved by employing sound-absorbing materials on walls and floors, alongside the strategic placement of quiet zones distanced from high-traffic areas. The overall aesthetic is one of minimal clutter and serene décor, utilizing simple design elements to prevent cognitive overload and create a calming environment. Furniture is carefully chosen for comfort and positioned to foster relaxation, complemented by views of nature or calming artwork that contribute to the tranquil atmosphere.

Technological integration also plays a significant role in fostering inclusivity within these sensory-sensitive designs. Interactive kiosks that provide information in multiple languages greatly benefit non-native speakers and those with hearing impairments. These kiosks can include headphone jacks for audio assistance and touch screens that allow adjustments in text size to accommodate those with visual impairments. Furthermore, mobile applications designed to assist with navigation can significantly aid individuals with cognitive impairments, the elderly, or first-time visitors who are not familiar with the hospital layout. These applications offer features like turn-by-turn directions, appointment reminders, and details about hospital services, enhancing independence and comfort for users navigating the hospital.

By embracing sensory design and technological advancements, healthcare facilities can address a wider array of patient needs, ensuring a respectful and considerate environment for all. This approach not only elevates the overall patient experience but also contributes to better health outcomes by reducing the stress and anxiety that often accompany hospital visits.

In conclusion, adopting inclusive sensory design underscores a healthcare facility's dedication to patient-centered care, acknowledging the necessity for medical environments to adapt to the varied needs of their patients. The subsequent discussion will explore methods for evaluating the effectiveness of these patient-centric design implementations. Continuous evaluation is essential for advancing the environment and confirming that it remains responsive to the changing needs of the patient community, thereby improving both the quality of care delivery and patient satisfaction.

Figure 4.10 illustrates the innovative integration of sensory-sensitive kiosks within a hospital environment, designed to provide an accommodating space for patients with diverse sensory needs. These kiosks feature adjustable interfaces that allow for customization of brightness, contrast, and volume to suit individual preferences, ensuring a user-friendly experience. Positioned in strategic locations, they minimize the sensory overload typically associated with high-traffic hospital areas. The incorporation of these kiosks reflects a deep commitment to sensory design, focusing on creating a calm and accessible atmosphere for all patients, particularly those who are overwhelmed by traditional bustling hospital environments. Such technological advancements not only facilitate smoother patient interactions but also enhance overall accessibility, demonstrating the hospital's dedication to inclusive design and patient-centered care.

## 4.6 Evaluation of Patient-Centric Design Implementations

The evaluation of patient-centric design implementations is crucial for understanding their effectiveness and identifying areas for improvement. This process involves assessing how well the physical environment and service designs meet the needs of patients and contribute to optimal health outcomes. Evaluating these designs helps ensure that the investments in hospital infrastructure truly enhance patient care and satisfaction.

Table 4.7 provides a summary of different evaluation methods used in healthcare settings, including surveys, focus groups, and observational studies, along with their specific outcomes. This table categorizes each method based on its effectiveness and

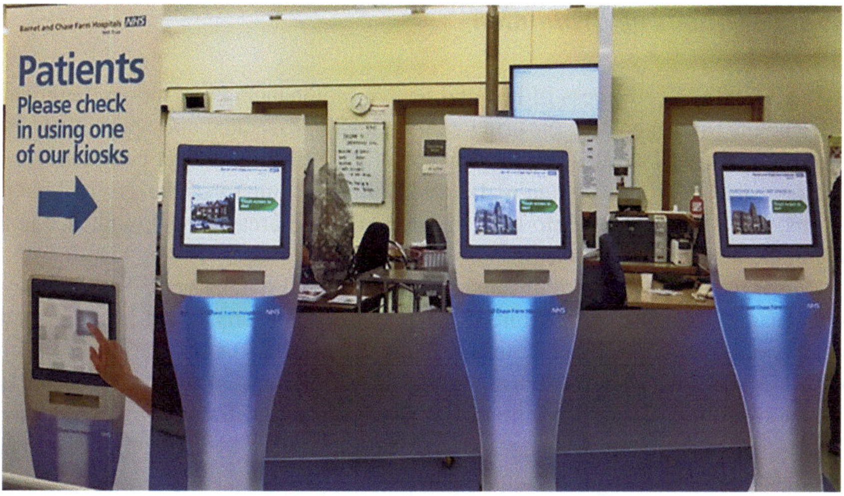


**Fig. 4.10** Sensory-adaptive self-check-in kiosks in hospital reception

primary areas of application, offering insights into how each approach contributes to understanding patient needs, staff efficiency, and system performance. By presenting a comparative analysis, the table aids healthcare administrators and researchers in selecting the most appropriate method for their specific evaluation goals, ensuring comprehensive and accurate data collection.

**Table 4.7** Summary of evaluation methods and outcomes

| Evaluation method | Primary application | Outcomes measured | Effectiveness |
|---|---|---|---|
| Surveys | Patient satisfaction | Patient comfort, satisfaction, feedback on services | High |
| Focus groups | Service improvement | In-depth insights into patient and staff experiences, ideas for improvement | Medium |
| Observational studies | Process optimization | Workflow efficiency, interaction quality, environmental impact | High |
| Electronic health records analysis | System performance | Treatment outcomes, patient follow-up rates, medication adherence | Very high |
| Patient interviews | Understanding individual needs | Detailed personal experiences, specific needs, and concerns | Medium |

### 4.6.1 Methods of Evaluation

Evaluating the effectiveness of hospital design in enhancing patient care and staff efficiency is fundamental to ensuring continuous improvement and optimal adaptation of healthcare facilities. Employing various methods to collect comprehensive data is crucial, as it reflects the real-world utility and impact of design implementations, providing insights into both the strengths and areas needing enhancement within the hospital environment.

One of the primary tools for evaluation is patient and staff surveys. These surveys offer quantifiable data that capture the experiences and satisfaction levels of both patients and hospital staff. They can be specifically tailored to assess various elements of the hospital environment, such as privacy measures, the comfort of patient rooms, and the overall navigability of the facility. Regular administration of these surveys enables hospitals to track changes over time, facilitating swift identification and rectification of areas requiring improvement.

In addition to surveys, focus groups provide a dynamic and interactive forum for deeper evaluation. These groups can consist of patients, their family members, or staff, discussing their experiences within a structured environment. This method allows for an in-depth exploration of specific concerns or positive aspects of hospital design that surveys might not fully capture. For instance, focus groups involving patients from a newly designed hospital wing can yield valuable insights into how the architectural changes have affected their recovery processes and overall experiences in the facility.

Observational studies also play a critical role in the evaluation process. This approach involves researchers or trained observers watching how individuals interact with the hospital environment without intervening. Such studies are especially effective in identifying design flaws that are not explicitly reported by the users but may obstruct efficient use of space. Observational data can highlight problems such as areas prone to congestion, underutilized spaces, or issues arising from the interaction with furniture or technological facilities.

For a thorough evaluation, hospitals often combine these methods, leveraging the unique strengths of each. Surveys provide a broad quantitative overview, focus groups add qualitative depth by exploring personal experiences and perceptions, and observational studies uncover subtle, unarticulated behaviors. Utilizing a blend of these evaluation tools offers a robust picture of how hospital designs impact patient outcomes and the daily workflows of staff.

In conclusion, systematic evaluation of hospital environments is crucial for ensuring that designs not only meet clinical needs but also enhance the overall care experience. By consistently collecting and analyzing feedback from various stakeholders through diverse methodologies, healthcare facilities can make well-informed decisions that lead to notable improvements in care delivery and patient satisfaction. This ongoing evaluation process not only aids in refining the physical aspects of hospital environments but also fosters a culture of care that values continuous feedback and strives for excellence in all facets of patient and staff experiences.

Equation 4.4 is designed to calculate the Impact Score for various design features within healthcare settings, directly correlating these features with patient satisfaction and recovery outcomes. This sophisticated formula is crucially positioned to quantify how specific architectural and interior design changes can be assessed for their effectiveness based on empirical data from survey responses and observational studies. The inclusion of this equation in discussions about healthcare facility design allows for a measurable, data-driven approach to enhancing the patient environment, thereby optimizing health outcomes.

$$\mathrm{IS} = \delta_0 + \delta_1 F_1 + \delta_2 F_2 + \cdots + \delta_n F_n \tag{4.4}$$

In this equation, IS represents the Impact Score, which is a quantitative measure of how effective different design features are in improving patient satisfaction and facilitating recovery. Each $F$ corresponds to a specific design feature (e.g., window size, art installations, room color), and $\delta$ values are coefficients that indicate the strength of each feature's impact on the overall patient experience. $\delta_0$ acts as the baseline impact score in the absence of these specified features.

This model empowers healthcare administrators and facility planners to implement targeted improvements within hospital environments. By calculating the Impact Score for each design feature, decision-makers can prioritize those modifications that have the most substantial evidence-backed benefits to patient care. This approach not only fosters a patient-centered design philosophy but also ensures that investments in facility upgrades are both strategic and effective. By leveraging Eq. 4.4, healthcare facilities can systematically enhance their environments, thereby directly contributing to improved patient outcomes and increased satisfaction. This rigorous quantitative assessment serves as a foundation for evidence-based design practices in the healthcare industry.

Figure 4.11 illustrates the systematic process involved in a mixed-methods evaluation approach used within healthcare settings. It begins with initiating the research, followed by designing the study framework, which sets the foundation for developing the necessary instruments, such as survey questions, focus group guidelines, and observational criteria. Subsequent stages include comprehensive data collection through surveys, focus groups, and observational studies. Each data collection method is meticulously analyzed to derive insightful results that are then integrated to form a cohesive understanding of the research question. The integration of findings culminates in reporting and disseminating the results, which not only provides valuable insights but also informs future practices. Feedback is actively sought after dissemination to refine and optimize the approach, ensuring continual improvement in healthcare research methodologies. This flowchart is essential for researchers and healthcare practitioners aiming to employ a holistic evaluation framework that captures the complexity and nuance of patient care and hospital administration.

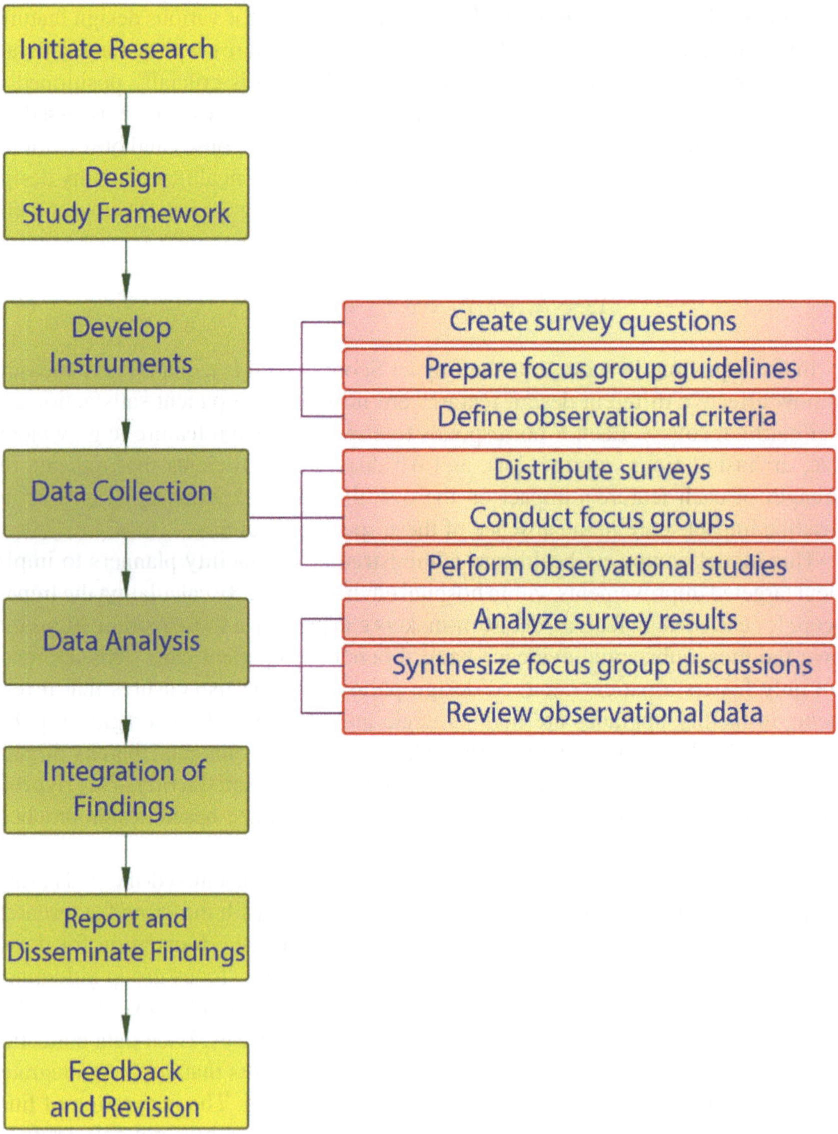


**Fig. 4.11** Mixed-methods evaluation approach in healthcare settings

### 4.6.2 *Performance Metrics*

To effectively measure the success of design interventions in healthcare environments, hospitals must establish specific and measurable performance metrics. These metrics serve not only to quantify the impact of design on patient care and facility operations but also provide essential data to inform ongoing improvements and innovations in hospital settings.

One critical set of performance metrics revolves around clinical outcomes. Metrics such as patient recovery times, infection rates, and the incidence of patient falls provide concrete evidence of design efficacy. For example, improvements in patient room designs such as enhanced lighting and reduced physical obstructions—can lead to a measurable decrease in patient falls, directly reflecting the success of these design changes. Similarly, tracking infection rates can help assess the effectiveness of updated environmental controls, like improved ventilation systems and more readily accessible hand-washing facilities, in enhancing patient safety.

Beyond clinical metrics, patient well-being and satisfaction indicators are also invaluable. These might include metrics such as the usage of medications for anxiety or assessments of patient mood, which help evaluate areas designed to reduce stress, such as therapeutic gardens or tranquil zones within the hospital. Additionally, patient satisfaction surveys are instrumental in measuring how design changes influence patient perceptions and comfort, providing a direct link between the physical environment and patient experience.

Operational efficiency metrics are equally important for understanding the impact of design on the functionality of healthcare facilities. These metrics may include the speed of service delivery, the efficiency of patient flow through different departments, and the utilization rates of various hospital areas. For instance, tracking the use of newly implemented patient-centered facilities, like family consultation rooms or multi-purpose spaces, can indicate whether these areas effectively meet their intended purposes.

Environmental and economic metrics also play a vital role in evaluating design success. Metrics such as energy consumption reflect the efficacy of sustainable design practices, while maintenance costs can provide insights into the durability and practicality of materials and designs in the challenging hospital environment.

To comprehensively assess the impact of design changes, hospitals should employ longitudinal studies and continuous monitoring of these metrics. This long-term approach enables the tracking of improvements over extended periods and facilitates the adjustment of strategies based on accumulated data. Continuous monitoring also aids in identifying trends that may not be immediately obvious, offering a deeper understanding of the long-term advantages or drawbacks associated with specific design choices.

In conclusion, by rigorously establishing and tracking a variety of performance metrics, hospitals can develop a robust framework for evaluating the effectiveness of design implementations. This data-driven methodology not only substantiates the benefits of thoughtful design but also pinpoints areas requiring further enhancement, ensuring that the hospital environment continuously evolves to better meet the needs of its users.

Table 4.8 provides empirical data demonstrating the relationship between specific design implementations in healthcare settings and measurable clinical outcomes, such as infection rates and recovery times. This table underscores the tangible benefits of targeted design interventions, showing how enhancements in facility layout,

Table 4.8 Correlation between design changes and clinical outcomes

| Design change | Clinical outcome measured | Pre-implementation rate | Post-implementation rate | Percentage improvement |
|---|---|---|---|---|
| Addition of HEPA filters | Infection rates | 1.5% | 0.7% | 53.3% |
| Improved lighting design | Recovery times for depression | 30 days | 22 days | 26.7% |
| Installation of ergonomic furniture | Staff injury rates | 2.0% | 1.2% | 40% |
| Enhanced patient room Acoustics | Patient stress levels | High | Moderate | Estimated 30% reduction |
| Introduction of green spaces | Length of stay in hospital | 5.0 days | 4.0 days | 20% |

material choices, and environmental conditions can directly influence patient health outcomes. It serves as a valuable resource for healthcare planners and administrators aiming to justify and guide investments in facility design based on evidence of clinical efficacy.

### 4.6.3 Continuous Improvement

Continuous improvement within healthcare environments is an essential process that ensures facilities remain adaptive and responsive to the dynamic landscape of medical care. The evaluation of hospital design and operation should be a perpetual activity, responding to shifts in patient demographics, technological advancements, and the evolving landscape of best practices in healthcare. This concept extends beyond mere troubleshooting; it necessitates a proactive approach to enhance and refine patient care and facility efficacy continuously.

To achieve continuous improvement, healthcare facilities must regularly assess the utilization and effectiveness of their spaces and services. This involves identifying areas within the hospital that are underutilized or ineffectively used and finding ways to transform these spaces to better serve patients and staff. For example, an underused ward might be converted into a more flexible space that can accommodate different types of care activities as needed, or it could be equipped with new technologies that streamline patient care processes.

In addition to reconfiguring physical spaces, continuous improvement also focuses on enhancing features within the healthcare facility that have received positive feedback from patients and staff. This could include expanding successful programs or services that improve patient satisfaction and care outcomes. For instance, if patients respond well to the integration of digital health tools that allow for easier access to their health information, the facility might consider expanding these technologies to more departments or enhancing their capabilities.

Integrating new technologies plays a critical role in continuous improvement by potentially transforming patient care delivery. As medical technology evolves, it offers new tools and systems that can significantly improve diagnostic accuracy, treatment options, and patient comfort. Hospitals must stay abreast of these developments and consider how best to incorporate them into their operations. This could mean adopting more sophisticated imaging technologies, implementing AI-driven diagnostic tools, or using telemedicine platforms to extend care reach.

The goal of continuous improvement in healthcare settings is to create an environment that is not only functional and efficient but also patient-centered and adaptive to future needs. By maintaining a commitment to ongoing assessment and adaptation, healthcare facilities can ensure that they provide the highest standard of care, remain competitive in the healthcare market, and most importantly, meet the changing needs of the communities they serve. This strategic approach to facility management and development is crucial for sustaining high performance and fostering innovation in healthcare delivery.

### 4.6.4 Stakeholder Involvement

Continuous improvement within healthcare settings is indispensable for adapting to the evolving needs of patients, advancements in medical technology, and changes in healthcare delivery models. This dynamic process ensures that hospitals not only rectify deficiencies but also continually innovate and enhance their offerings to maintain the highest standard of care.

The cornerstone of continuous improvement is the ongoing evaluation of facility performance and patient satisfaction. Regular feedback mechanisms, such as surveys, interviews, and focus groups, along with the application of performance metrics, enable healthcare facilities to systematically gather and analyze data. This consistent data collection helps identify areas where the design may not fully meet patient or staff needs or areas where it excels, thereby providing clear directives for future enhancements.

Adapting to changing demographics and needs is another critical aspect of continuous improvement. Hospitals cater to diverse populations, which may shift significantly over time. As such, healthcare facilities must modify their environments to align with these changing demographics. For example, as the population ages, enhancing design features to support mobility and accessibility becomes increasingly important. Similarly, growing cultural diversity might necessitate adjustments in design to enhance cultural sensitivity, such as including prayer rooms accommodating various religious practices or adapting dietary offerings in cafeteria services to reflect a broader range of cultural cuisines.

The rapid progression of medical and information technology also demands continual updates. Integrating cutting-edge technologies, such as telemedicine, augmented reality in surgical procedures, or advanced patient monitoring systems, necessitates ongoing revisions to both the physical and IT infrastructure of hospitals. These technologies not only elevate the quality of care but also improve patient experiences by making treatments less invasive and shortening recovery times.

Moreover, continuous improvement involves reassessing how space within the hospital is utilized to ensure it effectively meets current demands. Spaces that are consistently underutilized may require repurposing or redesigning to better fulfill the hospital's objectives. For instance, transforming seldom-used consultation rooms into versatile spaces for patient education or community health programs can enhance the hospital's role as a pivotal community health resource.

Innovation and experimentation are also integral to continuous improvement. Hospitals may experiment with new layouts, design elements, or patient care strategies on a small scale to determine their efficacy before wider implementation. This method allows for innovative solutions to be tested and refined in real-world settings, minimizing risks while maximizing potential benefits.

In conclusion, continuous improvement is an essential ongoing process that ensures healthcare facilities remain leaders in patient-centered care, safety, and operational efficiency. By fostering a culture of continual evaluation, adaptation, and innovation, hospitals can not only address current issues but also proactively enhance their environments to tackle future challenges. This commitment to perpetual assessment and adaptation is crucial for delivering the highest quality of care and maintaining a responsive, patient-focused healthcare environment.

### *4.6.5 Case Studies and Benchmarking*

Involving a diverse range of stakeholders in the evaluation and design processes of healthcare facilities is crucial for ensuring that the environments created effectively meet the varied needs of those who use them. Engaging patients, families, healthcare providers, and design professionals offers a comprehensive perspective that enhances the functionality of the facility and identifies potential areas for improvement.

The benefits of engaging diverse perspectives are manifold. Patients and their families, who are directly impacted by the design of the facility, often provide valuable insights into the usability and comfort of the spaces. Their firsthand experiences can uncover practical issues and benefits that may not be evident to staff or designers. Healthcare providers, including nurses, doctors, and support staff, contribute critical insights based on how design elements affect care delivery. They are uniquely positioned to identify design features that hinder efficient workflows or enhance patient care, given their daily interactions within these environments. Additionally, design professionals such as architects, interior designers, and engineers bring technical expertise to the table. They understand the feasibility of implementing changes and can anticipate potential challenges in modifying existing structures, ensuring that solutions are both practical and innovative.

Involving stakeholders in the evaluation process not only broadens the understanding of the facility's impact but also fosters a sense of ownership and commitment among those involved. When stakeholders are actively engaged in decision-making, they are more likely to support and advocate for the changes, thereby increasing the likelihood of successful implementation. This collaborative approach often leads to innovative solutions that might not have been considered without such diverse input.

Effective stakeholder engagement can be facilitated through various methods. Workshops and focus groups allow for dynamic discussions that yield in-depth feedback from different groups, revealing nuanced insights into the design's impact. Surveys and feedback forms, distributed regularly, provide ongoing data on stakeholder satisfaction and concerns, offering a quantifiable measure of design success and areas needing attention. Furthermore, including stakeholders in advisory capacities, such as on panels, ensures that ongoing decisions reflect the needs and preferences of those who frequently use the facility.

Establishing a continuous feedback loop by setting regular intervals for stakeholder feedback is essential. This process ensures that the evaluation is not only comprehensive at inception but remains ongoing and responsive to evolving healthcare practices, changing patient demographics, and technological advancements.

In conclusion, stakeholder involvement is a pivotal component of designing and maintaining healthcare facilities that genuinely cater to the needs of all users. By incorporating a wide range of perspectives in the evaluation process, hospitals can ensure that their environments are functional, effective, and inclusive. This strategy not only enhances the quality of care provided but also guarantees that facilities continue to evolve and adapt effectively over time, meeting the challenges of a dynamic healthcare landscape.

Table 4.9 compares the performance of our hospital against benchmarks from similar institutions, focusing on various operational and clinical metrics influenced by design features. This table illuminates areas where the hospital excels or requires improvement, offering a strategic lens through which to view the impact of design on healthcare delivery. Key metrics include patient satisfaction, recovery times, and infection rates, all of which reflect the hospital's functionality and responsiveness to patient needs. This comparative analysis not only underscores the benefits of stakeholder engagement in design processes but also highlights how diverse perspectives contribute to the development of facilities that better serve the community.

Table 4.9 Benchmarking results comparing similar healthcare facilities

| Metric | Our hospital | Benchmark average | Difference (%) | Design features impacting metric |
|---|---|---|---|---|
| Patient satisfaction | 92% | 87% | + 5.7 | Healing environments, patient-centric designs |
| Recovery times | 4.2 days | 5.0 days | −16 | Ergonomic patient care facilities, natural lighting |
| Infection rates | 0.9% | 1.2% | −25 | Advanced air filtration systems, surface materials |
| Staff turnover rate | 12% | 15% | −20 | Ergonomic workspaces, supportive staff environments |
| Operational efficiency | 85% | 78% | + 9 | Smart technology integrations, efficient layout designs |

## 4.7 Integration of Art and Culture in Healing Environments

The integration of art and culture in hospital environments plays a crucial role in enhancing the healing process and providing comfort to patients, families, and staff. This subsection explores how artistic and cultural elements can be effectively incorporated into healthcare settings to create environments that support emotional and psychological well-being alongside physical health.

### *4.7.1 Art as a Healing Tool*

Art has long been celebrated not just for its aesthetic value but also as a potent healing tool within healthcare settings, playing a pivotal role in the recovery processes of patients and enhancing the well-being of staff. The strategic integration of art into hospitals serves to transform sterile clinical environments into healing spaces, offering emotional, psychological, and physiological benefits that support the overall healing journey.

The therapeutic benefits of art are extensive. It has the power to evoke deep emotional responses, which can significantly alleviate stress, anxiety, and depression—common ailments among hospital patients. Research indicates that interaction with art can lead to a substantial reduction in perceived pain levels, possibly because it distracts patients from their discomfort and provides a mental diversion. The presence of paintings, sculptures, and murals depicting serene, natural landscapes can serve as virtual windows to the outside world, particularly beneficial in urban hospitals where actual nature views might be scarce. These representations of nature not only offer visual relief but also mimic the calming effects associated with time spent outdoors, promoting relaxation and mental peace.

Abstract art also contributes significantly by engaging the mind. The interpretation of abstract pieces can be a mentally stimulating exercise, particularly valuable for patients in long-term care. The inherent complexity and subjectivity of abstract art allows it to continuously engage viewers, offering new insights and distractions with each viewing, which is crucial for patients who spend significant amounts of time in recovery.

Implementing rotating art exhibitions in hospital corridors, waiting areas and patient rooms is another strategy to revitalize the hospital atmosphere and stimulate the minds of both patients and staff. These exhibitions can feature works by local artists or patients themselves, fostering a community spirit and making patients feel more involved and represented in their care environment. For hospital staff, the rotation of art can break the monotony and emotional drain often associated with clinical roles, providing fresh visual experiences that can rejuvenate and inspire.

Moreover, incorporating art that reflects the cultural diversity of the patient population can profoundly enhance feelings of comfort and belonging. Cultural representation in art can be particularly reassuring for patients who might feel vulnerable or isolated in the hospital setting, helping them feel respected and valued.

Beyond merely displaying art, some hospitals actively incorporate art therapy programs that engage patients in creating their artwork. This form of engagement not only offers emotional catharsis and a sense of achievement but also aids in psychological recovery. Art-making provides a powerful, non-verbal mode of expression, especially beneficial for patients who may struggle to express their feelings verbally.

In conclusion, the role of art in healthcare extends far beyond mere decoration. It is a multifaceted therapeutic tool that significantly impacts patient well-being and the overall care experience. By thoughtfully curating and rotating art collections, hospitals can create dynamic and supportive environments that facilitate healing and comfort, enhancing the quality of care and transforming hospitals into true sanctuaries of healing for both patients and staff.

### *4.7.2 Cultural Relevance and Sensitivity*

In today's increasingly globalized and diverse society, the cultural relevance and sensitivity of healthcare design have become paramount. Hospitals, serving a broad spectrum of populations, must integrate design elements that respect and reflect the diverse cultural identities of their patients. This not only honors individual identities but also fosters a supportive and welcoming environment, which is crucial for facilitating healing and comfort.

Artwork and decorations within hospital settings serve as powerful tools for expressing cultural diversity. By incorporating art that represents various cultures—such as paintings, sculptures, textiles, or crafts—hospitals can create spaces that mirror the multicultural makeup of their communities. This diverse display of cultural art does more than just beautify spaces; it sends a potent message of inclusivity and respect, helping patients from all backgrounds feel recognized and valued.

Another significant aspect of culturally sensitive design is incorporating design elements that resonate with local traditions and heritage. For instance, integrating colors, patterns, or materials significant to the local culture in the design of patient rooms, lobbies, or common areas can imbue the hospital environment with a sense of belonging and comfort. This approach is especially meaningful in areas with a prominent indigenous presence or distinctive local culture, as it helps connect patients with their community's roots and history, thereby enhancing their comfort and reducing stress.

The provision of culturally specific prayer rooms is essential in recognizing the spiritual needs of diverse patient populations. Hospitals can meet these needs by designing adaptable spaces that accommodate various faiths or by establishing several dedicated prayer rooms that cater to specific religious requirements, such as directionality in prayer, symbols, or cleanliness protocols. Such thoughtful design decisions emphasize the hospital's commitment to holistic patient care, respecting both spiritual and physical health.

Incorporating culturally specific motifs in the design of patient rooms and common areas can also significantly enhance the hospital's cultural sensitivity. These motifs can be expressed through wall art, fabric designs, or even architectural elements. However, these cultural elements must be incorporated respectfully and in consultation with community members to avoid stereotypes or cultural appropriation.

In conclusion, the integration of cultural relevance and sensitivity in hospital design transcends mere aesthetic considerations, aiming to create a healing environment that respects and celebrates the diversity of its users. Thoughtful integration of culturally reflective elements can significantly improve patient satisfaction and comfort, promoting a more inclusive atmosphere. Such an environment not only supports better health outcomes but also strengthens the bond between the hospital and the diverse community it serves, embodying a true commitment to patient-centered care.

Figure 4.12 illustrates a hospital corridor transformed by an art exhibition featuring diverse cultural artworks, symbolizing the hospital's commitment to cultural sensitivity and inclusivity. Such initiatives not only enrich the visual environment but also resonate deeply with patients from various cultural backgrounds, reinforcing a sense of belonging and respect. By integrating such culturally relevant art pieces, healthcare facilities can create a supportive and welcoming atmosphere that aids in healing and comfort, acknowledging the rich cultural diversity of the communities they serve.

### *4.7.3 Healing Gardens and Natural Elements*

The integration of healing gardens and landscaped areas within hospital settings underscores a profound commitment to holistic healing, offering patients, families, and staff a much-needed respite from the clinical environment. These natural sanctuaries are essential for enhancing both physical and mental recovery, providing a serene space where nature's calming effects can be fully appreciated.

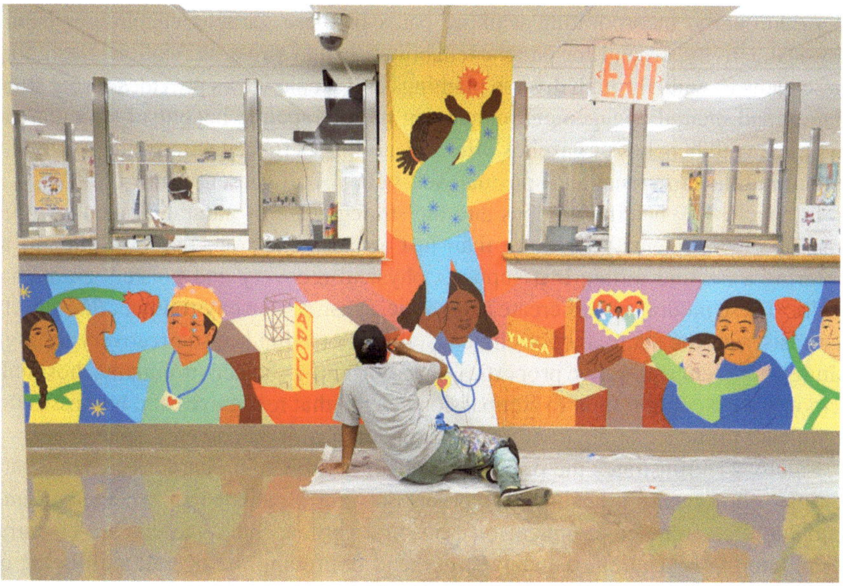


**Fig. 4.12** Enhancing cultural sensitivity through art in healthcare settings

Healing gardens are meticulously designed to engage the senses, promote relaxation, and alleviate stress. The presence of greenery, flowers, and natural landscapes has been scientifically proven to offer significant health benefits, including stress reduction, lower blood pressure, and improved emotional well-being. For patients undergoing treatment, these gardens serve as therapeutic havens that can potentially shorten recovery periods and decrease dependence on pain medications. The psychological lift provided by such serene settings often translates into tangible improvements in patient health outcomes.

When selecting plants and materials for these gardens, incorporating elements with cultural significance or therapeutic properties can greatly enhance their healing impact. For example, using medicinal plants known as traditional remedies can connect patients with their cultural heritage, offering comfort and familiarity. Moreover, opting for native plants not only helps maintain local biodiversity but also strengthens patients' connection to their surrounding natural environment, fostering a sense of peace and continuity.

Accessibility and enjoyment are critical considerations in the design of healing gardens. Paths should be wide enough to accommodate wheelchairs and walkers easily, ensuring that everyone can navigate space without difficulty. Water features like fountains or streams add a soothing auditory layer to the garden, helping to mask clinical noises and enhance the tranquil atmosphere. Adequate seating areas allow for rest and social interaction, which are vital for mental health and community feeling among patients, visitors, and staff.

Additionally, incorporating walking paths in these gardens encourages gentle physical activity, crucial for patient rehabilitation and stress alleviation. Designing these paths to wind through diverse plantings and scenic views provides visual stimulation and mental engagement, enhancing the therapeutic benefits. Quiet corners equipped with benches or gazebos offer private spaces for reflection and meditation, supporting the emotional and spiritual well-being of all garden visitors.

In conclusion, healing gardens and the strategic use of natural elements within hospital environments play more than just an aesthetic role; they are integral to fostering a healing milieu that supports physical recovery and bolsters emotional health. These natural spaces create a dynamic, sensory-rich environment that starkly contrasts with the static, indoor clinical settings, offering holistic recovery through nature's inherent restorative properties. By prioritizing such environments, hospitals demonstrate a commitment to compassionate care that recognizes and harnesses the significant influence of environmental factors on health outcomes.

Figure 4.13 illustrates the practical application of healing gardens in healthcare settings, showcasing a healthcare worker assisting a patient in a wheelchair, and enjoying the serene environment of a hospital garden. Such gardens are essential for promoting relaxation, reducing stress, and improving the overall well-being of patients, families, and staff. The design focuses on accessibility, with paths wide enough to accommodate mobility aids, while the natural setting helps alleviate the clinical atmosphere, offering a peaceful retreat that supports holistic healing and recovery.

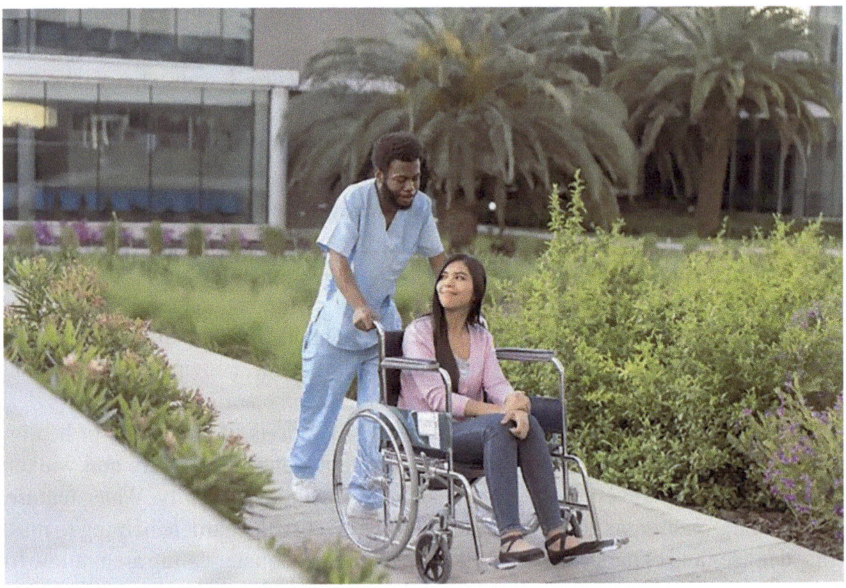

**Fig. 4.13** Enhancing recovery through healing gardens

### 4.7.4 *Performance Spaces*

Incorporating performance spaces within hospital settings is an innovative strategy that merges the arts with medicine, providing a dynamic and enriching dimension to the patient care experience. Live performances by musicians, storytellers, and dancers can be particularly therapeutic, bringing a unique form of emotional and psychological relief to patients, families, and staff. This approach is especially impactful in children's hospitals, where creating positive distractions can significantly affect treatment outcomes and the overall hospital experience for young patients.

Live performances introduce a dynamic energy that can transform the hospital atmosphere, instilling joy and providing a much-needed diversion from the clinical environment. For children, who may find the hospital setting intimidating, having access to live arts can make their experience far more positive, helping to alleviate fear and anxiety. The arts activate emotions and can also connect individuals with their cultural heritage, offering comfort and a semblance of normalcy during challenging times.

Designing flexible performance spaces in hospitals requires careful consideration to ensure they are accommodating and functional for various types of performances:

Flexibility: The space should be adaptable to host different performance types, from small acoustic sets to larger theatrical productions. Features like retractable seating, modular stages, and configurable lighting systems can transform the space to suit various events, accommodating diverse audience sizes and performance styles.

- **Acoustics**: Excellent acoustics are essential, ensuring sound quality is maintained at optimal levels. This is crucial in a hospital setting, where the clarity of sound contributes significantly to the enjoyment and impact of the performance. Using materials that provide sound insulation and investing in high-quality audio equipment will enhance the natural acoustics and minimize external noise.
- **Accessibility**: The performance area should be accessible to everyone, including performers with equipment and patients with mobility challenges. Easy access routes for performers and visibility considerations for patients, such as raised seating or strategically angled layouts, ensure that all attendees can enjoy the performances.
- **Comfort**: Comfortable seating is crucial, particularly since audience members may include patients dealing with various medical conditions. Accommodation for wheelchairs and other medical devices are necessary to provide a welcoming and inclusive environment for all attendees.

Beyond the physical design of performance spaces, hospitals should also consider the integration of performances into the daily life of the hospital. Scheduling regular performances can provide patients with enjoyable events to look forward to, enhancing their overall hospital experience. For those unable to attend in person, broadcasting performances to patient rooms can ensure that everyone can experience the therapeutic benefits of these cultural activities.

In conclusion, performance spaces within hospitals serve a critical role beyond entertainment; they are integral to a holistic healthcare approach that acknowledges the profound healing power of the arts. By seamlessly integrating these spaces into hospital designs, healthcare institutions not only support traditional medical treatments but also foster emotional and psychological recovery. This innovative inclusion enriches the hospital environment, enhancing the quality of care and bringing culture, art, and joy into the lives of patients and staff.

### *4.7.5 Interactive Art*

Interactive art installations in hospital settings are becoming increasingly recognized for their therapeutic value, offering patients a unique opportunity to engage actively with their environment. These installations do more than beautify the space; they empower patients by providing them with an active role in their interaction with the surroundings. This is especially crucial in a hospital setting where feelings of vulnerability and powerlessness are common. By participating in the creation or manipulation of art, patients gain a sense of agency and control, which can be profoundly beneficial for their emotional well-being and recovery process.

Interactive art serves multiple therapeutic roles: it empowers, entertains, and engages patients across various age groups, catering to their specific physical and cognitive needs. For instance, digital artworks that respond to movement or touch can transform a sterile hospital environment into a playful, engaging space. Such installations are particularly beneficial for pediatric patients, helping to alleviate the fear and anxiety associated with hospital stays. By interacting with art that changes in response to their actions, children can find joy and distraction, which eases their experience and promotes a positive mindset.

Moreover, interactive installations are not limited to digital mediums; tactile artworks encourage physical movement and mental engagement, offering a form of gentle physical therapy. These artworks are particularly useful for patients recovering from surgeries or injuries, as they promote gentle exercise and cognitive stimulation. For the elderly, installations designed with accessibility in mind—featuring tactile elements and gentle interactions—can provide cognitive and physical stimulation without being overwhelming.

The design of interactive art can also be inclusive, ensuring that patients with disabilities can engage fully. Installations might include features like tactile elements, voice-activated components, or other sensory-based interactions that are accessible to everyone, thereby enhancing inclusivity within the hospital environment. This aspect of interactive art is critical because it ensures that all patients, regardless of their physical capabilities, can benefit from the therapeutic advantages of engaging with art.

Interactive installations also play a vital role in fostering a sense of community and connection among patients, visitors, and hospital staff. Collaborative artworks, where people contribute collectively, help to reduce the isolation that can accompany

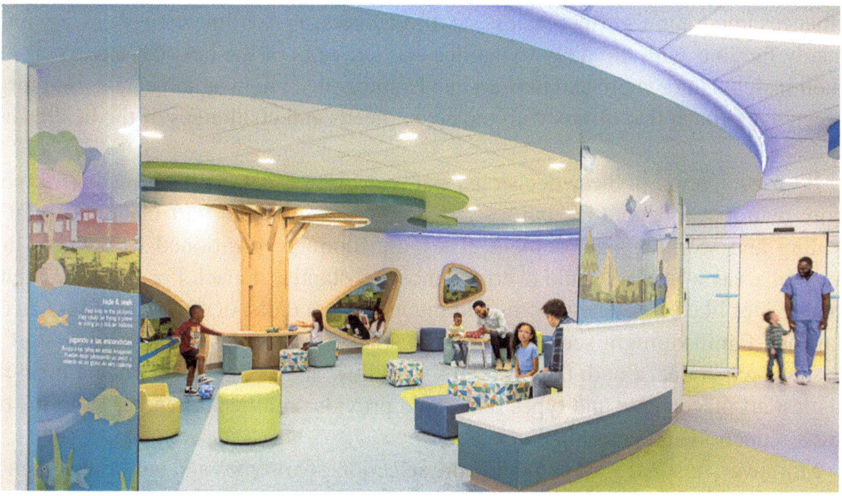

Fig. 4.14 Transforming pediatric care through interactive art

hospital stays. These shared artistic endeavors provide a non-clinical setting for social interaction, which can be uplifting and affirming for patients enduring long-term treatments.

In conclusion, interactive art installations in hospitals are more than aesthetic enhancements. They are integral to therapeutic strategies that aim to improve patient outcomes by providing mental and emotional relief, fostering a sense of community, and accommodating diverse patient needs holistically. Through these installations, hospitals can create a healing environment that actively supports the recovery process by involving patients in engaging and meaningful interactions.

Figure 4.14 showcases a pediatric area within a hospital designed to engage young patients through interactive art and thoughtful design. The environment features vibrant colors, playful furniture, and art installations that capture the imagination, turning what could be an intimidating hospital visit into an enjoyable experience. Interactive elements throughout space encourage play and exploration, which not only help in reducing anxiety and stress in young patients but also aid in their cognitive and physical development during recovery. Such dynamic environments are crucial in providing emotional comfort and promoting a positive mindset, integral to the healing process of children in healthcare settings.

### *4.7.6 Staff Involvement in Artistic Decisions*

Involving hospital staff in the decision-making process concerning artistic and cultural elements in healthcare settings is a strategic measure that significantly enhances integration's effectiveness. Staff members, as frontline caregivers, are intimately familiar with the daily dynamics and stressors of the hospital environment. Their insights into where and how art can be integrated are invaluable, allowing for a targeted approach that aligns with the actual needs and challenges of the healthcare setting.

By participating in these decisions, staff can ensure that the integrated art serves more than just a decorative purpose; it becomes an integral part of the hospital's healing and working environment. They are uniquely positioned to identify the hospital areas where art could exert the most substantial impact. For instance, high-stress zones like emergency rooms and intensive care units, or common areas where staff and families might seek solace and respite during stressful times. The staff's firsthand experiences with the hospital's workflow and stress points enable them to suggest art placements and types that could provide significant benefits in terms of stress reduction and emotional uplift.

Moreover, hospital staff can guide the selection of art types and cultural elements that resonate most effectively with the diverse demographics of both the patients served and the staff themselves. They might recommend artworks or cultural symbols that reflect the community's ethnic diversity, enhancing the sense of comfort and understanding patients feel within the hospital setting. Additionally, staff might advocate for interactive or engaging art pieces that not only serve as distractions for patients but also offer a momentary escape for caregivers themselves.

By incorporating staff feedback into the decision-making process, hospitals can create holistic healing environments that support clinical outcomes and cater to the emotional and psychological needs of patients and healthcare providers alike. This inclusive approach to environmental design recognizes the profound impact surroundings have on health and recovery. Environments that are visually appealing, culturally representative, and emotionally supportive not only nurture the spirit and body but also boost overall morale and job satisfaction among the staff.

Engaging staff in a continuous feedback loop regarding the art and cultural elements within the hospital ensures that the environment remains dynamic and responsive to evolving needs. As staff observe the impacts of these elements over time, they can provide ongoing insights into their efficacy, suggesting adjustments or new initiatives that might further enhance the healing environment.

In conclusion, the inclusion of hospital staff in the artistic and cultural decision-making process is critical for fostering a supportive and effective healing environment. This strategy ensures that the selected art and cultural representations are not only meaningful but also beneficial, reflecting the collective insights of those who interact with space daily. By valuing the perspectives of hospital staff, the environment becomes more cohesive and supportive, significantly enhancing the healing process and well-being of all involved.

Figure 4.15 delineates the systematic process of staff engagement in the selection and installation of art within hospital settings. Beginning with the initiation by hospital administration, the process involves forming a diverse art committee, gathering staff preferences through surveys, and democratically selecting art pieces. The

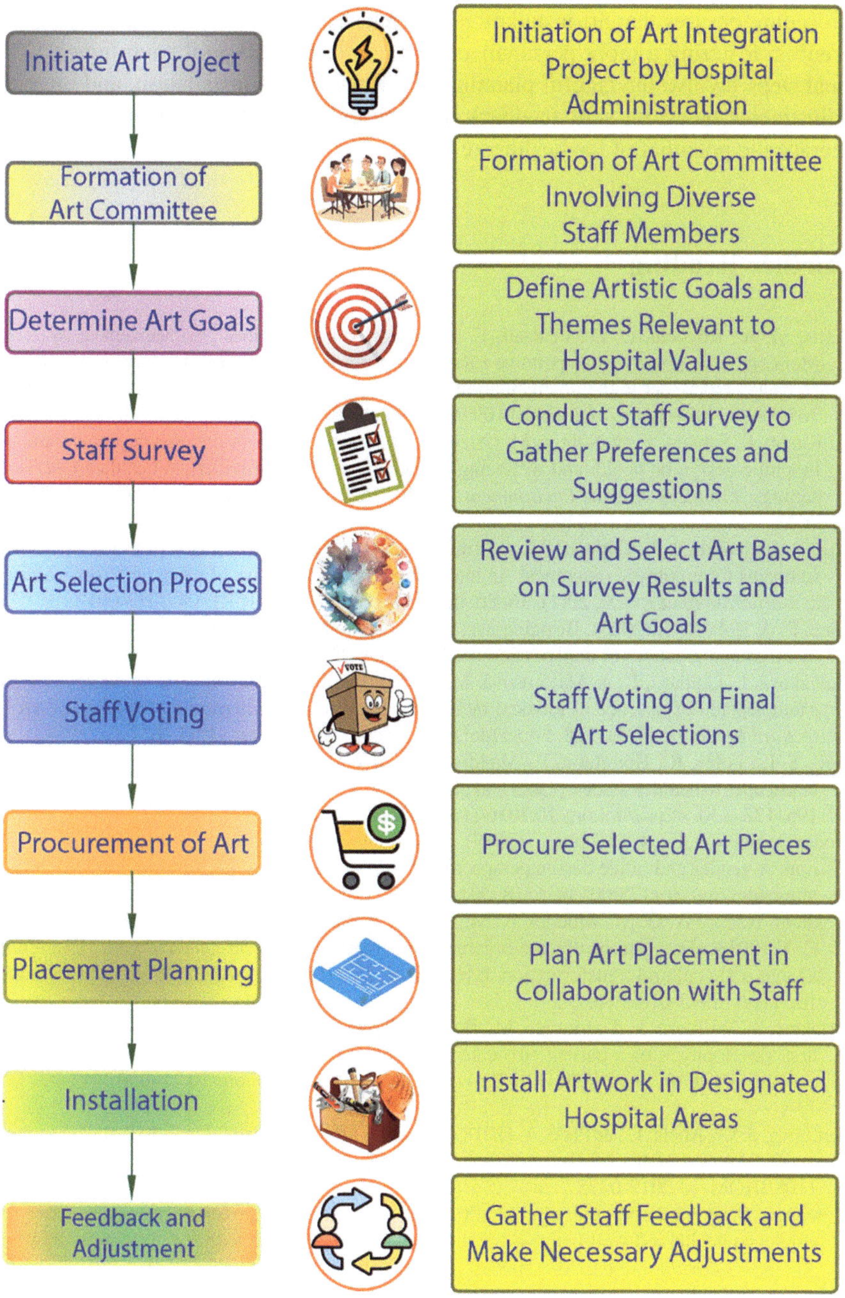


**Fig. 4.15** Staff involvement in art selection and placement at the hospital

procedure ensures that art installations resonate with the staff's and patients' preferences, fostering a more personalized and comforting hospital environment. The final steps involve the careful planning of art placement, installation, and ongoing adjustments based on staff feedback, thereby enhancing the aesthetic appeal and therapeutic potential of the healthcare facility.

## Further Reading


Aarts, M. A., Rotstein, O. D., Pearsall, E. A., Victor, J. C., Okrainec, A., McKenzie, M., ... & McLeod, R. S. (2018). Postoperative ERAS interventions have the greatest impact on optimal recovery: experience with implementation of ERAS across multiple hospitals. *Annals of Surgery*, 267(6), 992–997. https://doi.org/10.1097/SLA.0000000000002632

Al Khatib, I., Samara, F., & Ndiaye, M. (2024). A systematic review of the impact of therapeutical biophilic design on health and wellbeing of patients and care providers in healthcare services settings. *Frontiers in Built Environment*, 10, 1467692. https://doi.org/10.3389/fbuil.2024.1467692

Appel-Meulenbroek, R., Clippard, M., & Pfnür, A. (2018). The effectiveness of physical office environments for employee outcomes: An interdisciplinary perspective of research efforts. *Journal of Corporate Real Estate*, 20(1), 56–80. https://doi.org/10.1108/JCRE-04-2017-0012

Bate, P., & Robert, G. (2023). Bringing user experience to healthcare improvement: the concepts, methods and practices of experience-based design. CRC Press.

Belčáková, I., Galbavá, P., & Majorošová, M. (2018). Healing and therapeutic landscape design–examples and experience of medical facilities. *ArchNet-IJAR: International Journal of Architectural Research*, 12(3), 128. https://doi.org/10.26687/archnet-ijar.v12i3.1637

Bell, S. L., Foley, R., Houghton, F., Maddrell, A., & Williams, A. M. (2018). From therapeutic landscapes to healthy spaces, places and practices: A scoping review. *Social Science & Medicine*, 196, 123–130. https://doi.org/10.1016/j.socscimed.2017.11.035

Bodin Danielsson, C., & Theorell, T. (2019). Office employees' perception of workspace contribution: A gender and office design perspective. *Environment and Behavior*, 51(9-10), 995–1026. https://doi.org/10.1177/0013916518759146

Bokhour, B. G., Fix, G. M., Mueller, N. M., Barker, A. M., Lavela, S. L., Hill, J. N., ... & Lukas, C. V. (2018). How can healthcare organizations implement patient-centered care? Examining a large-scale cultural transformation. *BMC Health Services Research*, 18, 1-11. https://doi.org/10.1186/s12913-018-2949-5

Brickley, B., Williams, L. T., Morgan, M., Ross, A., Trigger, K., & Ball, L. (2021). Putting patients first: development of a patient advocate and general practitioner-informed model of patient-centred care. *BMC Health Services Research*, 21, 1-9. https://doi.org/10.1186/s12913-021-06273-y

Buttigieg, S. C., Abela, L., & Pace, A. (2018). Variables affecting hospital length of stay: a scoping review. *Journal of Health Organization and Management*, 32(3), 463–493. https://doi.org/10.1108/JHOM-10-2017-0275

Colenberg, S., Jylhä, T., & Arkesteijn, M. (2021). The relationship between interior office space and employee health and well-being–a literature review. *Building Research & Information*, 49(3), 352–366. https://doi.org/10.1080/09613218.2019.1710098

Costanzo, V., Fabbri, K., & Piraccini, S. (2018). Stressing the passive behavior of a Passivhaus: An evidence-based scenario analysis for a Mediterranean case study. *Building and Environment*, 142, 265–277. https://doi.org/10.1016/j.buildenv.2018.06.035

de Carvalho, E. C., Eduardo, A. H. A., Romanzini, A., Simão, T. P., Zamarioli, C. M., Garbuio, D. C., & Herdman, T. H. (2018). Correspondence between NANDA international nursing diagnoses and outcomes as proposed by the nursing outcomes classification. *International Journal of Nursing Knowledge*, 29(1), 66–78. https://doi.org/10.1111/2047-3095.12135

Dovjak, M., Shukuya, M., & Krainer, A. (2018). User-centred healing-oriented conditions in the design of hospital environments. *International Journal of Environmental Research and Public Health*, 15(10), 2140. https://doi.org/10.3390/ijerph15102140

Elsadek, M., Liu, B., & Xie, J. (2020). Window view and relaxation: Viewing green space from a high-rise estate improves urban dwellers' wellbeing. *Urban Forestry & Urban Greening*, 55, 126846. https://doi.org/10.1016/j.ufug.2020.126846

Engle, R. L., Mohr, D. C., Holmes, S. K., Seibert, M. N., Afable, M., Leyson, J., & Meterko, M. (2021). Evidence-based practice and patient-centered care: doing both well. *Health Care Management Review*, 46(3), 174–184. https://doi.org/10.1097/HMR.0000000000000254

Fabel, P. H., Wagner, T., Ziegler, B., Fleming, P. A., & Davis, R. E. (2019). A sustainable business model for comprehensive medication management in a patient-centered medical home. *Journal of the American Pharmacists Association*, 59(2), 285–290. https://doi.org/10.1016/j.hjdsi.2018.10.002

Fix, G. M., VanDeusen Lukas, C., Bolton, R. E., Hill, J. N., Mueller, N., LaVela, S. L., & Bokhour, B. G. (2018). Patient-centred care is a way of doing things: How healthcare employees conceptualize patient-centred care. *Health Expectations*, 21(1), 300–307. https://doi.org/10.1111/hex.12615

Gaminiesfahani, H., Lozanovska, M., & Tucker, R. (2020). A scoping review of the impact on children of the built environment design characteristics of healing spaces. *HERD: Health Environments Research & Design Journal*, 13(4), 98–114. https://doi.org/10.1177/1937586720903845

Halawa, F., Madathil, S. C., Gittler, A., & Khasawneh, M. T. (2020). Advancing evidence-based healthcare facility design: a systematic literature review. *Health Care Management Science*, 23, 453–480. https://doi.org/10.1007/s10729-020-09506-4

Jensen, A., & Bonde, L. O. (2018). The use of arts interventions for mental health and wellbeing in health settings. *Perspectives in Public Health*, 138(4), 209–214. https://doi.org/10.1177/1757913918772602

Jewkes, Y. (2018). Just design: Healthy prisons and the architecture of hope. *Australian & New Zealand Journal of Criminology*, 51(3), 319–338. https://doi.org/10.1177/0004865818766768

Johnson, J., Hall, L. H., Berzins, K., Baker, J., Melling, K., & Thompson, C. (2018). Mental healthcare staff well-being and burnout: A narrative review of trends, causes, implications, and recommendations for future interventions. *International Journal of Mental Health Nursing*, 27(1), 20–32. https://doi.org/10.1111/inm.12416

Ko, W. H., Schiavon, S., Zhang, H., Graham, L. T., Brager, G., Mauss, I., & Lin, Y. W. (2020). The impact of a view from a window on thermal comfort, emotion, and cognitive performance. *Building and Environment*, 175, 106779. https://doi.org/10.1016/j.buildenv.2020.106779

Kwame, A., & Petrucka, P. M. (2021). A literature-based study of patient-centered care and communication in nurse-patient interactions: barriers, facilitators, and the way forward. *BMC Nursing*, 20(1), 158. https://doi.org/10.1186/s12912-021-00684-2

Li, J., & Carayon, P. (2021). Health Care 4.0: A vision for smart and connected health care. *IISE* Transactions *on Healthcare Systems Engineering*, 11(3), 171–180. https://doi.org/10.1080/24725579.2021.1884627

Lygum, V. L., Poulsen, D. V., Djernis, D., Djernis, H. G., Sidenius, U., & Stigsdotter, U. K. (2019). Post-occupancy evaluation of a crisis shelter garden and application of findings through the use of a participatory design process. *HERD: Health Environments Research & Design Journal*, 12(3), 153–167. https://doi.org/10.1177/1937586718812444

Lyon, A. R., Munson, S. A., Renn, B. N., Atkins, D. C., Pullmann, M. D., Friedman, E., & Areán, P. A. (2019). Use of human-centered design to improve implementation of evidence-based psychotherapies in low-resource communities: protocol for studies applying a framework to assess usability. *JMIR Research Protocols*, 8(10), e14990. https://doi.org/10.2196/14990

Lytras, M. D., Papadopoulou, P., & Sarirete, A. (2020). Smart Healthcare: emerging technologies, best practices, and sustainable policies. *Innovation in Health Informatics*, 3-38. https://doi.org/10.1016/B978-0-12-819043-2.00001-0

Mahmood, F. J., & Tayib, A. Y. (2021). Healing environment correlated with patients' psychological comfort: Post-occupancy evaluation of general hospitals. *Indoor and Built Environment*, 30(2), 180–194. https://doi.org/10.1177/1420326X19888005

Mahrs Träff, A., Cedersund, E., & Abramsson, M. (2020). What promotes and what limits physical activity in assisted living facilities? A study of the physical environment's design and significance. *Journal of Aging and Environment*, 34(3), 291–309. https://doi.org/10.1080/02763893.2019.1683669

Marmo, R., Nicolella, M., Polverino, F., & Tibaut, A. (2019). A methodology for a performance information model to support facility management. *Sustainability*, 11(24), 7007. https://doi.org/10.3390/su11247007

Mayer, R. E. (2021). Evidence-based principles for how to design effective instructional videos. *Journal of Applied Research in Memory and Cognition*, 10(2), 229–240. https://doi.org/10.1016/j.jarmac.2021.03.007

McDonald, K., Escuro, K., & Edwards, S. (2024). Differences in design: How the care environment impacts healing and patient well-being. In *Handbook of Evidence-Based Inpatient Mental Health Programs for Children and Adolescents* (pp. 17–33). Cham: Springer International Publishing. https://doi.org/10.1007/978-3-031-62749-1_2

Nimlyat, P. S., Salihu, B., & Wang, G. P. (2024). The impact of indoor environmental quality (IEQ) on patients' health and comfort in Nigeria. *International Journal of Building Pathology and Adaptation*, 42(4), 510–530. https://doi.org/10.1108/IJBPA-06-2021-0089

Nosratabadi, S., Mosavi, A., Shamshirband, S., Zavadskas, E. K., Rakotonirainy, A., & Chau, K. W. (2019). Sustainable business models: A review. *Sustainability*, 11(6), 1663. https://doi.org/10.3390/su11061663

O'Brien, W., Wagner, A., Schweiker, M., Mahdavi, A., Day, J., Kjærgaard, M. B., ... & Berger, C. (2020). Introducing IEA EBC annex 79: Key challenges and opportunities in the field of occupant-centric building design and operation. *Building and Environment*, 178, 106738. https://doi.org/10.1016/j.buildenv.2020.106738

Oeljeklaus, L., Schmid, H. L., Kornfeld, Z., Hornberg, C., Norra, C., Zerbe, S., & McCall, T. (2022). Therapeutic landscapes and psychiatric care facilities: a qualitative meta-analysis. *International Journal of Environmental Research and Public Health*, 19(3), 1490. https://doi.org/10.3390/ijerph19031490

Paraskevopoulou, A. T., & Kamperi, E. (2018). Design of hospital healing gardens linked to pre-or post-occupancy research findings. *Frontiers of Architectural Research*, 7(3), 395–414. https://doi.org/10.1016/j.foar.2018.05.004

Perera, S., & Dabney, B. W. (2020). Case management service quality and patient-centered care. *Journal of Health Organization and Management*, 34(5), 551–568. https://doi.org/10.1108/JHOM-12-2019-0347

Poncette, A. S., Rojas, P. D., Hofferbert, J., Valera Sosa, A., Balzer, F., & Braune, K. (2020). Hackathons as stepping stones in health care innovation: case study with systematic recommendations. *Journal of Medical Internet Research*, 22(3), e17004. https://doi.org/10.2196/17004

Rodriguez-Suarez, C. A., Gonzalez-de la Torre, H., Hernandez-De Luis, M. N., Fernandez-Gutierrez, D. A., Martinez-Alberto, C. E., & Brito-Brito, P. R. (2023, September). Effectiveness of a standardized nursing process using NANDA international, nursing interventions classification and nursing outcome classification terminologies: a systematic review. *In Healthcare* (Vol. 11, No. 17, p. 2449). MDPI. https://doi.org/10.3390/healthcare11172449

Rojas-García, A., Turner, S., Pizzo, E., Hudson, E., Thomas, J., & Raine, R. (2018). Impact and experiences of delayed discharge: A mixed-studies systematic review. *Health Expectations*, 21(1), 41–56. https://doi.org/10.1111/hex.12619

Salingaros, N. A. (2019). The biophilic healing index predicts effects of the built environment on our wellbeing.
Sander, E. L. J., Caza, A., & Jordan, P. J. (2019). Psychological perceptions matter: Developing the reactions to the physical work environment scale. *Building and Environment*, 148, 338–347. https://doi.org/10.1016/j.buildenv.2018.11.020
Shanafelt, T., Trockel, M., Ripp, J., Murphy, M. L., Sandborg, C., & Bohman, B. (2019). Building a program on well-being: key design considerations to meet the unique needs of each organization. *Academic Medicine*, 94(2), 156–161. https://doi.org/10.1097/ACM.0000000000002415
Stoltz, J., & Grahn, P. (2021). Perceived sensory dimensions: An evidence-based approach to greenspace aesthetics. *Urban Forestry & Urban Greening*, 59, 126989. https://doi.org/10.1016/j.ufug.2021.126989
Totaforti, S. (2018). Applying the benefits of biophilic theory to hospital design. *City, Territory and Architecture*, 5, 1–9. https://doi.org/10.1186/s40410-018-0077-5
Vaartio-Rajalin, H., Santamäki-Fischer, R., Jokisalo, P., & Fagerström, L. (2021). Art making and expressive art therapy in adult health and nursing care: A scoping review. *International Journal of Nursing Sciences*, 8(1), 102–119. https://doi.org/10.1016/j.ijnss.2020.09.011
Van der Zwart, J., & Pilosof, N. P. (2020). Evidence-Based Design for Healthcare Work Environments. *Integrating the Organization of Health Services, Worker Wellbeing and Quality of Care: Towards Healthy Healthcare*, 245–262. https://doi.org/10.1007/978-3-030-59467-1_11
Vogus, T. J., Gallan, A., Rathert, C., El-Manstrly, D., & Strong, A. (2020). Whose experience is it anyway? Toward a constructive engagement of tensions in patient-centered health care. *Journal of Service Management*, 31(5), 979–1013. https://doi.org/10.1108/JOSM-04-2020-0095
Watts, P. I., McDermott, D. S., Alinier, G., Charnetski, M., Ludlow, J., Horsley, E., ... & Nawathe, P. A. (2021). Healthcare simulation standards of best practiceTM simulation design. *Clinical Simulation in Nursing*, 58, 14–21. https://doi.org/10.1016/j.ecns.2021.08.009
Weber, C., Monero Flores, V., Wheele, T. P., Miedema, E., & White, E. V. (2022). Patients' health & well-being in inpatient mental health-care facilities: a systematic review. *Frontiers in Psychiatry*, 12, 758039. https://doi.org/10.3389/fpsyt.2021.758039
Wild, D., Nawaz, H., Ullah, S., Via, C., Vance, W., & Petraro, P. (2018). Teaching residents to put patients first: creation and evaluation of a comprehensive curriculum in patient-centered communication. *BMC Medical Education*, 18, 1-9. https://doi.org/10.1186/s12909-018-1371-3
Werner, R. M., Coe, N. B., Qi, M., & Konetzka, R. T. (2019). Patient outcomes after hospital discharge to home with home health care vs to a skilled nursing facility. *JAMA Internal Medicine*, 179(5), 617–623. https://doi.org/10.1001/jamainternmed.2018.7998
Yap, S. F., Phillips, M., Hwang, E., & Xu, Y. (2022). Transforming healthcare service environments: a sensory-based approach. *Journal of Service Theory and Practice*, 32(5), 673–700. https://doi.org/10.1108/JSTP-02-2022-0033
Yang, T., White, M., Lipson-Smith, R., Shannon, M. M., & Latifi, M. (2024). Design Decision Support for Healthcare Architecture: A VR-Integrated Approach for Measuring User Perception. *Buildings*, 14(3), 797. https://doi.org/10.3390/buildings14030797
Zadeh, R. S., & Eshelman, P. (2019). Palliative design meets palliative medicine: a strategic approach to the design, construction, and operation of healthcare facilities to improve quality of life and reduce suffering for patients, families, and caregivers. *HERD: Health Environments Research & Design Journal*, 12(3), 179–186. https://doi.org/10.1177/1937586718820663

# Chapter 5
# Optimizing Hospital Layouts for Enhanced Staff Efficiency and Patient Care: The Role of Core Service Centralization

This chapter explores the strategic implementation of layout optimization to enhance staff efficiency within hospital settings, with a focus on the centralization of core services. By examining the impact of strategically centralized pharmacies, laboratories, and imaging centers, the chapter delves into how such configurations streamline workflows and reduce the time staff spend navigating between various departments, thereby significantly speeding up patient care processes. The analysis also covers the design of hospital pathways to optimize patient flow, aiming to reduce bottlenecks and improve overall patient satisfaction. Through case studies and theoretical frameworks, this chapter provides a comprehensive overview of how thoughtful architectural design can substantially increase operational efficiency, reduce staff stress, and foster a more efficient environment for patient care. The insights aim to guide healthcare administrators and designers in creating hospital layouts that not only meet medical needs but also enhance the operational dynamics of healthcare facilities.

## 5.1 Layout Optimization for Staff Efficiency

Optimizing the layout of a hospital is crucial for enhancing staff efficiency, which in turn can lead to better patient care and lower operational costs. A well-thought-out layout facilitates smooth workflows, minimizes unnecessary movement, and ensures that staff can perform their duties with maximum efficiency and minimum stress.



A. A. Firoozi and A. A. Firoozi, *Transformative Design in Healthcare*,
https://doi.org/10.1007/978-3-032-08019-6_5

### 5.1.1 Centralization of Core Services

In optimizing hospital layouts, the centralization of core services emerges as a pivotal strategy for enhancing operational efficiency and patient care. Core services, including pharmacies, laboratories, and imaging centers, are integral to the hospital's daily functions, heavily influencing patient treatment timelines and staff productivity. By strategically positioning these services in central locations within the hospital's layout, the design minimizes the need for extensive travel between different departments. This is particularly beneficial in acute care settings where the rapidity of response can significantly impact patient outcomes.

Centralization facilitates a streamlined workflow, enabling quicker transitions between diagnosis and treatment. For instance, with closer proximity to essential services, healthcare providers can speed up obtaining necessary medications from pharmacies or crucial diagnostic results from laboratories. This not only speeds up patient care but also enhances the ability of medical staff to manage multiple cases efficiently, reducing wait times and potentially lowering stress levels among both patients and healthcare professionals.

Moreover, this approach can contribute to a reduction in the risk of errors, as it minimizes the complexities associated with transporting samples or medications across long distances within the hospital. Centralizing these services can lead to better-coordinated care, as interdisciplinary teams can collaborate more effectively when they operate near one another. This is particularly critical in hospitals that handle a high volume of complex cases, where multi-specialty consultation is frequent and essential for effective patient management.

Ultimately, the centralization of core services within hospital layouts is not merely a matter of architectural design but a strategic decision that profoundly influences operational dynamics, enhances patient care, and optimizes resource allocation. As such, hospital administrators and design teams must consider these factors meticulously to ensure that the layout promotes not only medical efficiency but also a conducive environment for both patients and healthcare providers.

Equation 5.1, known as the "Efficiency Improvement Ratio," quantifies the efficiency enhancements in patient care delivery times resulting from the centralization of healthcare services. Positioned within the discussion on the benefits of centralization, this formula provides a crucial metric for evaluating how the strategic consolidation of services and resources across different departments can streamline operations, reduce necessary travel distances within facilities, and enhance inter-departmental collaboration.

$$\text{EIR} = \frac{T_{\text{before}} - T_{\text{after}}}{T_{\text{before}}} \times 100\% \tag{5.1}$$

In this equation, EIR represents the Efficiency Improvement Ratio, expressed as a percentage, which measures the relative improvement in operational efficiency due to centralization. $T_{\text{before}}$ is the average time taken for patient care delivery before centralization, including wait times and transit times between different service points.

**Table 5.1** Comparison of core service layouts

| Layout model | Staff travel time (min) | Response time (min) | Inter-departmental efficiency (%) |
|---|---|---|---|
| Centralized | 5 | 3 | 92 |
| Decentralized | 8 | 5 | 75 |
| Hybrid | 6 | 4 | 85 |

$T_{\text{after}}$ is the average time post-centralization, reflecting the optimized processes and reduced distances. The numerator $T_{\text{before}} - T_{\text{after}}$ calculates the absolute reduction in time while dividing this value by $T_{\text{before}}$ and multiplying by 100 converts this improvement into a percentage, which is easier to interpret and apply in decision-making processes.

This model serves a dual purpose: it not only demonstrates the tangible benefits of centralization in terms of time saved but also highlights enhanced collaborative efficiency that can be achieved when services are more integrated. By applying Eq. 5.1, healthcare administrators can quantitatively assess the impact of centralizing services, providing a strong basis for strategic planning and resource allocation. This approach ensures that centralization efforts are not just justified but are also aligned to enhance overall healthcare delivery, leading to more efficient patient care and potentially higher levels of patient and staff satisfaction.

Table 5.1 provides a comparative analysis of different layout models in healthcare facilities, specifically evaluating the effectiveness of centralization versus decentralized approaches. This table assesses key operational metrics such as staff travel time, response time, and inter-departmental efficiency, offering insights into how each layout impacts the functionality and performance of healthcare services. By presenting data on various layout configurations, the table aids administrators and designers in understanding the benefits and drawbacks of each approach, guiding decisions that optimize efficiency and service delivery in hospital settings.

### 5.1.2 *Streamlined Patient Flow*

The design of hospital pathways to streamline patient flow from admission through to discharge is a critical component in enhancing operational efficiency and patient satisfaction. This involves the careful planning of routes that guide patients logically through various service points, including outpatient services, diagnostic areas, treatment spaces, and inpatient wards. Such thoughtful design not only facilitates a smoother transition for patients across different stages of care but also optimizes the utilization of hospital resources and staff time.

Efficient patient flow is paramount in reducing bottlenecks which typically occur when too many processes converge in one area or when there is inadequate space to handle the volume of patient traffic. By creating well-defined and strategically located pathways, hospitals can ensure that patients move seamlessly through the care continuum without unnecessary delays. This not only minimizes cross-traffic—which can lead to operational inefficiencies—but also significantly decreases the risk of infection transmission, a critical consideration in hospital settings.

Moreover, streamlined patient flow is integral to enhancing the overall patient experience. By reducing the complexity and confusion often associated with navigating large hospital facilities, patients are likely to experience less stress and anxiety during their visits. This can lead to better patient outcomes, as a more relaxed patient is more likely to engage positively with healthcare providers and adhere more closely to treatment protocols.

In essence, the strategic design of patient pathways is not only about facilitating movement but about creating an environment that supports the health and well-being of patients by reducing waiting times, improving access to necessary services, and ensuring that the journey through the healthcare system is as efficient and stress-free as possible. This approach requires meticulous planning and understanding of the flow dynamics within hospital settings, making it an essential element of modern healthcare facility design.

### 5.1.3 Visibility and Accessibility

The optimization of hospital layouts significantly benefits from increasing visibility and accessibility between staff and patients, a principle that underpins many modern healthcare design strategies. One of the key aspects of achieving this is through the strategic placement of nursing stations. These stations are ideally positioned to allow for unobstructed views into patient rooms, facilitating immediate visual assessment and enabling staff to respond swiftly to patient needs. This layout not only ensures efficient patient monitoring but also enhances patient safety, as healthcare providers can quickly detect and address any emergent issues.

Moreover, the design of critical care units is another area where visibility and accessibility play crucial roles. These units are optimized to allow staff to monitor multiple patients simultaneously and efficiently. The architectural layout often includes large glass panes and minimal obstructions, allowing a clear line of sight to every patient from a centralized location. This arrangement is critical in environments where seconds can mean the difference between life and death. It allows healthcare providers to perform visual checks frequently and efficiently without necessarily entering each room, thus minimizing the risk of cross-contamination and saving valuable time in critical situations.

Additionally, enhanced visibility and accessibility facilitate better communication and coordination among medical staff, leading to more cohesive care delivery. It ensures that staff are readily available and easily accessible, not only to patients but also to each other, fostering a collaborative environment that is conducive to high-quality care. In designing these spaces, consideration is also given to the ease with which equipment can be moved and accessed, further streamlining the care process and reducing physical strain on healthcare workers.

In conclusion, the thoughtful consideration of visibility and accessibility in hospital design is not merely a matter of convenience but a critical component of effective healthcare delivery. By enabling better supervision, faster response times, and smoother operational flow, these design principles directly contribute to improved patient outcomes and overall hospital efficiency.

### *5.1.4 Integration of Technology*

The integration of technology into hospital layouts is a pivotal aspect of modern healthcare facility design, aimed at enhancing operational efficiency and improving patient care. Hospitals today leverage a range of technologies, with electronic health records (EHRs) being among the most critical. EHRs facilitate the seamless sharing of patient information across various departments, enhancing communication among healthcare providers and ensuring that patient data is up-to-date and easily accessible.

To support the effective use of such technologies, the physical layout of hospitals must be thoughtfully designed. This includes the strategic placement of workstations and areas specifically designated for digital access. These stations are equipped with the necessary hardware and connectivity options to access electronic systems efficiently. Moreover, the layout should accommodate the spatial needs of mobile computing solutions, such as tablets and mobile carts. These mobile devices are crucial for staff who require flexibility to move throughout the hospital while maintaining access to real-time data updates and patient records.

Furthermore, the integration of technology extends beyond just the placement of hardware; it also involves ensuring that the hospital infrastructure supports robust data security and connectivity. This includes providing secure Wi-Fi networks, sufficient data ports, and robust cybersecurity measures to protect sensitive patient information against unauthorized access. Additionally, hospitals must consider the ergonomic placement of technology to reduce physical strain on staff, such as adjustable amounts for screens and mobile units that adjust to different user heights.

Overall, the design of modern hospitals must be symbiotic with the technology used within them. By creating environments that enhance the accessibility and usability of technology, hospitals can not only streamline workflows but also significantly improve the quality of care provided to patients. This requires a meticulous approach to design that balances technological needs with the practical realities of healthcare delivery, ensuring that both staff and patients benefit from the advancements in medical technology.

Table 5.2 outlines a comprehensive checklist of essential technological features for modern hospitals, including hardware specifications, network requirements, and examples of effective integration points within hospital designs. This checklist serves as a valuable resource for healthcare administrators and IT professionals tasked with upgrading existing facilities or designing new ones. It ensures that all critical

**Table 5.2** Technology integration checklist

| Technology feature | Hardware specifications | Network requirements | Integration points in hospital design |
|---|---|---|---|
| Patient monitoring systems | High-resolution displays, wearable sensors | High-speed wireless connectivity | Patient rooms, ICU, outpatient areas |
| Electronic Health Records (EHR) | Secure servers, backup systems | Encrypted network access | Administrative areas, nurse stations |
| Telemedicine services | Cameras with HD video, microphones | Broadband internet connection | Patient rooms, emergency rooms |
| Mobile health applications | Compatible with iOS and Android | Cloud access, data synchronization | Throughout the hospital staff and patients |
| Automated pharmacy dispensers | Robotic dispensing systems | Real-time inventory management network | Pharmacy department |
| Smart building systems | IoT sensors for lighting, temperature | Integrated building management system | Common areas, patient rooms, operation theaters |

technological components are considered, from patient care systems to administrative support, enhancing overall functionality and future-proofing the healthcare environment.

### 5.1.5 *Flexible Workspaces*

Incorporating flexible workspaces into hospital layouts is essential for adapting to the dynamic nature of healthcare delivery and fluctuating hospital demands. These spaces, designed to be multipurpose, allow for a high degree of operational flexibility, enabling facilities to respond more effectively to varying patient loads and changing healthcare requirements.

Flexible workspaces in hospitals typically include rooms that can be easily converted for different uses, such as consultations, meetings, staff training, or even temporary patient care. This adaptability is crucial during peak times when hospital capacity is tested or in situations where there is a sudden need to isolate patients due to infectious disease outbreaks. For instance, a room normally used for staff meetings can quickly be transformed into additional patient care spaces with minimal adjustments needed, ensuring that the hospital can continue to operate efficiently without compromising patient care.

Moreover, these flexible spaces can enhance staff efficiency by providing areas where teams can collaborate on patient care plans, conduct interdisciplinary meetings, or engage in professional development activities without disrupting the hospital's daily operations. The design of these workspaces often includes movable furniture, adjustable lighting, and access to necessary technological support like video conferencing facilities, which further enhance their utility and responsiveness to hospital needs.

Furthermore, the psychological and social aspects of having adaptable work environments are significant. They contribute to a more satisfying work environment by reducing the monotony of a rigid workspace layout. This can lead to improved job satisfaction, lower turnover rates, and enhanced overall morale among staff.

Overall, the inclusion of flexible workspaces in hospital design is not merely a feature of modern architecture but a strategic element that enhances the resilience and adaptability of healthcare facilities. By planning for spaces that can serve multiple purposes, hospitals can ensure they remain prepared and responsive to both current and future healthcare challenges, thereby improving both staff well-being and patient care outcomes.

### 5.1.6 Efficient Use of Space

Maximizing the efficient use of space is paramount in hospital design, particularly given the limited physical footprint often available in healthcare settings. Hospitals must judiciously use every square foot to promote not only functionality but also a healing environment. This necessity drives the inclusion of multi-functional furniture, retractable work surfaces, and easily movable equipment, which collectively contributes to creating an adaptable and clutter-free workspace.

The incorporation of multi-functional furniture is a strategic choice that serves multiple purposes without occupying additional space. For example, seating that offers built-in compartments or work surfaces that fold away when not in use can significantly enhance the functionality of a space. Similarly, retractable or collapsible features in desks and shelving can transform a single-use space into a multi-use area, adapting quickly to the needs of the staff and patients alike.

Moreover, the mobility of medical equipment is another critical factor in efficient space utilization. Equipment on wheels, such as mobile medical carts or adjustable patient beds, can be moved as needed to wherever they are most required, thereby optimizing the workflow and minimizing the need for fixed, space-consuming installations. This mobility also aids in maintaining a clutter-free environment, which is essential not only for operational efficiency but also for maintaining cleanliness and preventing the spread of infections.

The goal of these strategies is to create an environment that feels spacious and adaptable, where staff can perform their tasks efficiently and effectively. By focusing on the smart integration of versatile furniture and equipment, hospitals can significantly enhance staff performance and, consequently, improve the quality of patient care.

In our subsequent discussions, we will delve deeper into the specific design considerations for diagnostic and treatment areas. This exploration will further illuminate how targeted design choices can optimize hospital operations and contribute to more effective healthcare delivery, underscoring the vital connection between spatial design and healthcare outcomes.

## 5.2 Designing for Diagnostic and Treatment Areas

The design of diagnostic and treatment areas is a critical component of hospital operations, significantly impacting the efficiency and effectiveness of medical care. These areas are hubs of activity where precision and timeliness are paramount, and the physical environment must support complex technical requirements and sensitive patient needs.

### *5.2.1 Functional Layouts*

The design of functional layouts in diagnostic and treatment areas is crucial for facilitating a streamlined and intuitive flow for both patients and healthcare providers. This strategic layout planning encompasses ensuring direct access from emergency entries to critical care areas, which is vital for rapid response in acute cases. Additionally, minimizing the physical distance between interconnected departments, such as radiology and surgery, enhances the efficiency of patient transfers and reduces the time needed for pre-operative preparations and post-operative care.

In a well-designed hospital, the logical sequencing of patient areas—from preparation to treatment and through to recovery—is meticulously mapped out to support a natural progression in patient care. This logical placement not only aids in reducing the physical and mental strain on patients and staff but also minimizes the likelihood of navigational errors and the associated risks of procedural delays or mistakes. For example, having pre-operative preparation areas adjacent to surgical suites allows for smooth transitions into and out of surgery, which can speed up moving patients through various treatment phases.

Moreover, these strategic layout decisions play a pivotal role in optimizing overall patient throughput. By reducing unnecessary waiting times and eliminating potential bottlenecks in patient flow, hospitals can handle a higher volume of cases more efficiently. This not only improves the capacity to deliver care but also enhances patient satisfaction and safety–critical metrics in healthcare quality assessments.

Ultimately, the thoughtful design of diagnostic and treatment areas, with an emphasis on functional layouts that promote efficiency and minimize error, is essential for maintaining a high standard of patient care. Such considerations ensure that hospitals are not only places of healing but also environments that operate with precision and effectiveness, directly contributing to improved health outcomes and operational success.

### 5.2.2 Equipment Integration

The integration of medical equipment into the design of modern diagnostic and treatment facilities requires meticulous planning to accommodate the complex technical demands of advanced medical technologies. These facilities are outfitted with a diverse array of sophisticated equipment, from MRI machines to robotic surgical systems, each with specific requirements for electrical and data connections, ventilation, and structural support. Effective space planning must, therefore, account not only for the operational footprint of such equipment but also for the necessary clearance areas that allow safe and efficient use and maintenance.

For instance, the installation of imaging equipment like CT scanners or MRI machines involves consideration for electromagnetic interference, requiring specialized shielding and isolated electrical circuits. Similarly, surgical suites equipped with robotic systems need reinforced flooring to support heavy equipment and flexible configurations to accommodate various procedural setups. Moreover, these areas must ensure easy accessibility for routine maintenance and emergency repairs, which are critical to the prolonged functionality and safety of medical devices.

The design must also incorporate a degree of flexibility to adapt to future technological advancements. The rapid pace of medical technology evolution means that diagnostic and treatment areas must be designed with an eye toward future growth and technology integration. This foresight allows for subsequent equipment upgrades or replacements without necessitating extensive structural renovations, thereby ensuring that facilities remain at the cutting edge of medical technology without disproportionate additional costs.

Ultimately, the successful integration of equipment into hospital design not only supports the operational needs of healthcare providers but also enhances patient care by ensuring that facilities can efficiently manage and maintain their technological assets. This integration is crucial for the longevity and adaptability of healthcare facilities, enabling them to continue delivering high-quality care in an ever-evolving technological landscape.

### 5.2.3 Environmental Controls

The meticulous control of environmental factors such as lighting, temperature, and acoustics is indispensable in the design of diagnostic and treatment areas within hospitals. These elements play a pivotal role in creating optimal conditions for medical procedures and enhancing patient comfort and recovery.

Lighting in medical facilities must be versatile to accommodate the varied requirements of different procedures. Adjustable lighting systems that can vary in intensity and color temperature are crucial. For instance, during intricate surgeries, bright, focused lights are necessary to provide clear visibility, whereas softer lighting may be used in patient recovery rooms to promote relaxation and healing. Moreover, the ability to control glare and shadow is essential in both diagnostic settings, such as radiology departments, and treatment areas to ensure accuracy and ease of operation.

Temperature control is equally critical, not only for patient comfort but also for the proper functioning of sensitive medical equipment. Many diagnostic tools, such as MRI machines and laboratory analyzers, require specific operating temperatures to function accurately and reliably. Thus, HVAC systems in hospitals must be precisely calibrated to maintain these conditions consistently. Additionally, patient areas need to be kept at comfortable temperatures to aid in recovery and ensure that patients remain relaxed during their treatments.

Acoustics also plays a fundamental role in the functionality of diagnostic and treatment areas. Soundproofing these spaces is essential to minimize distractions that could affect the concentration of medical staff performing complex tasks. Furthermore, reducing noise pollution contributes significantly to a patient's recovery environment, as excessive noise can increase stress and anxiety, potentially hindering recovery. In settings where clear communication is crucial, such as during surgical procedures or in critical care, effective acoustic design prevents miscommunication and enhances the overall efficiency of medical teams.

By integrating these environmental controls into the design of healthcare facilities, hospitals can provide environments that support high precision in medical tasks and offer comfort and tranquility to patients. These controlled settings not only optimize procedural outcomes but also enhance the overall experience for both patients and healthcare providers, ultimately contributing to better health outcomes and increased patient satisfaction.

### *5.2.4 Sterility and Infection Control*

In designing hospital spaces for surgeries and other invasive procedures, maintaining sterility and effectively controlling infections are paramount. The selection of appropriate materials and finishes in these critical areas demands careful consideration. Surfaces in surgical zones, for instance, must be non-porous to prevent the absorption and retention of microbes, and they should be easy to clean to uphold stringent hygiene standards. Seamless flooring is particularly advantageous as it lacks cracks and crevices that can harbor bacteria, thereby reducing the risk of hospital-acquired infections. Similarly, wall coverings and countertops are ideally designed with minimal joints and smooth finishes to avoid the accumulation of dust and microbes, which can be potential carriers of infection. Applying antimicrobial coatings on these surfaces can significantly enhance their sterility, adding an extra layer of protection against microbial contamination.

The architectural design of these spaces is equally critical in controlling infections. The design and choice of materials for ceilings should aim to prevent dust accumulation and facilitate easy cleaning—features essential in maintaining a sterile environment. For instance, materials that repel dust and architectural designs that minimize horizontal surfaces can greatly reduce the areas where dust and pathogens might collect. Moreover, the layout of the hospital should be strategically planned to ensure a clear segregation between clean and contaminated areas, which is crucial to prevent cross-contamination during and after medical procedures.

Furthermore, specialized ventilation systems play a vital role in sustaining sterility in surgical and invasive-procedure areas. Operating rooms and similar critical environments should be equipped with HVAC systems designed to maintain positive pressure. This configuration ensures that air is pushed out rather than allowing outside air in, which is crucial for preventing airborne contaminants, including pathogens, from entering sterile zones. Integrating High-Efficiency Particulate Air (HEPA) filters into these systems is essential as they can trap microorganisms, ensuring that the air remains clean and safe for both patients and medical staff.

By meticulously integrating appropriate materials and finishes, strategic architectural design, and advanced ventilation systems, healthcare facilities can effectively manage sterility and infection control. These elements are fundamental not only for the safety and health of patients but also for the well-being of medical staff and visitors. Effective design strategies contribute significantly to the efficacy and reputation of healthcare institutions by minimizing infection rates and enhancing patient outcomes, ultimately fostering environments that support optimal medical care and safety.

### *5.2.5 Patient Privacy and Comfort*

Ensuring patient privacy and comfort in the design of hospital diagnostic and treatment areas is as crucial as functional and technical considerations, necessitating a holistic approach to healthcare facility design that integrates elements catering to the emotional and psychological well-being of patients. This integration is foundational in creating a therapeutic environment conducive to healing.

Effective sound management is imperative to reduce noise pollution, a notable stressor in hospital settings that can significantly affect patients, especially in critical areas like diagnostic rooms and treatment facilities. Utilizing sound-absorbing materials such as acoustic ceilings, padded flooring, and noise-reducing wall panels is essential. These materials help in mitigating the propagation of noise, creating a quieter atmosphere that allows patients to feel more at ease, a quieter environment also facilitates clearer communication among healthcare providers, further enhancing the care delivery process.

Privacy is another cornerstone of patient-centered care, particularly in areas where patients are most vulnerable, such as waiting rooms, examination rooms, and recovery areas. The design of these spaces must prioritize the creation of private spaces that shield patients from exposure and prevent unintentional overhearing. Achieving this can involve strategic layout planning, including the installation of privacy curtains and the construction of structurally private rooms for consultations and recovery. Such measures not only uphold patient dignity but also ensure compliance with privacy laws and regulations.

Aesthetics also play a critical role in enhancing patient comfort and reducing anxiety within hospital environments. The choice of calming colors such as soft blues, greens, and earth tones can significantly alleviate stress, especially in treatment areas. Additionally, incorporating natural light and providing views of nature, whether through windows or digital displays, can further contribute to reducing patient stress and enhancing morale.

Moreover, ergonomics and accommodations for family members are vital for patients undergoing lengthy treatments. Ergonomically designed seating and adjustable beds can greatly increase physical comfort during extended medical procedures. Providing family members with comfortable seating areas in patient rooms or designated lounges acknowledges the critical role of the family in the healing process and enhances the overall patient experience by promoting a supportive environment.

By thoughtfully addressing these aspects of patient privacy and comfort, hospitals can forge environments that not only fulfill the technical and functional requirements of medical care but also foster a compassionate and supportive atmosphere. These considerations are integral to patient-centered care, aiding in faster recoveries, improving patient satisfaction, and embodying a more humane approach to healthcare delivery.

### *5.2.6 Adaptability*

Adaptability in the design of diagnostic and treatment areas is imperative to meet the dynamic demands of rapidly advancing medical technology and the evolving landscape of healthcare practices. As medical innovations continue to reshape the modalities of healthcare delivery, facilities must be designed with the flexibility to support these transformations efficiently, without necessitating extensive renovations or causing disruptions to ongoing services.

One practical approach to ensuring such adaptability is through the implementation of modular room designs. Modular systems offer the versatility to reconfigure or expand spaces in response to shifting needs or advancements in medical technology. This adaptability is particularly crucial in areas like radiology or surgical suites, where changes in equipment size and functionality might require different spatial configurations. By incorporating modular walls, interchangeable fixtures, and mobile cabinetry, these spaces can be easily adjusted or relocated, allowing hospitals to adapt quickly and economically to new requirements.

Furthermore, the underlying infrastructure of diagnostic and treatment areas must also be designed to support adaptability. This includes the installation of versatile utility frameworks capable of accommodating a wide range of medical technologies and services. For instance, electrical and data connections should be both accessible and upgradable to facilitate the seamless integration of emerging technological tools without significant infrastructural modifications. Additionally, structural supports designed to manage varying load requirements and duct systems adaptable to different ventilation needs are essential for future-proofing hospital spaces against forthcoming advancements in medical technology.

Designing with a focus on adaptability ensures that healthcare facilities are not only prepared to meet current operational standards but are also equipped to embrace future developments. This foresight significantly enhances the efficiency of medical procedures by minimizing the time and costs associated with frequent renovations. Moreover, it ensures that the latest healthcare innovations can be seamlessly integrated into the care delivery process, thereby improving patient outcomes.

As the discussion progresses into subsequent sections, the exploration will extend to how the integration of support and administrative spaces can further contribute to operational efficiencies in healthcare facilities. This includes examining how well-orchestrated behind-the-scenes workflows can bolster clinical functions, thus enhancing the overall functionality and effectiveness of healthcare environments.

## 5.3 Integration of Support and Administrative Spaces

The integration of support and administrative spaces within hospital design is crucial for maintaining efficient operations and supporting clinical functions. These areas serve as the backbone of hospital administration, handling everything from patient records and billing to staff scheduling and supply chain management. The thoughtful design of these spaces can significantly enhance productivity and streamline operations, which in turn supports better patient care.

### *5.3.1 Strategic Location*

The strategic location of administrative and support spaces within hospital layouts is critical for optimizing operations and maintaining an environment that supports high-quality patient care. Thoughtful positioning of these areas can significantly improve communication and workflow efficiency, which are essential to the effective delivery of healthcare services.

Proximity to clinical areas plays a crucial role in enhancing operational efficiency. By placing administrative offices near clinical zones, such as emergency departments or critical care units, hospitals can streamline communication between medical staff and administrative personnel. This proximity facilitates rapid decision-making and prompt administrative actions, allowing for quicker responses to clinical needs. For example, having administrative offices located near areas that handle patient admissions, transfers, or discharges enables smoother transitions and reduces bureaucratic delays that could negatively impact patient care. The ability to expedite processes such as patient admission, medical documentation, and discharge planning is vital in maintaining a seamless flow within the hospital, ensuring that both patient care and administrative duties are handled with minimal friction.

However, a balance must be maintained between interaction and isolation to ensure that the administrative presence does not disrupt the core patient care environment. While it is beneficial to position administrative offices close to clinical areas for operational reasons, these spaces should be designed to minimize any negative impact on patient spaces, particularly in terms of noise or increased foot traffic. Soundproofing materials should be used in the construction of office spaces to ensure that administrative activities do not create a noisy environment for patients. Additionally, buffer zones or architectural barriers between administrative and clinical areas can be employed to reduce interruptions, helping to maintain a peaceful and therapeutic atmosphere within patient care zones.

In addition to operational concerns, patient privacy and security are key factors in the strategic separation of administrative and clinical areas. Administrative offices often handle sensitive information, such as medical records, insurance details, and other confidential data. To ensure compliance with privacy regulations, these areas must be designed with secure entry and exit points, preventing unauthorized access. This strategic separation not only protects patient information but also ensures that confidential discussions and administrative activities remain private, further safeguarding the integrity of patient care processes.

By integrating these considerations into the design and layout of healthcare facilities, hospitals can achieve a balance that promotes administrative efficiency without compromising patient care, comfort, or privacy. Strategic positioning of administrative spaces contributes to the overall organization of the hospital, creating a more efficient, secure, and less intrusive environment that benefits both patients and healthcare providers. This thoughtful design supports the seamless management of healthcare operations while preserving the quality of the patient experience.

Table 5.3 offers a comparative analysis of different administrative layout strategies within healthcare facilities, evaluating their effectiveness in terms of communication efficiency, accessibility, and staff travel time. This table highlights the benefits of strategic placement of administrative units to optimize operations and support. By showcasing the impact of layout choices on the daily function of a hospital, this analysis assists facility managers and designers in making informed decisions that enhance overall efficiency and staff satisfaction.

**Table 5.3** Comparison of administrative layouts

| Layout strategy | Communication efficiency (%) | Accessibility score (out of 10) | Average staff travel time (min) |
|---|---|---|---|
| Centralized | 90 | 9 | 5 |
| Decentralized | 75 | 7 | 8 |
| Clustered | 85 | 8 | 6 |
| Open Plan | 80 | 9 | 7 |

### 5.3.2 *Efficient Space Utilization*

Efficient space utilization in administrative and support areas is critical for maximizing hospital resources, reducing operational costs, and creating a positive work environment. The thoughtful design of these spaces, prioritizing versatility and functionality, can significantly boost the productivity and well-being of staff, thereby enhancing the overall effectiveness of the healthcare facility.

One of the key strategies for optimizing space utilization is the implementation of multipurpose design. By creating spaces that can serve various functions, hospitals can make dynamic use of their footprint. For instance, a room initially designed to serve as a meeting space can also be equipped to host training sessions or function as a break area for staff. This multipurpose approach not only maximizes the use of available space but also reduces the necessity for additional areas dedicated solely to single-use functions. The use of flexible furniture and movable partitions enables these spaces to adapt quickly to the changing needs of the hospital. Such adaptability allows for the accommodation of different activities without the need for extensive reconfigurations, enhancing the hospital's ability to respond to varying demands with minimal disruption.

Incorporating ergonomic furniture into administrative areas is another essential factor in designing effective workspaces. Ergonomic chairs, adjustable desks, and supportive furnishings are vital to ensure that staff members can work comfortably over extended periods. This consideration helps reduce the risk of strain injuries and boosts job satisfaction, contributing to higher productivity and a more positive work environment. Moreover, well-organized storage solutions are crucial for maintaining an efficient workspace. Effective filing systems, accessible shelving, and ample storage areas are indispensable for keeping essential documents and supplies orderly and within easy reach. Such organizational enhancements streamline administrative processes, reducing the time staff spend searching for necessary materials, thus allowing them to focus more on their core responsibilities.

Furthermore, efficient space utilization fosters a cohesive team environment. Well-designed, multi-purpose spaces facilitate interaction among administrative and support staff, promoting collaboration and communication. This collaborative atmosphere can lead to enhanced problem-solving capabilities and spur innovation as team members engage more frequently and share ideas in a comfortable and accommodating setting.

By concentrating on these critical aspects of space utilization—multi-purpose functionality, ergonomic design, and effective organization, hospitals can craft administrative and support areas that not only meet the logistical demands of healthcare management but also improve the work experience for staff. This strategic approach to space planning ensures that the administrative backbone of the hospital operates with maximum efficiency and minimal disruption to clinical areas, supporting the institution's overarching goal of delivering exemplary patient care.

### 5.3.3 Technology Integration

The integration of technology into administrative and support areas is a pivotal aspect of modern hospital operations, facilitating efficient communication, data management, and overall functionality. As hospitals evolve into increasingly technologically advanced facilities, the design of these spaces must underscore the crucial role that technology plays in day-to-day operations.

A fundamental component of effective technology integration is the infrastructure required to support it. Administrative spaces must be equipped with necessary technological infrastructures, such as built-in data ports and multiple electrical outlets, strategically placed to support various technological devices and systems. Additionally, dedicated areas for servers and IT equipment are essential. These spaces should be designed to provide secure, climate-controlled environments that protect sensitive equipment from overheating and potential cyber threats. Proper infrastructure ensures that the backbone of hospital technology, its servers and data processing facilities—operates reliably and efficiently, safeguarding the integrity and accessibility of critical data.

Robust Wi-Fi coverage and mobile connectivity are also crucial for administrative and support areas. As healthcare systems increasingly rely on digital communications and cloud-based applications, seamless and uninterrupted internet access becomes essential. This level of connectivity supports the use of mobile devices such as tablets and smartphones, which play a significant role in various administrative functions, from scheduling and patient communication to resource management. Ensuring strong and consistent Wi-Fi coverage not only enhances the efficiency of administrative tasks but also allows staff the flexibility to work effectively from various locations within the facility.

Moreover, the design of administrative spaces should anticipate and adapt to future technological advancements. This may include installing flexible cable systems capable of handling upgrades to higher-speed internet and incorporating universal charging stations to accommodate new device standards as they emerge. Planning for future technological needs ensures that hospital spaces remain functional and relevant as the technological landscape evolves, supporting high levels of administrative performance without the need for disruptive and costly overhauls.

By prioritizing the integration of technology in the design of administrative and support areas, hospitals can significantly enhance operational efficiency and adaptability. This forward-thinking approach not only meets the current needs of hospital staff but also prepares the facility to seamlessly adopt future technological advancements, ensuring that administrative functions continue to operate at the forefront of healthcare management efficiency. This strategic integration forms the backbone of a technologically adept healthcare facility, poised to deliver top-tier medical care through enhanced communication and data management capabilities.

### 5.3.4 Support for Staff Needs

Catering to the needs of hospital staff in administrative and support areas is crucial for fostering a supportive and productive workplace. Given the demanding nature of healthcare professions, where staff often endure long and irregular hours, designing facilities with the well-being of personnel in mind can significantly enhance job satisfaction and retention.

One fundamental aspect of supporting staff needs is the provision of well-equipped break rooms. These spaces should offer a comfortable and relaxing environment where staff can genuinely unwind during breaks. Amenities such as comfortable seating, kitchen facilities, and entertainment options like televisions or games can make a substantial difference in staff morale. Moreover, creating spaces for meals that are pleasant and well-maintained encourages staff to take necessary breaks and recharge, which is essential for maintaining high levels of performance throughout their shifts.

Additionally, quiet areas designated for focused work are crucial in administrative and support spaces. These zones should be designed to minimize noise and interruptions, providing a sanctuary for staff who need to engage in tasks requiring high concentration, such as writing reports or managing complex data. Incorporating ergonomic furniture and adequate lighting in these spaces ensures that staff can work comfortably without physical strain, supporting their productivity and health.

Recognizing the high-stress nature of healthcare jobs, it is also beneficial to provide resources for stress relief and relaxation within the hospital. This could include dedicated areas for relaxation practices such as yoga or meditation, or even simpler additions like massage chairs. Offering programs or workshops focused on stress management techniques can further support staff well-being and might be integrated into the hospital's routine staff support services.

Furthermore, considering the unusual work hours of many hospital staff, including overnight or rotating shifts, it is important to design administrative and support areas to accommodate these patterns. Ensuring that facilities such as break rooms and quiet zones are accessible 24/7 and designed to provide comfort at any hour is crucial. Features like adjustable lighting that mimic natural rhythms can help regulate sleep cycles and enhance overall well-being, making these spaces functional and supportive regardless of the time of day.

By investing in the well-being of hospital staff through thoughtful design and facilities that cater to their comprehensive needs, hospitals can cultivate a more motivated and committed workforce. This attention to staff welfare not only improves individual job satisfaction but also enhances the overall quality of patient care. Well-rested and supported staff are better equipped to handle the challenges of their roles, ultimately leading to a more effective and compassionate healthcare environment.

Table 5.4 lists the various facilities and amenities provided in support areas specifically designed to contribute to staff well-being within healthcare settings. It details resources aimed at stress relief and relaxation, such as ergonomic furniture, relaxation resources, and quiet zones for focused work. This table underscores the importance

**Table 5.4** Amenities and facilities for staff support

| Facility/amenity | Description | Location | Accessibility |
|---|---|---|---|
| Relaxation rooms | Equipped with massage chairs, soft lighting, and quiet music | Near staff lounges | 24/7 access for all staff |
| Ergonomic workstations | Adjustable desks and chairs, supportive flooring | Throughout hospital | Available in all departments |
| Quiet zones | Areas designated for focused work without interruptions | Library, selected corridors | Open during hospital hours |
| Fitness center | Gym equipment, yoga mats, and space for exercise | Basement level | 24/7 access with staff ID |
| Outdoor gardens | Spaces with seating and greenery for relaxation | Adjacent to cafeteria | Daylight hours |

of staff support in healthcare environments by highlighting how these dedicated spaces can significantly enhance staff morale, reduce burnout, and improve overall job satisfaction and productivity.

### 5.3.5 *Security and Privacy*

In the design of hospital administrative and support areas, addressing security and privacy is paramount, especially considering the sensitive nature of the data handled in these spaces. Ensuring the confidentiality of patient information and financial records is crucial, not only for regulatory compliance, such as with HIPAA in the United States, but also as a fundamental component of maintaining patient trust and institutional integrity.

One primary design consideration for enhancing security is the implementation of controlled access points. Administrative areas, especially those where sensitive information is processed or stored, should be equipped with robust security measures like key card access systems or biometric scanners. This setup restricts entry strictly to authorized personnel, significantly reducing the risk of unauthorized access and potential data breaches. Such measures ensure that sensitive data remains secure from external threats and internal vulnerabilities, safeguarding the integrity of patient and institutional data.

Moreover, it is crucial to establish designated areas specifically designed for confidential discussions. These spaces should be acoustically treated to prevent eavesdropping and strategically located away from high-traffic areas to minimize the inadvertent overhearing of sensitive conversations. Ensuring privacy in these areas is essential for discussions that involve patient information, staff issues, or any other confidential matters. This design strategy not only protects privacy but also enhances the discretion necessary for handling sensitive information.

Incorporating surveillance and monitoring systems into the design of administrative areas can further bolster security. Strategically placed cameras act not only as a deterrent against unauthorized access and theft but also aid in upholding safety standards. However, it's important to carefully consider the placement of these cameras to strike a balance between security needs and privacy concerns. Surveillance should not intrude into areas where individuals expect a high degree of privacy, thus maintaining respect for personal spaces while still providing security coverage.

The physical layout of administrative areas should also be designed to facilitate natural surveillance and control flow paths. Positioning reception desks or security checkpoints at entry points allows for a clear line of sight into the area, enabling staff to effectively monitor comings and goings. Additionally, eliminating hidden corners and ensuring that hallways are well-lit can deter unauthorized access and enhance the overall security of space.

By integrating these security and privacy measures into the design of hospital administrative areas, facilities can effectively protect sensitive information while still maintaining a functional and welcoming environment for staff. This approach not only ensures compliance with legal requirements but also reinforces the hospital's commitment to safeguarding the personal and financial information of its patients and staff, thus strengthening institutional integrity and patient trust.

### *5.3.6 Flexibility for Future Changes*

In the ever-evolving landscape of the healthcare industry, marked by continuous technological advancements and shifts in administrative methodologies, designing administrative and support areas within hospitals with flexibility for future changes is essential. This foresight ensures that facilities can adapt to emerging needs without the need for extensive and costly renovations.

A practical approach to fostering this adaptability is the use of modular furniture in the design of flexible spaces. Modular furniture, which can be easily reconfigured or relocated, allows for quick adjustments to accommodate new technologies, enhance workflow, or refresh the space aesthetically. For example, desks and storage units with modular components can be rearranged to support more collaborative work environments or to increase privacy as required by changing work dynamics.

Additionally, adjustable room configurations are crucial for allowing spaces to be repurposed with minimal structural changes. The installation of movable walls or partition systems can transform a large meeting room into several smaller offices or vice versa, depending on the current demands. This versatility is invaluable in administrative areas where the scope of work and team sizes may vary over time.

Another key element is designing infrastructure to support technological growth. This includes installing sufficient conduits for cabling, providing multiple power sources, and ensuring that data ports and electrical outlets are plentiful and easily accessible. Such planning helps prevent the obsolescence of space as new technologies are introduced and become integral to hospital operations.

Beyond furniture and infrastructure, the overall design ethos should embrace adaptability. Creating inherently neutral spaces that can be easily adapted for various uses without specific design constraints can greatly enhance flexibility. For instance, designing rooms with generic dimensions that can serve multiple purposes—from offices to telemedicine centers—proves to be a practical and future-proof strategy.

By integrating such elements of flexibility into the design of support and administrative spaces, healthcare facilities not only enhance their current operational efficiency and staff productivity but also position themselves to seamlessly adapt to future challenges and changes in the healthcare landscape. As we delve into further sections, we will explore how adaptability and flexibility in design are integral not only in supporting current operations but also in preparing healthcare facilities for anticipated advancements and shifts in the industry. This approach ensures that hospitals remain responsive and resilient in the face of an uncertain and rapidly changing healthcare environment.

## 5.4 Adaptability and Flexibility in Design

Adaptability and flexibility in hospital design are critical for accommodating the rapidly changing landscape of healthcare delivery. These design principles ensure that healthcare facilities can respond effectively to evolving medical technologies, fluctuating patient volumes, and emergent health crises without requiring extensive overhauls or disruptions to service. Implementing adaptable and flexible design strategies not only enhances the ability of hospitals to provide high-quality care but also extends the lifecycle of the facility and maximizes return on investment.

### *5.4.1 Modular Design Concepts*

Modular design concepts are pivotal in creating adaptable healthcare facilities capable of responding effectively to dynamic needs and emergencies, such as pandemics. This approach ensures that hospitals can reconfigure spaces quickly and efficiently, meeting evolving medical demands with minimal disruption and enhancing overall healthcare delivery.

The essence of modular design in hospital settings lies in its ability to facilitate easy changes in the layout and functionality of spaces. Modular rooms, equipped with movable walls and flexible utilities like adjustable lighting, plumbing, and power sources, enable quick transformations of the environment. For instance, a space initially configured as an outpatient clinic can swiftly be converted into an inpatient care unit or a specialized treatment center when the need arises. This capability proves invaluable, especially during health crises or sudden influxes of patients, where rapid capacity expansion is critical.

Interchangeable fixtures and furniture further augment the versatility of healthcare spaces. Modular workstations, mobile storage units, and reconfigurable patient beds can be quickly rearranged or repurposed to suit varying medical functions and staff requirements. This adaptability not only addresses immediate space utilization concerns but also enhances the long-term usability of each piece, offering a cost-effective solution for hospitals operating with constrained budgets.

The advantages of modular design are particularly pronounced during emergencies that necessitate swift spatial adaptations. In scenarios like a pandemic, the ability to alter hospital spaces quickly is crucial for managing patient overflow, isolating contagious diseases, and providing dedicated treatment areas. Modular systems enable hospitals to scale up operations or pivot their service focus efficiently, without the extensive time and expense typically associated with traditional renovations.

Beyond its immediate adaptability, modular design also promotes sustainability and cost-effectiveness. By enabling the reuse and repurposing of spaces and materials, hospitals can minimize waste and reduce the financial burdens associated with constructing new structures. Moreover, the modular approach facilitates phased renovations or upgrades, which can be executed strategically over time without disrupting the hospital's overall operations.

Incorporating modular design principles into hospital architecture is not just about meeting the current demands of healthcare; it's also about preparing facilities to proactively face future challenges. This strategic approach significantly enhances a hospital's ability to deliver high-quality care under various conditions, ensuring readiness and resilience in facing changes, thereby securing its role as a critical infrastructure in the healthcare landscape.

### 5.4.2 *Scalable Infrastructure*

Designing hospital infrastructure with scalability at its core is essential for ensuring that healthcare facilities can meet evolving medical demands efficiently, without undergoing extensive and costly renovations. This forward-thinking approach not only enhances the functionality of a hospital but also secures its long-term viability and operational efficiency. A key component of this scalable design is the electrical system. Hospitals must plan electrical systems that not only meet current needs but also anticipate future growth and technological advancements. This could involve installing capacities that exceed immediate demands to accommodate new medical technologies or increased usage in the future. For example, incorporating spare conduits and additional panel capacities during the initial construction phase allows for easy integration of new technologies or expansion of existing systems with minimal disruption. This proactive planning prevents future logistical challenges and reduces downtime, which is critical in healthcare settings where continuous operation is essential.

In parallel, plumbing and ventilation systems in hospitals require a similar focus on scalability. This involves designing utility cores that can be easily expanded or modified as medical needs evolve. Accessible shafts that allow for the addition of extra piping or ductwork can significantly reduce time and expense when upgrading facilities. Moreover, ventilation systems, which play a crucial role in infection control and the management of airborne diseases, should be designed to accommodate increased capacity or enhanced filtration systems as requirements change over time. This foresight ensures that hospitals can adapt to changes in healthcare delivery and patient care standards without substantial structural changes.

Another critical aspect of scalability involves the physical layout and utilization of space within the hospital. It is essential to design spaces that can be easily modified, divided, or expanded according to changing needs. The use of demountable walls, for example, facilitates the quick reconfiguration of spaces to accommodate fluctuating patient volumes or to create new treatment areas as required. This level of flexibility is vital for hospitals to respond effectively to emergencies, epidemics, or changes in healthcare services.

Furthermore, as digital technology increasingly becomes integral to both clinical and administrative operations in healthcare, the scalability of IT infrastructure cannot be overlooked. Hospitals must ensure their data networks are robust enough to handle increased data flows and connectivity demands that come with digital expansion. This includes planning for additional data ports, wireless access points, and server capacity from the outset. Such preparations help prevent data bottlenecks and support the seamless integration of new digital tools and platforms, which are essential for modern healthcare delivery.

Overall, incorporating scalability into hospital design is not merely about anticipating the future; it's about creating a healthcare environment that can evolve and adapt with minimal disruption. This approach ensures that hospitals can continue to provide high-quality care efficiently and sustainably, adapting to new technologies, patient demands, and healthcare practices without the need for continuous physical expansions or overhauls.

### *5.4.3 Technology-Ready Spaces*

Designing hospitals to be technology-ready is imperative to ensure that these environments can seamlessly integrate rapid advancements in medical technology. This proactive approach in hospital design is crucial as it accommodates emerging diagnostic and treatment technologies without necessitating extensive modifications to the existing infrastructure. A fundamental element of creating technology-ready spaces involves the provision of adequate power and robust data connectivity. It is essential to install multiple power outlets and data ports throughout hospital rooms and corridors. This setup not only facilitates the easy addition of new equipment as it becomes available but also ensures that the electrical load capacity is sufficient to handle high-powered medical devices without the risk of overloading. Additionally, a state-of-the-art data infrastructure must be in place. This includes high-speed internet cables and strategically placed wireless access points to support large data transfers and meet the connectivity demands essential for modern medical technologies to function effectively.

Furthermore, the physical structure of the hospital must be equipped to support heavy medical equipment such as MRI machines and CT scanners, which are integral to advanced medical care but significantly heavy. Ensuring that floors and support systems can bear such heavy loads is critical. This might involve reinforced flooring or specialized mounting systems designed to securely hold and stabilize heavy equipment. By planning for these structural needs in advance, hospitals can significantly reduce the time and cost associated with retrofitting spaces when new technologies are introduced.

Incorporating flexible design elements into hospital spaces is another strategy to enhance their adaptability to new technologies. Adjustable lighting, movable walls, and modular furniture are examples of design choices that can transform spaces quickly and efficiently. These elements allow for the rapid reconfiguration of rooms to accommodate various types of medical equipment or changing functional needs without the need for major renovations.

Lastly, considering future expansion needs during the design phase is also strategic. Allocating extra space specifically for technological expansion, whether it's designing rooms with more square footage than currently necessary or including accessible utility rooms for additional technological infrastructure—can provide hospitals with the flexibility to adapt to new advancements as they arise.

By ensuring that hospital spaces are designed with future technology integration in mind, healthcare facilities position themselves at the forefront of medical care delivery. Technology-ready spaces not only facilitate the smooth adoption of cutting-edge medical equipment but also enhance continuity of care and operational efficiency, which are crucial for improving patient outcomes and streamlining healthcare delivery. This foresight in hospital design underscores the importance of adaptability in healthcare infrastructure, ensuring that facilities remain relevant and effective in the face of rapid technological evolution.

### *5.4.4 Responsive Layouts*

Incorporating responsive layouts in hospital design is a strategic move that significantly boosts the facility's capacity to adapt to fluctuating patient volumes and evolving medical needs. This flexibility is essential for maintaining efficient operations and upholding high standards of patient care, particularly during periods of increased demand. For example, versatile outpatient areas are a core feature of responsive layouts. These spaces are typically used as standard consultation rooms during regular hours. However, during peak times, such as flu season or in response to a public health crisis, they can be swiftly transformed into additional triage stations or extended waiting areas. Achieving such adaptability involves the use of movable partitions, foldable furniture, and modular design elements, enabling the space to be quickly and efficiently reconfigured according to the immediate needs.

Responsive layouts also extend to modifying operational workflows within the hospital. For instance, internal pathways can be designed to change based on the time of day or the specific requirements of the patients. This flexibility can be facilitated by adjustable signage, deployable barriers, and even temporary floor markings that direct foot traffic differently during emergencies or peak admission times. Such dynamic workflow adaptations ensure that the hospital environment can swiftly respond to various scenarios, enhancing both patient flow and safety.

The integration of technology further supports the concept of responsive layouts. Digital signage, for example, can be updated in real-time to redirect patients and staff according to the most pressing needs, whether it's directing them to different service areas or providing updates on waiting times. Additionally, automated systems can play a pivotal role in managing space usage effectively by tracking room occupancy and adjusting environmental controls like heating, ventilation, and lighting automatically, based on the current usage patterns.

Moreover, responsive layouts consider the need for quick expansion. In some hospital designs, non-medical areas such as conference rooms or cafeterias are equipped to double as patient care areas when necessary. The infrastructure in these spaces, including essential installations like oxygen lines and medical gas supplies, is integrated during initial construction but remains concealed until needed. This preparation allows these areas to be converted into functional medical spaces with minimal effort, ensuring that the hospital can increase its capacity promptly and efficiently in response to sudden increases in patient numbers.

By designing hospitals with responsive layouts, facilities are better equipped to manage varying patient volumes and adapt to diverse medical scenarios without significant disruptions. This approach not only enhances the functionality and operational efficiency of the hospital but also significantly improves the overall experience for both patients and staff. Ensuring that the hospital can meet both current and anticipated future healthcare demands with agility and effectiveness is a testament to the foresight and innovative thinking embedded in the concept of responsive layouts.

### 5.4.5 *Universal Rooms*

Designing universal rooms in hospitals represents a strategic approach aimed at enhancing operational flexibility and ensuring readiness for a variety of healthcare scenarios, from routine patient care to emergencies like infectious disease outbreaks. These multi-purpose rooms are equipped with adaptable features that enable them to swiftly transition between different medical needs without requiring extensive reconfiguration, thereby streamlining hospital operations and increasing their capacity to respond to various medical challenges.

One of the critical features of universal rooms is their adaptable airflow systems. These systems are engineered to modify the room's ventilation settings to create negative pressure environments, when necessary, which is essential for isolating airborne pathogens during infectious disease outbreaks. Conversely, the airflow can be adjusted back to standard settings for regular patient care, maintaining the room's functionality and safety for diverse uses. This flexibility in controlling the environmental conditions within the room is vital for both preventing cross-contamination and providing adaptable patient care settings.

Another key aspect of universal rooms is the use of easy-to-clean surfaces. The materials selected for these rooms typically include non-porous surfaces that prevent the absorption and retention of pathogens. These characteristics are crucial as they support rigorous cleaning protocols and withstand frequent sanitation with harsh chemicals, which is essential for both infectious disease control and routine patient care. The ability to maintain high standards of hygiene and cleanliness in these rooms is a cornerstone of their design, ensuring they can be quickly and effectively prepared for any patient scenario.

Furthermore, universal rooms are designed with flexible utility access. This includes easily accessible electrical and data ports, medical gas outlets, and modular furniture configurations that can be quickly rearranged or removed as needed. Such flexibility ensures that the room can be outfitted with various types of medical equipment or adapted for different treatment requirements swiftly, which is particularly important in situations requiring rapid changes in the setup of medical spaces.

Despite their functional versatility, it is crucial that universal rooms also prioritize patient comfort. This involves integrating elements such as adjustable lighting, temperature control, and soundproofing to make the environment as comfortable as possible for patients under various circumstances. Whether the room is used for intensive care, isolation, or standard patient stays, these features help maintain an environment conducive to healing and recovery.

The incorporation of advanced technology is also a hallmark of modern universal rooms. Features such as telemedicine capabilities allow for remote monitoring and consultations, reducing the need for physical contact during infectious disease outbreaks or when handling immuno-compromised patients. This technology not only enhances patient care but also extends the functionality of the room across broader clinical applications, enabling hospitals to leverage their resources more effectively.

By incorporating these versatile design elements, universal rooms significantly enhance a hospital's ability to efficiently respond to fluctuating medical demands. These rooms exemplify forward-thinking design, ensuring that hospitals can provide high-quality care under various conditions while maximizing the use of their physical space, ultimately contributing to better patient outcomes and more streamlined healthcare delivery.

### 5.4.6 Use of Demountable Partitions

Utilizing demountable partitions in hospital design represents a strategic choice that significantly enhances the adaptability and flexibility of healthcare facilities. This approach allows hospitals to respond swiftly and efficiently to changing clinical demands, which is especially beneficial in dynamic environments such as emergency departments or intensive care units (ICUs), where the needs of patients and room configurations can shift unexpectedly.

The ability to quickly reconfigure the size and layout of rooms is a primary benefit of demountable partitions. In emergency departments, for example, where the influx of patients can vary dramatically, these partitions allow for the expansion or compartmentalization of space depending on immediate needs. This flexibility is crucial during times of high demand, such as during mass casualty incidents or sudden outbreaks of infectious diseases. The quick adaptability provided by demountable partitions ensures that facilities can maintain operational efficiency and patient care quality without delay.

In the context of ICUs, the use of demountable partitions becomes even more critical. They enable the rapid creation of isolation rooms or the expansion of existing spaces to accommodate more medical equipment or personnel. This adaptability is vital for effective infection control and patient management, especially when managing highly infectious diseases or providing care to patients requiring intensive monitoring and complex setups. Additionally, the flexibility to adjust room sizes and layouts contributes to workflow efficiency and ensures that staff have adequate access to critical areas during high-stakes situations.

From a sustainability and economic perspective, demountable partitions offer considerable advantages. They allow for the reconfiguration of existing spaces rather than necessitating the construction of new structures or extensive renovations, thus avoiding the costs and environmental impacts associated with such activities. These partitions can be reused and repositioned as the needs of the facility evolve, supporting a more sustainable and cost-effective approach to hospital design.

Moreover, the use of demountable partitions contributes directly to the overall operational efficiency of healthcare facilities. By enabling quick adaptations to changing clinical demands without significant disruptions, these partitions help sustain continuous patient care, minimize downtime, and allow staff to operate more effectively under varying conditions.

As healthcare continues to evolve, the focus on adaptability and flexibility in hospital design, exemplified by the strategic use of demountable partitions, underscores a broader commitment to operational efficiency and responsive patient care. This approach not only meets the immediate needs of patients but also anticipates future demands, ensuring that hospitals remain equipped to provide high-quality care in an ever-changing healthcare landscape.

## 5.5 Enhancing Safety Through Architectural Choices

Enhancing safety through architectural choices is a fundamental aspect of hospital design that significantly impacts operational efficiency and patient care quality. Architectural strategies that promote safety help prevent accidents, reduce the spread of infections, and ensure a secure environment for both patients and staff. These choices are integral not only to meeting health standards but also to fostering a sense of trust and confidence in the healthcare facility.

Table 5.5 summarizes the key safety features stemming from architectural choices and their direct benefits to operational efficiency and patient care quality within healthcare facilities. This table highlights how specific design elements such as layout, material choices, and integrated technology can enhance both safety and functionality, contributing to more effective patient care and streamlined staff workflows. It serves as an essential reference for healthcare administrators and facility designers in understanding the profound impact of thoughtful architecture on healthcare delivery.

### *5.5.1 Infection Control Through Design*

The design of a hospital is crucial in controlling infections, with the integration of specific materials, layouts, and systems playing a key role in enhancing the facility's

**Table 5.5** Impact of architectural choices on safety and efficiency

| Architectural feature | Safety benefit | Impact on operational efficiency | Impact on patient care quality |
|---|---|---|---|
| Fire-resistant materials | Reduces risk of fire spread | Decreases downtime and maintenance needs | Enhance patient and staff safety |
| Barrier-free layouts | Prevents accidents and facilitates evacuation | Improves mobility and speeds up response times | Increases accessibility for patients with disabilities |
| Integrated monitoring systems | Allows for real-time tracking of patients and staff | Reduces response times to emergencies | Elevates patient security and monitoring |
| Natural lighting | Reduces eye strain and improves visibility | Lowers energy consumption and operational costs | Improves patient mood and recovery rates |
| Soundproofing | Minimizes noise pollution | Enhances workplace concentration and efficiency | Reduce stress and improve patient rest quality |

ability to minimize the spread of pathogens and ensure a safer environment for both patients and staff.

One of the fundamental aspects of infection control through design is the selection of appropriate materials used in hospital construction. Non-porous surfaces such as stainless steel, certain plastics, and treated wood are ideal because they do not allow bacteria and viruses to penetrate, making them easier to clean and disinfect. Additionally, the application of antimicrobial coatings on high-touch surfaces like doorknobs, light switches, and countertops actively reduces the presence of harmful microbes. These materials and coatings help cut down the transmission routes within the hospital, thereby reducing the likelihood of infection spread.

The strategic layout of the hospital is another critical factor in minimizing cross-contamination between different care areas. The design should segregate high-risk areas, such as surgical suites and intensive care units, from general patient areas to prevent the spread of infections. This includes creating dedicated pathways for the movement of staff and materials from sterile areas to contaminated zones without intersecting general corridors. Additionally, providing adequate space for the cleaning and sterilization of medical equipment and having designated areas for donning and doffing protective gear is vital in reducing the risk of healthcare-associated infections.

Advanced HVAC systems are also integral to infection control in hospitals. These systems are designed not only to maintain comfort but also to control infection risks. Sophisticated HVAC designs that include High-Efficiency Particulate Air (HEPA) filtration and ultraviolet (UV) light treatments can effectively remove pathogens from the air. Proper ventilation in patient rooms, operating rooms, and other critical areas involves controlling air pressures to create protective environments. For example, operating rooms typically maintain positive pressure to keep contaminated air out, whereas isolation rooms utilize negative pressure to contain harmful air particles within.

Implementing rigorous protocols for the monitoring and maintenance of these systems is equally important to ensure the optimal functioning of all design elements that contribute to a sterile environment. Regular checks and balances, including routine cleaning protocols, air quality tests, and surface swabbing, are essential to detect and address potential infection hotspots promptly.

By integrating these design principles, hospitals can significantly enhance their infection control measures. A well-thought-out hospital design not only supports the health and safety of patients but also protects healthcare workers, ultimately contributing to more effective and efficient healthcare delivery. This proactive approach in hospital design and operations is essential in establishing a robust defense against the spread of infections, thereby fostering a healthier hospital environment for everyone.

### 5.5.2 Fall Prevention

Fall prevention is an essential safety aspect in hospital design, particularly crucial for vulnerable patient populations such as the elderly or those with mobility impairments. Architectural and design elements are strategically employed to mitigate these risks and enhance safety across the facility, ensuring that patients can navigate the space without undue risk of falls.

One of the most effective architectural interventions for fall prevention is the installation of non-slip flooring materials. These surfaces provide enhanced traction underfoot, crucially reducing the likelihood of slips, especially in areas prone to wetness such as bathrooms, kitchens, and entryways. Textured vinyl, rubber, and certain types of treated ceramics are preferred for their ability to maintain grip even when wet, thereby significantly diminishing the risk of slip-related accidents.

Adequate lighting is another critical component in preventing falls within hospital settings. Proper illumination enables both patients and staff to see obstacles, floor-level changes, and other potential hazards. This is particularly important in corridors, bedside areas, and bathrooms—places where patients frequently move about unassisted, including during nighttime. Adjustable lighting options are advantageous, as they allow for the brightening of these spaces when necessary while ensuring that the light levels can be softened to avoid disturbing patients' rest.

The installation of handrails and other support structures throughout the hospital is vital for assisting patient mobility. Handrails should be installed in corridors, bathrooms, and bedside areas to provide necessary support for patients as they move around. These handrails need to be sturdy, well-positioned, and within easy reach to aid with balance and mobility. Furthermore, additional support structures such as grab bars next to toilets and in shower areas play a crucial role in providing stability and support for patients during movement and transfers.

Minimizing thresholds and removing floor-level changes are also effective strategies to reduce trip hazards. Design transitions should be as smooth as possible to avoid creating edges that could catch on feet or mobility devices. This includes the strategic design of doors to minimize threshold height and the installation of gently sloping, well-marked ramps where necessary.

Finally, the thoughtful layout of rooms and furniture contributes significantly to fall prevention. Patient rooms should be arranged to facilitate easy navigation, with ample space around beds and unobstructed pathways. Furniture placement is crucial, ensuring that it does not block movement and that essential items are within easy reach. In bathroom designs, ample turning space for mobility devices and the strategic placement of fixtures are critical to support safe and independent use.

By integrating these architectural and design elements into the structure of a hospital, the environment becomes safer for all occupants. This proactive approach not only improves the quality of care but also minimizes the potential for accidents that could worsen patients' conditions and extend their hospital stays, thereby contributing to a more effective and compassionate healthcare setting.

### *5.5.3 Emergency Preparedness*

Emergency preparedness is a crucial aspect of hospital design, ensuring rapid and effective responses to crises ranging from natural disasters to medical emergencies. The architectural planning and design elements dedicated to this end are pivotal in safeguarding both patients and staff, allowing for swift action when every second counts.

Clear and unobstructed pathways are fundamental to effective emergency preparedness. These evacuation routes must be logically designed to lead directly to emergency exits, minimizing confusion and delays during an evacuation. Flooring materials that enhance traction and visibility, along with signage like photoluminescent path markers, are essential, as they guide the way even during power failures, ensuring everyone knows the quickest route to safety.

The strategic placement of emergency exits is critical. To maximize safety, exits should be numerous and evenly distributed throughout the facility, ensuring that no area is too far from a safe escape route. Additionally, these exits must comply with accessibility laws, facilitating safe and efficient evacuation for everyone, including those with disabilities.

Robust structural design is another key component. Hospitals must be built to withstand events like earthquakes, floods, or hurricanes. This often involves structural reinforcements that exceed standard building codes, employing materials and construction techniques chosen for their durability and resilience. For example, seismic retrofitting is essential in earthquake-prone areas, while elevated structures may be necessary in flood-susceptible regions.

Designating specific areas for emergency response, such as triage zones, is also crucial. These areas should be strategically located to ensure they are quickly accessible. They need to be equipped to handle sudden increases in patient numbers, with flexible spaces that can be expanded or reconfigured as necessary. Utilities in these areas, such as power and water supplies, should have redundancies to ensure continued operation, even if primary systems fail.

Safety equipment placement must be carefully considered. Fire extinguishers, automated external defibrillators (AEDs), and first aid kits should be readily accessible throughout the hospital, clearly marked, and located in standardized spots familiar to all hospital personnel. This standardization helps ensure that during an emergency, time is not lost searching for these critical items.

Finally, effective communication systems are essential. These systems must be robust enough to function in various scenarios, including power outages and network disruptions. Backup power supplies, such as generators, and alternative communication methods, such as satellite phones, are vital to ensure that communication remains open between hospital staff and external emergency services.

By integrating these features into hospital design, facilities are better prepared to handle a range of emergencies. This preparedness not only enhances the safety and security of the hospital environment but also bolsters the facility's capacity to provide continuous care under adverse conditions, ensuring both patient and staff safety during critical times.

### *5.5.4 Security Measures*

In the realm of hospital design, security measures play a critical role in safeguarding both the physical safety of patients and staff and the integrity of sensitive medical information and valuable medical supplies. Effective security within a healthcare facility encompasses several key strategies, each contributing to a robust framework that protects against both internal and external threats.

One fundamental aspect of security is the implementation of controlled access points throughout the facility. These access points serve to regulate entry and exit, ensuring that only authorized individuals can access sensitive areas or handle high-value equipment and medications. This can be achieved through a combination of manned security posts and technologically advanced systems, such as card readers or biometric scanners, which provide a reliable means of access control. The careful management of these points is crucial not only for preventing unauthorized access but also for ensuring orderly patient discharges and visitor management, with all movements meticulously logged and monitored.

Surveillance plays a pivotal role in hospital security, with strategically placed cameras monitoring both public and restricted areas. These surveillance systems are essential in deterring potential thefts and ensuring constant vigilance in zones that are vulnerable to security breaches, such as medication storage rooms, equipment rooms, and parking areas. Modern surveillance technology may include motion sensors and real-time alerts, enhancing the system's overall effectiveness by providing immediate responses to unauthorized activities.

Equally important is the design of secure storage areas specifically for medications, sensitive equipment, and confidential patient records. These areas should be fortified with advanced locking systems and accessible only to authorized personnel to prevent any unauthorized access. For particularly sensitive items like medications and high-value medical supplies, systems such as automated dispensing cabinets and robust inventory control mechanisms are employed to meticulously track usage and prevent any form of pilferage.

The hospital design should also facilitate safe movement of personnel and patients, especially during off-hours. This includes installing adequate lighting, and clear signage, and ensuring that pathways are secure and continuously monitored by surveillance systems. Emergency call buttons and intercom systems in secluded areas further enhance safety by providing quick communication links for individuals needing assistance during less busy times.

Integrating security measures with the hospital's emergency protocols is crucial. For instance, during lockdown, access control systems should be capable of securing all doors automatically, while surveillance systems monitor the situation and guide emergency responses effectively.

Lastly, the human element of security cannot be overlooked—regular training and security drills are indispensable for ensuring that all hospital staff are prepared to respond adeptly in various security situations, from handling unauthorized entries to dealing with potential internal threats or emergencies.

By embedding these comprehensive security strategies into the architecture and operational protocols of healthcare facilities, hospitals can ensure a secure environment that protects not only their tangible assets but also the trust and safety of their patients and staff. This integration of security measures is foundational to maintaining the institution's reputation and effectiveness in delivering high-quality healthcare.

### *5.5.5 Visual Accessibility and Wayfinding*

Visual accessibility and effective wayfinding are paramount in hospital design, directly enhancing safety by facilitating easy and efficient navigation for patients, visitors, and staff within the facility. Effective wayfinding is particularly vital in complex environments like hospitals, where confusion and stress are common and can significantly contribute to safety incidents and accidents.

A cornerstone of effective wayfinding is the implementation of clear and consistent signage throughout the hospital. Signage plays a critical role in guiding individuals smoothly through the facility, reducing potential disorientation. These signs must be designed with easy-to-read fonts and utilize high-contrast colors to accommodate those with visual impairments. Strategic placement of signage at key decision points such as entrances, elevator banks, and corridor intersections ensures a smooth flow of traffic and helps individuals make intuitive navigation decisions.

The architectural layout of the hospital also significantly impacts the effectiveness of wayfinding. Hospitals should aim for an intuitive design that facilitates straightforward paths from one area to another, minimizing the labyrinthine complexity often associated with hospital corridors. Clear sightlines and the inclusion of distinctive landmarks, such as art installations or uniquely designed areas, can greatly aid in orientation by providing easy-to-recognize reference points that assist in navigation.

Utilizing color coding and icons enhances traditional wayfinding strategies, making navigation more intuitive. Different colors can denote specific wings or floors, allowing individuals to quickly ascertain their location and direction. Icons are equally important; they provide universal cues that transcend language barriers, clearly marking essential services and facilities like bathrooms, pharmacies, and emergency exits.

The integration of technology offers further enhancements to traditional wayfinding systems. Modern hospitals can implement digital solutions such as interactive kiosks or mobile apps that provide real-time navigation assistance, facility updates, and important alerts about any operational changes. These digital tools can also cater to specific navigational needs, such as offering routes optimized for wheelchair access, thereby personalizing the navigation experience to accommodate all users.

Beyond physical and digital aids, the training of hospital staff in wayfinding assistance is crucial. The knowledgeable staff can not only provide directions but also escort patients as needed, contributing to a hospitable environment that mitigates stress and makes the facility seem less daunting, particularly for those who are visually impaired or new to the hospital.

By emphasizing visual accessibility and effective wayfinding in hospital design, healthcare facilities can significantly improve safety and operational efficiency. This focus not only aids in reducing the risks associated with disorientation and confusion but also fosters a more organized and patient-centric healthcare environment, ultimately enhancing the overall experience for all users of the facility.

### *5.5.6 Natural Disaster Resilience*

Incorporating resilience into hospital architecture is essential to ensure that healthcare facilities can continue to operate effectively during and after natural disasters. This resilience is critical not only for safeguarding the safety of patients and staff but also for maintaining the continuity of vital healthcare services when they are most needed.

One of the primary considerations in building disaster-resilient hospitals is adopting flood-resistant construction techniques, especially in areas known to be susceptible to flooding. This can involve elevating the structure above anticipated flood levels to prevent water ingress, utilizing waterproof materials in the construction of lower levels, and installing physical barriers such as water barriers and sump pumps to manage and mitigate floodwater. Additionally, thoughtful landscaping can contribute to resilience by sculpting the surrounding terrain to divert water away from the building, further protecting the facility from flood damage.

In regions prone to seismic activity, earthquake resilience is paramount. Hospitals in these areas should incorporate advanced seismic design principles that allow structures to withstand tremors. This might include the use of flexible building materials that can absorb and withstand shaking, the installation of base isolators to dampen ground motion, and reinforcing the building's overall structural integrity to prevent collapse. Staying up-to-updated with the latest advancements in technology and materials is crucial to enhancing and maintaining seismic resilience over time.

Another cornerstone of disaster preparedness is ensuring the availability of emergency power systems to sustain critical hospital operations during power outages, which are common during natural disasters. This typically involves setting up backup generators that can operate independently from the grid. Hospitals must ensure that these generators are well-maintained and have sufficient fuel storage on-site to remain operational throughout a disaster.

Safe evacuation routes and designated safe areas within the hospital are essential for effective disaster response. These routes and areas should be clearly marked, easily accessible, and structurally reinforced to provide shelter during emergencies.

They should also be equipped with essential emergency supplies, first aid kits, and other resources needed to support patients and staff during a crisis.

Effective communication systems are vital in maintaining coordination and management during a disaster. Hospitals should have diverse and redundant communication systems, including landlines, mobile devices, and radios, to ensure connectivity even if one system fails. These systems must facilitate robust internal communication among hospital staff and enable seamless external communication with emergency services, government agencies, and other critical stakeholders.

By prioritizing these safety-focused architectural and operational elements, hospitals can enhance their disaster resilience, reducing risk and disruption during crises. These strategic choices in hospital design not only safeguard critical healthcare infrastructure but also reinforce the role of hospitals as pivotal community resources, ensuring their operational effectiveness and reliability in times of natural disasters. This commitment to resilience is fundamental to improving patient outcomes and maintaining community trust and safety during critical times.

Table 5.6 lists various architectural strategies employed to enhance resilience against natural disasters within healthcare facilities. This table details specific materials and design techniques used to withstand events such as floods and earthquakes, ensuring the continuity of care and safety of patients and staff during such crises. By outlining these robust construction practices and innovative materials, the table provides vital information for architects, engineers, and facility managers aiming to build or retrofit healthcare facilities to meet high standards of disaster resilience.

Next, we will conduct a comparative analysis of operational models, exploring how different design strategies are implemented in various healthcare settings and their impact on operational effectiveness. This analysis will highlight the diversity of approaches in hospital design and their tailored adaptations to meet specific environmental challenges and healthcare needs.

**Table 5.6** Architectural strategies for disaster resilience

| Disaster type | Architectural strategy | Materials used | Design techniques |
|---|---|---|---|
| Floods | Elevated structures | Waterproof concrete, marine-grade steel | Raised platforms, sloped landscaping |
| Earthquakes | Seismic retrofitting | Flexible joint systems, shock absorbers | Base isolation, flexible utility connections |
| Hurricanes | Wind-resistant design | Impact-resistant glass, reinforced concrete | Aerodynamic building shapes, secure roof designs |
| Fires | Fire-resistant construction | Fire-resistant drywall, intumescent paint | Compartmentalization, fire barriers |
| Tornadoes | Reinforced safe rooms | Steel reinforcements, concrete walls | Heavy-duty door and window locks, underground bunkers |

## 5.6 Comparative Analysis of Operational Models

In the dynamic field of healthcare, different operational models can greatly influence the design and functionality of hospital facilities. A comparative analysis of these models provides insights into how design choices can enhance or impede operational efficiency and patient care. This section examines several key operational models within healthcare settings and how architectural designs support or challenge their effectiveness.

### *5.6.1 Traditional Hospital Model*

The traditional hospital model, which has formed the backbone of healthcare infrastructure for decades, is defined by its centralized services and hierarchical departmental structures. This model is built on a systematized approach to healthcare delivery, organizing distinct medical specialties into separate departments. While this structure supports specialized care and aims to maximize resource utilization and operational efficiency, it also presents several architectural and operational challenges that can impact patient experience and adaptability to changing healthcare demands.

Centralization of key services such as diagnostics, laboratories, and pharmacies is a hallmark of the traditional hospital model. By centralizing these critical services, the model seeks to streamline operations and enhance efficiency within individual departments. However, this centralization often requires patients to traverse various parts of the hospital to access different types of care, which can be both time-consuming and stressful. This movement can detract from patient experience, adding a layer of complexity to patient journeys that are already under the strain of medical concerns.

The architectural design of traditional hospitals typically mirrors the hierarchical nature of their operations, with clear demarcations between different specialties and services. While this separation facilitates specialized care by grouping expertise and resources, it may hinder the integration of services. Coordinating care for patients requiring multidisciplinary interventions can become challenging, creating bottlenecks and potential gaps in the continuum of care.

Patient flow within traditional hospitals is often designed to optimize the movement of patients through various service areas, leading to complex layouts with long hallways and multiple wings. Such designs, necessary to segregate different care types (e.g., inpatient versus outpatient), result in expansive facilities that can be difficult to navigate. For patients and visitors, especially those dealing with stress or health anxieties, these sprawling complexes can be intimidating and confusing, exacerbating an already tense experience.

A significant limitation of the traditional hospital model is its lack of flexibility and adaptability. The physical layout, characterized by fixed and specialized spaces, poses challenges when adapting to shifts in healthcare demand or integrating new technologies. This rigidity can be particularly problematic during emergencies or outbreaks that necessitate rapid reconfiguration of spaces, leading to inefficiencies and potential disruptions in service delivery.

Moreover, while the traditional model is designed to support high-volume patient care, it often struggles with effective cross-departmental interactions. The physical and operational separation between departments can create silos, fostering communication barriers that may lead to delays in care delivery and sometimes disjointed patient experiences.

Despite its challenges, the traditional hospital model has provided a stable foundation for delivering specialized, high-volume care. However, the evolving landscape of healthcare, characterized by increasing integration of technology and a greater emphasis on patient-centered approaches, calls for modifications to this model. Modern hospital designs are increasingly reflecting these needs, focusing on modularity, integrated care teams, and patient-centric services to enhance flexibility, improve patient experiences, and facilitate smoother cross-departmental collaboration. This evolution in hospital architecture and operations is essential to meet the contemporary demands of healthcare delivery effectively.

### 5.6.2 *Integrated Care Model*

The Integrated Care Model represents a paradigm shift in healthcare delivery, emphasizing a holistic approach that not only addresses acute medical issues but also incorporates preventive care and chronic disease management. This model seeks to provide comprehensive, continuous care that effectively coordinates all aspects of a patient's health needs. The architectural design of facilities within this model plays a crucial role in enhancing its functionality and effectiveness by creating environments that support and facilitate this integrated approach to healthcare.

Community-based facilities are a defining feature of the Integrated Care Model, contrasting sharply with the large, centralized buildings typical of the traditional hospital model. These smaller, localized facilities are embedded within the communities they serve, making healthcare services more accessible and less daunting for patients. This proximity encourages regular engagement with healthcare services, fostering ongoing health management and preventive care, which are central to the model's objectives.

Architectural designs in this model prioritize creating welcoming, non-clinical environments that reduce the institutional feel of healthcare settings. This is achieved by incorporating design elements commonly found in residential or hospitality settings, such as natural lighting, comfortable seating, and inviting décor. The aim is to make these spaces feel less like traditional hospitals and more like part of the community fabric, which can significantly enhance patient comfort and willingness to engage in regular health management practices.

Flexibility is another cornerstone of architectural design in the Integrated Care Model. Facilities must be adaptable to a wide array of healthcare functions, from conventional medical treatments and diagnostics to preventive care and community health initiatives. This flexibility often manifests in modular designs that allow spaces to be reconfigured according to changing needs, multi-purpose rooms, and areas that facilitate community involvement and interaction. Such versatile spaces are essential for supporting the model's comprehensive approach to health and wellness.

Despite their community-based orientation, these facilities are not isolated but are seamlessly integrated into larger health systems. Architectural designs facilitate this integration through advanced technology and infrastructure that support robust communication and data exchange across different settings. Features might include designated areas for telemedicine, enabling consultations with specialists who are not physically present, and integrated information systems that maintain continuity in patient records and care management across multiple locations.

The architectural design under the Integrated Care Model also supports a broad spectrum of health services beyond traditional medical care. Facilities are equipped to handle mental health counseling, nutritional advising, physical therapy, and social services, reflecting the model's holistic view of health. This requires architectural planning that accommodates a variety of activities, including group programs, private consultations, and community educational seminars.

By aligning architectural design with the principles of holistic, patient-centered care, the Integrated Care Model significantly enhances patient outcomes and satisfaction. It shifts healthcare from a reactive to a proactive stance, focusing on the prevention and effective management of chronic conditions. Consequently, these facilities do more than just meet the immediate health needs of their communities; they become pivotal in fostering long-term health and well-being, effectively transforming the landscape of healthcare delivery.

### *5.6.3 Patient-Centered Medical Home (PCMH)*

The Patient-Centered Medical Home (PCMH) model marks a significant evolution in healthcare delivery, emphasizing the development of strong, enduring relationships between patients and their healthcare providers. This model is designed to increase accessibility, enhance the quality of care, and improve the efficiency of healthcare services by focusing on the individual needs and conditions of each patient. The architectural design of facilities implementing the PCMH model is critical, as it directly influences the effectiveness of these interactions by creating environments that are conducive to personalized and comfortable exchanges between patients and their care teams.

Central to the PCMH model is personalized care spaces, which are typically smaller and more intimate than those found in traditional hospital settings. These spaces are meticulously designed to foster close interactions between patients and their caregivers, enhancing the personal connection and understanding that are vital for effective care. The design often features rooms that feel less like clinical settings and more like welcoming, calming spaces. This approach not only helps in reducing patient anxiety but also promotes open communication, which is crucial for accurate diagnosis and effective treatment planning.

Privacy is another cornerstone of the PCMH model, necessitating the inclusion of private consultation areas within the architectural design. These areas provide a confidential space where patients can discuss their health concerns openly and without fear of being overheard. The emphasis on privacy helps in building trust between patients and healthcare providers, which is fundamental to the patient-centered approach.

The design of waiting areas in PCMH facilities also reflects the model's emphasis on patient comfort and stress reduction. Waiting rooms are equipped with comfortable seating, plenty of natural light, and amenities such as Wi-Fi, refreshments, and reading materials. These features make the waiting experience less stressful and more pleasant for patients and their families, contributing to a positive overall healthcare experience.

Recognizing the integral role of family in the healthcare process, PCMH facilities often include dedicated areas designed to facilitate family involvement. These spaces allow family members to actively participate in discussions and decision-making about the patient's care plan. Architectural considerations might include family-friendly waiting areas and consultation rooms that are spacious enough to accommodate multiple family members comfortably. Some facilities also offer amenities for extended stays, recognizing the supportive role families play in the care continuum.

Accessibility is a critical aspect of the PCMH model, ensuring that all patients, including those with disabilities, have easy access to medical care. Facilities are designed with features such as wheelchair-accessible entrances, corridors, rooms, lower countertops, and specialized bathroom facilities. The goal is to eliminate physical barriers to healthcare services, ensuring that every patient receives the care they need in an environment that respects their dignity and enhances their ability to receive care.

By integrating these architectural elements, facilities operating under the PCMH model can truly focus on patient-centered care. This environment not only promotes health but also enhances overall well-being by supporting the model's goals of continuous, coordinated care. It fosters the establishment of trusting relationships between patients and providers, which are essential for delivering effective, compassionate, and personalized healthcare.

### *5.6.4 Specialized Service Centers*

Specialized service centers, such as those dedicated to cancer treatment or pediatric healthcare, play a critical role in addressing specific medical needs through a combination of expertise and cutting-edge technology. The architectural design of these facilities is pivotal, not only in accommodating specialized treatment protocols but also in creating environments conducive to both patient recovery and professional healthcare delivery. One of the foremost considerations in the design of such centers is the integration of advanced medical technologies. For instance, cancer treatment facilities often require space specifically constructed to house complex imaging equipment like PET scanners and radiation therapy machines, necessitating the incorporation of radiation-shielding materials to ensure safety. Pediatric hospitals, in contrast, might focus on integrating technology that supports interactive treatment and monitoring, which can help in managing the unique needs of young patients.

Moreover, the architecture of specialized service centers extends beyond mere functionality to incorporate therapeutic elements designed to enhance patient well-being and expedite recovery. This includes the strategic use of calming colors, natural lighting, and views that connect patients with the outside environment. Pediatric centers, for example, often employ bright, vibrant colors and themes to create a friendly and less intimidating atmosphere for children. Cancer centers, on the other hand, might include tranquil spaces designed to offer patients privacy and peace during stressful treatment periods.

The design must also consider the needs of immunocompromised patients, common in settings like oncology centers where patients might be undergoing treatments such as chemotherapy. To protect these vulnerable groups, the centers are equipped with controlled environment features like HEPA-filtered air systems, antimicrobial surfaces, and layouts that minimize the risk of infection by reducing exposure to pathogens.

Another significant architectural consideration is the integration of clinical and research spaces, particularly in facilities like cancer treatment centers where the synergy between research and clinical applications is tight. By situating laboratories adjacent to or near treatment areas, these centers foster a seamless collaboration between researchers and clinicians, thereby facilitating the rapid translation of research breakthroughs into clinical practices and enhancing patient access to pioneering treatments.

Recognizing the essential role of family in the healthcare journey, specialized service centers often include facilities designed to support family members. These might encompass private waiting areas, accommodations for extended stays, and amenities like play areas for siblings of pediatric patients, as well as counseling and recreational facilities. Such provisions help alleviate the emotional burden on families, thereby contributing positively to the patient's care experience.

Lastly, given the complex nature of services offered and the variety of patients catered to, these centers prioritize clear and intuitive navigation within their premises. Effective wayfinding systems are crucial, involving straightforward signage and thoughtfully designed layouts that aid patients and their families in navigating the facilities effortlessly.

By addressing these comprehensive design considerations, specialized service centers can ensure they not only meet the intricate demands of treatment protocols but also significantly enhance the overall healthcare experience for patients and their families, thereby fulfilling their mission to deliver specialized, high-quality care in a supportive and efficient environment.

### 5.6.5 High-Efficiency Operational Model

The High-Efficiency Operational Model in healthcare facilities, particularly in ambulatory surgical centers, is crafted to optimize speed and efficiency, enhancing patient throughput without compromising the quality of care. This model is ideally suited for procedures that permit patients to be discharged on the same day, and the architectural design of these facilities is a critical component in achieving this efficiency.

Architecturally, the layout of these facilities is strategically designed to support streamlined workflows, facilitating logical progression from patient check-in, through various treatment stages, to discharge. This design minimizes unnecessary movement and physical barriers, which are crucial in enabling swift transitions at each phase of the patient's visit. Central to this layout is the arrangement of treatment rooms around core support areas, which significantly reduces the distance staff need to travel to access equipment and supplies, thus speeding up service delivery.

In terms of aesthetics, these facilities often embrace a minimalist design to expedite maintenance and cleaning, supporting rapid turnover rates. Surfaces are designed to be smooth and easy to clean, with minimal crevices where bacteria or dirt could accumulate, thereby reducing cleaning time and enhancing hygiene. The decor balances functionality with appeal, creating an environment that is both professional and calming, which is essential for efficient operation.

The patient flow within these centers is meticulously organized to enhance efficiency. The placement of registration, pre-operative areas, operating rooms, and recovery zones follows a natural sequence that minimizes backtracking and cross-traffic, streamlining the patient experience from entry to exit. This strategic setup not only accelerates the process but also minimizes the likelihood of errors and reduces patient waiting times, which is crucial for facilities that handle a high volume of patients.

Quick and unobstructed access to treatment areas is another priority in the high-efficiency model. Clear signage, well-organized rooms, and unimpeded hallways are designed to reduce the time it takes for a patient to move from entry to treatment. Similarly, discharge processes are optimized to ensure rapid transitions out of the facility. These areas are equipped with streamlined procedures for post-treatment reviews and instructions, facilitated by technology solutions like automated checkout systems that manage prescriptions, billing, and scheduling of follow-up appointments efficiently.

Technology plays a pivotal role in these operational models, with innovations such as automated patient check-in systems, electronic medical records accessible from multiple points, and real-time patient tracking systems to monitor progress through various care stages. These technological integrations are crucial in reducing manual tasks, decreasing the possibility of human error, and enhancing overall efficiency.

By focusing on these key architectural and operational strategies, facilities employing the high-efficiency model can significantly improve their capacity to handle a larger number of patients effectively. This approach not only elevates patient satisfaction by minimizing waiting times and reducing the length of stay but also optimizes the facility's overall capacity and resource utilization, ultimately enhancing the quality and efficiency of healthcare delivery.

### 5.6.6 Sustainable Model

The Sustainable Model in healthcare design emphasizes environmental responsibility while enhancing the well-being of both patients and staff, aligning with global sustainability goals and responding to the growing awareness among healthcare providers and patients about the benefits of eco-friendly facilities. This model integrates several architectural and operational strategies aimed at reducing environmental impact and promoting a healthy setting for recovery and work.

Key to this model is the use of green building materials. Facilities designed under the sustainable model prioritize environmentally friendly and resource-efficient construction materials. This includes the use of recycled materials, locally sourced supplies to minimize transportation emissions, and non-toxic finishes that improve indoor air quality. Such choices not only reduce the environmental footprint of the facilities but also promote a healthier environment for patients and staff, mitigating issues related to off gassing and poor air quality that can affect vulnerable populations.

Natural lighting and ventilation are also cornerstone elements of sustainable healthcare design. By maximizing the use of natural light through the strategic placement of windows and skylights, these facilities reduce reliance on artificial lighting, which decreases energy consumption and enhances the aesthetic and therapeutic quality of the environment. Additionally, natural ventilation systems help reduce the need for air conditioning, improving energy efficiency and creating a more comfortable and health-promoting indoor climate.

The implementation of energy-efficient systems is another hallmark of the sustainable model. Advanced HVAC systems designed to minimize energy use, the incorporation of LED lighting, and the utilization of smart-building technologies that adjust power consumption based on real-time occupancy and environmental conditions are all measures that not only reduce the facility's carbon footprint but also lead to significant operational cost savings over time.

Therapeutic green spaces within these facilities provide substantial health benefits. Gardens, green roofs, and indoor plantings not only serve as areas for relaxation and mental health relief but also enhance biodiversity and manage stormwater runoff. These spaces offer patients and staff a sanctuary for relaxation and stress relief, essential for overall well-being and mental health recovery, contributing to the sustainability of the facility's surroundings.

While sustainable design often involves higher initial investments, the long-term savings achieved through reduced energy costs, combined with the health benefits of a better environment, render this model financially viable. Moreover, sustainable facilities tend to experience benefits in staff retention and patient satisfaction, as these eco-conscious, visually appealing environments are attractive places to work and heal.

Many sustainable healthcare facilities aim for certifications like LEED (Leadership in Energy and Environmental Design), which not only boost their marketability but also ensure compliance with stringent global standards. Such certifications necessitate rigorous adherence to sustainability criteria, which encourages ongoing improvements in healthcare architectural and operational practices.

By adopting these sustainable practices, healthcare facilities not only significantly reduce their environmental impact but also foster environments that promote healing and well-being. This approach underscores the need for healthcare architectures that balance operational efficiency with environmental responsibility, ensuring that contemporary healthcare settings can meet current demands while being well-prepared for future challenges and trends in healthcare provision.

## 5.7 Enhancing Staff Well-Being and Productivity

The design of healthcare facilities profoundly impacts on the well-being and productivity of staff. Incorporating design elements that promote efficient workflows and reduce stress can significantly improve job satisfaction and the overall quality of care provided. This subsection explores various strategies to enhance operational efficiencies specifically aimed at supporting staff in healthcare environments.

### *5.7.1 Ergonomic Design*

Ergonomic design in healthcare facilities plays a crucial role in ensuring the well-being and efficiency of staff, especially in environments demanding prolonged and often physically taxing activities. By implementing proper ergonomic practices, healthcare facilities can significantly reduce the physical strain on staff, prevent injuries, and thereby enhance productivity while reducing absenteeism associated with work-related ailments.

One of the foundational elements of ergonomic design in healthcare settings is the provision of adjustable furniture. This includes adjustable-height desks, which allow staff members to alternate between sitting and standing positions. Such flexibility is instrumental in alleviating back pain and minimizing the risk of chronic musculoskeletal disorders. Additionally, ergonomic chairs that support proper posture are vital, particularly for roles that require extensive sitting, such as administrative tasks or monitoring stations. These chairs should offer adjustable features such as armrests, seat depth, and lumbar support to cater to the individual physical needs of healthcare staff.

The strategic positioning of workstations is another critical aspect of ergonomic design. It is essential to set up workstations so that monitors are at eye level and keyboards and other tools are within easy reach to prevent awkward postures. Proper workstation setup helps ensure that staff do not need to stretch excessively or twist their bodies to access necessary equipment, significantly reducing the risk of strains and injuries.

In healthcare, the diversity of tasks requires equipment to not only be ergonomically designed but also customizable to different users and situations. For instance, mobile workstations on wheels (WOWs) used by nurses need to be easily adjustable and lightweight to maneuver without significant physical effort. Similarly, handheld devices used for recording patient data should be designed to minimize strain and enhance ease of use.

Specific ergonomic considerations are also necessary for different hospital areas, depending on the tasks performed. For example, surgical theaters require ergonomically designed surgical tools and adjustable operating tables to help surgeons maintain optimal posture during lengthy procedures. In laboratory settings, considerations might include the appropriate height for microscopes and the use of anti-fatigue matting where staff may need to stand for extended periods.

The effectiveness of ergonomic design is further reinforced through proper training and awareness programs for staff. These training sessions are crucial in educating staff about the importance of ergonomics, demonstrating how to adjust their workstations correctly, and sharing best practices for minimizing physical strain during their shifts.

Lastly, a successful ergonomic initiative should include continuous assessment and feedback mechanisms. Regular evaluations of ergonomic solutions and active solicitation of staff feedback are essential to identify areas needing improvement and ensure that the ergonomic designs effectively meet the workforce's needs.

By incorporating these comprehensive ergonomic design principles, healthcare facilities can create safer, more comfortable, and supportive work environments. This commitment not only leads to a more satisfied and healthier workforce but also elevates the overall quality of care provided to patients, showcasing the profound impact of ergonomics in healthcare settings.

### 5.7.2 *Efficient Workflow Layouts*

Efficient workflow layouts are fundamental in healthcare settings, playing a pivotal role in enhancing the speed and quality of patient care as well as improving staff satisfaction. Optimizing the physical layout of a hospital can significantly streamline staff movements, reduce fatigue, and minimize the time dedicated to non-clinical tasks, thereby fostering a more effective and responsive healthcare environment.

One of the primary strategies in developing efficient workflow layouts is the creation of clear, unobstructed pathways throughout the hospital. This design principle ensures that staff can move quickly and safely between different areas, which is especially crucial in emergencies. By minimizing obstacles and designing corridors that can accommodate high traffic volumes without congestion, hospitals can substantially decrease the time it takes for staff to reach their destinations, enhancing the overall speed and efficiency of care delivery.

Another critical aspect is the separation of different traffic flows within the hospital to streamline operations and reduce disruptions. By designing separate pathways for public and service operations, hospitals can minimize noise and disturbances around critical care areas. For instance, service corridors might be utilized solely by staff for logistical operations, such as transporting equipment, meals, or laundry, thereby keeping these activities out of patient and public view and reducing interruptions to patient care.

Strategic placement of essential service areas like nursing stations, supply rooms, and other critical resources is also vital. For example, centralizing nursing stations within patient wards can significantly enhance nurses' ability to monitor and access their patients swiftly. Additionally, positioning supply rooms close to points of care can reduce the time staff spend retrieving necessary supplies, thereby increasing the time they can dedicate to patient care.

Visibility and accessibility in the design of work areas are paramount for maintaining effective patient monitoring. Implementing design features such as glass walls or strategically placed observation windows allows staff to oversee patient areas without the need to remain at the bedside continuously. Such considerations can lead to faster response times and heightened patient safety, especially in critical care settings.

The incorporation of modular and adaptable design elements also contributes to workflow efficiency. Facilities that include flexible room configurations, movable workstations, and adjustable storage solutions can adapt more readily to changing needs and workflows. This flexibility is particularly beneficial in healthcare settings where the demand for different services can fluctuate.

Finally, the integration of technology into the workflow layout is essential for modern healthcare facilities. Utilizing electronic health record systems accessible from multiple locations within a unit, centralized control rooms for monitoring patient vitals, and mobile devices for data entry on the go can all streamline processes and reduce redundant movements.

By adhering to these design principles, hospitals can create environments that not only promote efficient workflows and reduce unnecessary staff movement but also improve the overall quality of care provided to patients and the working conditions for staff. This holistic approach to layout planning supports both the current operational needs and the adaptability required to meet future changes in healthcare delivery.

### *5.7.3 Restorative Spaces for Staff*

Creating restorative spaces for staff in healthcare settings is crucial for sustaining their well-being and enhancing overall job performance. These designated areas provide sanctuaries where staff can physically and mentally recharge, offering a necessary reprieve from the rigorous demands of patient care. Thoughtfully designing these spaces can yield significant improvements in staff satisfaction, retention, and productivity, contributing to a more harmonious work environment.

To be effective, restorative spaces must be distinctly separated from patient care zones. Locating these areas away from the clinical environment helps to provide a genuine break, reinforcing a mental and physical separation from work-related stress. This separation allows staff to relax and decompress themselves in an environment devoid of professional reminders, fostering a true sense of respite and recuperation.

The ambiance of restorative spaces is another critical factor. These areas should be designed to cultivate a tranquil environment, utilizing soothing colors, comfortable furniture, and soft lighting to create a peaceful and calming atmosphere. Providing a variety of seating options, including loungers, bean bags, and quiet booths—can accommodate different relaxation preferences among staff, ensuring that everyone can find a space that suits their needs for unwinding.

Incorporating access to natural light and, where feasible, views of the outdoors are highly beneficial. Natural light has been shown to boost mood, enhance alertness, and assist in regulating circadian rhythms, which is particularly advantageous for staff working long or irregular hours. Additionally, if the facility's location and space permit, access to outdoor areas such as gardens or terraces can further amplify the restorative quality of these spaces, offering staff a connection to the outside world and a place to breathe fresh air during breaks.

The inclusion of natural elements within restorative spaces can significantly elevate their calming effect. Indoor plants, small water features such as fountains or aquariums, and the use of natural materials like wood and stone not only enhance the aesthetic appeal but also improve indoor air quality and provide a tactile and visual connection to nature. These elements serve to create a more organic and serene setting, ideal for relaxation and stress relief.

Moreover, designing restorative spaces with zones dedicated to quiet relaxation and separate areas for entertainment and social interaction can address varying needs among staff. Quiet zones equipped with comfortable chairs are perfect for reading, meditation, or napping, while entertainment areas might feature amenities such as televisions, games, or spaces for informal social gatherings, catering to those looking for active relaxation or social interaction during their breaks.

Allowing staff to personalize these spaces or contribute to their design can further enhance their utility and appeal. Personal touches such as artwork, photographs, or a selection of books chosen by staff can make these areas feel more inviting and personalized, encouraging regular use.

By prioritizing and investing in the creation of restorative spaces, healthcare organizations can demonstrate a profound commitment to staff welfare. Such spaces acknowledge the inherently challenging nature of healthcare work and play a vital role in supporting the physical and mental health of staff, ultimately leading to a healthier, more productive workplace environment.

### 5.7.4 Access to Technology

Access to advanced technology is a critical factor in the efficiency and effectiveness of modern healthcare facilities. The implementation of state-of-the-art technological solutions in hospitals not only enhances patient care but also boosts the job satisfaction and productivity of healthcare professionals. Seamless integration of hospital systems is a foundational aspect, allowing for the effective connection of electronic health records (EHRs), patient management systems, and diagnostic tools. Such integration ensures that patient information is readily accessible across various departments, significantly reducing the time needed to retrieve data from multiple sources and minimizing the risks associated with data inconsistencies.

Moreover, equipping healthcare staff with mobile technology solutions, such as tablets and smartphones connected to the hospital's central systems, empowers them with the ability to access and update patient data in real-time from any location within the facility. This capability is particularly crucial for medical professionals who require immediate access to patient records to make informed decisions quickly. Additionally, mobile devices facilitate telemedicine consultations, enabling specialists to offer their expertise remotely, thus optimizing the efficiency of healthcare delivery and expanding the reach of specialized medical care.

It is also essential that the technology adopted by healthcare institutions features user-friendly interfaces that are straightforward to navigate, even in high-pressure environments typical of healthcare settings. Simplified interfaces reduce the required training time and accelerate the adoption rate among staff, allowing them to concentrate more on patient care rather than technology management. In addressing the challenges of data security, hospitals must implement robust security measures to protect against cyber threats. This includes encrypted data storage, secure login procedures, and regular software updates. Adherence to stringent compliance regulations like HIPAA in the United States is also critical to ensure the privacy of patient data and maintain trust in healthcare services.

Furthermore, technology in healthcare should enhance communication and collaboration among medical staff. Secure messaging apps, integrated communication systems, and collaborative care planning platforms are essential in maintaining a cohesive care team. Such tools ensure that everyone, from doctors to nurses to specialists, is well-informed of the patient's care progress, fostering a unified approach to healthcare delivery.

Lastly, the success of technology implementation is heavily dependent on the availability of continuous training and robust support systems. Ongoing education and support for healthcare technologies are imperative to build confidence among staff, ensuring they are competent in utilizing these tools effectively. This comprehensive approach not only facilitates superior patient care but also nurtures a culture of innovation and ongoing enhancement in healthcare practices, leading to broader improvements in patient outcomes and staff morale.

### 5.7.5 Noise Reduction Measures

Noise reduction is an essential element in hospital settings, aiming to create a serene environment conducive to the recovery of patients and the optimal functioning of staff. Excessive noise not only contributes to stress and fatigue among healthcare providers but also impedes effective communication, which is critical in medical settings. Implementing strategic noise reduction measures can significantly enhance the quality of care and the overall hospital atmosphere.

A fundamental method for controlling noise is the integration of soundproofing materials within the hospital infrastructure. Utilizing acoustic panels, specialized insulation, and double-paned glass in walls, floors, and ceilings effectively dampens external noises and those from different sections of the hospital. This approach not only reduces the transmission of noise but also contributes to a more controlled acoustic environment.

Additionally, incorporating soft-closing mechanisms on doors, drawers, and cabinets can drastically lower ambient noise levels. These features are particularly beneficial during nighttime or in areas meant for patient rest, preventing disruptive slamming sounds that can startle and disturb patients and staff alike. Designing quiet zones within the hospital is another proactive strategy. These areas provide a sanctuary from the bustling activity typical of hospital operations, offering both patients and staff a space to retreat to a quieter, more peaceful environment. Strategically locating these zones away from high-traffic areas and reinforcing them with extra soundproofing measures ensures their effectiveness.

The layout of the hospital also plays a crucial role in noise management. Thoughtfully designing the hospital layout to place inherently noisy departments—such as emergency rooms and areas for children's activities—away from zones requiring quiet, like intensive care units and patient recovery areas, helps prevent noise crossover. Introducing buffer zones such as storage and administrative spaces between noisy and quiet areas further aids in mitigating noise propagation.

Operational protocols that are conscious of noise levels are equally important. Implementing policies that reduce the need for overhead paging, encouraging the use of low-volume personal communication devices, and fostering a culture of quietness among staff in sensitive areas are all effective strategies. Moreover, regular maintenance of medical and mechanical equipment ensures they operate quietly and efficiently, preventing noise disturbances due to malfunctions or wear and tears.

By adopting these comprehensive noise reduction measures, hospitals can significantly improve the working conditions for their staff, boosting focus and productivity, while also cultivating a tranquil and healing atmosphere for patients. This holistic approach to noise management underscores the commitment to enhancing both the efficacy of medical care and the overall experience within healthcare facilities.

### 5.7.6 Visual Connectivity

Visual connectivity in hospital design is pivotal in boosting operational efficiency, facilitating staff collaboration, and enhancing patient safety. Through carefully planned architectural and interior design elements that promote clear lines of sight in patient care areas, hospitals can create environments that enable seamless communication and constant, yet unobtrusive, supervision.

Incorporating strategic architectural features such as glass walls and large windows in essential areas like nurse stations or control centers is critical. This design choice allows healthcare staff to maintain visual contact with patients while managing other duties, facilitating continuous monitoring without constant physical presence. Such configurations not only alleviate the physical demands on staff but also improve response times to patient needs, enhancing the overall efficacy of the healthcare delivery process.

The hospital layout also significantly influences visual connectivity. Open floor plans with minimal visual obstructions ensure that patient beds are aligned with the nurse's station's line of sight or that consultation rooms are arranged around a central observation point. This setup allows staff to oversee multiple areas simultaneously, fostering efficient monitoring and a sense of openness and accessibility throughout the facility.

Additionally, the strategic use of reflective surfaces and adequate lighting can extend visual connectivity to areas where direct lines of sight are compromised. Mirrors can be strategically placed to reflect views into otherwise obscured sections of a hospital, while proper lighting ensures that these areas are free of shadows and dim lighting, which could hinder visibility.

Interactive communication points, such as digital screens synchronized with the hospital's management systems, also play a crucial role. Placed in locations visible to staff moving through interconnected areas, these screens provide real-time updates on patient status or urgent alerts, enhancing communication and response efficiency.

For units requiring intensive monitoring, such as ICUs or emergency departments, designated observation areas equipped with technology to monitor vital signs and video feeds allow comprehensive patient oversight without the need to enter rooms. This setup not only bolsters patient safety but also supports stringent infection control protocols.

Furthermore, integrating visual connectivity with the external environment by incorporating views of nature through windows or internal gardens can significantly impact staff morale and patient recovery. Exposure to natural views has been associated with decreased stress levels and enhanced mental well-being, benefiting both patients and healthcare providers.

By emphasizing visual connectivity in the design of healthcare facilities, hospitals can craft spaces that optimize healthcare logistics while also supporting the well-being and efficiency of their staff. This holistic approach not only streamlines operational processes but also indirectly raises the quality of patient care, showcasing the profound impact of thoughtful design in healthcare settings.

# Further Reading

Ahsan, M. M. (2020). Strategic decisions on urban built environment to pandemics in Turkey: Lessons from COVID-19. *Journal of Urban Management*, 9(3), 281–285. https://doi.org/10.1016/j.jum.2020.07.001

Allen, J., Hutchinson, A. M., Brown, R., & Livingston, P. M. (2018). User experience and care for older people transitioning from hospital to home: Patients' and carers' perspectives. *Health Expectations*, 21(2), 518–527. https://doi.org/10.1111/hex.12646

Almansour, I., Seymour, J. E., & Aubeeluck, A. (2019). Staff perception of obstacles and facilitators when providing end of life care in critical care units of two teaching hospitals: A survey design. *Intensive and Critical Care Nursing*, 53, 8–14. https://doi.org/10.1016/j.iccn.2019.04.003

Anthony Jnr, B. (2022). An exploratory study on academic staff perception towards blended learning in higher education. *Education and Information Technologies*, 27(3), 3107–3133. https://doi.org/10.1007/s10639-021-10705-x

Bates, D. W., & Singh, H. (2018). Two decades since to err is human: an assessment of progress and emerging priorities in patient safety. *Health Affairs*, 37(11), 1736–1743. https://doi.org/10.1377/hlthaff.2018.0738

Bianchi, M., Bagnasco, A., Bressan, V., Barisone, M., Timmins, F., Rossi, S., ... & Sasso, L. (2018). A review of the role of nurse leadership in promoting and sustaining evidence-based practice. *Journal of Nursing Management*, 26(8), 918-932. https://doi.org/10.1111/jonm.12638

Birken, S., Clary, A., Tabriz, A. A., Turner, K., Meza, R., Zizzi, A., ... & Charns, M. (2018). Middle managers' role in implementing evidence-based practices in healthcare: a systematic review. *Implementation Science*, 13, 1-14. https://doi.org/10.1186/s13012-018-0843-5

Bogaert, B. (2022). Moving toward person-centered care: Valuing emotions in hospital design and architecture. *HERD: Health Environments Research & Design Journal*, 15(2), 355–364. https://doi.org/10.1177/19375867211062101

Bomhof-Roordink, H., Gärtner, F. R., Stiggelbout, A. M., & Pieterse, A. H. (2019). Key components of shared decision making models: a systematic review. *BMJ open*, 9(12), e031763. https://doi.org/10.1136/bmjopen-2019-031763

Brambilla, A., Rebecchi, A., & Capolongo, S. (2019). Evidence based hospital design. A literature review of the recent publications about the EBD impact of built environment on hospital occupants' and organizational outcomes. *Annali di Igiene Medicina Preventiva e di Comunita*, 31(2), 165–180. https://hdl.handle.net/11311/1074776

Britto, M. T., Fuller, S. C., Kaplan, H. C., Kotagal, U., Lannon, C., Margolis, P. A., ... & Seid, M. (2018). Using a network organisational architecture to support the development of Learning Healthcare Systems. *BMJ quality & safety*, 27(11), 937–946. https://doi.org/10.1136/bmjqs-2017-007219

Bruns, E. J., Parker, E. M., Hensley, S., Pullmann, M. D., Benjamin, P. H., Lyon, A. R., & Hoagwood, K. E. (2019). The role of the outer setting in implementation: associations between state demographic, fiscal, and policy factors and use of evidence-based treatments in mental healthcare. *Implementation Science*, 14, 1–13. https://doi.org/10.1186/s13012-019-0944-9

Capolongo, S., Gola, M., Brambilla, A., Morganti, A., Mosca, E. I., & Barach, P. (2020). COVID-19 and healthcare facilities: a decalogue of design strategies for resilient hospitals. *Acta Bio Medica: Atenei Parmensis*, 91(9-S), 50. https://doi.org/10.23750/Fabm.v91i9-S.10117

Carmona, M. (2019). Place value: Place quality and its impact on health, social, economic and environmental outcomes. *Journal of urban design*, 24(1), 1–48. https://doi.org/10.1080/13574809.2018.1472523

Cignitas, C. P., Torrents Arévalo, J. A., & Vilajosana Crusells, J. (2022). Positive management and the balanced scorecard: A successful strategy for organizations. *Journal of Positive School Psychology*, 6(3), 2606–2627. http://hdl.handle.net/2117/367242

Clarke, D. J., Burton, L. J., Tyson, S. F., Rodgers, H., Drummond, A., Palmer, R., ... & Forster, A. (2018). Why do stroke survivors not receive recommended amounts of active therapy? Findings from the ReAcT study, a mixed-methods case-study evaluation in eight stroke units. *Clinical rehabilitation*, 32(8), 1119-1132. https://doi.org/10.1177/0269215518765329

Colenberg, S., Jylhä, T., & Arkesteijn, M. (2021). The relationship between interior office space and employee health and well-being–a literature review. *Building Research & Information*, 49(3), 352-366. https://doi.org/10.1080/09613218.2019.1710098

Dietz, L., Horve, P. F., Coil, D. A., Fretz, M., Eisen, J. A., & Van Den Wymelenberg, K. (2020). 2019 novel coronavirus (COVID-19) pandemic: built environment considerations to reduce transmission. *Msystems*, 5(2), 10–1128. https://doi.org/10.1128/msystems.00245-20

Dong, B., Prakash, V., Feng, F., & O'Neill, Z. (2019). A review of smart building sensing system for better indoor environment control. *Energy and Buildings*, 199, 29-46. https://doi.org/10.1016/j.enbuild.2019.06.025

DuBose, J., MacAllister, L., Hadi, K., & Sakallaris, B. (2018). Exploring the concept of healing spaces. *HERD: Health Environments Research & Design Journal*, 11(1), 43–56. https://doi.org/10.1177/1937586716680567

Escoffery, C., Lebow-Skelley, E., Haardoerfer, R., Boing, E., Udelson, H., Wood, R., ... & Mullen, P. D. (2018). A systematic review of adaptations of evidence-based public health interventions globally. *Implementation Science*, 13, 1–21. https://doi.org/10.1186/s13012-018-0815-9

Escoffery, C., Lebow-Skelley, E., Udelson, H., Böing, E. A., Wood, R., Fernandez, M. E., & Mullen, P. D. (2019). A scoping study of frameworks for adapting public health evidence-based interventions. *Translational behavioral medicine*, 9(1), 1–10. https://doi.org/10.1093/tbm/ibx067

Farokhzadian, J., Dehghan Nayeri, N., & Borhani, F. (2018). The long way ahead to achieve an effective patient safety culture: challenges perceived by nurses. *BMC health services research*, 18, 1–13. https://doi.org/10.1186/s12913-018-3467-1

Fatima, T., Malik, S. A., & Shabbir, A. (2018). Hospital healthcare service quality, patient satisfaction and loyalty: An investigation in context of private healthcare systems. *International journal of quality & reliability management*, 35(6), 1195-1214. https://doi.org/10.1108/IJQRM-02-2017-0031

Fischer, M., Safaeinili, N., Haverfield, M. C., Brown-Johnson, C. G., Zionts, D., & Zulman, D. M. (2021). Approach to human-centered, evidence-driven adaptive design (AHEAD) for health care interventions: a proposed framework. *Journal of General Internal Medicine*, 36, 1041–1048. https://doi.org/10.1007/s11606-020-06451-4

Gaminiesfahani, H., Lozanovska, M., & Tucker, R. (2020). A scoping review of the impact on children of the built environment design characteristics of healing spaces. *HERD: Health Environments Research & Design Journal*, 13(4), 98–114. https://doi.org/10.1177/193758672090 3845

Geerligs, L., Rankin, N. M., Shepherd, H. L., & Butow, P. (2018). Hospital-based interventions: a systematic review of staff-reported barriers and facilitators to implementation processes. *Implementation Science*, 13, 1–17. https://doi.org/10.1186/s13012-018-0726-9

Graban, M. (2018). Lean hospitals: improving quality, patient safety, and employee engagement. Productivity Press.

Halawa, F., Madathil, S. C., Gittler, A., & Khasawneh, M. T. (2020). Advancing evidence-based healthcare facility design: a systematic literature review. *Health Care Management Science*, 23, 453–480. https://doi.org/10.1007/s10729-020-09506-4

Henrique, D. B., & Godinho Filho, M. (2020). A systematic literature review of empirical research in Lean and Six Sigma in healthcare. *Total Quality Management & Business Excellence*, 31(3-4), 429-449. https://doi.org/10.1080/14783363.2018.1429259

Hick, J. L., Hanfling, D., Wynia, M. K., & Pavia, A. T. (2020). Duty to plan: health care, crisis standards of care, and novel coronavirus SARS-CoV-2. *Nam Perspectives*, 2020. https://doi.org/10.31478/F202003b

Hill, C., Knafl, K. A., & Santacroce, S. J. (2018). Family-centered care from the perspective of parents of children cared for in a pediatric intensive care unit: an integrative review. *Journal of pediatric nursing*, 41, 22–33. https://doi.org/10.1016/j.pedn.2017.11.007

Holeman, I., & Kane, D. (2020). Human-centered design for global health equity. *Information technology for development*, 26(3), 477–505. https://doi.org/10.1080/02681102.2019.1667289

Isokääntä, S., Koivula, K., Honkalampi, K., & Kokki, H. (2019). Resilience in children and their parents enduring pediatric medical traumatic stress. *Pediatric Anesthesia*, 29(3), 218–225. https://doi.org/10.1111/pan.13573

Jain, N., Burman, E., Stamp, S., Shrubsole, C., Bunn, Rn., Oberman, T., ... & Davies, M. (2021). Building performance evaluation of a new hospital buildig in the UK: Balancing indoor environmental quality and energy performance. *Atmosphere*, 12(1), 115. https://doi.org/10.3390/atmos12010115

Jewkes, Y. (2018). Just design: Healthy prisons and the architecture of hope. *Australian & New Zealand Journal of Criminology*, 51(3), 319-338. https://doi.org/10.1177/0004865818766768

Kärmeniemi, M., Lankila, T., Ikäheimo, T., Koivumaa-Honkanen, H., & Korpelainen, R. (2018). The built environment as a determinant of physical activity: a systematic review of longitudinal studies and natural experiments. *Annals of behavioral medicine*, 52(3), 239–251. https://doi.org/10.1093/abm/kax043

Kelly, J. T., Campbell, K. L., Gong, E., & Scuffham, P. (2020). The Internet of Things: Impact and implications for health care delivery. *Journal of medical Internet research*, 22(11), e20135. https://doi.org/10.2196/20135

Lawati, M. H. A., Dennis, S., Short, S. D., & Abdulhadi, N. N. (2018). Patient safety and safety culture in primary health care: a systematic review. *BMC family practice*, 19, 1–12. https://doi.org/10.1186/s12875-018-0793-7

Lee, D. (2019). A model for designing healthcare service based on the patient experience. *International Journal of Healthcare Management*, 12(3), 180-188. https://doi.org/10.1080/20479700.2017.1359956

Li, J., & Carayon, P. (2021). Health Care 4.0: A vision for smart and connected health care. *IISE Transactions on Healthcare Systems Engineering*, 11(3), 171–180. https://doi.org/10.1080/24725579.2021.1884627

Liu, X., Zheng, J., Liu, K., Baggs, J. G., Liu, J., Wu, Y., & You, L. (2018). Hospital nursing organizational factors, nursing care left undone, and nurse burnout as predictors of patient safety: A structural equation modeling analysis. *International journal of nursing studies*, 86, 82–89. https://doi.org/10.1016/j.ijnurstu.2018.05.005

Manzoor, F., Wei, L., Hussain, A., Asif, M., & Shah, S. I. A. (2019). Patient satisfaction with health care services; an application of physician's behavior as a moderator. *International journal of environmental research and public health*, 16(18), 3318. https://doi.org/10.3390/ijerph16183318

Marques, B., Freeman, C., & Carter, L. (2021). Adapting traditional healing values and beliefs into therapeutic cultural environments for health and well-being. *International journal of environmental research and public health*, 19(1), 426. https://doi.org/10.3390/ijerph19010426

Moore, T. H., Kesten, J. M., López-López, J. A., Ijaz, S., McAleenan, A., Richards, A., ... & Audrey, S. (2018). The effects of changes to the built environment on the mental health and well-being of adults: Systematic review. *Health & place*, 53, 237-257. https://doi.org/10.1016/j.healthplace.2018.07.012

Mouratidis, K. (2021). Urban planning and quality of life: A review of pathways linking the built environment to subjective well-being. *Cities*, 115, 103229. https://doi.org/10.1016/j.cities.2021.103229

Mujan, I., Anđelković, A. S., Munćan, V., Kljajić, M., & Ružić, D. (2019). Influence of indoor environmental quality on human health and productivity-A review. *Journal of cleaner production*, 217, 646–657. https://doi.org/10.1016/j.jclepro.2019.01.307

Nagasawa, Y. (2020). Global Hospitals in 2050—a review of the historical development of hospital building studies from a global perspective. *Japan Architectural Review*, 3(1), 5–24. https://doi.org/10.1002/2475-8876.12131

Oh, D. L., Jerman, P., Silvério Marques, S., Koita, K., Purewal Boparai, S. K., Burke Harris, N., & Bucci, M. (2018). Systematic review of pediatric health outcomes associated with childhood adversity. *BMC pediatrics*, 18(1), 1–19. https://doi.org/10.1186/s12887-018-1037-7

Oben, P. (2020). Understanding the patient experience: a conceptual framework. *Journal of patient experience*, 7(6), 906–910. https://doi.org/10.1177/2374373520951672

Pieterse, A. H., Stiggelbout, A. M., & Montori, V. M. (2019). Shared decision making and the importance of time. *Jama*, 322(1), 25–26. https://doi.org/10.1001/jama.2019.3785

Quesado, P. R., Aibar Guzmán, B., & Lima Rodrigues, L. (2018). Advantages and contributions in the balanced scorecard implementation. *Intangible capital*, 14(1), 186-201. https://doi.org/10.3926/ic.1110

Romito, B., Jewell, J., Jackson, M., Ernst, K., Hill, V., Hsu, B., ... & Vinocur, C. (2021). Child life services. *Pediatrics*, 147(1). https://doi.org/10.1542/peds.2020-040261

Sadek, A. H., & Willis, J. (2020). Are we measuring what we ought to measure? A review of tools assessing patient perception of the healthcare built environment and their suitability for oncology spaces. *Journal of Environmental Psychology*, 71, 101486. https://doi.org/10.1016/j.jenvp.2020.101486

Scholl, I., LaRussa, A., Hahlweg, P., Kobrin, S., & Elwyn, G. (2018). Organizational-and system-level characteristics that influence implementation of shared decision-making and strategies to address them—a scoping review. *Implementation Science*, 13, 1–22. https://doi.org/10.1186/s13012-018-0731-z

Shannon, M. M., Nordin, S., Bernhardt, J., & Elf, M. (2020). Application of theory in studies of healthcare built environment research. *HERD: Health Environments Research & Design Journal*, 13(3), 154–170. https://doi.org/10.1177/1937586719901108

Shelton, R. C., Cooper, B. R., & Stirman, S. W. (2018). The sustainability of evidence-based interventions and practices in public health and health care. *Annual review of public health*, 39(1), 55–6. https://doi.org/10.1146/annurev-publhealth-040617-014731

Sturm, H., Rieger, M. A., Martus, P., Ueding, E., Wagner, A., Holderried, M., … & WorkSafeMed Consortium, 2019. Sturm, H., Rieger, M. A., Martus, P., Ueding, E., Wagner, A., Holderried, M., ... & WorkSafeMed Consortium. (2019). Do perceived working conditions and patient safety culture correlate with objective workload and patient outcomes: A cross-sectional explorative study from a German university hospital. *PloS one*, 14(1), e0209487. https://doi.org/10.1371/journal.pone.0209487

Sufian, S., Blackie, M., Michel, J., & Garden, R. (2020). Centering patients, revealing structures: the health humanities portrait approach. *Journal of Medical Humanities*, 41, 459-479. https://doi.org/10.1007/s10912-020-09640-8

Uslu, B. Ç., Okay, E., & Dursun, E. (2020). Analysis of factors affecting IoT-based smart hospital design. *Journal of Cloud Computing*, 9(1), 67. https://doi.org/10.1186/s13677-020-00215-5

Wadsworth, K. H., Archibald, T. G., Payne, A. E., Cleary, A. K., Haney, B. L., & Hoverman, A. S. (2019). Shared medical appointments and patient-centered experience: a mixed-methods systematic review. *BMC Family Practice*, 20, 1–13. https://doi.org/10.1186/s12875-019-0972-1

Watts, P. I., McDermott, D. S., Alinier, G., Charnetski, M., Ludlow, J., Horsley, E., ... & Nawathe, P. A. (2021). Healthcare simulation standards of best practiceTM simulation design. *Clinical Simulation in Nursing*, 58, 14–21. https://doi.org/10.1016/j.ecns.2021.08.009

Zhang, Z., Zou, Y., & Xiao, W. (2023). Exploration of a virtual restoration practice route for architectural heritage based on evidence-based design: a case study of the Bagong House. *Heritage Science*, 11(1), 35. https://doi.org/10.1186/s40494-023-00878-8

# Chapter 6
# Revolutionizing Healthcare Architecture Through Digital Innovation: The Integration of BIM, Real-Time Monitoring, and Predictive Analytics

This chapter delves deeply into the transformative impact of digital technologies on healthcare architecture. It highlights how tools like Building Information Modeling (BIM) revolutionize the planning and design processes, enhancing accuracy and operational efficiency from construction through long-term management. The chapter further explores real-time monitoring and responsive environments that adapt dynamically to enhance patient comfort and facility efficiency. A significant focus is placed on personalizing patient spaces with technologies that allow environmental and entertainment customization, contributing to improved patient experiences. Additionally, the integration of wearable technologies is examined for its ability to streamline patient data integration and enhance clinical decision-making. The chapter also addresses the critical need for enhanced security measures to protect sensitive data within these advanced digital frameworks. Lastly, it emphasizes the role of data analytics in optimizing facility design and operations, underscoring the potential for predictive analytics to refine energy management and patient care pathways. This comprehensive analysis not only underscores the current integration of digital advancements in healthcare facilities but also paves the way for future innovations that promise to further enhance the efficacy and patient-centric focus of healthcare environments.

## 6.1 The Impact of Digital Technologies on Healthcare Architecture

Digital technologies are profoundly transforming healthcare architecture, pushing the boundaries of what is possible in the design and functionality of medical facilities. The integration of these technologies affects nearly every aspect of healthcare delivery, from patient diagnostics and treatment to facility management and operations.

 
A. A. Firoozi and A. A. Firoozi, *Transformative Design in Healthcare*,
https://doi.org/10.1007/978-3-032-08019-6_6

### 6.1.1 Enhanced Design and Planning Tools

The integration of digital technology into healthcare architecture, notably through Building Information Modeling (BIM) and other advanced software tools, has fundamentally transformed the design and construction of medical facilities. BIM represents a revolutionary shift in architectural design, especially within the intricate domain of healthcare architecture. This technology facilitates the creation of digital models that encapsulate both the physical and functional characteristics of a facility. Unlike traditional 2D drafting, BIM incorporates 3D modeling and extends it with dimensions like time management, cost estimation, and operational specifics. For healthcare facilities, this means that all involved—from architects to engineers—can visualize and iterate on the building's lifecycle, from initial conception through to maintenance, long before ground is broken.

The precision and detail afforded by BIM and similar tools are unparalleled. These technologies allow architects to craft highly accurate and detailed models that present stakeholders with a three-dimensional view of the proposed space, significantly enhancing the understanding of spatial dynamics and interactions. Such detailed visualization is particularly crucial in healthcare settings where the layout directly influences functionality and patient flow—factors that are vital for efficient and effective medical care.

Moreover, these advanced tools enable the simulation of various building systems which is invaluable in a healthcare context. For instance, using BIM to model airflows can ensure that ventilation systems in surgical suites or isolation rooms are optimally designed to mitigate infection risks. Lighting simulations can also be employed to create environments that support the well-being of patients and the operational efficiency of medical staff, aligning with healthcare standards and best practices.

Efficiency in managing traffic flow and space utilization is another significant benefit of these tools. In the context of a hospital, where the smooth movement of patients, staff, and materials is crucial, the ability to analyze and optimize foot traffic and layouts can prevent bottlenecks. This optimization is critical not only for day-to-day operations but also in emergency scenarios that require clear and unobstructed pathways.

BIM and similar digital tools also facilitate enhanced collaboration among all parties involved in the creation of a healthcare facility. Through a shared digital environment, architects, engineers, medical professionals, and other stakeholders can collaboratively refine the facility's design during the planning phase, ensuring that the final construction meets all necessary criteria and functional requirements comprehensively.

Lastly, the economic and temporal efficiencies brought by BIM cannot be overstated. Early detection of design conflicts allows for adjustments before construction, substantially reducing the necessity for costly modifications during the building phase. Moreover, BIM can expedite the regulatory approval process by providing clear, detailed, and comprehensible representations of the facility, thereby streamlining critical administrative procedures.

**Table 6.1** Comparison of traditional versus BIM-enhanced design processes

| Design process | Time to completion | Cost efficiency | Error reduction (%) |
|---|---|---|---|
| Traditional | 24 months | $10M | 5 |
| BIM-enhanced | 18 months | $8M | 25 |

In conclusion, the advent of enhanced design and planning tools like BIM is revolutionizing healthcare architecture. These tools not only elevate the accuracy of designs but also enhance the overall functionality of medical facilities. They optimize both the construction process and the ongoing management of the facility, thereby delivering long-term benefits to healthcare providers and patients alike.

Table 6.1 provides a comparative analysis of traditional versus Building Information Modeling (BIM)-enhanced design processes in healthcare construction. This table lists key metrics such as time to completion, cost efficiency, and error reduction, drawing on the latest research and case studies to demonstrate the transformative impact of BIM on architectural practices. By highlighting these metrics, the table showcases how BIM not only streamlines the design process but also leads to significant improvements in accuracy, efficiency, and overall project outcomes, making it an indispensable tool in modern healthcare facility design.

### *6.1.2 Real-Time Monitoring and Responsive Environments*

The advent of digital technologies in healthcare architecture has catalyzed the development of responsive environments that adapt dynamically to both user needs and environmental conditions. These smart environments, equipped with an array of sensors and automated systems, enable real-time monitoring and adjustments, thereby enhancing patient comfort, operational efficiency, and overall sustainability.

Healthcare facilities are increasingly being outfitted with diverse sensors that monitor environmental conditions, occupancy, and usage patterns. These sensors are foundational to creating environments that can automatically respond to changes without human intervention. For instance, occupancy sensors can detect whether a room is in use and adjust lighting and HVAC (Heating, Ventilation, and Air Conditioning) systems accordingly to save energy. This capability is crucial in managing the energy consumption of large medical facilities and reduces unnecessary utility use when spaces are unoccupied.

Automated systems in these environments take the data collected by sensors and use it to adjust settings such as lighting, temperature, air quality, and window shading to maintain optimal conditions. This level of automation ensures that the environment for both patients and staff are consistently comfortable, which can significantly aid in patient recovery and enhance staff performance. For example, lighting systems that mimic natural daylight patterns help regulate circadian rhythms, potentially improving patient sleep quality and recovery rates.

In more sophisticated applications, responsive environments may also feature architectural elements that can be automatically reconfigured. Movable walls or partitions can alter room sizes and layouts depending on current needs, such as accommodating different group sizes for therapy sessions or providing more private spaces for consultations. This flexibility enhances the facility's ability to cater to various medical activities and patient privacy requirements efficiently.

From an operational standpoint, these technologies significantly contribute to energy efficiency and cost reduction. Smart HVAC systems, for example, can adjust airflow and temperature in real-time based on detailed data like occupancy levels and external weather conditions, ensuring that energy is not wasted on heating or cooling unoccupied spaces. Such optimizations not only reduce operational costs but also align with the broader environmental sustainability goals of the healthcare sector.

Security and safety in healthcare facilities are also enhanced through real-time monitoring technologies. Integrated systems that manage automated door locks, surveillance cameras, and emergency response mechanisms can respond swiftly to any security breaches or safety incidents. This centralized control ensures a high level of protection for both patients and staff, enabling immediate lockdowns or emergency responses when necessary.

Additionally, the extensive data collected from these smart environments serves as a valuable resource for analytics, offering insights into trends in energy usage, space utilization, and patient activity. This information can inform long-term strategies to improve facility management and patient care, driving continual refinements in healthcare delivery and facility design.

Responsive environments in healthcare facilities represent a transformative progression in architectural design, emphasizing adaptability, efficiency, and patient-centric care. By harnessing digital technologies for real-time monitoring and automated system adjustments, these environments not only optimize day-to-day operations but also support the long-term objectives of sustainability and improved healthcare outcomes.

Equation 6.1 is designed to calculate the energy savings achieved using automated adjustments in HVAC (Heating, Ventilation, and Air Conditioning) and lighting systems, incorporating occupancy sensors. This equation is carefully placed within the discussion on how such automated systems optimize energy usage based on real-time room occupancy, significantly reducing unnecessary power consumption and enhancing overall energy efficiency in healthcare facilities.

$$E_{\text{savings}} = (E_{\text{standard}} - E_{\text{automated}}) \times H \times D \tag{6.1}$$

In this formula, $E_{\text{savings}}$ represents the total energy savings achieved through automation. $E_{\text{standard}}$ is the energy consumption under standard operation without occupancy-based adjustments, typically measured in kilowatt-hours (kWh). $E_{\text{automated}}$ is the reduced energy consumption when systems are adjusted based on actual occupancy, also measured in kWh. $H$ denotes the number of operational hours per day that the automated systems are active, and $D$ represents the number of days the systems are used annually.

By implementing Eq. 6.1, healthcare facility managers and sustainability officers can quantify the direct benefits of integrating smart technology into building management systems. This calculation provides a clear metric of energy conservation, justifying the initial investments in smart technologies by showcasing the potential reductions in energy costs and environmental impact. The use of occupancy sensors to tailor the functioning of HVAC and lighting dramatically decreases the operational energy footprint of healthcare environments, promoting a sustainable approach that aligns with modern energy management standards and healthcare goals.

### *6.1.3 Personalization of Patient Spaces*

The personalization of patient spaces through digital technology marks a significant evolution in healthcare delivery, focusing on enhancing patient comfort and empowerment. As technological capabilities expand, they offer unprecedented opportunities for customizing patient environments to individual preferences and specific medical requirements, significantly improving the overall patient experience during hospital stays.

Smart patient rooms exemplify this shift, as they are increasingly equipped with sophisticated systems that provide patients the autonomy to adjust their surroundings to their liking. These rooms often feature integrated control systems—accessible via tablets, smartphones, or bedside controls—that allow adjustments to lighting, window shades, room temperature, and bed positioning. This capability not only elevates comfort but also plays a crucial role in recovery by enabling patients to tailor their environment to what feels most conducive to their well-being.

Entertainment and connectivity options have also been customized to enhance patient experience. Smart TVs, on-demand streaming services, and robust Wi-Fi connectivity allow patients to access a wide range of entertainment options suited to their tastes—crucial during extended stays. These technologies help patients maintain connections with the outside world, facilitating communication with family and friends, attendance at virtual meetings, or continuation of work, all of which help mitigate feelings of isolation and maintain a sense of normalcy.

Voice-activated technology further personalizes the patient's space by enhancing accessibility. Patients with mobility challenges or other disabilities can use voice commands to control room features such as lights, televisions, and even summon nursing staff without physically having to press buttons. This not only fosters independence but also upholds dignity in patient care.

Additionally, adaptive lighting and acoustic management systems can modify environmental conditions based on time of day or specific patient needs. Lighting systems that emulate natural daylight patterns can support healthy sleep cycles, while acoustics can be fine-tuned to create soothing environments that encourage relaxation and aid in recovery.

The integration of health monitoring technologies in smart patient rooms offers another layer of personalized care. These systems continuously track patient vitals and conditions, integrating data directly into medical records accessible by healthcare providers. This integration supports ongoing monitoring and enables timely medical interventions, ensuring that patient care is both immediate and appropriately responsive.

Moreover, smart systems can gather feedback from patients on their experiences and preferences. This feedback is invaluable as it helps refine the personalization of care and environment for future admissions, whether for the same patient or to enhance protocols for patient care more broadly.

By embedding these digital innovations into patient space design, healthcare facilities can craft environments that are not only comfortable and conducive to healing but also finely attuned to the unique needs and preferences of each patient. This patient-centered approach does not just improve patient satisfaction; it also drives better clinical outcomes, as a comfortable and empowered patient is more likely to experience a quicker and more effective recovery.

Table 6.2 details the features of smart patient rooms, including voice-activated technology, adjustable beds, and entertainment options, and their impact on patient satisfaction. This table illustrates the integration of advanced technology in personalizing patient spaces to enhance comfort, accessibility, and overall patient experience. By listing these features alongside their benefits, the table demonstrates how technological advancements contribute to a more patient-centered approach in healthcare facilities, leading to increased patient satisfaction and improved care outcomes.

**Table 6.2** Features of smart patient rooms

| Feature | Description | Impact on patient satisfaction |
|---|---|---|
| Voice-activated technology | Allow patients to control lighting, temperature, and call nurses without physical effort | Increases comfort and autonomy, leading to higher satisfaction |
| Adjustable beds | Beds that can be easily adjusted to various positions for optimal comfort and medical needs | Enhances comfort, facilitates better rest and recovery |
| Personal entertainment systems | Integrated systems providing access to TV, internet, and games directly from the patient's bed | Improves mood, reduces feelings of isolation |
| Remote health monitoring | Devices that monitor patient vitals and alert staff to any needs without constant physical check-ins | Increase safety, ensures timely medical response |
| Environmental controls | Patient-controlled window shades, lighting, and thermostat | Empowers patients, tailored comfort increases satisfaction |

### 6.1.4 Integration of Wearable Technologies

The integration of wearable technologies into the architecture of healthcare facilities represents a transformative advancement in patient monitoring and data-driven healthcare. Wearable devices, such as fitness trackers, smartwatches, and specialized medical wearables, play a pivotal role in continuous health monitoring. This is particularly beneficial for patients managing chronic conditions, undergoing rehabilitation, or in recovery from surgery or acute illnesses.

Healthcare facilities are increasingly being designed to support the seamless integration of data from wearable technologies into hospital information systems. This integration necessitates a robust IT infrastructure capable of handling significant volumes of real-time data efficiently. By ensuring that data from wearables is directly incorporated into electronic health records (EHRs), healthcare providers gain access to timely information on patient vitals and conditions. This immediate access to data enhances the responsiveness and personalization of care, allowing for quicker and more precise medical interventions.

The use of wearable technologies enables continuous, non-invasive monitoring of patients. Devices that can track heart rate, blood pressure, glucose levels, and other vital signs in real-time are particularly beneficial in environments where constant monitoring is critical. This capability allows healthcare professionals to detect potential health issues before they escalate into serious complications, thus improving patient safety and reducing the risk of hospital remissions.

Furthermore, the continuous data stream from wearable technologies provides a richer, more detailed view of a patient's health over time. This comprehensive data enables clinicians to make more informed decisions, tailor treatments to individual patient needs, adjust medications based on up-to-date health data, and evaluate the effectiveness of medical interventions. This approach not only improves the precision of healthcare but also enhances its effectiveness by allowing for adjustments based on real-time insights.

Wearable technology also empowers patients by involving them more actively in their care management. Through patient portals and mobile applications, patients can access their health data, gain insights into their health conditions, and track progress toward recovery. This accessibility promotes greater patient engagement and accountability, encouraging adherence to treatment plans and motivating patients to make informed health and lifestyle choices.

In addition to enhancing patient care and empowerment, the integration of wearable technologies facilitates significant contributions to medical research and development. The vast amounts of data collected can be anonymized and utilized in clinical studies to advance medical knowledge, develop new treatments, identify disease patterns, and improve overall healthcare outcomes on a broader scale.

Lastly, with increased reliance on data from wearable technologies, ensuring data security becomes paramount. Healthcare facilities must incorporate secure networks and robust data protection measures in their architecture. This includes employing encryption, implementing secure access controls, and maintaining rigorous cybersecurity protocols to safeguard sensitive patient information.

By embedding wearable technology integration into the design of healthcare facilities, the healthcare sector can not only enhance the quality and efficiency of care provided but also pave the way for future innovations in personalized medicine. This strategic integration supports immediate clinical needs and fosters a proactive, data-driven approach to healthcare management.

Equation 6.2 is a sophisticated model that illustrates how data collected from wearable technologies can be analyzed to predict patient health events or outcomes, thereby informing and influencing clinical decisions. This equation is integral to discussions about the growing importance of continuous, real-time health monitoring through wearables in modern healthcare settings. By leveraging the predictive analytics derived from these devices, healthcare providers can achieve a more proactive approach to patient care, potentially preventing adverse events and optimizing treatment strategies.

$$P_{\text{event}} = \sigma(\beta_0 + \beta_1 X_1 + \beta_2 X_2 + \cdots + \beta_n X_n) \tag{6.2}$$

In this equation, $P_{\text{event}}$ represents the probability of a specific health event or outcome, such as a cardiac incident or a sudden decline in patient stability. The function $\sigma$ is a logistic function that maps the linear combination of inputs to a probability between 0 and 1, making it suitable for predicting binary outcomes (events happen or does not happen). Each $X_i$ represents different metrics collected from wearable devices, such as heart rate, blood pressure, activity level, or sleep quality. The coefficients $\beta_i$ quantify the impact of these variables on the likelihood of the health event, with $\beta_0$ acting as the intercept or baseline risk without any other influencing factors.

By applying Eq. 6.2, clinicians can harness the power of data to enhance their understanding of patient health trends and potential risks. This model not only facilitates a more informed clinical decision-making process but also underscores the critical role of technology in advancing patient care. Continuous monitoring through wearables offers a dynamic and responsive approach to healthcare, allowing for earlier interventions and tailored treatment plans based on individual patient data profiles. This proactive model of care, underpinned by predictive analytics, represents a significant shift towards personalized and precision medicine in the healthcare industry.

### 6.1.5 *Enhanced Security Measures*

In the evolving landscape of healthcare, the imperative for integrating robust security measures within the architectural design of medical facilities is crucial. The surge in digital technology reliance for both operational and clinical purposes underscore the necessity for heightened security to protect both hardware and sensitive patient information. Effective architectural strategies must therefore encompass a dual approach: securing physical infrastructure and enhancing cybersecurity protocols.

To fortify physical infrastructure, contemporary architectural design must prioritize the protection of critical hardware elements. This involves the strategic placement of servers and data centers in secure locations within the facility, shielded by advanced access control systems such as biometric scanners and keycard entries. Such measures are essential to deter unauthorized access and ensure the integrity of the hardware. Additionally, the integration of comprehensive surveillance systems and the deployment of trained security personnel are vital. These measures collectively contribute to a robust framework that monitors and secures the physical assets critical to the facility's operations.

Parallel to physical security, architectural design must also support an advanced cybersecurity infrastructure. This encompasses the deployment of secure network configurations that include state-of-the-art firewalls, intrusion detection systems, and encrypted data transmission protocols. Such technologies are crucial for the protection of data integrity and confidentiality. Furthermore, data storage solutions—whether on-site or cloud-based—must adhere to stringent data protection standards and comply with relevant laws and regulations, such as the Health Insurance Portability and Accountability Act (HIPAA) in the United States. Ensuring these cybersecurity measures are intrinsic to architectural design is fundamental in safeguarding sensitive information from cyber threats.

While implementing these security measures, they must be integrated seamlessly into the healthcare facility's design to maintain aesthetic coherence and functional efficiency without compromising the comfort and privacy of patients and staff. For instance, the strategic positioning of security cameras and access control systems should enhance security while blending unobtrusively with the facility's design ethos, thereby maintaining a welcoming environment for all users.

Moreover, the resilience and redundancy of security systems are critical considerations in architectural planning. Designing for redundancy involves ensuring that there are adequate backup systems and alternative data storage and network pathways, which can be activated during a security breach or system failure to maintain continuous operations. This aspect of design not only enhances the robustness of security measures but also ensures that the facility can sustain operations under various contingencies.

Finally, a comprehensive security strategy must include regular training for all facility staff on the best practices for data security and the proficient use of digital systems. This training should be accompanied by routine drills and updates to keep security protocols current and top-of-mind. Additionally, integrating these security systems with broader hospital management systems can enhance operational efficiency and enable swift responses to security incidents. Such holistic integration facilitates centralized monitoring and management, bolstering the overall security framework of the facility.

By embedding enhanced security measures into the architectural design, healthcare facilities can create a secure environment that protects against both physical and digital threats. This proactive approach not only safeguards sensitive data but also fosters trust among patients and staff, thereby supporting the delivery of high-quality healthcare.

Table 6.3 presents a detailed checklist of required security protocols and compliance measures for healthcare facilities, emphasizing the importance of adhering to data protection laws and regulations. This checklist serves as a comprehensive guide for healthcare administrators and IT professionals to ensure that their facilities meet stringent security standards to protect patient information and comply with legal requirements. The list includes key protocols and practices that mitigate risks associated with data breaches and unauthorized access, essential for maintaining trust and integrity within healthcare services.

In the next section, we will explore how data analytics is being utilized to further optimize healthcare facility design, leveraging insights from vast data sets to make informed decisions that enhance the efficacy and efficiency of healthcare environments.

**Table 6.3** Security protocols and compliance checklist

| Security protocol/measure | Description | Compliance requirement |
|---|---|---|
| Data encryption | Encrypt patient data both at rest and in transit to ensure confidentiality | Mandatory under HIPAA and GDPR |
| Multi-factor authentication (MFA) | Require multiple forms of verification for accessing sensitive information | Recommended by cybersecurity best practices; required by some regulations |
| Regular security audits | Conduct comprehensive reviews of security infrastructure and policies | Required annually by HIPAA; recommended best practice |
| Access control policies | Define and enforce who can access specific types of data and systems | Mandatory under data protection laws |
| Incident response plan | Establish a plan for responding to data breaches, including notification procedures | Required under HIPAA and GDPR |
| Patient data anonymization | Remove identifying information from patient data used for research | Necessary for compliance with privacy laws |
| Network security measures | Implement firewalls, intrusion detection systems, and anti-virus software | Essential for protecting against external and internal threats |

## 6.2 Role of Data Analytics in Design Optimization

Data analytics is revolutionizing the field of healthcare architecture by enabling more informed and precise design decisions that enhance operational efficiency and patient care. The ability to collect, analyze, and interpret vast amounts of data from healthcare operations provides architects and facility planners with insights that were previously unattainable, leading to optimized design solutions that directly address the specific needs of healthcare facilities.

### *6.2.1 Optimizing Patient Flows*

The integration of data analytics into the design and operation of healthcare facilities significantly enhances the efficiency and effectiveness of patient flows, thereby optimizing both the patient experience and facility operations. Through the strategic analysis of patient movement data, healthcare administrators and architects can identify and rectify inefficiencies within the hospital layout that impede smooth patient navigation and service delivery.

Data analytics serves as a pivotal tool in pinpointing bottlenecks and inefficiencies within healthcare settings. By meticulously analyzing the data collected from patient movements and interactions within the facility, these tools can highlight congested areas such as overcrowded waiting rooms, sluggish registration processes, and sub optimally located key services. Identifying these problem areas allows for targeted architectural and operational adjustments. Redesigning the layout or repositioning services can significantly reduce congestion, thereby streamlining patient flow and minimizing the potential stress and confusion associated with navigating complex hospital environments.

Moreover, the insights garnered from data analytics empower architects to design intuitive and patient-centric hospital layouts. These designs often involve the strategic placement of frequently accessed services like emergency care, radiology, and laboratories in proximity to each other to minimize travel times within the facility. This thoughtful placement not only enhances operational efficiency but also significantly improves the experience of patients and staff by reducing the fatigue associated with long distances and complex navigation paths.

Additionally, the application of data analytics extends to the optimization of waiting times and the overall patient scheduling process. Analyzing data on peak times and patient volumes enables the design of waiting areas that are better equipped to handle varying patient numbers comfortably. Similarly, registration processes can be streamlined to enhance efficiency based on the insights from patient flow data, significantly reducing waiting times and improving patient satisfaction.

Infection control is another critical aspect that benefits from data analytics. By studying patient and staff movement patterns, facility layouts can be optimized to minimize cross-traffic, thereby reducing the risk of infection transmission, which is paramount in areas treating immuno-compromised individuals. Strategic design can ensure that pathways for vulnerable patient populations do not intersect with high-traffic public areas, thus enhancing safety and infection control.

The use of advanced simulation and modeling tools in data analytics also plays a crucial role. These tools allow healthcare administrators to simulate various layout scenarios and their impact on patient flow before implementing any physical modifications. This predictive approach can identify potential issues and areas for improvement, enabling cost-effective planning without the need for extensive physical trial-and-error.

Ongoing improvement is a further advantage provided by data analytics. Continuous monitoring and analysis of patient flow and facility usage offer real-time insights that can lead to immediate adjustments in layouts, signage, and operational protocols. This dynamic approach allows healthcare facilities to remain adaptable and responsive to evolving needs, ensuring that the infrastructure continually supports optimal patient care and operational efficiency.

By harnessing the power of data analytics to optimize patient flows, healthcare facilities can not only improve operational efficiencies and patient care quality but also ensure that the infrastructure effectively supports the delivery of medical services in a safe and patient-friendly environment. This analytical approach ensures that the healthcare facility is a nexus of efficiency and patient-centered care, continually adapting to meet the demands of its users.

Table 6.4 provides a before-and-after analysis of how changes in the layout of healthcare facilities have impacted patient flow, demonstrating reductions in waiting times and improvements in traffic patterns. This table illustrates the effectiveness of intuitive layouts and strategic redesigns in enhancing operational efficiency and patient experience. By quantifying the effects of these layout changes, the table highlights their significance in creating a smoother, more efficient healthcare environment that minimizes stress and maximizes care delivery.

### *6.2.2 Energy Efficiency and Sustainability*

The role of advanced data analysis in enhancing energy efficiency and sustainability of healthcare facilities is increasingly crucial. By meticulously examining patterns of energy consumption and harnessing data-driven insights, architects and engineers are empowered to develop strategies that effectively reduce operational costs while simultaneously contributing to environmental sustainability. These strategies are not only fiscally prudent but also ethically aligned with global sustainability efforts, demonstrating the healthcare sector's commitment to responsible environmental stewardship.

Optimizing the use of natural light through data analytics is a primary strategy for reducing the dependence on artificial lighting, which is a significant energy drain in hospital settings. Detailed analysis of sunlight patterns and the architectural orientation of buildings can guide the strategic placement of windows and the selection

**Table 6.4** Impact of layout changes on patient flow

| Department | Before change (waiting time in minutes) | After change (waiting time in minutes) | Improvement (%) | Traffic flow improvement (description) |
|---|---|---|---|---|
| Emergency | 30 | 18 | 40 | Increased triage efficiency, reduced bottlenecks |
| Outpatient | 25 | 15 | 40 | Streamlined check-in process, better signage |
| Radiology | 20 | 10 | 50 | Improved scheduling, dedicated patient pathways |
| Pharmacy | 15 | 8 | 46.7 | Optimized layout for faster service, reduced congestion |
| Admissions | 18 | 12 | 33.3 | Direct routing, separate entrance for admissions |

of appropriate glazing materials. These materials are chosen to maximize daylight exposure while minimizing thermal gain, thereby reducing the need for both lighting and air conditioning. This not only decreases energy consumption but also enhances indoor environmental quality for both patients and staff, contributing to better health outcomes and greater comfort.

In addition to natural lighting, the selection of building materials plays a critical role in the energy efficiency of healthcare facilities. Data analytics aids in choosing materials with high thermal mass suitable for environments with significant temperature variations, helping to stabilize interior temperatures naturally. Advanced insulation materials are also selected based on performance characteristics analyzed through data models, ensuring they meet the highest standards of energy efficiency. This careful material selection helps maintain internal climate control with minimal energy expenditure, enhancing sustainability.

The integration of smart building systems represents another significant application of data analytics in healthcare facilities. These systems use sensors and real-time data to adjust heating, ventilation, air conditioning, and lighting based on actual conditions and occupancy levels. For example, HVAC systems can be optimized to modify temperature settings automatically in unoccupied spaces or adjust operation based on external weather conditions and internal thermal loads. Such dynamic adjustments lead to substantial energy savings and operational efficiencies.

Ongoing monitoring of energy consumption is essential for maintaining and improving energy efficiency over time. Through continuous data analysis, healthcare facilities can pinpoint inefficient operations or outdated equipment that may require upgrades. This persistent monitoring allows for the adaptation of operational practices and the integration of new technologies, ensuring sustained energy efficiency regardless of changes in facility usage or technological advancements.

Beyond energy, data analytics play a pivotal role in the sustainable management of other resources like water and waste. Analyzing water usage patterns can identify inefficiencies, such as leaks, and lead to the implementation of solutions like gray water recycling for non-potable uses, which further enhances the facility's sustainability profile. Additionally, data analytics supports compliance with environmental regulations and facilitates detailed reporting on energy savings and sustainability initiatives, which is crucial for meeting regulatory standards and reinforcing the facility's commitment to environmental sustainability.

Through these comprehensive strategies, data analytics enable healthcare facilities to significantly enhance their energy efficiency and sustainability. This not only leads to reduced operational costs but also minimizes the environmental footprint of healthcare operations, aligning them with broader global sustainability goals and fostering a healthy, supportive environment for patients and staff. These efforts underscore the healthcare sector's role as a leader in sustainable practices, setting a benchmark for other industries to follow.

Equation 6.3 is designed to quantify energy savings achieved through the strategic use of natural light and the integration of smart building systems. This formula is strategically placed in the context of how optimizing natural light and employing smart technologies contribute significantly to energy efficiency in modern buildings. By calculating energy savings, facility managers can assess the effectiveness of these environmentally conscious strategies in reducing power consumption and operational costs.

$$E_{\text{savings}} = (E_{\text{old}} - (E_{\text{natural}} + E_{\text{smart}})) \times T \tag{6.3}$$

In this formula, $E_{\text{savings}}$ represents the total energy savings. $E_{\text{old}}$ is the energy usage before implementing natural light optimization and smart systems, measured in kilowatt-hours (kWh). $E_{\text{natural}}$ and $E_{\text{smart}}$ represent energy consumption after these implementations have been made, respectively. $E_{\text{natural}}$ accounts for the reduced energy usage due to increased reliance on natural light during daylight hours, while $E_{\text{smart}}$ reflects further reductions from smart system integrations, which adjust lighting, HVAC, and other systems based on real-time occupancy and ambient conditions. $T$ denotes the total operating hours in which these systems are active, providing a multiplier that extends the savings calculation over a given period, typically a month or a year.

By utilizing Eq. 6.3, building administrators can effectively monitor and verify the energy savings resulting from eco-friendly design choices and technology upgrades. This equation highlights the dual benefits of enhancing natural light utilization and implementing smart building technologies, both of which serve to decrease energy demands and promote sustainability. Such strategic integrations not only lead to significant cost savings but also support broader environmental goals by minimizing the carbon footprint of building operations.

### 6.2.3 Enhanced Patient Experience

Data analytics is increasingly recognized as a transformative tool for enhancing patient experience within healthcare facilities. By systematically analyzing patient feedback, behavior patterns, and various data points, healthcare providers gain valuable insights that drive decisions to improve comfort, convenience, and overall satisfaction. This approach ensures that patient care is not only responsive but also anticipates needs, fostering a more nurturing and supportive healthcare environment.

Collecting and analyzing patient feedback is a fundamental aspect of data analytics. Utilizing surveys, digital platforms, and direct comments, healthcare providers can gather significant insights into areas where patient satisfaction may be lacking. For instance, recurring issues such as prolonged wait times or complex navigation within facilities can be systematically addressed. Analyzing this feedback allows for targeted operational changes or physical layout modifications to enhance patient convenience and reduce any potential stress or discomfort associated with the healthcare environment.

Monitoring behavior patterns within the facility through data analytics is another crucial strategy. Using sensor data to track movement can reveal high-traffic areas or bottlenecks that negatively affect patient comfort and the efficiency of the facility. Understanding these patterns enables healthcare managers to make informed decisions on modifying physical layouts—such as widening corridors, repositioning furniture, or redesigning the flow of movement—to alleviate crowded areas and enhance the overall ease of movement within the facility.

Acoustic comfort is a significant component of the patient's experience, as excessive noise can impede recovery and lead to dissatisfaction. Through data analytics, healthcare facilities can identify areas where noise levels are problematic, determine the sources, and understand when disruptions are most frequent. Solutions may include enhancing acoustic isolation in walls, redesigning areas where noise is generated or adjusting the operational times of noisy equipment to less sensitive hours. Such targeted interventions can substantially improve the acoustic environment, contributing positively to patient recovery and comfort.

Customizing patient rooms based on data-driven insights allows for accommodations tailored to the specific needs or preferences of different patient groups. Analysis might reveal that certain room orientations or layouts are associated with better patient outcomes or higher satisfaction. Leveraging this information, rooms can be designed to optimize natural light, offer pleasant views, or provide easier access for caregivers—each adjustment aimed at enhancing patient well-being and comfort.

Enhancing wayfinding within healthcare facilities is another area where data analytics can make a significant impact. Large medical facilities can be challenging to navigate, leading to patient frustration. Data analytics can identify common points of confusion and implement effective wayfinding solutions such as improved signage, interactive digital maps, or enhanced staff training to assist patients and visitors more efficiently.

Lastly, optimizing environmental controls in patient care areas through smart systems can dramatically improve the patient's experience. Data analytics enables the customization of environmental conditions—such as temperature, humidity, and lighting—according to individual patient feedback and preferences. These systems can adjust environmental parameters in real-time, ensuring optimal comfort and fostering a conducive healing environment.

By weaving data analytics into the fabric of healthcare facility design and management, healthcare providers can craft a more patient-centered experience. This data-driven approach not only elevates the quality of care but also ensures that healthcare facilities continuously evolve to meet the dynamic needs and expectations of patients, ultimately enhancing health outcomes and boosting overall patient satisfaction.

Table 6.5 outlines common patient complaints and the data-driven improvements made in response, along with the resulting outcomes in terms of patient satisfaction scores. This table highlights how targeted changes based on patient feedback and analytics can significantly enhance the patient experience within healthcare facilities. By documenting specific complaints, the corresponding interventions, and their impacts on satisfaction, the table demonstrates the value of a responsive and adaptable healthcare service that prioritizes patient needs and comfort.

### 6.2.4 Facility Management and Maintenance

The integration of data analytics into facility management and maintenance marks a pivotal advancement in the operational efficiency and sustainability of healthcare facilities. Utilizing predictive analytics, healthcare organizations can revolutionize how they maintain their critical infrastructures, preempt equipment failures, and optimize resource allocation, ultimately fostering a safer and more efficient environment for both patients and staff.

Predictive maintenance stands out as a cornerstone benefit of data analytics in facility management. By leveraging continuous monitoring of equipment performance and data from sensors and Internet of Things (IoT) devices, analytics platforms are adept at detecting patterns or anomalies that may indicate impending equipment failures. This predictive capability allows facility managers to proactively address potential issues before they escalate into costly breakdowns, thereby minimizing downtime and reducing the financial burden associated with emergency repairs. This not only ensures uninterrupted service but also extends the lifespan of vital medical equipment.

Resource optimization is another significant advantage offered by data analytics. By analyzing usage patterns across various systems and areas within the healthcare facility, managers can optimize the deployment of resources ranging from energy to human capital. For instance, insights into energy consumption patterns can inform more efficient adjustments to heating, ventilation, and air conditioning (HVAC) systems, balancing cost savings with optimal comfort and safety. Similarly, data-driven insights can streamline staff scheduling, ensuring that personnel

**Table 6.5** Summary of patient feedback improvements

| Patient complaint | Improvement made | Outcome | Patient satisfaction score (before) (%) | Patient satisfaction score (after) (%) |
|---|---|---|---|---|
| Long wait times | Streamlined scheduling and increased staffing during peak hours | Reduced average wait time by 30% | 70 | 85 |
| Lack of privacy | Redesigned patient areas to enhance privacy with soundproofing and better partitioning | Increased perception of privacy | 65 | 82 |
| Poor communication | Implemented a new digital communication system to keep patients informed about their care process | Improved understanding and engagement | 60 | 78 |
| Uncomfortable bedding | Upgraded mattresses and adjustable beds in all patient rooms | Enhanced comfort during recovery | 55 | 75 |
| Insufficient parking | Expanded parking facilities and improved signage | Easier access and reduced stress on arrival | 50 | 70 |

are efficiently allocated according to actual needs, and enhancing productivity while maintaining high standards of patient care.

Lifecycle management of infrastructure components and medical equipment is greatly enhanced through data analytics. With detailed insights into the usage and wear of equipment, facility managers can accurately plan for future expenditures such as replacements and upgrades. This foresight helps in budgeting and ensures that facilities remain equipped with the latest technologies and comply with regulatory standards, avoiding unexpected expenses and operational disruptions.

Enhanced safety protocols are critical in healthcare settings, and data analytics significantly bolsters this aspect by analyzing incident reports and maintenance records to identify patterns indicative of safety risks. This proactive approach not only helps in averting potential accidents but also ensures compliance with stringent health and safety regulations, thereby safeguarding both patients and staff.

Environmental monitoring through data analytics offers real-time oversight of critical parameters such as air quality, temperature, and humidity. This capability allows facility managers to maintain environmental conditions within the optimal range for patient health and comfort. Adjustments can be made dynamically in response to changes in external weather conditions or internal factors, ensuring that the facility environment consistently supports patient recovery and staff well-being.

Automated reporting and compliance, facilitated by data analytics, streamline the maintenance of regulatory documentation. Automated systems can generate detailed reports on aspects such as energy consumption, safety measures, and maintenance activities, ensuring that healthcare facilities remain prepared for inspections and audits. This automation not only enhances accuracy but also saves valuable time and resources, allowing staff to focus more on patient care rather than administrative tasks.

By embracing a data-driven approach in facility management and maintenance, healthcare organizations can enhance operational efficiencies, extend the longevity of their infrastructure, improve safety standards, and ensure a comfortable environment for patients and staff. This strategic use of data analytics not only addresses current operational needs but also strategically positions healthcare facilities to meet future challenges and embrace technological advancements.

Equation 6.4 provides a robust model for predicting the likelihood of equipment failure based on maintenance data analytics, a cornerstone of effective preventive maintenance scheduling. Positioned within the discussion on predictive maintenance strategies, this formula is vital for facilities that aim to maximize the operational reliability and longevity of their equipment through data-driven decision-making. Predictive maintenance, unlike reactive maintenance, utilizes historical and real-time data to preemptively identify and address potential equipment issues before they result in costly downtime.

$$P_{\text{failure}} = 1 - e^{-\lambda t} \tag{6.4}$$

In this equation, $P_{\text{failure}}$ represents the probability of equipment failure within a given timeframe. The variable $\lambda$ (lambda) is the failure rate, determined from historical maintenance and operational data of the equipment. $t$ represents the time elapsed since the last maintenance activity. The exponential term $e^{-\lambda t}$ describes the decreasing probability of survival (or increasing probability of failure) as time progresses, which is a common statistical pattern in the lifespan of mechanical systems.

The use of Eq. 6.4 allows maintenance managers to strategically schedule inspections, repairs, and replacements based on the calculated risk of failure, rather than on a fixed schedule or in response to actual breakdowns. This proactive approach not only ensures higher availability and reliability of equipment but also optimizes maintenance costs by preventing unnecessary repairs and reducing unexpected breakdowns. By integrating predictive analytics into maintenance planning, facilities can achieve greater operational efficiency and extend the lifespan of critical equipment, making this equation a key tool in modern maintenance strategies.

### 6.2.5 *Staffing and Resource Optimization*

Data analytics has revolutionized staffing and resource optimization in healthcare facilities by aligning human resources and facility operations efficiently with patient needs and operational demands. This strategic implementation allows hospitals to adapt to varying demand patterns dynamically, enhancing the quality of care provided and overall operational efficiency.

One of the pivotal applications of data analytics in healthcare is the development of predictive staffing models. These models utilize historical data, seasonal trends, and local events that could influence patient influx—such as public gatherings or flu seasons—to forecast demand accurately. This predictive capability enables hospital administrators to tailor staffing levels precisely, ensuring that the facility is adequately staffed to handle the anticipated patient load without incurring the high costs associated with overstaffing. By matching staffing levels to patient demand, hospitals can maintain a high level of care and efficiency while controlling operational costs.

Data analytics also plays a crucial role in dynamic resource allocation. By monitoring usage data of critical hospital facilities such as operating rooms, imaging equipment, and patient beds, analytics help in the efficient scheduling and utilization of these resources. For example, if the data indicates a surge in demand for certain procedures at specific times, the scheduling for operating rooms can be adjusted to match this demand, thereby maximizing use and minimizing idle times. This not only enhances the productivity of healthcare facilities but also ensures that resources are available when patients need them the most.

Furthermore, optimizing staffing levels and resource allocation directly impacts the efficiency of care delivery. Through analytics, patterns in patients flow across various departments can be identified, enabling administrators to adjust staff assignments strategically to reduce wait times and eliminate bottlenecks. This streamlined approach ensures that patients receive prompt and effective care, which is crucial for patient satisfaction and positive health outcomes.

An essential benefit of effective staffing optimization is the reduction of staff fatigue. Data analytics identify times and areas within the hospital that experience high stress, allowing management to implement strategic staff rotations and provide additional support where necessary. This not only helps maintain a high standard of care but also safeguards healthcare workers from burnout, fostering a healthier work environment.

Integrating data analytics with human resources systems further enhances staffing management. This integration offers deep insights into staff availability, qualifications, and training needs, facilitating informed decisions regarding hiring, ongoing training, and personnel development. This cohesive approach ensures that the hospital staff is well-equipped to meet the demands of their roles, enhancing overall service delivery.

Moreover, advanced data analytics and artificial intelligence tools enable real-time adjustments to staffing and resource deployment. This immediacy is invaluable

**Table 6.6** Efficiency gains from staff and resource optimization

| Metric | Before implementation | After implementation | Improvement (%) |
|---|---|---|---|
| Patient wait times | 45 min | 30 min | 33.3 |
| Staff satisfaction | 65% | 85% | 30.8 |
| Resource utilization | 70% | 90% | 28.6 |
| Bed turnover rate | 4 h | 3 h | 25 |
| Emergency response time | 15 min | 10 min | 33.3 |

in emergencies or unexpected surges in patient intake, where rapid decision-making is crucial to effectively managing the situation.

As technology evolves, the integration of data analytics into healthcare operations is expected to deepen, providing even more sophisticated tools for managing staffing and resources. This ongoing advancement not only enhances the operational facets of healthcare delivery but also significantly impacts strategic planning and the long-term sustainability of healthcare facilities. By leveraging these technologies, healthcare providers can ensure that they are well-prepared to meet both current and future challenges, thereby improving the overall healthcare system's efficiency and effectiveness.

Table 6.6 illustrates the tangible benefits of data-driven staffing solutions by comparing key metrics such as patient wait times, staff satisfaction, and resource utilization before and after implementation. This table demonstrates how strategic adjustments in staff scheduling and resource management, guided by analytics, significantly enhance operational efficiency and satisfaction levels among both patients and staff. It serves as a compelling showcase of how modern data tools can optimize hospital operations, leading to more efficient and effective healthcare delivery.

Next, we will explore future directions in medical technology integration, examining how upcoming innovations may further influence the design and operation of healthcare facilities, potentially revolutionizing how care is delivered and managed.

## 6.3 Future Directions in Medical Technology Integration

The integration of medical technology into healthcare facilities is an evolving field that promises to reshape the landscape of healthcare delivery. As technological advancements continue to accelerate, future trends in medical technology integration will focus on enhancing efficiency, accuracy, and personalization of care. These innovations will require thoughtful architectural design to ensure that facilities are not only equipped to handle new technologies but are also adaptable to future developments.

### 6.3.1 *Robotics and Automation*

The integration of robotics and automation within healthcare settings marks a significant shift in the operational and care delivery models of medical facilities. As these technologies advance, they necessitate thoughtful architectural adaptations to accommodate their unique requirements while ensuring safe and efficient interactions among machines, patients, and healthcare professionals.

A primary architectural consideration in accommodating robotics is the design of efficient pathways for robotic systems. These pathways are crucial, particularly for robots tasked with logistics operations, such as the delivery of medications, supplies, or food. To facilitate seamless robotic navigation without impeding human traffic, healthcare facilities might need to incorporate dedicated lanes specifically for robotic use. Alternatively, more innovative solutions, such as overhead systems or floor-embedded guidance tracks, could be employed to optimize space and enhance efficiency.

Furthermore, healthcare facilities must consider the storage, charging, and maintenance needs of robotic devices. Designating specific areas for these functions is essential to ensure that robots are readily accessible, fully charged, and properly maintained without causing service delays. These areas should be strategically located to facilitate quick deployment and efficient use of robotic services, and they must be equipped with the necessary infrastructure to handle routine technical repairs and software updates.

Robotic operations require robust support from the facility's infrastructure, particularly in terms of electrical and data systems. The integration of high-capacity power supplies is critical to meet the changing requirements of various robotic devices, while advanced data networks are essential for managing communications and control systems efficiently. Proper integration of these infrastructural elements is crucial to the functionality and reliability of robotic technologies within the healthcare environment.

Another vital design aspect is ensuring the safety and coexistence of robots and humans within shared spaces. Treatment and communal areas must be designed to allow the safe operation of robots alongside human staff, with adequate space to prevent collisions and ensure seamless workflows. Safety protocols, such as emergency stop mechanisms and advanced sensor systems, should be integrated into the facility design to mitigate the risk of accidents and ensure a safe environment for all users.

The design of treatment areas also needs to be flexible to accommodate both traditional human-delivered care and robotic interventions. For example, surgical rooms may require additional space to accommodate the physical dimensions and operational range of surgical robots, beyond what is typically necessary for human surgeons. Diagnostic areas might also need adaptations to facilitate the use of diagnostic robots, ensuring that these spaces can support a hybrid model of care delivery effectively.

Privacy and security are paramount, especially with the increased integration of robotics in healthcare. Architectural designs should ensure that data transmissions by robots are secure and that areas where sensitive interactions occur are adequately shielded from unauthorized access by both robotic and human entities.

As robotics and automation increasingly permeate the healthcare sector, the architecture of medical facilities must evolve to support these technologies effectively. This evolution will enhance operational efficiencies and expand the possibilities for patient care, paving the way for more innovative, responsive, and efficient healthcare environments. The ongoing architectural challenge will be to harmonize the technical requirements of robotic systems with the comfort and safety needs of human users, ensuring that the healthcare facilities of the future can fully capitalize on the advancements offered by these transformative technologies.

Table 6.7 provides a detailed checklist of necessary architectural adjustments for integrating robotics in healthcare facilities, covering aspects such as storage, maintenance, and safety features. This table serves as a comprehensive guide for healthcare administrators and facility designers to prepare their environments for the effective use of robotics, ensuring that all logistical and safety considerations are met to optimize the benefits of this technology in patient care and hospital operations.

**Table 6.7** Robotics integration checklist

| Requirement | Description | Importance level | Architectural adjustment needed |
|---|---|---|---|
| Storage facilities | Secure and accessible areas for storing robots when not in use | High | Designated robotics storage rooms with charging stations |
| Maintenance areas | Dedicated spaces for routine checks and repairs | High | Maintenance workshops equipped with necessary tools and diagnostic equipment |
| Safety features | Measures to ensure safe interaction between robots and humans | Critical | Installation of emergency stop buttons, sensor-based collision avoidance systems |
| Navigation pathways | Clear and wide pathways for robot movement | Medium | Modification of hallways and access points to accommodate robotic dimensions |
| Power supply | Reliable and uninterrupted power sources for continuous operation | High | Enhanced electrical infrastructure with backup power solutions |
| Integration with building systems | Seamless interaction with existing hospital systems | High | Digital integration with hospital's network for control and monitoring |

### *6.3.2 Advanced Imaging Technologies*

The rapid evolution of imaging technology in healthcare demands significant adaptations in the architectural design of medical facilities. These adaptations are crucial to accommodate the installation and safe operation of increasingly sophisticated and heavier imaging equipment, such as MRI machines and multi-slice CT scanners. As these technologies provide more detailed and rapid insights into the human body, the facilities that house them must evolve to meet these new challenges effectively.

One of the primary architectural considerations for modern imaging technologies is the need for reinforced flooring and structural support. The significant weight of advanced imaging machines often necessitates the reinforcement of flooring to prevent structural damage and ensure safety. Moreover, the overall structural design of the building may need adjustments to support the heavy load of these devices without compromising the building's integrity. This reinforcement is critical not only for the installation phase but also for the long-term sustainability of the facility.

Power supply is another crucial aspect of accommodating advanced imaging technologies. These machines typically require more power than standard medical equipment, necessitating enhanced power supplies and dedicated electrical circuits designed to handle the increased demand. Facilities must ensure these power systems are not only capable of supporting high-energy equipment but are also reliable enough to provide uninterrupted service, essential for the precision and safety of diagnostic procedures.

Radiation shielding is also a vital component, especially for imaging modalities that use radiation, such as X-rays and CT scans. To protect patients and staff from exposure, facilities must incorporate radiation shielding in the design of imaging rooms. This often involves constructing lead-lined walls, doors, and windows, tailored to meet stringent regulatory standards for radiation protection. The effectiveness of this shielding is crucial for safety and compliance with health regulations.

Efficient workflow layout within the facility is equally important. The placement of imaging equipment should facilitate a seamless flow from patient intake to diagnosis and treatment. Strategic planning of the layout is essential to minimize the physical distance patients travel, which is particularly important for those in discomfort. This consideration helps streamline operations and reduces the time patients and staff spend moving between different areas of the facility.

Patient comfort and privacy must also be prioritized in the design of advanced imaging areas. Facilities should provide calm and welcoming waiting areas to help reduce patient anxiety, incorporating soothing aesthetics and privacy measures. Design elements such as private changing areas, secure storage for personal belongings, and tranquil decor can significantly enhance the patient experience during potentially stressful procedures.

Additionally, integration with data systems is critical, given the vast amounts of data produced by modern imaging technologies. Architectural designs should include provisions for robust data management infrastructure, including secure data centers and high-capacity network systems to handle the storage, processing, and secure transfer of medical images.

Finally, adaptability for future technological advancements is a key consideration. Imaging facilities should be designed with flexibility in mind, allowing for easy updates or replacements of equipment as new technologies and improvements emerge. Modular designs or accessible equipment rooms can facilitate these technological upgrades without the need for extensive renovation.

By addressing these critical factors, healthcare facilities can ensure that their imaging technologies not only meet the current diagnostic and operational needs but are also well-prepared to adapt to future advancements. This forward-thinking approach not only enhances the capability to provide top-tier patient care but also positions the facilities at the forefront of medical technology innovation.

Equation 6.5 is essential for calculating the required thickness and type of materials needed for effective radiation shielding in imaging rooms, where the control of exposure is critical to ensure the safety of both patients and healthcare staff. Positioned within the context of discussing radiation shielding requirements, this formula helps in designing facilities that adhere to safety standards, minimizing radiation risks associated with diagnostic imaging procedures such as X-rays and CT scans.

$$T = \frac{\log(I_0/I)}{\mu} \tag{6.5}$$

In this equation, $T$ represents the required thickness of the shielding material. $I_0$ is the initial intensity of the radiation without any shielding, and $I$ is the desired intensity of radiation after it passes through the shielding material, indicating the level of radiation that is considered safe. $\mu$ is the linear attenuation coefficient of the shielding material, which quantifies the material's ability to absorb or block radiation. This coefficient varies depending on the material's composition and the type of radiation being shielded.

The logarithmic function $\log(I_0/I)$ calculates the ratio of the initial to the desired intensity, providing a measure of how much the radiation needs to be reduced to meet safety standards. Dividing this value by $\mu$ gives the thickness of the material needed to achieve this reduction. This calculation ensures that imaging rooms are equipped with adequate shielding to protect against harmful radiation exposure, facilitating the safe use of radiological imaging technologies.

By using Eq. 6.5, healthcare facility designers and radiologists can determine the most appropriate materials and thicknesses for constructing radiation barriers, tailored to the specific radiation sources and energies used in their imaging practices. This targeted approach not only upholds safety regulations but also optimizes the protection of health and well-being in medical imaging environments.

### 6.3.3 Telemedicine and Remote Monitoring

As telemedicine and remote monitoring technologies increasingly become staples in healthcare delivery, the architectural design of healthcare facilities needs to adapt to support these innovations effectively. This evolution involves creating spaces specifically designed for virtual interactions and integrating the necessary infrastructure for high-quality remote care.

One of the fundamental adaptations is the inclusion of dedicated telemedicine rooms within healthcare facilities. These rooms should be equipped with high-quality video and audio systems to facilitate clear and effective communication between healthcare providers and patients. The design of these rooms should prioritize optimal lighting, acoustics, and camera placement to enhance the quality of virtual interactions, aiming to replicate the experience of in-person consultations as closely as possible.

Additionally, healthcare facilities should provide private patient areas designed for attending virtual appointments. This is especially important in outpatient clinics or hospitals where patients may not have the means to engage in telemedicine from their homes. These spaces should be designed with soundproofing to ensure privacy and confidentiality and equipped with user-friendly technology to accommodate patients of varying ages and levels of tech proficiency.

The increasing reliance on data transmission in telemedicine and remote monitoring necessitates the establishment of secure and robust data centers within healthcare facilities. These centers should be capable of managing large volumes of sensitive patient data securely. It is imperative to incorporate advanced cybersecurity measures to safeguard patient information against potential breaches. The architectural design should also include provisions for the physical security of these areas, ensuring that access is restricted to authorized personnel only.

Remote monitoring technologies, which continuously track patient health metrics, require seamless integration into the facility's IT infrastructure. This integration should include network enhancements to support the real-time transmission of data from patients' homes or wearable devices to healthcare providers. Facilities must ensure that their network systems are robust and reliable, capable of handling high data loads to facilitate uninterrupted monitoring.

Healthcare facilities must also be designed with adaptability in mind to accommodate the rapid evolution of telemedicine and remote monitoring technologies. Design strategies might include modular rooms that can be easily updated or reconfigured as technology advances and patient needs change. Such flexibility allows for the timely updating of both hardware and software systems without significant disruptions to facility operations.

Furthermore, the successful adoption of telemedicine and remote monitoring technologies depends on the proficiency of healthcare staff. Designing specific areas within the facility for training on the use of these technologies can significantly enhance staff adoption rates and effectiveness. These training areas should be

**Table 6.8** Telemedicine impact analysis

| Metric | Before telemedicine integration | After telemedicine integration | Improvement (%) |
|---|---|---|---|
| Patient access (number of patients) | 1000 | 1500 | 50 |
| Average patient satisfaction score | 75 | 85 | 13.3 |
| Wait times for consultation (minutes) | 30 | 10 | 66.7 |
| Follow-up compliance rate | 60% | 80% | 33.3 |
| Reach to remote areas (number of remote patients served) | 200 | 400 | 100 |

equipped with the same technologies that are used in actual telemedicine interactions to provide hands-on learning experiences.

By integrating these design elements, healthcare facilities can effectively leverage the benefits of telemedicine and remote monitoring. This approach not only meets the current demands of healthcare delivery but also prepares facilities for future advancements. It enhances patient access to care, improves the efficiency of healthcare delivery, and expands the capability of facilities to meet a broader range of patient needs, thus enriching the overall healthcare experience.

Table 6.8 showcases before and after statistics on patient access and satisfaction following the integration of telemedicine facilities. This table highlights the transformative effects of telemedicine on healthcare delivery, demonstrating how remote monitoring and virtual consultations have expanded access to medical care and improved patient satisfaction rates. The data emphasizes the critical role of telemedicine in modernizing healthcare practices, making care more accessible and convenient, particularly in underserved areas.

### *6.3.4 Personalized Medicine*

The paradigm shift towards personalized medicine, particularly in fields like genomic medicine and tailored therapies, calls for a profound transformation in the architectural design of healthcare facilities. These advanced fields require highly specialized laboratories and data analysis facilities that can handle sophisticated processes and technologies essential for personalized treatment approaches.

Sophisticated laboratory spaces are fundamental to the advancement of personalized medicine. These laboratories need to be equipped with state-of-the-art genomic sequencing and bioinformatics technologies. The architectural design must cater to the specific needs of this sensitive equipment, which includes maintaining strict controls over temperature and humidity, ensuring vibration isolation, and preventing contamination. Additionally, the layout of these laboratories should be meticulously planned to support complex workflows typical of genomic studies. This involves arranging spaces to minimize cross-contamination and facilitate a logical sequence of processes from one stage of analysis to the next, enhancing efficiency and accuracy.

The design of these facilities must also be flexible and scalable to keep pace with rapid technological advancements in genomic medicine. Modular layouts that can be easily adjusted or expanded are essential, allowing for the swift integration of new technologies and equipment as they become available. This adaptability prevents the need for extensive renovations or disruptions, ensuring that the facilities remain at the cutting edge of personalized medicine without significant downtime.

Advanced data analysis facilities are also crucial due to the massive amounts of data generated by personalized medicine practices. Architectural designs should include dedicated spaces for data scientists and bioinformaticians, equipped with high-performance computing resources and secure data storage facilities. These areas must be designed to support intensive computational work and comply with stringent data protection regulations, facilitating robust data analysis capabilities within the healthcare facility.

Moreover, integrating clinical and research areas within healthcare facilities can significantly enhance the efficacy of personalized medicine. Such integration allows for the swift translation of research findings into clinical applications, benefiting patients more directly and rapidly. Architectural considerations for this integration should include the strategic placement of research labs near patient care areas and the creation of shared spaces that encourage collaboration between clinicians and researchers.

Given the sensitive nature of genomic data and personalized therapies, enhanced security measures are imperative. Facilities must incorporate advanced security systems to protect these areas, including surveillance, biometric access controls, and secure data networks. These measures help safeguard patient information and maintain the integrity of the research and treatments being conducted.

Finally, personalized medicine often requires in-depth consultations between patients and healthcare providers regarding genetic information and treatment options. Therefore, facilities should include private consultation areas that offer a secure and comfortable environment for these sensitive discussions. These areas should be designed to promote a sense of privacy and trust, facilitating open communication between patients and their caregivers.

By addressing these specific architectural and design needs, healthcare facilities can effectively support the growing field of personalized medicine. This not only enhances the facility's capacity for cutting-edge research and customized patient care but also positions it as a leader in medical science, ready to embrace future advancements in treatment technologies and methodologies.

Equation 6.6 serves as a critical tool for performing a cost–benefit analysis of incorporating advanced genomic facilities into healthcare infrastructure, particularly within discussions on the long-term benefits and cost-effectiveness of investing in personalized medicine. This equation enables healthcare administrators and policymakers to quantify the financial returns on investments made in personalized medicine, which often includes sophisticated genomic testing and tailored treatment plans that can significantly improve patient outcomes and potentially reduce overall healthcare costs.

$$\text{ROI} = \frac{(B - C)}{C} \times 100\% \tag{6.6}$$

In this formula, ROI represents the Return on Investment, expressed as a percentage, which measures the efficiency and profitability of the investment. $B$ denotes the total benefits realized from the investment, which in the context of personalized medicine, can include improved patient outcomes, reduced hospital stays, and decreased long-term healthcare costs due to more effective treatments. $C$ represents the initial and ongoing costs associated with setting up and maintaining genomic facilities, including equipment, staffing, training, and operational expenses.

By applying Eq. 6.6, healthcare decision-makers can systematically evaluate the financial implications of adopting personalized medicine approaches. The equation highlights not only the direct financial gains but also the broader economic impact through enhanced patient care and potentially lower treatment costs over time. Such a cost–benefit analysis is crucial for justifying the upfront investments in advanced medical technologies and for planning sustainable, patient-centered healthcare services that leverage the latest advancements in genomic medicine. This approach encourages a strategic investment in healthcare infrastructure that aligns with the goals of improving treatment precision and patient satisfaction, while also considering the long-term financial health of the institution.

### *6.3.5 Wearable Health Technology*

The rapid proliferation of wearable health technologies represents a significant leap forward in continuous patient monitoring and personalized care. As these devices increasingly become a staple in modern medical practice, healthcare facilities must adapt their architectural and infrastructural designs to effectively support the integration and utilization of the data generated by wearables.

One critical adaptation for healthcare facilities is the integration of dedicated monitoring centers. These centers would be specially equipped to manage the continuous streams of data produced by wearable devices, providing real-time health monitoring and data analysis. Staffed by professionals trained to interpret this data, these centers would act as vital hubs, alerting medical staff to changes in patient conditions and enabling swift and informed medical interventions.

To handle the voluminous and complex data generated by wearables, healthcare facilities must enhance their data reception capabilities. This enhancement includes upgrading IT infrastructure to securely and efficiently manage high volumes of real-time data. Facilities will need to implement high-speed internet connections, secure wireless access points throughout the facility, and integrate scalable cloud technologies for robust data storage and processing capabilities.

Securing integration of data from wearable devices into existing Electronic Health Records (EHR) systems is another essential requirement. This integration demands robust cybersecurity measures and the development of standardized protocols to ensure seamless merging of data from wearables with traditional medical records. Such integration enhances the continuity and comprehensiveness of patient care documentation, making it a critical aspect of modern healthcare facility operations.

Given the increased capture and storage of data from wearables, addressing privacy and security concerns is paramount. Architectural designs should incorporate secure areas for data management that safeguard against unauthorized physical and digital access. Implementing encryption, secure servers, and adherence to stringent health data regulations, such as HIPAA in the United States, are vital measures to protect patient information.

Furthermore, the effective use of wearable health technologies requires dedicated spaces within healthcare facilities for training and support. These areas should be equipped with the necessary technologies and designed to facilitate training healthcare providers in the use of new systems and the interpretation of data from wearables. Ongoing education and support are crucial as these technologies evolve and become more ingrained in medical practice.

Finally, healthcare facilities should also consider establishing patient interaction zones. These areas would serve as informational points where patients can learn about wearable technologies, understand how to incorporate these devices into their health management and interact with technology specialists. Providing access to educational materials and supporting demonstrations of new devices in these zones can significantly enhance patient engagement and compliance with wearable technology use.

By making these architectural and operational adjustments, healthcare facilities can effectively leverage the capabilities of wearable health technologies. This proactive approach not only enhances patient care by providing continuous, data-driven insights but also positions healthcare facilities at the forefront of medical technology and patient engagement, paving the way for a more data-informed and patient-centric healthcare environment.

Table 6.9 presents data on the adoption rates and health outcome improvements associated with the use of wearable technology in healthcare settings. This table details the significant role wearable devices play in enhancing patient monitoring by providing continuous health data, facilitating early detection of potential health issues, and improving chronic disease management. The statistics highlight the growing reliance on these technologies and their effectiveness in elevating the quality of care and patient autonomy in health management.

**Table 6.9** Wearable technology usage statistics

| Wearable technology | Adoption rate (%) | Health outcome improvements (%) | Key benefits observed |
|---|---|---|---|
| Fitness trackers | 40 | 20 | Improved patient engagement in fitness |
| ECG monitors | 25 | 35 | Early detection of cardiac issues |
| Blood pressure monitors | 30 | 25 | Better hypertension management |
| Glucose monitoring devices | 15 | 40 | Enhanced diabetes control |
| Sleep trackers | 20 | 15 | Improved sleep quality management |

### *6.3.6 Interactive Patient Environments*

Interactive patient environments herald a transformative shift in healthcare facility design, merging cutting-edge technology with patient care to forge spaces that are not only functionally robust but also conducive to healing. Envisioning the future of hospital rooms, one imagines spaces equipped with dynamic walls and surfaces that adapt to the needs and preferences of their occupants. The integration of advanced display technologies is pivotal in this context. These technologies could convert ordinary walls and surfaces into interactive panels that display real-time health data, care schedules, or even therapeutic imagery customized to a patient's specific treatment requirements or personal tastes. The seamless incorporation of such technology is crucial, ensuring that it enhances the overall aesthetic of the room while maintaining a tranquil atmosphere that does not overwhelm or unsettle the patient.

Equally vital is the role of embedded sensor systems in creating responsive interactive environments. These sensors could monitor various patient metrics, such as movement, vital signs, or stress levels, and adapt the environment accordingly. For instance, they might adjust lighting, temperature, or even the visuals on interactive surfaces to better suit the patient's current state. This responsive capability underscores the importance of ensuring these environments are attuned to the nuances of patient and staff needs.

The effectiveness of these technologies also hinges on the intuitiveness of the user interfaces. These interfaces should be accessible to patients of varying ages and technological adeptness, incorporating touch-responsive surfaces, voice-controlled mechanisms, or gesture recognition technologies that not only facilitate ease of use but also minimize the risk of infection transmission through physical contact. Furthermore, the integration of these interactive systems with the hospital's electronic medical records is crucial. Such integration ensures that the displayed information is consistently updated, providing both patients and healthcare providers with the latest data on treatment plans, medication schedules, and other essential health details.

However, the incorporation of technology in patient environments also necessitates stringent privacy and security measures. Architectural and technological designs must include robust data networks and secure storage solutions to safeguard sensitive patient information. Moreover, it is imperative to ensure that the information displayed in these interactive settings is accessible only to authorized personnel to protect patient privacy.

Beyond the technological aspects, the architectural design of these spaces must also incorporate therapeutic elements. This involves the strategic use of colors, textures, and materials that foster relaxation and well-being. Technology itself should be capable of displaying calming scenes or personalized content that aids in stress reduction and facilitates the healing process.

As the scope of these technologies extends to telemedicine and remote care, architectural designs will increasingly need to accommodate virtual interactions without compromising the quality of care or the patient experience. These spaces must be designed to facilitate easy access to telemedicine technologies while ensuring adequate privacy and maintaining a tangible connection between patients and healthcare providers, even in their physical absence.

The exploration of how these advanced technologies reshape patient interactions within healthcare systems illuminates the pressing need for innovative architectural solutions. These solutions must support a digital and interconnected approach to healthcare, emphasizing environments that are not only technologically advanced but also deeply attuned to the therapeutic needs of patients.

Equation 6.7 encapsulates a set of environmental adjustment algorithms designed to automatically fine-tune room conditions such as lighting, temperature, and display settings based on real-time sensor data from the patient's environment. This equation is crucial in discussions about interactive environments where dynamic adjustments are essential to create a responsive and comfortable setting for patients, enhancing their experience and potentially improving health outcomes.

$$C_{\text{new}} = C_{\text{current}} + k(\Delta S) \tag{6.7}$$

- $C_{\text{new}}$ represents the new settings for room conditions (light, temperature, displays).
- $C_{\text{current}}$ is the current setting before adjustment.
- $\Delta S$ denotes the change needed based on sensor inputs, which measure factors such as ambient light levels, room temperature, and patient activity or presence.
- $k$ is a sensitivity factor that determines how much the settings should adjust in response to a given change in sensor data.

This algorithm allows systems within a patient's room to adapt automatically, ensuring optimal comfort and environmental conditions at all times. The sensitivity factor κ κ can be tailored to different environments and patient needs, making it a flexible and powerful tool in healthcare settings where personalized care is increasingly important. By using Eq. 6.7, healthcare facilities can implement smart environmental controls that not only respond in real-time to changes but also anticipate needs based on patterns and specific patient preferences, thus elevating the standard of care and leveraging technology to foster better patient outcomes.

## 6.4 Designing for Telemedicine and Remote Care

The rapid expansion of telemedicine and remote care is reshaping how healthcare services are delivered, emphasizing the need for healthcare facilities to adapt their designs to support these technologies effectively. Telemedicine not only extends the reach of medical services to remote or underserved populations but also reduces the burden on traditional healthcare settings by managing routine consultations and follow-ups online. This shift requires thoughtful architectural considerations to accommodate the unique needs of telemedicine within healthcare facilities.

### *6.4.1 Dedicated Telemedicine Spaces*

As telemedicine increasingly becomes a cornerstone of modern healthcare delivery, the architectural design of healthcare facilities is compelled to adapt, incorporating dedicated spaces that facilitate virtual consultations. These telemedicine spaces, essential for the clear and private exchange between patients and healthcare providers, must be meticulously designed to uphold the integrity of remote consultations. Central to these spaces is the incorporation of high-quality video and audio equipment. Such spaces require professional-grade cameras, microphones, and speakers to ensure high-quality, uninterrupted communication. The objective is to replicate the clarity and immediacy of in-person consultations, facilitating a seamless exchange of medical advice and information.

Acoustic integrity is another critical factor in the design of telemedicine spaces. Effective soundproofing is paramount to maintain privacy and prevent the intrusion of external noise typical of hospital environments. Utilizing materials such as acoustic panels, sound-absorbing wall treatments, and double-paned glass can significantly enhance the acoustic isolation of these rooms. Such measures ensure that consultations are not only private but free from disruptive noise that could compromise the quality of the interaction.

The physical layout and ergonomics of telemedicine spaces also demand careful consideration. The furniture and spatial arrangement should prioritize comfort and functionality to accommodate prolonged consultations. Adjustable seating, adequate desk space for necessary equipment, and optimally positioned lighting that avoids glare and shadows are essential components. These elements collectively contribute to a setting that supports the physical well-being of both patients and healthcare providers during remote interactions.

Privacy features are integral to the design of telemedicine spaces. Consultations must be conducted without the risk of eavesdropping or unintended observations. Incorporating privacy curtains, frosted glass, and secure doors can safeguard the confidentiality of the interactions taking place within these spaces. Moreover, the technical infrastructure of telemedicine spaces must be robust, featuring high-speed internet connections with redundancy measures to prevent downtime, secure networks for the protection of patient data, and ample power outlets to support all necessary equipment.

Furthermore, the design of telemedicine spaces should be both adaptable and scalable to keep pace with evolving technologies and healthcare practices. Employing modular furniture and flexible room configurations allows for future modifications and upgrades without extensive renovations. Additionally, ensuring accessibility is crucial. These spaces must be designed to accommodate patients with disabilities, adhering to established accessibility standards in terms of space dimensions, furniture design, and equipment placement.

By thoughtfully integrating these elements, healthcare providers can ensure that dedicated telemedicine spaces within facilities not only enhance the effectiveness of virtual consultations but also align with broader healthcare objectives of improving accessibility and efficiency in medical service delivery. Such spaces are instrumental in advancing the reach and quality of healthcare, adapting to the shifting paradigms of patient interaction in the digital age.

Table 6.10 provides a comprehensive checklist that outlines all necessary technical and ergonomic features for setting up dedicated telemedicine rooms in healthcare facilities. This checklist serves as a practical guide to ensure that telemedicine spaces are well-equipped to facilitate efficient and effective virtual consultations, enhancing patient and provider experience. It details both the hardware requirements and the environmental considerations that contribute to a functional and comfortable telemedicine interaction.

### *6.4.2 Technical Infrastructure*

Robust technical infrastructure is the cornerstone of effectively deploying and operating telemedicine services within healthcare facilities. As telemedicine increasingly becomes a staple in healthcare delivery, the infrastructure supporting it must guarantee seamless, secure, and uninterrupted services. A high-speed internet connection forms the backbone of this infrastructure, enabling real-time video conferencing and swift data transmission without lags or disruptions. Healthcare facilities must prioritize investments in high-bandwidth connections while ensuring network redundancy to circumvent service interruptions. This may include securing services from multiple internet providers or installing dedicated lines that maintain connectivity even when one source fails, thus ensuring continuous operation during critical telemedicine sessions.

**Table 6.10** Checklist for telemedicine room requirements

| Requirement category | Specific requirement | Importance level |
|---|---|---|
| Technical equipment | High-definition video cameras | Essential |
| | High-quality microphones | Essential |
| | Dual monitors for multitasking | Recommended |
| Connectivity | High-speed internet connection | Essential |
| | Secure network infrastructure | Essential |
| Furniture and ergonomics | Adjustable seating for providers | Essential |
| | Ergonomic desks with easy cable management | Essential |
| | Soundproofing to ensure privacy | Essential |
| Lighting and acoustics | Soft, non-glare lighting to reduce eye strain | Essential |
| | Acoustic panels to minimize echo | Recommended |
| Accessibility | Wheelchair-accessible furniture and doorways | Essential |
| Decor | Calm and neutral color schemes | Recommended |
| Additional features | Remote-controlled room environment settings (e.g., temperature, lighting) | Recommended |
| | Emergency communication systems | Recommended |

Security is paramount, given the sensitive nature of patient data exchanged during telemedicine consultations. Networks must be fortified with encrypted connections, virtual private networks (VPNs), and secure sockets layer (SSL) protocols to protect data in transit. Beyond encryption, segmenting the internal network is crucial, ensuring that sensitive data and telemedicine transmissions are isolated from the network's less secure sections. This stratification helps prevent potential breaches and safeguard patient information effectively.

Furthermore, continuity of care is critical, necessitating robust backup systems. These systems should seamlessly take over in the event of a primary system failure, thus ensuring uninterrupted telemedicine services. Facilities should implement backup power solutions, like uninterruptible power supplies (UPS) and generators, alongside redundant data storage solutions to mitigate data loss risks. Compliance with data protection regulations, such as HIPAA in the United States or GDPR in Europe, is also crucial. Healthcare facilities must not only deploy secure technological solutions but also establish stringent policies and protocols regulating access to and usage of patient data, thereby ensuring both privacy and regulatory compliance.

The technical infrastructure must also be scalable and flexible to adapt to evolving telemedicine technologies and expanding service demands. This might involve modular network designs that facilitate easy expansion or upgrades, cloud-based solutions for dynamic resource allocation based on demand, and systems that can integrate seamlessly with emerging technologies. Training and technical support are integral to the effective utilization of telemedicine technologies. Facilities should ensure that healthcare providers and technical support staff receive ongoing training to handle the technology adeptly. A dedicated technical support team should be in place to address issues swiftly, minimizing downtime and allowing healthcare providers to concentrate on patient care rather than technical glitches.

Regular monitoring and maintenance of the technical infrastructure are essential to preempt potential issues that could impact service quality. This includes routine network performance checks, security audits, and timely updates to software and hardware components, ensuring the system remains secure and functions optimally.

By embedding these elements into the architectural and operational fabric of healthcare facilities, telemedicine services can be delivered reliably and securely, significantly enhancing patient care and broadening the scope of medical services. This infrastructure not only supports current practices but also establishes a solid foundation for the anticipated advancements in digital healthcare delivery.

Equation 6.8 is specifically designed to calculate the minimum data throughput needed to effectively support simultaneous telemedicine consultations, ensuring that these critical communications occur without any lag. This calculation is particularly relevant in discussions highlighting the necessity of high-speed internet connections in healthcare settings, where reliable and seamless digital communication is crucial for delivering telemedicine services efficiently.

$$D_{\text{min}} = N \times B \times R \tag{6.8}$$

In this formula:

- $D_{\text{min}}$ represents the minimum data throughput required, measured in megabits per second (Mbps).
- $N$ is the number of simultaneous telemedicine consultations that the system needs to support.
- $B$ denotes the bandwidth required per consultation, which includes both the upload and download speeds necessary to ensure high-quality video and audio streaming without interruptions.
- $R$ is a redundancy factor that provides additional bandwidth to handle potential fluctuations in internet speed and ensures a buffer for maintaining smooth communication even under less ideal conditions.

By applying Eq. 6.8, healthcare administrators can determine the necessary internet bandwidth to support telemedicine operations without disruptions. This ensures that telemedicine consultations are conducted smoothly, enhancing the patient and provider experience. Proper planning of network capabilities using this equation helps healthcare facilities avoid common pitfalls related to inadequate bandwidth, which can lead to decreased care quality during remote consultations.

### *6.4.3 Patient and Provider Comfort*

Ensuring comfort for both patients and healthcare providers is paramount in designing telemedicine spaces within healthcare facilities. This focus on comfort not only enriches the telemedicine experience but also enhances the effectiveness of consultations and overall satisfaction for everyone involved. To support healthcare providers who often engage in lengthy telemedicine sessions, ergonomic workstations are indispensable. These workstations should include adjustable chairs that promote proper posture, desks set at an optimal height, and monitor stands adjusted to eye level to prevent physical strain. The design of these spaces should also facilitate natural movement and include periodic breaks, incorporating accessories like footrests and ergonomic keyboards or mice to minimize the risk of repetitive strain injuries.

For patients accessing telemedicine services, the provision of private booths or rooms that are both comfortable and secure is crucial. These spaces should be designed to be welcoming and equipped with appropriate technology, ensuring confidentiality and comfort. Soundproofing is essential to maintain privacy, while comfortable seating and soft, calm interior designs can significantly alleviate anxiety, helping patients feel more relaxed during consultations. Furthermore, the technology available in these spaces must be user-friendly to accommodate individuals of varying ages and technological proficiency. This includes simple interfaces, clear instructions, and for those with disabilities, accessibility features like screen readers, large text options, and easily navigable touchscreens.

Acoustic design also plays a critical role in ensuring that conversations during telemedicine sessions are clear and undisturbed by external noise. The use of sound-absorbing materials in the construction of walls, floors, and ceilings, as well as designs that minimize echo, can enhance the auditory clarity of consultations. Lighting, too, is essential; it should be designed to prevent glare on screens and provide a flattering and clear view for video communications. Adjustable lighting options can accommodate different times of the day and variations in natural light, ensuring that both participants appear visibly comfortable and professional during video calls.

Ventilation and climate control are further aspects that influence comfort in telemedicine spaces. Maintaining a comfortable temperature and ensuring good air quality is particularly important in enclosed spaces designated for private consultations. Effective climate control systems can prevent the environment from becoming stifling, thereby keeping both patients and providers comfortable throughout their interactions.

By meticulously considering these design elements, healthcare facilities can establish telemedicine spaces that significantly boost the comfort and efficacy of remote healthcare delivery. This detailed approach ensures that telemedicine is not only a functional alternative to traditional in-person visits but also a preferred option for many due to its convenience and enhanced comfort.

Table 6.11 rates various ergonomic and acoustic features based on user feedback, highlighting their impact on comfort and productivity in healthcare environments. This table provides a detailed assessment of different comfort-related aspects, from seating to noise levels, quantifying their benefits to help healthcare facility managers and designers make informed decisions. By focusing on the user experience, this table underscores the importance of comfort in promoting both staff efficiency and patient satisfaction in healthcare settings.

**Table 6.11** Comfort features rating

| Comfort feature | User comfort rating (1–10) | Productivity impact rating (1–10) | Notes on user feedback |
|---|---|---|---|
| Ergonomic seating | 9 | 8 | "Significantly reduces back pain during long shifts" |
| Adjustable lighting | 8 | 7 | "Helps reduce eye strain, especially in overnight shifts" |
| Acoustic panels | 7 | 8 | "Noticeable reduction in stress due to lower noise levels" |
| Climate control | 9 | 7 | "Optimal temperature settings boost morale and focus" |
| Anti-fatigue flooring | 8 | 9 | "Greatly appreciated in areas where staff stand for long periods" |

### *6.4.4 Integration with Traditional Care*

Integrating telemedicine with traditional care practices is crucial for crafting a unified healthcare delivery system that maximizes the advantages of both modalities. Achieving this integration demands careful architectural and operational planning to ensure healthcare providers can smoothly transition between virtual and in-person care. One effective strategy is the development of hybrid examination rooms. These rooms are designed to support both physical examinations and telemedicine consultations, equipped with the requisite medical tools for standard in-person exams and the technological setup necessary for conducting virtual consultations, such as high-quality video and audio systems. This dual functionality allows healthcare providers to use the same space for various types of consultations, thus enhancing flexibility and operational efficiency.

Furthermore, technology integration points are essential for facilitating a seamless transition between virtual and in-person care. These should be strategically distributed throughout the facility, including in examination rooms, waiting areas, and patient rooms, equipped with computer stations, tablets, or mobile devices. This setup ensures that digital patient data is readily accessible and can be updated swiftly, irrespective of whether the care provided is remote or face-to-face. Additionally, the integration process is supported by flexible scheduling systems capable of managing both telemedicine and in-person appointments. These systems should be capable of real-time updates and accessible to all relevant staff, ensuring efficient resource and space allocation.

Dedicated areas for staff training are also vital, providing the necessary technology and environment for staff to familiarize themselves with telemedicine platforms and integrate these tools into their regular practice effectively. Effective communication infrastructure is another cornerstone, enabling coordination across different sections of the healthcare facility. This might include unified communications platforms that allow staff to quickly send and receive information, whether they are engaged in telemedicine or direct patient care.

Patient flow management should also be a priority in architectural design, facilitating the navigation of both telemedicine and in-person care pathways within the facility. Clear signage and guidance systems can direct patients to the appropriate areas based on their consultation type, reducing confusion and enhancing the patient experience. Finally, ensuring privacy and security in environments where digital and physical care modalities intersect is paramount. Facilities must incorporate secure areas for storing both physical and digital patient records and design private spaces for consultations to occur without the risk of eavesdropping or data breaches.

By thoughtfully designing healthcare facilities that integrate telemedicine with traditional care practices, providers can offer more comprehensive, flexible, and efficient services. This integrated approach not only meets a wide range of patient needs but also elevates the overall quality of care, ensuring that patients receive the most appropriate treatment at the right time and in the most suitable format.

Equation 6.9 is formulated to evaluate the efficiency of space utilization in hybrid rooms, considering both their physical and virtual care capacities. This is essential in discussions about the need for healthcare facilities to have adaptable and scalable spaces that can efficiently transition between in-person and virtual care modalities. Such adaptability is increasingly important in modern healthcare environments, where the flexibility of room functions can enhance overall operational efficiency and patient satisfaction.

$$\mathrm{SUE} = \frac{(P + V)}{A} \tag{6.9}$$

In this equation:

- SUE represents Space Utilization Efficiency, which is a measure of how effectively the physical space of a room is used for both physical and virtual healthcare activities.
- $P$ denotes the capacity of the room for providing physical care, measured in the number of patients that can be treated in the space per day.
- $V$ represents the virtual care capacity, indicating how many virtual consultations or remote interactions can be conducted from this space each day.
- $A$ is the total area of the room in square meters, providing a baseline for measuring space efficiency.

By utilizing Eq. 6.9, healthcare planners and administrators can quantify how effectively space is being used in hybrid environments, supporting decisions about room designs and layouts that accommodate both direct patient care and telehealth services. This approach ensures that healthcare facilities can maximize their space efficiency, adapting dynamically to the varying demands of patient care, thereby optimizing both the physical and virtual capabilities of their environments. This model encourages thoughtful design that supports a seamless blend of care delivery methods, crucial for the evolving landscape of healthcare services.

### 6.4.5 Adaptability for Future Technologies

As telemedicine continues to evolve, the necessity for healthcare facilities to adapt to emerging technologies grows increasingly vital. Designing for adaptability is crucial, as it allows facilities to integrate new advancements seamlessly, maintaining their relevance and efficacy in the swiftly evolving healthcare technology landscape. One effective approach is the incorporation of modular design elements. By utilizing modular rooms, walls, and furniture, facilities can be reconfigured or updated as needed to accommodate new equipment or different layouts. This modularity enables healthcare facilities to respond promptly to technological advancements or shifts in medical practice without substantial downtime or disruption.

Building scalable infrastructure is another foundational strategy for future-proofing healthcare facilities. This includes the installation of conduit systems that can accommodate additional wiring or fiber optics as technological demands grow. Similarly, power supplies and network systems should be designed with the capability of being easily upgraded to manage increased loads or more sophisticated equipment. The concept of universal spaces is also instrumental in enhancing a facility's adaptability. Designing rooms to be suitable for both in-person and telemedicine interactions, with built-in technological hookups and flexible furniture arrangements, allows these spaces to easily adapt to shifts in healthcare delivery methods.

Facilities should also designate technology-ready areas pre-equipped for rapid integration of new technologies. These could feature universal docking stations, high-speed data ports, and adjustable power supplies, enabling quick and efficient setup of new devices and systems. The choice of materials and finishes plays a significant role in adaptability. Opting for durable, easy-to-clean surfaces that can withstand frequent updates or modifications helps preserve the facility's functionality and aesthetic appeal over time.

Fostering innovation within healthcare facilities can be achieved by designing collaborative spaces where staff can explore new technologies and treatment methodologies. These innovation hubs can position healthcare providers at the cutting edge of medical technology, facilitating the development and testing of new approaches in a controlled environment. Lastly, the integration of robust digital infrastructure is imperative. Facilities must have IT systems capable of interfacing with a variety of devices and platforms, alongside secure, scalable cloud storage solutions for efficient data management.

**Table 6.12** Technology adaptation index

| Healthcare facility | Technological infrastructure score (out of 10) | Staff training score (out of 10) | Institutional adaptability score (out of 10) | Overall adaptation index (out of 100) |
|---|---|---|---|---|
| Hospital A | 8 | 7 | 9 | 85 |
| Hospital B | 6 | 5 | 7 | 65 |
| Clinic C | 9 | 8 | 8 | 85 |
| Health center D | 7 | 9 | 6 | 75 |
| Specialty clinic E | 8 | 6 | 8 | 75 |

By adhering to these design principles, healthcare facilities can ensure they remain adaptable and responsive to the demands of modern healthcare delivery. This adaptability not only enhances the facility's capacity to integrate new telemedicine technologies but also ensures the continued provision of high-quality care as the scope of digital health expands.

Table 6.12 assesses healthcare facilities on their readiness and adaptability to integrate future telemedicine technologies. This table provides a benchmark index that evaluates various facilities based on factors such as existing technological infrastructure, staff training levels, and institutional willingness to adopt new technologies. The index helps to highlight where healthcare facilities stand in their preparedness to embrace upcoming innovations in telemedicine, which is crucial for maintaining competitive and cutting-edge patient care services.

### 6.4.6 *Accessibility Considerations*

Ensuring accessibility in telemedicine systems is paramount for creating an inclusive and equitable healthcare experience. Both digital and physical aspects must cater comprehensively to all users, particularly those with disabilities. The design of digital platforms for telemedicine is a critical area that requires meticulous attention to adhere to various accessibility standards. This commitment ensures that individuals with visual, hearing, motor, or cognitive disabilities can effectively utilize these services. Features such as voice recognition technologies provide an alternative for those unable to use traditional input devices like keyboards. For users with visual impairments, ensuring compatibility with screen readers and offering functionalities like adjustable text sizes and contrast settings are essential. Moreover, including captions in video content addresses the needs of the hearing impaired. The overall interface design should focus on simplicity to facilitate ease of navigation for users with varying levels of technological proficiency, enhancing their ability to engage with telemedicine services confidently.

On the physical front, the accessibility of telemedicine facilities within healthcare environments is equally crucial. Key elements include wheelchair-accessible entryways and consultation rooms, along with appropriately designed furniture to accommodate diverse patient needs. Compliance with standards such as the Americans with Disabilities Act (ADA) is non-negotiable, ensuring that physical spaces feature wide doors, sufficient room for maneuverability, and accessible seating arrangements. These modifications are fundamental not only to meet legal requirements but also to foster a genuinely accommodating environment for all patients.

Training and support for both patients and staff play a vital role in maximizing the accessibility of telemedicine. Educating healthcare providers and patients on how to effectively utilize telemedicine technologies, especially considering the needs of those with disabilities, is crucial. Providing robust support services ensures that any issues encountered during telemedicine interactions are promptly addressed, making help readily available through various means, whether a phone call or online assistance. This proactive approach to training and support underscores the commitment to accessible healthcare delivery.

Moreover, the integration of telemedicine systems with various assistive technologies is essential. Healthcare facilities should ensure their systems are compatible with specialized equipment used by patients, such as alternative input devices or software that adapts the user interface for easier access. This integration extends the usability of telemedicine solutions, accommodating a broader range of disabilities and user preferences.

Continuous testing and improvement based on feedback from users with disabilities are vital for refining telemedicine services. This iterative process helps identify and rectify accessibility barriers, ensuring the system evolves to meet user needs effectively. Such ongoing enhancements are crucial as technological advances, requiring telemedicine services to adapt continuously to new standards and user expectations.

By emphasizing these accessibility considerations, healthcare facilities can ensure their telemedicine services are genuinely accessible, thereby advancing broader healthcare goals of inclusivity and equity. As we continue to embrace technological advancements, the focus on refining accessibility in telemedicine will remain a critical, ongoing endeavor to ensure that all patients receive the care they need in a manner that respects their circumstances and needs.

In the subsequent section, we will delve into the concept of smart hospitals and explore how the integration of advanced technologies into healthcare facilities can revolutionize patient care, enhance operational efficiency, and offer a more responsive and personalized healthcare environment.

Table 6.13 provides a comprehensive checklist to ensure that telemedicine facilities meet all required accessibility standards. This checklist serves as a guideline for compliance verification, helping healthcare providers ensure that their telemedicine services are accessible to all patients, including those with disabilities. The list covers various aspects of physical and digital accessibility, promoting an inclusive approach to healthcare that adheres to legal and ethical standards.

**Table 6.13** Accessibility compliance checklist

| Accessibility requirement | Checklist item | Compliance needed (yes/no) | Notes |
|---|---|---|---|
| Physical access | Wheelchair-accessible entrances and facilities | Yes | Ensure all entry and exit points are barrier-free |
| Visual accessibility | High-contrast visuals and signage for telemedicine interfaces | Yes | Critical for patients with visual impairments |
| Auditory accessibility | Availability of hearing aid-compatible devices | Yes | Essential for patients with hearing difficulties |
| Technological compatibility | Support for assistive technologies like screen readers | Yes | Allows seamless use by patients with special tech needs |
| Language accessibility | Multilingual support in telemedicine applications | Yes | Important for non-English speaking patients |
| Cognitive accessibility | Simplified user interfaces and instructions | Yes | Helps patients with cognitive disabilities navigate services easily |

## 6.5 Smart Hospitals: A Forward-Looking Approach

Smart hospitals represent the pinnacle of integrating technology and healthcare, utilizing advanced digital systems, artificial intelligence, and the Internet of Things (IoT) to create a highly efficient, responsive, and patient-centered environment. This forward-looking approach not only enhances the quality of care provided but also dramatically improves operational efficiencies and patient outcomes.

### *6.5.1 Integrated Digital Infrastructure*

The transition toward smart hospitals is significantly underpinned by the establishment of an integrated digital infrastructure, which serves as the nerve center for orchestrating seamless interactions between various technological systems within the facility. This digital backbone is indispensable for facilitating efficient operational workflows, delivering high-quality patient care, and enabling informed, real-time decision-making processes. At the heart of a smart hospital lies its robust IT infrastructure. Architectural considerations are crucial to accommodate dedicated IT spaces designed to securely house servers, data processing units, and other essential hardware. These facilities are constructed to ensure maximum security, with controlled access to safeguard the systems and the sensitive data they manage.

In the realm of connectivity, the reliance on cloud-based solutions and the necessity for real-time data exchange dictate the need for ultra-high-speed internet connections. This infrastructure is meticulously designed to include redundant pathways to mitigate downtime and ensure uninterrupted service, essential for operations ranging from telemedicine consultations to intricate data analyses. Moreover, the network infrastructure is tailored to efficiently handle high volumes of data traffic, thereby eliminating bottlenecks and facilitating a smooth flow of information across the hospital's systems.

Data security is a cornerstone in the architecture of smart hospitals. Secure data centers are integrated into the hospital design, featuring advanced security measures such as fire suppression systems, climate control, and stringent physical security protocols. These measures are in place to protect sensitive patient information and proprietary data from unauthorized access or breaches, with compliance with national and international data security standards being a top priority.

The scalability and flexibility of the digital infrastructure are critical to accommodate future technological advancements and expanding operational needs. The network is designed with modular components that allow for easy upgrades or replacements, ensuring the longevity and adaptability of the system. This flexibility is crucial for integrating emerging technologies seamlessly, without necessitating extensive overhauls.

Integration across various systems, including electronic health records (EHR), building management systems (BMS), and real-time health monitoring devices, is meticulously planned to enable efficient data exchange and interoperability. This interconnectedness supports the deployment of advanced analytics and machine learning tools, which enhance operational efficiency and elevate patient care quality by providing deep, actionable insights.

Moreover, the design of smart hospitals includes comprehensive redundancy and backup systems to guarantee reliability and continuity of service. Critical systems, especially those related to power supply and data storage, are equipped with redundancies such as backup generators, uninterruptible power supplies (UPS), and duplicate data storage solutions. These systems are crucial for maintaining hospital operations during unexpected power failures or system malfunctions.

Energy efficiency is also a key component of the infrastructure design in smart hospitals, reflecting the significant technological reliance. Energy-efficient cooling systems for IT equipment, the integration of renewable energy sources, and the employment of smart building technologies are pivotal in optimizing energy consumption throughout the facility.

By fostering a robust and integrated digital infrastructure, smart hospitals are positioned to revolutionize healthcare delivery. This infrastructure not only supports the immediate needs of patient care and facility operations but also lays the groundwork for continuous innovation and adaptation to meet future healthcare challenges. This dynamic framework promises to enhance the responsiveness and personalization of healthcare, making it more attuned to the evolving needs of patients and healthcare professionals alike.

**Table 6.14** IT infrastructure components

| IT component | Specifications | Security features | Purpose of smart hospital |
|---|---|---|---|
| Servers | High-performance CPUs, scalable memory of up to 1 TB | Firewalls, Intrusion Detection Systems (IDS) | Manage and process patient data, support applications |
| Data centers | Redundant power supplies, HVAC cooling systems | Physical access controls, Video surveillance | House critical hardware and software, data storage |
| Network devices | High-speed routers and switches, Wi-Fi 6 capability | Encryption, Secure VPN access | Facilitate communication and data transfer across hospital |
| Cloud storage solutions | Hybrid cloud environment | Data encryption at rest and in transit | Offsite data storage and disaster recovery |
| Security appliances | Unified Threat Management (UTM) devices | Anti-virus, Anti-spyware, Content filtering | Protect network from external threats, manage web traffic |
| Backup systems | Onsite and offsite backup options | Encrypted backups, Automated backup schedules | Ensure data integrity and availability during failures |

Table 6.14 lists the critical components of IT infrastructure required in smart hospitals, including servers, data centers, and network devices, complete with specifications and security features. This detailed overview helps healthcare administrators understand the robust setup necessary to support a smart hospital's operations, emphasizing the importance of high-capacity, secure, and reliable IT systems. The table serves as a comprehensive guide to constructing or upgrading IT infrastructure to meet the advanced demands of modern healthcare technology, ensuring efficiency and data protection.

### *6.5.2 Automation and AI*

In the evolving landscape of smart hospitals, the role of automation and artificial intelligence (AI) is becoming increasingly central, fundamentally transforming how healthcare services are delivered. The incorporation of these technologies into the healthcare environment necessitates a thoughtful architectural approach to ensure optimal functionality and integration within the setting. To accommodate this technological shift, smart hospitals are incorporating designated areas specifically tailored for automated equipment. For instance, spaces dedicated to robotic pharmacies are crucial as they automate medication dispensing, requiring precise design considerations to enhance their efficiency and safety. Similarly, automated laboratory systems that handle patient samples also demand meticulously planned environments tailored to their operational needs.

Moreover, the interaction between AI-driven systems and human-operated functions is a critical aspect of modern healthcare facilities, requiring the creation of integration zones. These areas are designed to facilitate seamless collaboration between technology and healthcare professionals. For example, AI-assisted diagnostic centers are environments where technologies like advanced imaging software are used alongside human expertise, such as that of radiologists, to analyze patient images. These zones must support both the technological infrastructure and the professionals, ensuring a harmonious workflow that enhances diagnostic accuracy and efficiency.

The backbone of AI operations in healthcare is a robust infrastructural setup that can handle substantial computational demands and data processing. Facilities must be equipped with advanced data centers, featuring powerful servers and high-speed networking capabilities that are scalable to future advancements in AI technology. This setup ensures that AI systems can operate efficiently and evolve without the need for extensive overhauls.

Architectural flexibility is another cornerstone in the design of smart hospitals. As AI and automation technologies continue to advance, hospital spaces must be adaptable to accommodate these changes. This adaptability can be achieved through modular design elements and reconfigurable spaces that allow for easy modifications to integrate new or updated technologies as they become available.

Security is paramount, particularly with the increased reliance on data-driven technologies. Smart hospitals must incorporate enhanced data security measures within their architectural designs. This includes secure areas for IT operations that protect against both physical and cyber threats, such as secure server rooms and restricted access to critical technology zones. Integrating advanced cybersecurity measures directly into the building's network is also crucial to safeguarding sensitive patient data.

Energy management is a significant consideration, given the energy-intensive nature of AI and automation systems. Smart hospitals should employ energy management systems that optimize power usage and improve efficiency. Implementations might include smart grids, the use of renewable energy sources, and energy recovery systems, all aimed at minimizing environmental impacts and operational costs.

Despite the technological emphasis, the comfort and well-being of patients and staff remain a priority. The design of hospital environments should ensure that the integration of technology does not detract from the human-centric nature of healthcare. This includes creating spaces that are welcoming and comfortable, utilizing elements like natural light, comfortable seating areas, and interactive displays that provide both informational and entertainment value to patients and visitors.

By strategically designing for the integration of automation and AI, smart hospitals can forge environments that not only harness cutting-edge technology for operational efficiency and enhance patient care but also maintain adaptability and security as the paradigms of healthcare technology continue to evolve. These thoughtfully designed spaces ensure that the technology serves to augment the healthcare experience rather than overshadow the human elements critical to patient care and comfort.

Equation 6.10 is developed to estimate the computational load required for AI operations in healthcare settings. This formula is integral to understanding and planning for the technological infrastructure needed to support advanced AI applications, from diagnostic algorithms to patient management systems. These applications are critical in modern healthcare, where AI significantly enhances efficiency and accuracy across various operations.

$$\mathrm{CL} = \frac{D \times P}{S} \tag{6.10}$$

In this equation:

- CL stands for computational load, which measures the total computational power required to perform AI operations, expressed in terms of operations per second or gigaflops.
- $D$ represents the total data input size that needs to be processed by the AI system, measured in gigabytes (GB), which includes patient data, imaging files, and real-time health monitoring data.
- $P$ denotes the complexity of the processing tasks involved in the AI operations, which could be a factor based on the type and depth of analysis, or the complexity of algorithms being used. This might include layers in neural networks, the number of parameters in a machine learning model, or the computational steps required for processing.
- $S$ is the speed of the processing unit in the AI system, measured in gigaflops (billion floating-point operations per second), which indicates how quickly the system can process the incoming data.

By applying Eq. 6.10, healthcare facility administrators and IT departments can calculate the necessary computational resources needed to ensure that AI systems operate efficiently without lag or delay. This calculation helps in scaling up infrastructure appropriately to meet the demands of AI-driven applications, ensuring that healthcare providers can leverage the full potential of AI to improve patient care, enhance diagnostic accuracy, and optimize operational workflows. Proper planning using this equation supports the strategic implementation of AI technologies, crucial for driving innovation and efficiency in healthcare services.

### *6.5.3 Responsive Environments*

Responsive environments in smart hospitals epitomize a significant shift in architectural design, where spaces dynamically adapt to enhance the comfort, safety, and efficiency of both patients and staff. By integrating advanced technologies, these environments adjust automatically to various situational demands without the need for manual intervention, thereby optimizing the healthcare experience.

One of the key technologies pivotal to creating responsive environments is smart glass. This technology allows for the modulation of window transparency and light transmittance, which can be adjusted automatically to suit patient preferences or specific medical requirements. For instance, windows can become opaque to ensure privacy during medical procedures or can be cleared to let in natural light during recovery periods, fostering a better healing environment through mood enhancement. The ability of smart glass to respond to different needs without human input exemplifies the shift towards more patient-centered, responsive hospital designs.

Advanced HVAC systems are another cornerstone of responsive environments, crucial for maintaining optimal conditions within healthcare facilities. These systems are enhanced with IoT sensors that adjust settings based on room occupancy, external weather conditions, and specific healthcare needs, such as maintaining certain temperature and humidity levels to mitigate the spread of airborne pathogens. This capability ensures that the environment remains comfortable and safe for patients and staff, adapting in real-time to changes within the facility.

Moreover, the integration of IoT-connected fixtures extends the responsive capabilities of smart hospitals. Lighting, window shades, and even hospital beds that connect to the Internet of Things can be adjusted remotely to better serve the needs of the facility. For example, lighting intensity can adapt throughout the day to support the natural circadian rhythms of patients or be brightened during medical examinations or procedures. Window shades can automatically adjust to the sun's position to maintain optimal indoor lighting conditions, enhancing both energy efficiency and patient comfort.

Energy management systems play a crucial role in optimizing power usage across the hospital. By monitoring real-time energy consumption, these systems can dynamically adjust the operation of various components, like dimming lights in unoccupied rooms or modulating HVAC output depending on the area's usage, thereby reducing energy waste and promoting sustainability.

The deployment of a comprehensive sensor network is essential in responsive environments. These sensors track a variety of parameters including motion, light, noise, and air quality, providing the necessary data for the environment to adapt accordingly. The architectural design must consider the strategic placement and integration of these sensors to ensure they function effectively without compromising the aesthetic or therapeutic nature of space.

Flexibility in design is also crucial for accommodating future technological advancements. Smart hospitals should be built with modular components and infrastructure that can be easily updated or replaced, allowing for the seamless integration of emerging technologies without the need for extensive renovations.

While the functionality of these responsive systems is paramount, their aesthetic and functional integration within the hospital setting is equally important. Architectural designs should ensure that these technological elements are discreetly incorporated, maintaining a tranquil and healing atmosphere that supports patient recovery without the overt presence of technology.

**Table 6.15** Sensor and IoT device specifications

| Sensor/IoT device | Specifications | Use cases | Integration points in hospital |
|---|---|---|---|
| Wearable health monitors | Heart rate, oxygen saturation monitoring | Real-time patient health monitoring | Patient rooms, home health care |
| Environmental sensors | Temperature, humidity sensors | Monitor and adjust hospital environment | HVAC systems, operating rooms, patient rooms |
| Location tracking sensors | RFID, infrared sensors | Track equipment, staff, and patient locations | Emergency departments, general wards |
| Smart beds | Weight sensors, motion sensors | Monitor patient movement, vital signs | Patient rooms, ICU |
| Hand hygiene compliance sensors | Alcohol sensor, motion detection | Ensure compliance with hygiene protocols | Entrances, corridors, near patient rooms |
| Air quality sensors | $CO_2$, particulate matter sensors | Monitor and manage indoor air quality | All patients and public areas |

By implementing these design principles, smart hospitals with responsive environments can significantly enhance patient care and operational efficiency. These spaces not only adapt to the immediate needs of those within them but also contribute to the overarching goals of sustainability and efficacy in healthcare delivery, setting a new standard for future hospital designs.

Table 6.15 details the specifications and use cases of various sensors and IoT devices used in responsive hospital environments, highlighting how these technologies contribute to patient care, facility management, and medical procedures. This table provides healthcare administrators and technology teams with a clear, detailed breakdown of the types of sensors and devices available, their capabilities, and their integration within hospital systems to enhance efficiency, safety, and patient engagement.

### *6.5.4 Patient Interaction and Engagement*

In smart hospitals, the enhancement of patient interaction and engagement through technological integration is pivotal for elevating healthcare delivery and increasing patient satisfaction. The strategic placement of smart devices within patient rooms is instrumental in granting patients' greater control over their care and environment, thereby fostering a sense of autonomy and active participation in their health management.

The cornerstone of this approach is the integration of smart devices within patient rooms. These devices, such as tablets or bedside monitors, are designed to provide intuitive and accessible interfaces that allow patients of all ages and levels of tech-savviness to easily communicate with their healthcare providers. Patients can use these devices to schedule follow-up appointments, request assistance, or even order meals, thus enhancing their control and involvement in the care process.

Additionally, patients should have the ability to adjust environmental settings directly from their beds or through a handheld device. This integration should be seamless, incorporating controls that are easily accessible and simple to operate, thereby enhancing patient comfort and contributing to a more conducive recovery environment. Being able to adjust lighting, temperature, and window shades independently empowers patients, making their stay more comfortable and personalized.

Entertainment and educational access is also a critical component of patient rooms in smart hospitals. Smart televisions or display screens should provide a dual function—offering entertainment options like television programming and movies, along with internet access and tailored educational content pertinent to the patient's health condition. This not only aids in keeping patients entertained during their hospital stay but also plays a crucial role in keeping them informed and engaged in their recovery process.

Voice-activated technology is another innovative feature that can significantly enhance the patient's experience, especially for those with mobility challenges. By incorporating systems that respond to voice commands, patients can effortlessly control room settings, summon assistance, or navigate entertainment and educational content, thus adding a layer of convenience and accessibility to their hospital stay.

With the integration of these advanced technologies, the security and privacy of patient data become paramount. It is essential to implement secure networks that guard against potential data breaches and ensure that all communications and data transmissions within and out of the patient rooms are kept confidential.

From an ergonomic and aesthetic standpoint, the physical setup of technology in patient rooms must be carefully considered. Devices and screens should be positioned at appropriate heights and angles for easy use, ensuring that they are accessible and comfortable for all patients. Moreover, the technological elements should be designed to harmonize with the room's aesthetics, creating an environment that feels less clinical and more healing, thereby enhancing the overall experience of patients.

Lastly, technology should facilitate improved interactions between the healthcare staff and patients. Rooms equipped with systems that alert staff to patient needs or provide quick health status updates can significantly enhance the efficiency and effectiveness of care delivery, ensuring that medical personnel can respond swiftly to any patient's requirements.

By adhering to these design principles, smart hospitals can craft patient rooms that leverage technology not only to improve health management and comfort but also to augment the overall patient experience. This holistic approach supports better clinical outcomes and reinforces the hospital's reputation as an innovative, patient-centered institution committed to delivering superior healthcare services.

Table 6.16 evaluates various technologies based on their impact on patient satisfaction and engagement, offering an analytical view of how these tools enhance patient interaction within healthcare settings. This assessment highlights the effectiveness of

**Table 6.16** Patient engagement technologies assessment

| Technology | Impact on patient satisfaction | Impact on patient engagement | Overall effectiveness score (out of 10) |
|---|---|---|---|
| Interactive patient portals | High | High | 9 |
| Virtual reality therapy | Moderate | High | 8 |
| Mobile health apps | High | Moderate | 8 |
| Telehealth services | High | High | 9 |
| Automated patient reminders | Moderate | Moderate | 7 |
| Wearable fitness trackers | Low | High | 7 |

each technology in improving the quality of care and the patient experience, aiding healthcare providers in selecting the most impactful solutions for their facilities.

### 6.5.5 *Security and Privacy*

In the modern landscape of healthcare, where digital technologies pervade every facet of operations, the security and privacy of sensitive information are paramount. Smart hospitals, at the intersection of technology and healthcare, are particularly susceptible to a range of security threats that can compromise patient safety and privacy. To mitigate these risks, a robust security framework is essential. This framework should include advanced cybersecurity measures tailored to the healthcare context. Smart hospitals should deploy comprehensive cybersecurity protocols, such as firewalls and intrusion detection systems, to safeguard against potential breaches. Encryption protocols must be implemented to ensure that data, whether in transit or at rest, is protected from unauthorized access. The architecture of these hospitals should be designed to facilitate not just cybersecurity but also physical security measures, creating a safe environment for both patients and healthcare providers.

The network infrastructure of a smart hospital is a critical component that requires meticulous design to enhance security. Hospitals must adopt a segmented network architecture to effectively isolate patient data from operational data, thereby reducing the risk of cross-contamination in the event of a data breach. The use of secure, encrypted wireless networks is crucial to maintaining the integrity and confidentiality of sensitive information while supporting the mobility needs of staff and patients. Regular updates and security audits are also vital to adapt to the evolving landscape of cyber threats, ensuring that the hospital's defenses remain impervious to new vulnerabilities.

Beyond digital security, physical security measures are imperative to prevent unauthorized access to critical areas such as server rooms, data centers, and areas where patient records are stored. Access to these sensitive areas should be controlled through advanced authentication systems such as biometrics, key cards, or PIN codes, which not only restrict entry but also monitor it through log entries. These logs are crucial for identifying potential security breaches and initiating prompt corrective actions. Additionally, surveillance systems equipped with strategically placed cameras can monitor and record activities around high-risk areas and common spaces, providing real-time security data that is crucial for the prevention of unauthorized activities.

Privacy concerns extend into the physical layout and design of patient care areas. Hospitals must prioritize the confidentiality of patient interactions and treatments. This can be achieved through architectural and interior designs that use soundproofing materials to prevent eavesdropping, and layouts that minimize public exposure to treatment areas. Privacy-enhancing features like screens or smart glass, which can be adjusted to become opaque, ensure that patient dignity is maintained during sensitive procedures.

An often-overlooked aspect of security is the human element. Regular training and awareness programs are essential for hospital staff and patients. These programs should focus on secure data handling practices, the identification of phishing attempts, and the security of personal and hospital devices. Furthermore, smart hospitals must have comprehensive emergency response and disaster recovery plans. These plans should include secure, redundant data backups, off-site storage solutions, and well-equipped emergency operations centers to ensure continuity of care and data integrity in the event of major incidents or disasters.

By integrating these multifaceted security measures into their architectural and operational strategies, smart hospitals can provide a secure environment that safeguards patient data and health information against both cyber and physical threats. These practices not only meet legal and ethical standards but also enhance patient trust, a crucial element in the healthcare provider-patient relationship, particularly in an era where data privacy concerns are at the forefront of patient's minds.

Equation 6.11 is a sophisticated risk assessment model designed to quantify the risk levels of data breaches in healthcare settings by evaluating system vulnerabilities and the probabilities of potential threats. This model is crucial in discussions about cybersecurity in healthcare, where the protection of sensitive patient data is paramount. Implementing a quantitative approach to assessing risk levels enables healthcare organizations to prioritize resources and strengthen their defenses against data breaches effectively.

$$R = V \times T \times I \tag{6.11}$$

In this equation:

- $R$ represents the overall risk level of a data breach occurring, providing a quantitative measure that can guide decision-making in cybersecurity strategies.

- $V$ denotes the vulnerability score, which quantifies the susceptibility of the system to security threats. This score is derived from factors such as outdated software, lack of proper encryption, and insufficient access controls.
- $T$ is the threat probability, indicating the likelihood of a potential security breach attempt. This probability is influenced by factors like the value of the stored data, the profile of the healthcare institution, and the current threat landscape.
- $I$ stands for impact, which measures the potential severity and consequences of a data breach occurring. This includes considerations such as the type and amount of data at risk, compliance implications, and potential damage to the organization's reputation.

By utilizing Eq. 6.11, healthcare IT security teams can effectively assess and model the risk associated with various data security vulnerabilities and external threats. This model assists in systematically identifying high-risk areas, enabling targeted improvements in security protocols and measures. Prioritizing security efforts based on quantified risk levels helps ensure that resources are allocated efficiently, enhancing the protection of critical healthcare data against breaches. This rigorous approach to risk assessment is essential for maintaining trust and ensuring the integrity of patient information within the healthcare sector.

### 6.5.6 Sustainability

The pursuit of sustainability in smart hospitals encompasses more than environmental stewardship; it is about integrating efficiency, enhancing patient health, and optimizing cost-effectiveness into every facet of their operation and design. Smart hospitals, by embedding sustainable practices into their core functionalities, not only bolster their commitment to ecological responsibility but also foster environments that promote healthier outcomes for patients and more economically viable operations for healthcare systems.

Incorporating energy-efficient technologies stands as a cornerstone of sustainable smart hospital design. Utilizing advanced HVAC systems that adapt to changes in environmental conditions and occupancy levels, implementing LED lighting equipped with motion sensors and automatic dimming, and deploying comprehensive energy management systems that monitor and regulate energy consumption throughout the facility are essential strategies. These systems not only reduce energy usage but also cut down on operational costs, making healthcare facilities more sustainable.

The choice of building materials also plays a crucial role in the sustainability of smart hospitals. Materials selected should be sustainable, have minimal impact on indoor air quality, and contribute to the overall healthiness of the environment. This includes opting for building materials that emit low levels of volatile organic compounds (VOCs), are sourced through responsible practices, and are recyclable at their lifecycle's end. Durability is also a key consideration, as materials that last longer reduce the need for frequent replacement, thereby minimizing waste.

The architectural design of smart hospitals should emphasize maximizing natural light, which serves multiple functions. Natural light not only reduces dependency on artificial lighting, thus conserving energy but also significantly improves the health and well-being of both patients and staff. Studies have shown that natural light can enhance mood, boost sleep quality, and even potentially reduce the length of hospital stays. Features like large windows, skylights, and strategically designed internal courtyards are instrumental in achieving these benefits.

Further integrating sustainability, smart hospitals can adopt green roofs and living walls. These features contribute to a building's energy efficiency by enhancing insulation and reducing heat absorption, which lowers cooling costs. They also improve air quality and provide accessible green spaces for patients and staff, which can be therapeutic and contribute to the overall wellness environment of the facility.

Water conservation is another critical aspect of sustainable hospital operations. Implementing systems that reduce water use, such as low flow plumbing fixtures, and promoting the recycling of water, like using rainwater for non-potable purposes, are vital. These measures not only lessen the water footprint of the facility but also align with broader environmental conservation goals.

Effective waste management strategies are imperative in healthcare settings, which are known for significant waste generation. Smart hospitals should establish rigorous protocols for waste segregation, promote recycling initiatives, and ensure the safe disposal of medical waste. Moreover, reducing the use of single-used items and replacing them with reusable alternatives wherever feasible can substantially decrease the volume of waste produced.

Lastly, sustainable design in smart hospitals also means ensuring that the infrastructure is resilient and adaptable to various climate conditions and environmental stresses. This includes building considerations that can withstand extreme weather events such as floods or hurricanes, enabling hospitals to maintain operations during such crises without compromising safety or quality care.

As smart hospitals evolve, their dedication to sustainable practices sets a new benchmark for healthcare infrastructure globally. Not only does this approach mitigate immediate environmental impacts, but it also ensures the long-term sustainability and resilience of healthcare facilities, thereby equipping them to face future challenges and healthcare crises effectively. This commitment to sustainability is integral as healthcare facilities prepare to navigate the complexities of future healthcare needs, focusing on resilience and adaptability to continue delivering high-quality care under diverse conditions.

Table 6.17 lists key sustainability metrics for smart hospitals, including Energy Use Intensity (EUI), water reuse rates, and waste recycling percentages. This table provides a metrics-based assessment of sustainability practices in healthcare facilities, helping to quantify their environmental impact and effectiveness in achieving sustainability goals. By benchmarking these metrics, the table aids healthcare administrators in evaluating their facilities against industry standards and identifies areas for improvement in their sustainability initiatives.

**Table 6.17** Sustainability metrics and benchmarks

| Sustainability metric | Description | Current performance | Industry benchmark | Target goal |
|---|---|---|---|---|
| Energy use intensity (EUI) | Measures energy consumption per square foot | 150 $kWh/m^2$ | 130 $kWh/m^2$ | 120 $kWh/m^2$ |
| Water reuse rate | Percentage of wastewater recycled for use | 40% | 50% | 60% |
| Waste recycling rate | Percentage of waste materials recycled | 30% | 45% | 50% |
| Renewable energy utilization | Proportion of energy from renewable sources | 20% | 35% | 50% |
| Green building certification | Percentage of buildings certified (e.g., LEED) | 50% | 60% | 75% |

## 6.6 Preparing for Future Healthcare Challenges

As healthcare continues to evolve, facilities must be designed with the foresight to handle future challenges, whether they arise from demographic shifts, technological advancements, or global health threats. The adaptability and resilience of healthcare infrastructure are crucial in ensuring that facilities can continue to provide high-quality care under varying conditions and in response to unforeseen crises.

### *6.6.1 Designing for Epidemic and Pandemic Preparedness*

Designing healthcare facilities with a proactive approach to epidemic and pandemic preparedness is essential in today's increasingly interconnected world. The architecture and operational strategies of such facilities must be agile, allowing for quick adaptation to the heightened demands and shifting healthcare needs that arise during public health emergencies. This requires a thorough integration of design elements that can accommodate rapid changes in function and capacity while maintaining strict infection control protocols.

One of the fundamental aspects of designing for pandemic readiness involves creating modular and convertible spaces within healthcare facilities. These spaces should be capable of serving multiple purposes, easily transitioning into isolation wards or intensive care units as needed during an outbreak. To facilitate these transitions with minimal disruption, such spaces should have ready access to necessary utilities and be constructed to allow quick modifications without significant structural changes. Convertible ventilation systems are particularly crucial, as they enable rooms to alter air pressure settings to create negative pressure environments, a key strategy in preventing the spread of airborne diseases.

Dedicated isolation areas are also a core component of pandemic-preparedness design. These areas should be distinctly segregated from general patient care zones to prevent cross-contamination. They need to feature independent access and exit routes, specialized waste disposal systems, and dedicated ventilation that adheres to the highest standards of infection control. Designing these areas to be self-contained allows for effective management of infectious diseases while protecting both patients and healthcare staff from potential exposure.

To handle the logistical demands of public health crises, facilities should incorporate scalable spaces designed for mass testing or vaccination. These areas must be configured to optimize the flow of individuals and prevent overcrowding, with potential extensions into outdoor spaces or repurposed areas like parking lots during emergency scenarios. Additionally, robust infection control measures are essential. This includes the use of antimicrobial surfaces in high-contact areas and the strategic placement of sanitation stations throughout the facility, especially in entry points and high-traffic zones.

The integration of telehealth is another critical design consideration, enhancing the facility's capacity to manage patient flow and provide remote consultations during surges in demand. Spaces specifically designed for telehealth should be equipped with the necessary technological infrastructure, allowing healthcare providers to offer continuous and effective care remotely.

Properly addressing the needs of healthcare staff during emergencies is equally important. Facilities should include designated support areas where staff can rest and recuperate. These spaces are vital for maintaining staff well-being and operational efficiency during prolonged response efforts. They should be equipped with amenities for sleeping, eating, and personal downtime, designed to offer comfort and reduce stress.

Lastly, the infrastructure of the facility must be inherently flexible to support rapid adjustments in utility demands, such as increased electrical and water usage during a crisis. This includes provisions for backup power and additional water supplies, along with HVAC systems that can adapt to a range of medical scenarios and environmental conditions.

By embedding these design principles, healthcare facilities can ensure they are well-prepared to respond effectively to epidemics and pandemics. This approach not only facilitates rapid adaptation to emergency conditions but also underscores a commitment to safety, efficiency, and the continued well-being of both patients and healthcare workers. This preparation is crucial for building resilient healthcare infrastructures capable of facing the challenges posed by future public health crises.

Table 6.18 lists essential features and systems necessary for epidemic readiness in healthcare facilities, including advanced ventilation systems, isolation wards, and comprehensive emergency protocols. This checklist serves as a comprehensive reference for designing and equipping facilities to effectively manage and respond to epidemic outbreaks. By outlining critical infrastructure and procedural requirements,

**Table 6.18** Checklist for epidemic preparedness

| Feature/system | Description | Importance level | Notes |
|---|---|---|---|
| Advanced ventilation systems | Systems that enhance air filtration and circulation to control the spread of airborne diseases | Essential | Must meet or exceed HEPA filter standards |
| Isolation wards | Dedicated areas designed to treat infectious patients separately from other patients | Essential | Should be equipped with separate air-handling systems |
| Emergency protocols | Comprehensive plans for handling sudden outbreaks, including staff roles and communication strategies | Essential | Must be regularly updated and practiced |
| Personal protective equipment (PPE) stocks | Adequate reserves of PPE for healthcare workers | Essential | Regular checks to ensure supplies are not expired and are sufficient for extended outbreaks |
| Infection control training | Ongoing training for all healthcare staff on infection prevention and control best practices | High | Training should include updates on the latest guidelines and practices |
| Rapid testing facilities | On-site capabilities for quick testing to identify and manage infections effectively | High | Integration with local health authorities for real-time data sharing |
| Telehealth services | Systems to provide remote care and consultations to minimize contact and spread | High | Should be capable of handling increased loads during epidemics |

the table aims to ensure healthcare environments are prepared to handle infectious diseases safely and efficiently, protecting both patients and healthcare workers.

## 6.6.2 *Accommodating an Aging Population*

As demographic trends shift towards a growing elderly population, healthcare facilities are increasingly tasked with adapting their environments to meet the unique needs of older adults. This demographic often grapples with chronic diseases and mobility limitations, necessitating thoughtful design choices that emphasize accessibility, safety, and comprehensive care. Such adaptations are crucial for providing effective, compassionate, and accessible healthcare to elderly patients, ensuring they receive the support they need in a comfortable and conducive setting.

To begin with, healthcare facilities must prioritize the navigability of their spaces. This includes designing clear and unobstructed pathways that accommodate mobility aids such as wheelchairs and walkers. Floors should be constructed with non-slip materials, and corridors should be sufficiently wide to support the safe passage of mobility aids. Furthermore, the accessibility of elevators and ramps is crucial—they should be conveniently located and easy to use for all patients, regardless of mobility level. Visual aids are also essential; signage should be prominently displayed and feature large, legible fonts to aid those with visual impairments.

The integration of assistive technologies plays a pivotal role in enhancing the care provided to elderly patients. Innovative solutions such as wearable devices can monitor vital signs and detect falls, alerting healthcare providers to potential emergencies. Smart beds, which adjust automatically, can help optimize comfort for patients and prevent the development of bedsores. Additionally, voice-activated technology in patient rooms can allow individuals to control environmental factors such as lighting and temperature without physically moving, which can be particularly beneficial for those with limited mobility.

Telehealth is another critical component in modernizing care for the aging population. By integrating telehealth capabilities, healthcare facilities can drastically reduce the need for physical visits, which can be challenging and exhausting for older patients. Telehealth allows for the remote monitoring of chronic conditions and enables virtual consultations, providing consistent and accessible care from the comfort of the patient's home. Facilities should be equipped with dedicated spaces for telehealth that ensure privacy and are outfitted with the necessary technological infrastructure to facilitate these digital interactions effectively.

Adapting patient rooms to meet the diverse needs of elderly patients is also vital. This includes the provision of adjustable beds, accessible personal care areas, and bathrooms that accommodate mobility aids. Installing safety features such as handrails, emergency call buttons, and anti-scald devices on taps can further promote independence while ensuring patient safety.

Creating therapeutic environments within healthcare facilities can significantly enhance the well-being of elderly patients. Access to natural light, serene outdoor spaces, and designated quiet zones can provide spaces for relaxation and stress reduction, which are especially beneficial for this demographic. Additionally, social spaces that encourage interaction can play a crucial role in improving mental health and overall well-being for older adults. These areas should be designed to foster comfortable and safe social engagements, promoting a sense of community and support among patients.

Lastly, the importance of specialized staff training cannot be overstated. Healthcare facilities should incorporate dedicated areas for staff training specifically focused on geriatric care. This training should encompass the unique physical, psychological, and social challenges faced by elderly patients, equipping healthcare providers with the knowledge and skills needed to deliver compassionate and effective care.

By embedding these design principles into the architecture and operations of healthcare facilities, we can ensure that the needs of the aging population are met comprehensively. These facilities will not only be well-equipped to provide medical support but will also enhance the quality of life for elderly patients, making healthcare environments more inclusive, supportive, and effective for all.

Equation 6.12 is devised to calculate the efficiency of ergonomic designs in healthcare settings, particularly focusing on reducing mobility-related accidents among the elderly. This mathematical model is critical in discussions about the importance of ergonomics in the design of medical facilities, ensuring they are tailored to support the mobility needs of older adults. By enhancing the ergonomic features of a facility, healthcare providers can significantly lower the incidence of accidents, thereby improving safety and overall patient care quality.

$$\mathrm{EDE} = 1 - \frac{A_{\text{after}}}{A_{\text{before}}} \tag{6.12}$$

In this equation:

- EDE stands for ergonomic design efficiency, which measures the effectiveness of ergonomic improvements in reducing accidents. This efficiency is expressed as a ratio reflecting the reduction in accidents due to ergonomic interventions.
- $A_{\text{before}}$ represents the number of mobility-related accidents recorded before implementing ergonomic design changes. This figure serves as a baseline to assess the initial risk level within the environment.
- $A_{\text{after}}$ denotes the number of mobility-related accidents observed after the ergonomic design interventions have been put in place, illustrating the impact of these changes.

By applying Eq. 6.12, healthcare facility designers and administrators can quantitatively evaluate the success of ergonomic enhancements in their buildings. This assessment allows them to verify the practical benefits of such designs in terms of reducing falls and other mobility-related incidents among elderly patients. The resulting data can be used to justify further investments in ergonomic improvements, support modifications in facility design, and contribute to the development of safer healthcare environments. This approach not only enhances patient safety but also aligns with broader healthcare goals of improving quality of life and care for elderly populations.

### *6.6.3 Technological Scalability*

In the rapidly evolving landscape of medical technology, healthcare facilities must be designed for adaptability to accommodate new advancements without necessitating extensive structural changes. This foresight in design ensures that healthcare institutions can continuously offer state-of-the-art care by integrating the latest technological innovations efficiently and effectively.

The key to achieving this is the creation of flexible and modular spaces within the healthcare facility. These spaces are designed to be easily reconfigured or expanded to incorporate new technologies or equipment as they become available. The use of modular designs, including movable walls, interchangeable fixtures, and modular furniture systems, allows specific areas of a facility to be updated or altered without impacting the overall infrastructure. This flexibility is crucial for the seamless integration of new medical equipment and technologies, ensuring that updates can be made with minimal disruption to the facility's operations.

Infrastructure that supports high-speed digital connectivity is another critical component. Facilities must be equipped with robust data cabling, extensive wireless access points, and the capacity to upgrade these systems to accommodate faster and more reliable technologies as they emerge. Additionally, the infrastructure should have adequate data processing capabilities to handle the large volumes of information generated by new medical technologies, ensuring that data management is efficient and secure.

For the integration of future technologies, healthcare facilities should include easily accessible utility channels. These channels, potentially implemented within raised floors or ceiling spaces, should be designed to accommodate additional electrical and data cables easily. Such designs allow for quick access to maintenance, upgrades, and the seamless incorporation of new equipment, thus minimizing operational disruptions.

Considering the substantial power requirements and heat generation of advanced medical equipment, healthcare facilities must also plan for scalable power supply and cooling systems. This involves installing electrical systems that can be scaled up to meet increased demand and HVAC systems capable of adjusting to the additional heat load introduced by new technologies. This foresight in utility planning is crucial for maintaining environmental control within clinical spaces, ensuring both patient comfort and the optimal functioning of medical equipment.

Predefined integration points for new technologies within the facility's design can further streamline the adaptation process. These points, equipped with connections for power, data, and other necessary utilities, facilitate the quick and efficient installation of new systems, reducing the time and cost associated with technological upgrades.

To foster continuous innovation, including designated spaces such as innovation labs within the facility, is beneficial. These labs serve as incubators for new ideas and technologies, allowing medical staff to engage with upcoming advancements in a controlled environment. This proactive approach not only aids in the smooth transition of technologies from experimental to clinical use but also ensures staff are well acquainted with new tools before they are fully integrated into the healthcare setting.

Lastly, the inclusion of training and simulation rooms equipped with the latest technology is essential for ensuring that staff can effectively utilize new advancements. These rooms provide a dynamic environment where healthcare professionals can receive training on new procedures and technologies, keeping them adept at the forefront of medical practice.

By embedding technological scalability into the architecture and operational philosophy of healthcare facilities, institutions can ensure they remain flexible and responsive to the rapid advancements in medical technology. This strategic foresight not only enhances operational efficiency and future-proofs the facility but also

**Table 6.19** Technology upgrade impact assessment

| Technology upgrade | Estimated cost | Expected downtime | Space requirements | Impact on operations | Cost–benefit ratio |
|---|---|---|---|---|---|
| MRI machine upgrade | $2 million | 2 weeks | 200 sq. ft | High improvement in diagnostic precision | High |
| Electronic health records system | $500,000 | 3 days | No additional space needed | Increasing data accessibility and security | Moderate |
| Telemedicine expansion | $300,000 | 1 week | 100 sq. ft | Extends care reach, reduces patient in-flow | High |
| Patient monitoring system | $600,000 | 5 days | 150 sq. ft | Enhance real-time patient care and monitoring | High |
| Robotics for surgery | $1.5 million | 3 weeks | 300 sq. ft | Improves surgical outcomes, reduces human error | High |

guarantees that patients have access to the most advanced medical care available as technology continues to progress.

Table 6.19 evaluates the impact of potential technological upgrades on existing healthcare facilities, considering factors like cost, downtime, and space requirements. This assessment aids decision-makers in understanding the practical implications of integrating new technologies and balancing the benefits against the operational disruptions they may cause. By quantifying these factors, the table provides a strategic framework to guide investments in technology, ensuring they are both effective and feasible within the current facility constraints.

### *6.6.4 Sustainability and Environmental Resilience*

Sustainability and environmental resilience are increasingly vital in the design and operation of healthcare facilities, especially as the effects of climate change intensify. Modern healthcare facilities must be built to withstand diverse climatic conditions while minimizing environmental footprints, ensuring they remain safe, operational, and responsible stewards of environmental resources. Here's how these objectives can be seamlessly integrated into the design and functioning of healthcare facilities:

Incorporating energy-efficient materials and systems is foundational in reducing a healthcare facility's carbon footprint. High-performance insulation, energy-efficient windows, and advanced HVAC systems designed to consume less energy while optimizing indoor air quality are essential. Furthermore, the adoption of LED lighting, energy-efficient appliances, and smart building systems capable of adjusting energy consumption based on real-time occupancy and needs can drive significant reductions in energy use. These technologies not only lower energy costs but also contribute to the broader goal of environmental sustainability.

The integration of green roofs is another effective strategy for enhancing the ecological footprint of healthcare facilities. Green roofs provide excellent insulation, helping to reduce the costs associated with heating and cooling. They also play a crucial role in stormwater management by absorbing rainwater, thereby mitigating runoff and the associated risk of flooding. Moreover, these green spaces contribute to air quality improvement and offer therapeutic benefits to patients and staff, fostering better mental and physical health through accessible, natural settings.

Energy recovery systems represent a smart investment for enhancing energy efficiency within healthcare settings. By capturing and reutilizing waste heat from the building's exhaust systems to warm incoming fresh air, these systems reduce the energy demand associated with heating and ventilation. This is particularly beneficial in healthcare facilities, where maintaining precise control over air quality and temperature is crucial for patient care and comfort.

Water conservation is another pillar of sustainable healthcare facility design. Installing low-flow fixtures, optimizing laundry and sterilization processes, and implementing systems for rainwater collection and reuse can dramatically reduce water usage. Additionally, choosing landscaping that includes native plant species can decrease the overall water requirements for maintaining facility grounds, further reducing the environmental impact.

Designing with climate resilience in mind means preparing for local climate conditions and potential natural disasters. This could involve elevating structures in flood-prone areas, utilizing materials that can withstand extreme weather conditions, and designing shading features to mitigate the effects of rising temperatures. It is also critical to have robust contingency plans in place for maintaining water and power supplies during severe weather events to ensure continuous operation.

The construction phase itself presents an opportunity to reinforce sustainability commitments. Employing sustainable construction practices—such as managing construction waste effectively, using locally sourced materials to minimize transportation emissions, and selecting sustainably produced or recycled materials—can significantly reduce the environmental impact of building a new healthcare facility.

Finally, healthcare facilities must consider their broader environmental impact. This encompasses reducing emissions from facility operations, managing hazardous materials and waste responsibly, and initiating comprehensive recycling programs.

By embracing these sustainable practices and resilience strategies, healthcare facilities not only mitigate their environmental impact but also enhance their capacity to adapt to changing climate conditions. Such efforts not only align with global sustainability goals but also boost the facility's operational efficiency and reputation within the community, creating a healthier, more sustainable environment for both patients and staff. This proactive approach in healthcare facility design and operation serves as a model for responsible, future-focused healthcare delivery.

Equation 6.13 is designed to calculate the Sustainability Performance Index (SPI) for healthcare facilities, offering a comprehensive measure based on critical environmental factors such as energy consumption, waste management, and water use. This formula plays a pivotal role in discussions about the environmental impact of healthcare operations and underscores the importance of sustainability practices within these settings. By quantifying sustainability performance, healthcare facilities can assess their environmental footprint and identify areas for improvement.

$$\mathrm{SPI} = \frac{E_{\mathrm{eff}} + W_{\mathrm{eff}} + H_{\mathrm{eff}}}{3} \tag{6.13}$$

In this equation:

- SPI stands for the sustainability performance index, which provides an overall score reflecting the environmental sustainability of the facility.
- $E_{\mathrm{eff}}$ is the energy efficiency score, calculated based on the facility's energy consumption relative to its size and the services it provides. This score considers factors like the use of renewable energy sources, energy-saving technologies, and the overall energy management practices of the facility.
- $W_{\mathrm{eff}}$ represents the waste management efficiency score, which evaluates how effectively the healthcare facility manages its waste. This includes the reduction, recycling, and safe disposal of medical and non-medical waste, aiming to minimize the environmental impact.
- $H_{\mathrm{eff}}$ denotes the water use efficiency score, assessing the facility's water conservation and management strategies. This score reflects the efforts to reduce water consumption, recycle water when possible, and implement sustainable water management practices.

By calculating the SPI using Eq. 6.13, healthcare facilities can gain a clear understanding of their sustainability performance across multiple dimensions. This index helps in benchmarking against best practices and regulatory standards, driving improvements in environmental sustainability. The SPI not only serves as an indicator of current performance but also guides strategic planning for future sustainability initiatives, helping healthcare facilities reduce their ecological footprint while enhancing their operational efficiency and compliance with environmental regulations. This approach supports the broader goals of promoting environmental responsibility within the healthcare sector.

### 6.6.5 Mental Health Considerations

As mental health increasingly becomes recognized as a crucial aspect of comprehensive healthcare, the design of healthcare facilities plays a pivotal role in fostering environments conducive to mental and emotional well-being. Thoughtful architectural and interior design choices can significantly enhance patient recovery and improve the overall well-being of both patients and staff. Here's how healthcare facilities can integrate these essential mental health considerations into their design to create spaces that support healing and comfort:

Creating calming environments is fundamental to a mental health-friendly facility. The use of natural light, which has been scientifically shown to elevate mood and increase energy levels, should be maximized in patient areas and staff workspaces. Color schemes also play a crucial role; soft, soothing colors can create a tranquil atmosphere that promotes calmness and relaxation. Additionally, incorporating natural elements such as indoor plants, water features, and artwork that depicts serene landscapes can help reduce stress and anxiety, making the healthcare environment more welcoming and less institutional.

Access to outdoor spaces like gardens or courtyards is another critical element. These spaces provide a crucial escape from the clinical environment, allowing patients and staff to connect with nature, which has been shown to have therapeutic effects. Such areas should be designed for easy accessibility, ensuring they are secure and private to allow individuals to relax and rejuvenate without concerns about safety or privacy.

Quite zones within healthcare facilities can serve as vital retreats from the bustling activity and noise typically associated with medical settings. These areas should be strategically insulated from the sounds of hospital operations and designed with comfort in mind—featuring comfortable seating, soft lighting, and perhaps elements of nature and art to enhance the peaceful ambiance. Such spaces are essential for patients and staff alike, providing a sanctuary for relaxation or contemplation away from the stresses of healthcare environments.

Privacy is especially crucial in areas designated for mental health treatment. Patient rooms and consultation areas should be designed to feel secluded and secure, with soundproofing features to minimize external noise, thus creating a sense of tranquility and isolation necessary for effective therapy. Spaces for one-on-one counseling or group therapy should be discreet, comfortable, and designed to maintain confidentiality, facilitating open communication and effective treatment.

The integration of mental health services into general medical areas can play a significant role in destigmatizing mental health care. By designing spaces within primary care settings specifically for mental health services, facilities can make these services more accessible and visible, encouraging holistic care and collaboration between mental health professionals and medical staff.

Facilities should also incorporate flexible spaces that can be used for a variety of therapeutic activities such as group therapy sessions, workshops, or relaxation

**Table 6.20** Mental health design integration

| Design feature | Mental health benefit | Patient recovery improvement (%) |
|---|---|---|
| Natural lighting | Reduces depression and anxiety symptoms | 30 |
| Quiet zones | Lowers stress and agitation levels | 25 |
| Green spaces | Enhances mood and cognitive function | 40 |
| Art installations | Boosts positivity and reduces perceived pain | 20 |
| Ergonomic furniture | Improves comfort and reduces physical stress, aiding mental recovery | 15 |

practices like yoga and meditation. These spaces need to be versatile and adaptable, equipped to accommodate different activities and group sizes with appropriate furnishings and resources.

Finally, wayfinding and navigation within healthcare facilities should be straightforward and stress-free. For individuals, dealing with mental health challenges, complex layouts, and confusing navigation can be sources of significant anxiety. Implementing clear, simple wayfinding systems, including intuitive signage and logical paths, can help mitigate stress and enhance the overall experience for patients, visitors, and staff.

By embedding these mental health considerations into the design of healthcare facilities, architects and planners can create environments that support both the physical and mental health needs of individuals. These facilities become places of healing and comfort, reflecting a holistic approach to health that recognizes the critical importance of mental and emotional well-being in overall health outcomes.

Table 6.20 lists various design features alongside their mental health benefits, incorporating findings from recent studies and statistics on patient recovery improvements. This table emphasizes the critical role of thoughtful design in supporting mental health within healthcare environments. Each feature is analyzed for its impact on reducing stress, enhancing patient comfort, and promoting overall mental well-being, providing a valuable resource for architects and healthcare providers aiming to create supportive spaces for mental health recovery.

### *6.6.6 Community Integration and Accessibility*

The incorporation of healthcare facilities into their communities is essential, transforming them into accessible, welcoming, and inclusive spaces that foster a sense of belonging and support. This integration is crucial for making sure that such spaces serve the community effectively, catering to diverse needs while promoting a supportive environment.

To begin with, ensuring physical accessibility is paramount. Healthcare facilities must be designed to be accessible to all individuals, including those with disabilities. This entails the provision of barrier-free access points such as ramps, sufficiently wide doors, and elevators that are equipped to accommodate wheelchairs and other mobility aids. The interior layout should be intuitive, allowing easy navigation throughout the facility. This is facilitated by clear, prominently displayed signage in multiple languages to serve a diverse population. Such measures ensure that no individual faces barriers to accessing care due to physical limitations.

Furthermore, reflecting on the cultural diversity of the community is vital in the design and operation of healthcare facilities. This can be achieved by integrating multicultural elements into the design of the facility, offering services in various languages, and ensuring that the staff is trained in cultural competence. The inclusion of artwork, decor, and programming that celebrate local traditions and cultures plays a significant role in making all community members feel represented and valued. Healthcare facilities must be seen as spaces where cultural diversity is not only acknowledged but actively celebrated.

In addition to accommodating cultural and physical needs, healthcare facilities should offer a broad spectrum of services tailored to the community's varied needs. This includes not only standard medical care but also preventive care, mental health services, chronic disease management, and rehabilitation services. By extending their services to include community health initiatives—like nutrition classes, fitness programs, and health screening facilities can weave themselves into the very fabric of the community they serve.

Creating community spaces within healthcare facilities further supports community integration. Spaces designated for community events such as health fairs, educational seminars, and support groups, which are easily accessible and adaptable for various uses, encourage active community participation. These activities not only bring people into the facility for health-related events but also foster an ongoing engagement with the community, enhancing the facility's role as a community hub.

Accessibility extends beyond the physical design of the facility. Ensuring easy access through public transportation, adequate parking, and pedestrian-friendly pathways is essential for community integration. Facilities must also embrace digital connectivity, allowing people to access healthcare resources remotely, which is increasingly important in today's digital age.

The economic impact of healthcare facilities on their communities is also significant. By hiring local staff, sourcing supplies from local businesses, and engaging with local organizations, facilities can stimulate the local economy and build community goodwill. This economic integration helps in creating sustainable community relationships and positions the healthcare facility as a key stakeholder in the community's economic health.

Lastly, the design and operation of healthcare facilities with an eye toward environmental sustainability is increasingly important. Utilizing green building techniques and sustainable materials and installing systems that reduce energy and water consumption not only addresses environmental concerns but also underscores a commitment to the long-term health and well-being of the community.

By meticulously addressing these diverse but interconnected elements, healthcare facilities can ensure their successful integration into the communities they serve. This holistic approach not only enhances the care provided but also bolsters the overall health and cohesion of the community, securing the facility's role as a trusted and integral resource. As societal needs evolve and new technologies emerge, maintaining flexibility and foresight in design and function will be essential to adapt and continue meeting the community's health demands effectively.

## 6.7 Technological Failures and Mitigation Strategies

While the integration of advanced technologies in healthcare facilities offers significant benefits, it also introduces potential risks and failures that can adversely affect operations and patient care. This subsection addresses common technological failures in healthcare settings and outlines effective mitigation strategies to ensure continuous, reliable, and safe technology use.

### *6.7.1 Identifying Common Technological Failures*

In the modern digital environment of healthcare, technological failures represent a significant threat to the continuity and safety of patient care and facility operations. Proactively identifying and mitigating such failures is essential to maintain the reliability and security of healthcare services. Effective strategies can significantly reduce the likelihood of disruptions and protect sensitive data, ultimately ensuring a high standard of patient care.

One of the most critical challenges is managing system downtimes, which can disrupt patient care and block access to vital information. To address this, healthcare facilities must invest in redundant systems and reliable backup solutions capable of seamlessly taking over in case of system failure. Additionally, facilities should conduct regularly scheduled maintenance and updates during off-peak hours to minimize the impact on patient care and facility operations. This proactive approach ensures that the system remains robust, and that patient care continues uninterrupted even during technical difficulties.

Another significant concern is the threat of data breaches, which can have dire consequences due to the sensitive nature of patient data stored within healthcare systems. To safeguard against breaches, facilities must implement comprehensive cybersecurity measures, including robust firewalls, advanced encryption protocols, multi-factor authentication, and regular security audits. It is equally crucial to train staff on best data security practices to mitigate risks associated with human error, thereby fortifying the security framework from within.

Equipment malfunctions also pose a considerable risk, potentially delaying critical diagnoses and treatments, which could endanger patient health. To prevent such scenarios, healthcare facilities should enforce regular maintenance schedules for all medical equipment, carried out by certified technicians. The availability of rapid response repair teams and backup equipment is vital to minimize downtime and maintain continuity of care when primary equipment fails.

Software errors are another area of concern, as they can lead to the mishandling of patient data or misinterpretations of clinical information, which could result in clinical errors. Healthcare facilities must ensure that all software systems undergo rigorous testing before they are deployed. Regular updates should be managed meticulously to maintain system compatibility and functionality. Continuous monitoring for software operational anomalies is crucial in detecting and rectifying issues promptly before they impact patient care.

Furthermore, the importance of training and awareness cannot be overstated. Regular, ongoing training for healthcare staff on the proper use of technology and potential technological issues is fundamental. Facilities should also conduct regular drills and simulations of technological failures to prepare staff for actual incidents, enhancing their readiness and response capabilities.

Developing a comprehensive incident response plan is essential for effectively managing technological failures. Such a plan should outline clear procedures for rapidly addressing various types of technological issues, mitigating their impact, communicating with all stakeholders involved, and swiftly restoring normal operations.

Finally, fostering strong collaborations with technology providers is crucial. These relationships ensure that healthcare facilities can quickly access support and expertise when issues arise. Holding technology providers to high standards of service and responsiveness should be a stipulated part of contract agreements.

By proactively addressing these common technological failures through comprehensive planning, regular training, and robust system design, healthcare facilities can ensure more reliable and secure operations. These measures not only protect the facility and its patients but also contribute to the overall effectiveness and efficiency of healthcare delivery, fostering a resilient healthcare environment.

Table 6.21 provides a detailed checklist of required accessibility features for inclusivity in healthcare facilities, including specifications for ramps, door widths, and signage. This guide is essential for architects and planners tasked with designing environments that accommodate all individuals, regardless of physical ability. By outlining these specific features, the checklist ensures that healthcare settings are accessible and welcoming to everyone, promoting equality and compliance with legal accessibility standards.

**Table 6.21** Accessibility features checklist

| Accessibility feature | Specification | Importance level | Compliance standard |
|---|---|---|---|
| Ramps | Slope ratio of 1:12, at least 36 inches wide | Essential | ADA requirements |
| Door widths | Minimum of 32 inches clear width for accessibility | Essential | ADA requirements |
| Signage | High contrast, tactile text with Braille | Essential | ADA requirements |
| Elevators | Minimum dimensions of 54 inches by 80 inches, with accessible controls | Essential | ADA requirements |
| Bathroom facilities | ADA-compliant grab bars, turning space of 60 inches diameter | Essential | ADA requirements |
| Parking | Designated accessible parking spots, adequately sized and marked | Essential | ADA requirements |
| Hallway width | Minimum of 42 inches to allow wheelchair accessibility | Essential | ADA requirements |
| Emergency systems | Visual and audio emergency alerts, accessible call buttons | High | ADA requirements |

### *6.7.2 Redundancy Systems*

Implementing redundancy systems within healthcare facilities is an essential measure to guarantee uninterrupted service and continuity of care during any form of technological or systemic failure. Redundancy ensures that backup systems or processes can seamlessly take over the critical functions of the healthcare facility without any disruption, thus maintaining operational integrity and ensuring continuous patient care.

One of the foundational elements of redundancy in healthcare facilities is the setup of backup power supplies. Critical areas such as operating rooms, intensive care units, and emergency departments should be equipped with uninterruptible power supplies (UPS) and backup generators that automatically activate during a power outage. This setup is crucial as it ensures that life-supporting equipment and essential medical devices continue to operate without interruption, thereby safeguarding patient health during power fluctuations or outages.

Data redundancy also plays a vital role in protecting electronic health records and other critical medical data. Healthcare facilities should implement robust data storage solutions that automatically duplicate and mirror data to multiple locations. This can be achieved through redundant servers located on-site or through cloud-based solutions that ensure real-time data backup in geographically dispersed data centers.

Such measures prevent data loss and allow quick data recovery, thus maintaining the integrity and availability of medical records.

Communication is another critical area that requires redundancy. In emergencies where regular communication networks may fail, alternative communication systems such as satellite phones, two-way radios, or dedicated internet lines should be available. These systems should be capable of operating independently of the primary network, ensuring that staff can maintain effective coordination and communication during critical times.

Additionally, the installation of redundant networking hardware is necessary to support continual network connectivity. This includes having duplicate switches, routers, and wireless access points, which should be configured to automatically switch to backup hardware without manual intervention if the primary hardware fails. Such arrangements ensure that there is no interruption in network service, which is crucial for accessing patient information, communicating within the facility, and utilizing digital health tools.

Critical software applications used in patient management, diagnostics, and medication dispensing should be equipped with failover systems. These systems enable the software to switch instantly to a backup system in the event of a failure, minimizing downtime and ensuring that the management of patient care remains unaffected.

To ensure that these redundancy systems are always operational and effective, regular testing and maintenance are imperative. Healthcare facilities should conduct routine drills to test the transition to backup systems and perform regular updates to both hardware and software components to keep all systems running efficiently and securely.

Lastly, staff must be thoroughly trained on the operational procedures related to backup systems. They should be familiar with how to switch to and operate these systems during failures. Proper training can significantly reduce response times and enhance the facility's ability to maintain continuous patient care during unforeseen disruptions.

By integrating these comprehensive redundancy strategies, healthcare facilities can greatly reduce the risks associated with system failures. This proactive approach not only preserves the operational integrity of the facility but also ensures that the quality of patient care and safety remains uncompromised, regardless of any external or internal disruptions.

Equation 6.14 is developed to estimate the efficiency and required capacity of redundancy systems in healthcare settings, ensuring that critical systems have the necessary backup to maintain operations during unexpected failures. This calculation is pivotal in the context of healthcare facilities where continuous operation is essential for patient safety and care, such as in electrical systems, medical gas, and data storage.

$$\mathrm{RE} = \frac{C_{\text{actual}}}{C_{\text{required}}} \times 100\% \tag{6.14}$$

In this equation:

- RE stands for redundancy efficiency, expressed as a percentage, which measures the effectiveness and adequacy of redundancy systems in place.
- $C_{\text{actual}}$ is the actual capacity of the redundancy systems installed, measured in terms of power, data storage, or any other relevant unit, depending on the system being evaluated.
- $C_{\text{required}}$ represents the estimated capacity needed to ensure uninterrupted operation during primary system failures. This capacity is based on the critical needs of the healthcare facility, including peak demand scenarios and essential safety margins.

By applying Eq. 6.14, healthcare administrators and facility managers can assess whether their redundancy systems are adequate to handle emergencies and system failures without affecting the facility's operations. This calculation helps in ensuring that all critical systems are backed up sufficiently to handle unexpected disruptions, thereby safeguarding patient care and facility operations. The efficiency score derived from this equation also aids in identifying areas where improvements are needed, allowing for targeted upgrades to enhance system reliability and patient safety. This proactive approach to redundancy planning is crucial for maintaining high standards of care and operational continuity in healthcare environments.

### 6.7.3 Regular Maintenance and Updates

Regular maintenance and timely updates are fundamental to the smooth and efficient operation of healthcare facilities, playing a crucial role in minimizing the risk of technological failures that could potentially compromise patient care. Establishing robust practices for the upkeep and modernization of equipment and systems is essential for maintaining high standards of healthcare delivery.

To begin, healthcare facilities should set up a structured maintenance schedule for all technological equipment and systems, including medical devices, computer systems, and networking hardware. This schedule should align with the manufacturer's recommendations and adhere to industry's best practices. Regular maintenance not only prolongs the lifespan of the equipment but also ensures that it operates at peak efficiency. This systematic approach helps in avoiding unexpected malfunctions that could disrupt healthcare services.

In addition to scheduled maintenance, implementing preventive maintenance programs is critical. These programs proactively address potential issues before they escalate into significant problems. Routine inspections, tests, and servicing of equipment should be carried out regularly to ensure everything is in optimal condition. Preventive measures help identify and rectify signs of wear and tear, outdated software, or hardware beginning to fail, thus maintaining continuous operation and reliability.

Keeping software systems up to date is equally important. Regularly applying updates and patches to all software, including operating systems, management software, and specialized healthcare applications, is vital. These updates often include critical security enhancements that protect systems from vulnerabilities, as well as functional improvements and bug fixes that boost overall performance.

Effective documentation and tracking of all maintenance activities and updates are indispensable. Detailed records should be kept, documenting the service date, maintenance or update nature, identified or resolved issues, and the personnel involved. This documentation is crucial for future maintenance planning and provides a valuable resource for troubleshooting if recurrent issues arise.

Educating healthcare staff about the importance of maintenance and updating is also essential. Training should cover recognizing signs of equipment malfunction, protocols for reporting issues, and understanding the critical role that regular maintenance plays in patient safety and quality of care. Ensuring that all staff members are aware and informed can help in the early detection of potential problems, thus preventing them from affecting patient care.

Building strong relationships with equipment vendors and service providers is another key aspect. Service agreements should encompass timely maintenance and support, and leveraging vendor expertise ensures that maintenance is carried out to the highest standards. This can also facilitate access to specialized knowledge and replacement parts necessary for sophisticated medical equipment.

Lastly, having protocols in place for emergencies or unscheduled maintenance is crucial. Facilities must have rapid response plans that ensure immediate action is taken when critical equipment fails, preventing any adverse impact on patient care.

By adopting these comprehensive strategies, healthcare facilities can significantly mitigate the risks associated with technological failures. Regular maintenance and timely updates are not merely about keeping systems operational; they reinforce the facility's commitment to safety and quality care, ensuring that technology continues to serve as a reliable asset in delivering exceptional healthcare.

### *6.7.4 Comprehensive Training Programs*

In the realm of healthcare, where technology rapidly evolves and directly impacts patient outcomes, the implementation of comprehensive training programs for healthcare staff is indispensable. These programs are foundational in maximizing the effectiveness of new technologies while minimizing the risks associated with user errors and technological mishaps. The cornerstone of such initiatives is to ensure that all staff are not only familiar with the basic operation of new equipment but are also proficient in its advanced functionalities and its integration with existing systems. To achieve this, healthcare facilities must embark on meticulously designed initial training sessions that introduce the staff to new technologies. These sessions should be robust, covering all necessary operational details and ensuring staff are well-versed in the technology from the outset. This foundational training is crucial as it sets the stage for the staff's confidence and competence, significantly reducing the likelihood of operational errors.

Moreover, given the continuous advancements in healthcare technology, ongoing education becomes critical. Healthcare facilities should institute regular training updates to inform and educate staff about the latest technological upgrades and modifications. These updates should not only cover new features but should also reinforce fundamental usage protocols and best practices. Additionally, training programs must include comprehensive troubleshooting guides and emergency protocols, equipping the staff with the knowledge to handle common technical issues and to act swiftly in case of technological failures. Understanding whom to contact for immediate support, how to switch to backup systems, and how to manually perform tasks when technology fails, are all essential components of a well-rounded training program. This ensures that in times of technological disruption, the impact on patient care is minimal, and normal operations can be restored swiftly.

Hands-on practice sessions form another pillar of effective training, providing staff with the opportunity to engage in new technologies in a controlled, non-critical environment. Simulation-based training is particularly beneficial as it mimics real-life scenarios, allowing staff to practice and refine their responses to various challenges in a risk-free setting. Following these practical experiences, assessments should be conducted to evaluate and ensure that the staff have thoroughly understood and absorbed the training. These assessments can lead to certification, which serves as a benchmark for proficiency and aid in maintaining high operational standards across the healthcare facility.

Cross training is another strategic element that enhances operational flexibility and fosters a broader understanding of the technology throughout the facility. By training multiple staff members across different roles or departments, facilities can ensure that there is always a proficient operator available, especially during shifts or emergencies. Finally, integrating feedback mechanisms into the training programs is vital for continuous improvement. Feedback collected from staff routinely using the technology offers valuable insights into the effectiveness of the training and highlights areas needing further enhancement or clarification.

By prioritizing these critical aspects of training, healthcare facilities can ensure that their staff are not only equipped to use technology effectively but are also ingrained with a culture of continuous learning and adaptability. Such a culture is essential in the dynamic field of healthcare technology, where being ahead of the curve can significantly enhance patient care and operational efficiency.

### *6.7.5 Robust Cybersecurity Measures*

In the digital age, healthcare facilities increasingly rely on electronic systems to manage sensitive patient information, making robust cybersecurity measures essential to safeguard against data breaches and cyber threats. To protect this vital data, healthcare organizations must employ a multifaceted approach to cybersecurity, starting with the encryption of patient data. Encryption should be standard both in transit—when data is being transferred—and at rest—when data is stored. This ensures that sensitive information remains unintelligible to unauthorized users throughout its lifecycle.

Equally critical is the implementation of secure access controls. These controls should be stringent, allowing only authorized personnel access to sensitive data. Facilities should employ strong password policies, incorporate biometric authentication, and use multi-factor authentication (MFA) systems to enhance security. The principle of least privilege should govern access rights, granting individuals the minimum level of access required for their job functions, thereby limiting potential exposure of sensitive information.

Regular security audits are vital in maintaining a secure information environment. These audits should be conducted by independent security experts and include comprehensive penetration testing and vulnerability assessments, along with reviews of existing policies and procedures. Such audits help in identifying and mitigating potential vulnerabilities before they can be exploited.

Additionally, deploying firewalls and intrusion detection systems (IDS) provides a necessary barrier against unauthorized access and monitors network traffic for suspicious activity, respectively. These systems are crucial for providing real-time protection and for early detection of potential threats that could compromise the network.

Ongoing staff training in cybersecurity best practices is another pillar of a robust cybersecurity strategy. Employees should be educated about the risks of common cyber threats such as phishing, social engineering, and malware, and trained in the proper handling of sensitive information. Regular awareness programs can equip staff with the knowledge needed to prevent breaches and to respond appropriately to security incidents.

A well-defined incident response plan is imperative for a rapid and effective reaction to security breaches. This plan should detail the procedures for containment, notification of affected parties, and system restoration. Regular drills should be conducted to ensure that the response team is prepared to manage incidents swiftly and efficiently.

Continuous monitoring of all systems is essential for detecting and responding to threats in real-time. Healthcare facilities must ensure that their software and systems are regularly updated with the latest security patches to protect against newly emerging vulnerabilities. Moreover, robust data backup and recovery procedures must be established to ensure that critical data can be restored in the event of data loss or corruption. Backups should be frequent, comprehensive, and stored securely offsite to safeguard against both cyber and physical threats.

By implementing these robust cybersecurity measures, healthcare facilities can significantly enhance their defense against the myriads of cyber threats present in today's digital landscape. This not only preserves the integrity and confidentiality of patient data but also maintains the trust placed in healthcare providers, ensuring that patient care continues to be delivered safely and effectively within a secure digital environment.

### *6.7.6 Developing a Response Plan for Technological Failures*

Developing a comprehensive response plan for technological failures is a crucial strategy for healthcare facilities aiming to minimize disruptions and maintain continuity of care. Such a plan is vital as it prepares the staff to handle unexpected technological breakdowns effectively, thereby safeguarding patient care and operational stability. The first step in crafting such a plan involves the identification of potential technological failures that could adversely affect the facility. These can range from hardware malfunctions and software crashes to data breaches and power outages. Understanding the varied nature of these failures is fundamental in formulating tailored responses that address the specific needs and risks associated with each type of failure.

Once potential failures are identified, healthcare facilities must develop detailed response procedures for each scenario. These procedures should outline the immediate steps required to manage and mitigate the effects of the failure, specifying essential actions such as contacting IT support, engaging third-party vendors, switching to backup systems, and implementing measures to secure data and system integrity. The clarity and precision of these procedures are critical, as they must be straightforward enough to facilitate quick and decisive action during an emergency.

An effective response plan also involves the designation of specialized response teams. These teams should consist of individuals who are not only trained in the facility's technological systems but are also familiar with the specific processes needed for recovery and continuity. Clearly defined roles and responsibilities are essential to prevent overlap and confusion during a crisis, ensuring each team member knows exactly what is expected of them.

Additionally, a robust communication plan is an integral part of the response strategy. This plan should detail the protocols for informing staff, patients, and other stakeholders about the technological failure and the steps being taken to resolve it. It should specify who is authorized to communicate critical information externally, the methods by which updates will be communicated, and the platforms that will be used, such as internal messaging systems or public announcements.

To ensure the effectiveness of the response plan, regular training sessions and simulation drills are necessary. These drills, designed to mimic various types of technological failures, provide staff with practical experience and help to reinforce the response procedures. The feedback collected from these simulations is invaluable, offering insights that can be used to refine both the response plan and the associated training programs.

Moreover, the response plan should ensure that all critical systems are backed by reliable and readily accessible alternative solutions. This includes maintaining redundant hardware, alternative software solutions, and manual processes that can take over the critical functions of the healthcare facility until the primary systems are restored. Regular reviews and updates of the response plan are also necessary to accommodate new technological deployments, changes in operational procedures, or insights gained from past incidents. This iterative process ensures that the response plan remains relevant and effective.

Lastly, maintaining comprehensive documentation of the response plan is imperative. This documentation should include up-to-date contact lists, detailed procedure manuals, and access credentials for backup systems. It should be securely stored yet readily accessible to authorized personnel, ensuring that critical information is available when needed.

By meticulously developing and regularly updating a detailed response plan for technological failures, healthcare facilities can enhance their preparedness for unexpected technological disruptions. This proactive approach not only protects patient care during such incidents but also bolsters the overall resilience and reliability of the facility's technological infrastructure.

### *6.7.7 Vendor Support and Service Agreements*

Establishing robust relationships with technology vendors and securing comprehensive service agreements are essential strategies for ensuring that healthcare facilities receive timely and effective support in the event of technological failures. These partnerships are critical in maintaining the reliability and efficiency of the technology infrastructure that underpins crucial healthcare operations. Effective management of these vendor relationships and service agreements is foundational to ensure that healthcare facilities can rely on their technological systems to function optimally without disruption.

The selection of the right vendors is the initial step in this process. Healthcare facilities should prioritize vendors known for their reliability and excellent service within the healthcare sector. Assessing a vendor's reputation, financial stability, customer service reviews, and compliance with healthcare standards is imperative. This evaluation helps in choosing vendors that not only meet but exceed the expectations and needs of healthcare operations, ensuring that they can reliably support the critical functions these facilities perform.

Negotiating service agreements with these vendors is equally crucial. These agreements should be comprehensive and specifically tailored to the needs of the healthcare facility. Critical elements to negotiate include guaranteed response times to ensure that any issues are addressed swiftly, thereby minimizing the impact on operations. Round-the-clock technical support is essential as healthcare facilities operate 24/7, requiring immediate attention to issues regardless of the time. The availability of necessary replacement parts, stipulated to be readily accessible and quickly delivered, minimizes equipment downtime and expedites repairs. Regular maintenance checks and updates should be included in the agreements to prevent failures and ensure that all technologies are equipped with the latest software, which may include crucial security enhancements.

Service agreements should also include detailed performance metrics with defined penalties for service lapses. These metrics hold vendors accountable and ensure that they adhere to the service levels agreed upon, with penalties serving as a deterrent against substandard service delivery. Moreover, the agreement should offer scalability and flexibility, allowing the healthcare facility to adjust services or add new technologies as its needs evolve without necessitating a renegotiation of the entire agreement.

Regular reviews and audits of the vendor's services are necessary to ensure that the agreed standards are maintained over time. These assessments help in identifying any discrepancies or areas for improvement, facilitating timely adjustments to the service agreements. Establishing strong communication channels with vendors is also critical. Effective and open communication ensures that any technological issues are promptly identified and addressed and that the vendor is aware of the evolving needs of the healthcare facility.

By focusing on these critical aspects of vendor support and service agreements, healthcare facilities can ensure robust backing in managing technological issues. This proactive approach not only safeguards the functionality and security of crucial healthcare systems but also supports the overarching mission of delivering high-quality patient care. These strategic partnerships and carefully crafted agreements are indispensable in the modern healthcare landscape, where technology plays a pivotal role in day-to-day operations.

## Further Reading

Aceto, G., Persico, V., & Pescapé, A. (2018). The role of Information and Communication Technologies in healthcare: taxonomies, perspectives, and challenges. *Journal of Network and Computer Applications*, 107, 125–154. https://doi.org/10.1016/j.jnca.2018.02.008

Al-Emran, M., Mezhuyev, V., & Kamaludin, A. (2018). Technology Acceptance Model in M-learning context: A systematic review. *Computers & Education*, 125, 389–412. https://doi.org/10.1016/j.compedu.2018.06.008

Albahri, O. S., Zaidan, A. A., Zaidan, B. B., Hashim, M., Albahri, A. S., & Alsalem, M. A. (2018). Real-time remote health-monitoring Systems in a Medical Centre: A review of the provision of healthcare services-based body sensor information, open challenges and methodological aspects. *Journal of medical systems*, 42, 1–47. https://doi.org/10.1007/s10916-018-1006-6

Alfian, G., Syafrudin, M., Ijaz, M. F., Syaekhoni, M. A., Fitriyani, N. L., & Rhee, J. (2018). A personalized healthcare monitoring system for diabetic patients by utilizing BLE-based sensors and real-time data processing. *Sensors*, 18(7), 2183. https://doi.org/10.3390/s18072183

Amri, M. M., & Abed, S. A. (2023). The data-driven future of healthcare: a review. *Mesopotamian Journal of Big Data*, 2023, 68–74. https://doi.org/10.58496/MJBD/2023/010

Anikwe, C. V., Nweke, H. F., Ikegwu, A. C., Egwuonwu, C. A., Onu, F. U., Alo, U. R., & Teh, Y. W. (2022). Mobile and wearable sensors for data-driven health monitoring system: State-of-the-art and future prospect. *Expert Systems with Applications*, 202, 117362. https://doi.org/10.1016/j.eswa.2022.117362

Burton-Jones, A., Akhlaghpour, S., Ayre, S., Barde, P., Staib, A., & Sullivan, C. (2020). Changing the conversation on evaluating digital transformation in healthcare: Insights from an institutional analysis. *Information and Organization*, 30(1), 100255. https://doi.org/10.1016/j.infoandorg.2019.100255

Cascini, F., Santaroni, F., Lanzetti, R., Failla, G., Gentili, A., & Ricciardi, W. (2021). Developing a data-driven approach in order to improve the safety and quality of patient care. *Frontiers in public health*, 9, 667819. https://doi.org/10.3389/fpubh.2021.667819

Chang, K. M., Dzeng, R. J., & Wu, Y. J. (2018). An automated IoT visualization BIM platform for decision support in facilities management. *Applied sciences*, 8(7), 1086. https://doi.org/10.3390/app8071086

Chau, K. Y., Lam, M. H. S., Cheung, M. L., Tso, E. K. H., Flint, S. W., Broom, D. R., ... & Lee, K. Y. (2019). Smart technology for healthcare: Exploring the antecedents of adoption intention of healthcare wearable technology. *Health psychology research*, 7(1). https://doi.org/10.4081/Fhpr.2019.8099

Chen, L. K., Yuan, R. P., Ji, X. J., Lu, X. Y., Xiao, J., Tao, J. B., ... & Jiang, L. Z. (2021). Modular composite building in urgent emergency engineering projects: A case study of accelerated design and construction of Wuhan Thunder God Mountain/Leishenshan hospital to COVID-19 pandemic. *Automation in Construction*, 124, 103555. https://doi.org/10.1016/j.autcon.2021.103555

Chen, M., & Decary, M. (2020, January). Artificial intelligence in healthcare: An essential guide for health leaders. In Healthcare management forum (Vol. 33, No. 1, pp. 10–18). *Sage CA: Los Angeles, CA: SAGE Publications*. https://doi.org/10.1177/0840470419873123

Chen, S., Guo, X., Wu, T., & Ju, X. (2020). Exploring the online doctor-patient interaction on patient satisfaction based on text mining and empirical analysis. *Information Processing & Management*, 57(5), 102253. https://doi.org/10.1016/j.ipm.2020.102253

Costin, A., Adibfar, A., Hu, H., & Chen, S. S. (2018). Building Information Modeling (BIM) for transportation infrastructure–Literature review, applications, challenges, and recommendations. *Automation in construction*, 94, 257–281. https://doi.org/10.1016/j.autcon.2018.07.001

Dugstad, J., Eide, T., Nilsen, E. R., & Eide, H. (2019). Towards successful digital transformation through co-creation: a longitudinal study of a four-year implementation of digital monitoring technology in residential care for persons with dementia. *BMC health services research*, 19, 1–17. https://doi.org/10.1186/s12913-019-4191-1

Dwivedi, R., Mehrotra, D., & Chandra, S. (2022). Potential of Internet of Medical Things (IoMT) applications in building a smart healthcare system: A systematic review. *Journal of oral biology and craniofacial research*, 12(2), 302–318. https://doi.org/10.1016/j.jobcr.2021.11.010

Fennelly, O., Cunningham, C., Grogan, L., Cronin, H., O'Shea, C., Roche, M., ... & O'Hare, N. (2020). Successfully implementing a national electronic health record: a rapid umbrella review. International *Journal of Medical Informatics*, 144, 104281. https://doi.org/10.1016/j.ijmedinf.2020.104281

Fang, D., Pan, S., Li, Z., Yuan, T., Jiang, B., Gan, D., ... & Liu, Z. (2020). Large-scale public venues as medical emergency sites in disasters: lessons from COVID-19 and the use of Fangcang shelter hospitals in Wuhan, China. *BMJ global health*, 5(6), e002815. https://doi.org/10.1136/bmjgh-2020-002815

Godfrey, A., Hetherington, V., Shum, H., Bonato, P., Lovell, N. H., & Stuart, S. (2018). From A to Z: Wearable technology explained. *Maturitas*, 113, 40–47. https://doi.org/10.1016/j.maturitas.2018.04.012

Granić, A., & Marangunić, N. (2019). Technology acceptance model in educational context: A systematic literature review. *British Journal of Educational Technology*, 50(5), 2572–2593. https://doi.org/10.1111/bjet.12864

Halawa, F., Dauod, H., Lee, I. G., Li, Y., Yoon, S. W., & Chung, S. H. (2020). Introduction of a real time location system to enhance the warehouse safety and operational efficiency. *International Journal of Production Economics*, 224, 107541. https://doi.org/10.1016/j.ijpe.2019.107541

Hathaliya, J. J., & Tanwar, S. (2020). An exhaustive survey on security and privacy issues in Healthcare 4.0. *Computer Communications*, 153, 311–335. https://doi.org/10.1016/j.comcom.2020.02.018

Hathaliya, J. J., Tanwar, S., Tyagi, S., & Kumar, N. (2019). Securing electronics healthcare records in healthcare 4.0: A biometric-based approach. *Computers & Electrical Engineering*, 76, 398–410. https://doi.org/10.1016/j.compeleceng.2019.04.017

Hayn, D., Veeranki, S., Kropf, M., Eggerth, A., Kreiner, K., Kramer, D., & Schreier, G. (2018). Predictive analytics for data driven decision support in health and care. *it-Information Technology*, 60(4), 183–194. https://doi.org/10.1515/itit-2018-0004

Hong, H., & Oh, H. J. (2020). The effects of patient-centered communication: exploring the mediating role of trust in healthcare providers. *Health communication*, 35(4), 502–511. https://doi.org/10.1080/10410236.2019.1570427

Hwang, M., & Jiang, Y. (2023). Personalization in digital health interventions for older adults with cancer: A scoping review. *Journal of Geriatric Oncology*, 14(8), 101652. https://doi.org/10.1016/j.jgo.2023.101652

Jagadeeswari, V., Subramaniyaswamy, V., Logesh, R. T. A., & Vijayakumar, V. (2018). A study on medical Internet of Things and Big Data in personalized healthcare system. *Health information science and systems*, 6(1), 14. https://doi.org/10.1007/s13755-018-0049-x

Jayaraman, P. P., Forkan, A. R. M., Morshed, A., Haghighi, P. D., & Kang, Y. B. (2020). Healthcare 4.0: A review of frontiers in digital health. *Wiley Interdisciplinary Reviews: Data Mining and Knowledge Discovery*, 10(2), e1350. https://doi.org/10.1002/widm.1350

Jiang, S. (2020). The relationship between face-to-face and online patient-provider communication: examining the moderating roles of patient trust and patient satisfaction. *Health communication*. https://doi.org/10.1080/10410236.2018.1563030

Kalid, N., Zaidan, A. A., Zaidan, B. B., Salman, O. H., Hashim, M., & Muzammil, H. J. J. O. M. S. (2018). Based real time remote health monitoring systems: A review on patients prioritization and related" big data" using body sensors information and communication technology. *Journal of medical systems*, 42, 1–30. https://doi.org/10.1007/s10916-017-0883-4

Keshta, I., & Odeh, A. (2021). Security and privacy of electronic health records: Concerns and challenges. *Egyptian Informatics Journal*, 22(2), 177–183. https://doi.org/10.1016/j.eij.2020.07.003

Kumar, M., Kumar, A., Verma, S., Bhattacharya, P., Ghimire, D., Kim, S. H., & Hosen, A. S. (2023). Healthcare Internet of Things (H-IoT): Current trends, future prospects, applications, challenges, and security issues. *Electronics*, 12(9), 2050. https://doi.org/10.3390/electronics12092050

Lyles, C. R., Nelson, E. C., Frampton, S., Dykes, P. C., Cemballi, A. G., & Sarkar, U. (2020). Using electronic health record portals to improve patient engagement: research priorities and best practices. *Annals of internal medicine*, 172(11_Supplement), S123-S129. https://doi.org/10.7326/M19-0876

Mannino, A., Dejaco, M. C., & Re Cecconi, F. (2021). Building information modelling and internet of things integration for facility management—Literature review and future needs. *Applied Sciences*, 11(7), 3062. https://doi.org/10.3390/app11073062

Mathews, S. C., McShea, M. J., Hanley, C. L., Ravitz, A., Labrique, A. B., & Cohen, A. B. (2019). Digital health: a path to validation. *NPJ digital medicine*, 2(1), 38. https://doi.org/10.1038/s41746-019-0111-3

Meena, J. S., Choi, S. B., Jung, S. B., & Kim, J. W. (2023). Electronic textiles: New age of wearable technology for healthcare and fitness solutions. *Materials Today Bio*, 19, 100565. https://doi.org/10.1016/j.mtbio.2023.100565

Mele, C., Marzullo, M., Di Bernardo, I., Russo-Spena, T., Massi, R., La Salandra, A., & Cialabrini, S. (2022). A smart tech lever to augment caregivers' touch and foster vulnerable patient engagement and well-being. *Journal of Service Theory and Practice*, 32(1), 52–74. https://doi.org/10.1108/JSTP-12-2020-0292

Moro Visconti, R., & Morea, D. (2020). Healthcare digitalization and pay-for-performance incentives in smart hospital project financing. *International journal of environmental research and public health*, 17(7), 2318. https://doi.org/10.3390/ijerph17072318
Nasr, M., Islam, M. M., Shehata, S., Karray, F., & Quintana, Y. (2021). Smart healthcare in the age of AI: recent advances, challenges, and future prospects. *IEEE Access*, 9, 145248–145270. https://doi.org/10.1109/ACCESS.2021.3118960
Nguyen, M., Waller, M., Pandya, A., & Portnoy, J. (2020). A review of patient and provider satisfaction with telemedicine. *Current allergy and asthma reports*, 20, 1-7. https://doi.org/10.1007/s11882-020-00969-7
Niknejad, N., Ismail, W. B., Mardani, A., Liao, H., & Ghani, I. (2020). A comprehensive overview of smart wearables: The state of the art literature, recent advances, and future challenges. *Engineering Applications of Artificial Intelligence*, 90, 103529. https://doi.org/10.1016/j.engappai.2020.103529
Nikou, S., Agahari, W., Keijzer-Broers, W., & de Reuver, M. (2020). Digital healthcare technology adoption by elderly people: A capability approach model. *Telematics and Informatics*, 53, 101315. https://doi.org/10.1016/j.tele.2019.101315
Ometov, A., Shubina, V., Klus, L., Skibińska, J., Saafi, S., Pascacio, P., ... & Lohan, E. S. (2021). A survey on wearable technology: History, state-of-the-art and current challenges. Computer Networks, 193, 108074. https://doi.org/10.1016/j.comnet.2021.108074
Opoku, D. G. J., Perera, S., Osei-Kyei, R., & Rashidi, M. (2021). Digital twin application in the construction industry: A literature review. *Journal of Building Engineering*, 40, 102726. https://doi.org/10.1016/j.jobe.2021.102726
Palanisamy, V., & Thirunavukarasu, R. (2019). Implications of big data analytics in developing healthcare frameworks–A review. *Journal of King Saud University-Computer and Information Sciences*, 31(4), 415–425. https://doi.org/10.1016/j.jksuci.2017.12.007
Preum, S. M., Munir, S., Ma, M., Yasar, M. S., Stone, D. J., Williams, R., ... & Stankovic, J. A. (2021). A review of cognitive assistants for healthcare: Trends, prospects, and future directions. *ACM Computing Surveys (CSUR)*, 53(6), 1–37. https://doi.org/10.1145/3419368
Rahimi, B., Nadri, H., Afshar, H. L., & Timpka, T. (2018). A systematic review of the technology acceptance model in health informatics. *Applied clinical informatics*, 9(03), 604–634. https://doi.org/10.1055/s-0038-1668091
Rodgers, M. M., Alon, G., Pai, V. M., & Conroy, R. S. (2019). Wearable technologies for active living and rehabilitation: Current research challenges and future opportunities. *Journal of rehabilitation and assistive technologies engineering*, 6, 2055668319839607. https://doi.org/10.1177/2055668319839607
Salleh, M. I. M., Abdullah, R., & Zakaria, N. (2021). Evaluating the effects of electronic health records system adoption on the performance of Malaysian health care providers. *BMC medical informatics and decision making*, 21, 1–13. https://doi.org/10.1186/s12911-021-01447-4
Secundo, G., Shams, S. R., & Nucci, F. (2021). Digital technologies and collective intelligence for healthcare ecosystem: Optimizing Internet of Things adoption for pandemic management. *Journal of Business Research*, 131, 563–572. https://doi.org/10.1016/j.jbusres.2021.01.034
Senbekov, M., Saliev, T., Bukeyeva, Z., Almabayeva, A., Zhanaliyeva, M., Aitenova, N., ... & Fakhradiyev, I. (2020). The recent progress and applications of digital technologies in healthcare: a review. *International journal of telemedicine and applications*, 2020(1), 8830200. https://doi.org/10.1155/2020/8830200
Shanafelt, T., Trockel, M., Ripp, J., Murphy, M. L., Sandborg, C., & Bohman, B. (2019). Building a program on well-being: key design considerations to meet the unique needs of each organization. *Academic Medicine*, 94(2), 156–161. https://doi.org/10.1097/ACM.0000000000002415
Signorelli, G. R., Lehocki, F., Mora Fernández, M., O'Neill, G., O'Connor, D., Brennan, L., ... & Garcia-Gomez, J. M. (2019). A research roadmap: connected health as an enabler of cancer patient support. *Journal of Medical Internet Research*, 21(10), e14360. https://doi.org/10.2196/14360

Sun, Y., Lo, F. P. W., & Lo, B. (2019). Security and privacy for the internet of medical things enabled healthcare systems: A survey. *IEEE Access*, 7, 183339–183355. https://doi.org/10.1109/ACCESS.2019.2960617

Talal, M., Zaidan, A. A., Zaidan, B. B., Albahri, A. S., Alamoodi, A. H., Albahri, O. S., ... & Mohammed, K. I. (2019). Smart home-based IoT for real-time and secure remote health monitoring of triage and priority system using body sensors: Multi-driven systematic review. *Journal of medical systems*, 43, 1–34. https://doi.org/10.1007/s10916-019-1158-z

Tapuria, A., Porat, T., Kalra, D., Dsouza, G., Xiaohui, S., & Curcin, V. (2021). Impact of patient access to their electronic health record: systematic review. *Informatics for Health and Social Care*, 46(2), 194–206. https://doi.org/10.1080/17538157.2021.1879810

Tian, S., Yang, W., Le Grange, J. M., Wang, P., Huang, W., & Ye, Z. (2019). Smart healthcare: making medical care more intelligent. *Global Health Journal*, 3(3), 62–65. https://doi.org/10.1016/j.glohj.2019.07.001

Tortorella, G. L., Saurin, T. A., Fogliatto, F. S., Rosa, V. M., Tonetto, L. M., & Magrabi, F. (2021). Impacts of Healthcare 4.0 digital technologies on the resilience of hospitals. *Technological Forecasting and Social Change*, 166, 120666. https://doi.org/10.1016/j.techfore.2021.120666

Tsai, C. H., Eghdam, A., Davoody, N., Wright, G., Flowerday, S., & Koch, S. (2020). Effects of electronic health record implementation and barriers to adoption and use: a scoping review and qualitative analysis of the content. *Life*, 10(12), 327. https://doi.org/10.3390/life10120327

Tuli, S., Tuli, S., Wander, G., Wander, P., Gill, S. S., Dustdar, S., ... & Rana, O. (2020). Next generation technologies for smart healthcare: challenges, vision, model, trends and future directions. *Internet technology letters*, 3(2), e145. https://doi.org/10.1002/itl2.145

Tyler, J., Choi, S. W., & Tewari, M. (2020). Real-time, personalized medicine through wearable sensors and dynamic predictive modeling: a new paradigm for clinical medicine. Current opinion in systems biology, 20, 17–25. https://doi.org/10.1016/j.coisb.2020.07.001

Wowak, K. D., Handley, S., Kelley, K., & Angst, C. M. (2024). Strategic sourcing of multicomponent software systems: The case of electronic medical records. *Decision Sciences*, 55(3), 227–244. https://doi.org/10.1111/deci.12576

Xie, Y., Lu, L., Gao, F., He, S. J., Zhao, H. J., Fang, Y., ... & Dong, Z. (2021). Integration of artificial intelligence, blockchain, and wearable technology for chronic disease management: a new paradigm in smart healthcare. *Current Medical Science*, 41(6), 1123–1133. https://doi.org/10.1007/s11596-021-2485-0

Zhang, J., Yang, M., Ge, Y., Ivers, R., Webster, R., & Tian, M. (2022). The role of digital health for post-surgery care of older patients with hip fracture: a scoping review. *International Journal of Medical Informatics*, 160, 104709. https://doi.org/10.1016/j.ijmedinf.2022.104709

# Chapter 7
# Foundations of Safety and Compliance in Healthcare Architecture: Navigating Building Codes, Accreditation Standards, and Sustainability Practices

This chapter, meticulously examines the comprehensive landscape of standards and guidelines crucial for hospital design and operation, ensuring adherence to safety and quality care. It discusses the intricate roles of building codes, healthcare accreditation standards, and specialized facility requirements, each integral to maintaining hospitals that are safe, accessible, and functional. This includes a focus on fire safety, universal accessibility, and the infrastructural demands of specialized units like surgical suites and ICUs, highlighting the necessity for rigorous stakeholder involvement in formulating these regulations. Furthermore, the chapter delves into sustainability standards, advocating environmentally responsible practices in healthcare architecture, such as energy efficiency and the use of sustainable materials. It also addresses the impact of local and national health regulations, which govern aspects from medical waste handling to the storage of pharmaceuticals, emphasizing their importance in the daily operational efficiency and compliance of healthcare facilities. Through a detailed narrative, this chapter underscores the pivotal role of regulatory frameworks in shaping healthcare facilities that not only meet current health demands but are also prepared for future challenges, thus ensuring ongoing patient safety and care quality.

## 7.1 Current Standards and Guidelines in Hospital Design

Understanding the current standards and guidelines in hospital design is essential for architects, planners, and healthcare administrators to ensure that facilities not only meet the required legal and safety standards but also provide high-quality care. These standards and guidelines are developed to standardize practices, enhance patient safety, and ensure facilities are equipped to handle the healthcare needs of the population.



A. A. Firoozi and A. A. Firoozi, *Transformative Design in Healthcare*,
https://doi.org/10.1007/978-3-032-08019-6_7

### 7.1.1 Building Codes and Safety Standards

The robustness of hospital infrastructure critically hinges on the meticulous adherence to building codes and safety standards, which serve as the cornerstone for hospital design. These regulations are carefully crafted to ensure the structural integrity, fire safety, and accessibility of healthcare facilities, thereby safeguarding both patients and healthcare professionals. Adherence to these codes is not merely a legal formality but a fundamental aspect of healthcare ethics and patient safety.

Building codes encompass a broad spectrum of design and construction standards, which include the stipulation of fire-resistant materials and the integration of advanced fire suppression systems. Such materials and systems are selected based on their proven efficacy to withstand extreme conditions, ensuring that the hospital remains a haven during emergencies. Moreover, these codes mandate the design of accessible exit routes, which are crucial in facilitating the swift and safe evacuation of occupants in the event of a fire or other emergencies.

In addition to fire safety, building codes also emphasize universal accessibility. This is vital in a hospital setting where patients may have varying levels of mobility. Regulations specify the width of doorways, the design of entryways, and the installation of ramps and elevators, ensuring that all individuals, regardless of physical ability, can navigate the facility safely and independently.

The formulation of these codes involves a rigorous process of stakeholder consultation, which includes medical professionals, architects, engineers, and advocacy groups for the disabled. This collaborative approach ensures that the codes are comprehensive, addressing the nuanced needs of all users of medical facilities. By adhering to these standards, hospitals not only comply with legal requirements but also demonstrate a commitment to the highest standards of patient care and safety. This adherence is crucial, as it directly impacts the hospital's ability to provide a secure and functional environment conducive to healing and recovery.

Table 7.1 provides a comprehensive comparison of key building codes from different regions, including the International Building Code (IBC), Eurocode, and others, with a focus on requirements specific to healthcare facilities. This table highlights variations in structural safety, fire protection, accessibility, and emergency systems across different codes to give healthcare facility planners and architects a clear, comparative view of international standards. Understanding these differences is crucial for ensuring global best practices in the construction and renovation of healthcare environments.

### 7.1.2 Healthcare Accreditation Standards

In the intricate ecosystem of healthcare, the role of healthcare accreditation standards is pivotal, serving as a comprehensive framework that guides the operational and design aspects of medical facilities. These standards, meticulously set by reputable

**Table 7.1** Comparison of International Building Codes

| Building code | Region | Structural safety standards | Fire protection requirements | Accessibility specifications | Emergency system guidelines |
|---|---|---|---|---|---|
| International Building Code (IBC) | USA | High; includes detailed seismic regulations | Comprehensive; includes requirements for fire barriers, alarms, and sprinkler systems | Detailed; includes guidelines on door widths, ramp slopes, and restroom design | Require detailed evacuation and lockdown plans |
| Eurocode | Europe | Very high; includes extensive load calculations and materials testing | Stringent; fire resistance and containment measures | Comprehensive; covers accessibility in public and private buildings | Mandates emergency escape routes and external firefighting access points |
| National Building Code of Canada (NBC) | Canada | High; includes climatic considerations and seismic assessments | Detailed; includes provisions for fire detection, suppression, and compartmentalization | Includes accessibility measures like ADA | Includes requirements for emergency power and lighting systems |
| British Standards (BS) | United Kingdom | High; includes specific regulations for healthcare buildings | Stringent; like Eurocode with additional local requirements | Detailed; aligned with European norms for accessibility | Comprehensive emergency and fire safety measures required |
| Australian Building Codes Board (ABCB) | Australia | High; includes earthquake and cyclone considerations | Comprehensive; fire safety strategies are integral | Robust emphasizes access to and within buildings for all people | Requires installation of emergency warning and intercommunication systems |

healthcare accreditation organizations, are instrumental in ensuring that hospitals not only provide safe and efficient care but also foster an environment that respects patient privacy, minimizes infection risks, and maintains high levels of environmental health.

For instance, in the United States, the Joint Commission, a preeminent body in healthcare accreditation, stipulates a set of rigorous criteria that hospitals must meet to earn and maintain accreditation. These criteria are far-reaching, influencing everything from the spatial configuration of patient rooms to the systems governing air quality and noise control. The layout of patient rooms, as guided by these standards, focuses on optimizing patient recovery while safeguarding privacy and dignity. This involves considerations such as the placement of beds, the design of visitor areas, and the integration of bathroom facilities that are accessible and conducive to patient needs.

Furthermore, infection control is a critical component of accreditation standards, necessitating the implementation of advanced ventilation systems that filter and purify air to meet stringent health and safety benchmarks. These systems are essential in preventing the spread of airborne diseases—a paramount concern in hospital settings. Additionally, the standards dictate specific materials and surfaces that are easier to clean and maintain, thereby reducing the potential for contamination.

Environmental health also receives considerable attention under these standards, encompassing aspects such as noise levels and lighting. The control of noise within hospital premises is crucial, as excessive noise can adversely affect patient recovery and staff performance. Accreditation standards, therefore, mandate the use of sound-absorbing materials and the design of buffer zones to minimize the intrusion of external noise into sensitive areas like patient rooms and intensive care units.

Compliance with healthcare accreditation standards is not merely about fulfilling regulatory requirements; it represents a deep-seated commitment to quality assurance and continuous improvement. Hospitals that meet these standards demonstrate their dedication to providing a therapeutic environment that enhances patient outcomes and staff well-being. This compliance is continually assessed through regular reviews and audits, ensuring that hospitals not only meet these standards at a point in time but sustain them through ongoing enhancements and responsive adaptations to emerging healthcare challenges.

Table 7.2 lists major healthcare accreditation bodies such as the Joint Commission and details their core standards related to hospital design and operations. This table serves as a vital resource for healthcare administrators and facility planners to understand the specific standards emphasized by each accreditation body. By adhering to these standards, healthcare facilities can ensure they meet quality benchmarks that impact patient safety, care quality, and operational efficiency.

**Table 7.2** Key healthcare accreditation standards by organization

| Accreditation body | Region | Core standards related to design | Core standards related to operations |
|---|---|---|---|
| The Joint Commission (TJC) | United States | Safety and accessibility of physical facilities | Emergency preparedness, patient flow, and staff safety |
| Commission on Accreditation of Rehabilitation Facilities (CARF) | International | Accessibility, and environmental safety for rehabilitation | Patient care processes, continuity of care, outcomes evaluation |
| Det Norske Veritas (DNV) | International | Building safety, environmental sustainability | Patient safety, risk management, and quality improvement processes |
| Australian Council on Healthcare Standards (ACHS) | Australia | Compliance with Australian building standards, patient safety in design | Clinical governance, patient care processes, healthcare outcomes |
| National Accreditation Board for Hospitals & Healthcare Providers (NABH) | India | Infrastructure norms, infection control in design | Quality benchmarks in clinical services, patient rights, and education |

### *7.1.3 Specialized Facility Requirements*

Within the complex architecture of a hospital, specialized facilities such as surgical suites, intensive care units, and imaging centers are governed by precise requirements that ensure the safety and efficacy of medical procedures. These requirements, intricately defined by both best medical practices and regulatory guidelines, are critical for accommodating the sophisticated demands of modern healthcare services.

Surgical suites, for example, are designed with stringent standards regarding room size and layout to accommodate a range of surgical procedures and the necessary medical equipment. The spatial dimensions must be sufficient not only to house advanced surgical tools but also to ensure unobstructed workflow for medical teams. This includes adequate space for the movement of surgical staff and the positioning of patient monitoring devices. Furthermore, these rooms require specialized air handling systems capable of maintaining high air exchange rates, essential for minimizing the risk of infection. The air quality in surgical suites is meticulously controlled, removing contaminants and maintaining a sterile environment crucial for patient safety during operations.

Intensive care units (ICUs) also have unique design specifications that cater to the critical nature of care provided in these settings. ICUs are equipped with enhanced electrical and gas hook-ups to support multiple high-dependency machines per patient. The layout is optimized to allow constant visual monitoring of patients by medical staff, facilitating immediate intervention when necessary. Additionally, these units often feature isolation capabilities to prevent the spread of infections, particularly for patients with compromised immune systems.

Imaging centers, housing sophisticated diagnostic equipment such as MRI machines, CT scanners, and X-rays, require specific design considerations to accommodate the size and operational requirements of these technologies. This includes reinforced flooring to support heavy machinery, shielded walls to prevent radiation exposure, and controlled access to ensure patient privacy and safety. The design also considers the ergonomic needs of radiologists and technicians, integrating features that reduce the risk of repetitive strain injuries and enhance overall workflow efficiency.

The design of these specialized facilities further extends to the ergonomic considerations for medical staff. Ergonomics is pivotal in preventing work-related injuries and enhancing the efficiency of medical operations. This involves the strategic placement of equipment and furniture, the selection of appropriate flooring materials to reduce fatigue and the design of workstations that align with natural body movements and frequent tasks. These considerations not only improve the safety and comfort of the staff but also contribute to better patient care by enabling medical professionals to perform at their best.

In essence, the specialized requirements of different hospital areas are a blend of advanced engineering, medical insights, and ergonomic design, all aimed at optimizing the safety, effectiveness, and efficiency of healthcare delivery. These standards are not static but evolve with advancements in medical technology and practices, ensuring that hospital facilities remain at the forefront of patient care and safety.

Equation 7.1 outlines the calculation for determining the Air Changes per Hour (ACH) in surgical suites to maintain sterile conditions. This is crucial for ensuring an environment conducive to minimizing the risk of infections during surgical procedures. The equation for ACH is designed to specify the number of times the air within a room should be replaced each hour, which is vital in settings where sterility is paramount. The calculation of ACH is particularly important in surgical suites where maintaining a sterile environment reduces the risk of postoperative infections, a critical component in patient care and recovery.

The ACH in surgical suites is influenced by various factors including the size of the room, the number of personnel, the type of surgery, and the efficiency of the air filtration system. Industry standards typically recommend a higher ACH rate in rooms used for surgeries to ensure a clean air supply that minimizes the presence of airborne contaminants. The equation is structured to integrate these variables effectively:

$$\text{ACH} = \left(\frac{\text{Volume of Air Filtered } (V)}{\text{Volume of the Room } (R)}\right) \times 60 \tag{7.1}$$

where

- $V$ is the volume of air filtered per minute,
- $R$ is the volume of the room, and
- The factor of 60 converts the air changes from minutes to hours.

This formula ensures that surgical suites are equipped with enough air changes to maintain a sterile environment, thus safeguarding patient health and optimizing surgical outcomes. Implementing this calculation supports the operational standards necessary for infection control in medical facilities, contributing to overall healthcare quality and effectiveness.

### *7.1.4 Sustainability Standards*

The integration of sustainability standards into hospital design represents a critical evolution in the approach to healthcare infrastructure, aligning the demands of medical facilities with the imperatives of environmental stewardship and public health promotion. As awareness of environmental impacts grows, healthcare facilities are increasingly expected to lead by example, adopting sustainable practices that mitigate their ecological footprint while enhancing patient care environments.

Among the most recognized frameworks guiding these efforts is the Leadership in Energy and Environmental Design (LEED) certification system. LEED provides a comprehensive set of criteria for evaluating the environmental performance of a building, focusing on key aspects such as energy efficiency, water usage, carbon emissions, and the quality of the indoor environment. Hospitals pursuing LEED certification are required to demonstrate adherence to these criteria through various means, including the utilization of energy-efficient heating, ventilation, and air conditioning (HVAC) systems, the implementation of advanced water recycling processes, and the choice of low-emitting materials to improve indoor air quality.

Moreover, the selection of materials in the construction of healthcare facilities is profoundly influenced by sustainability standards. Sustainable materials are those that are sourced responsibly, have a reduced impact on the environment, and contribute to the health and well-being of building occupants. For hospitals, this means choosing materials that do not emit harmful chemicals, that are durable and recyclable, and that contribute to the thermal efficiency of the building, thus reducing overall energy consumption.

Designing with the local climate and ecology in mind is another pivotal aspect of sustainable hospital design. This approach, often referred to as "bioclimatic design," involves understanding and utilizing local environmental conditions to naturally enhance building performance. For example, in warmer climates, hospitals might incorporate shading devices and orientation strategies that minimize heat gain and reliance on air conditioning, while in cooler regions, they might maximize solar gain and thermal mass to reduce heating requirements.

The benefits of incorporating sustainability standards extend beyond environmental impacts. Hospitals designed with these principles in mind can also promote better health outcomes by providing a healing environment that maximizes natural light, ensures excellent air quality, and incorporates views of nature, which have been shown to speed recovery times and reduce stress among patients and staff alike.

**Table 7.3** Sustainability metrics and targets for hospitals

| Sustainability metric | Current average | Target | Standard/benchmark |
|---|---|---|---|
| Energy consumption (kWh per sq m) | 250 | 150 | LEED, Green Building Council |
| Water usage (liters per day per bed) | 1000 | 800 | EPA WaterSense, LEED |
| Waste recycling rate (%) | 45 | 75 | Zero Waste International Alliance |
| Greenhouse gas emissions (kg $CO_2$ eq per sq m) | 50 | 30 | Paris Agreement, Green Building Standards |
| Renewable energy use (%) | 20 | 50 | Renewable Energy Directives |

As the healthcare sector continues to evolve, the push towards more sustainable and environmentally friendly building practices will likely intensify. Hospitals that embrace these standards not only contribute to a more sustainable planet but also position themselves as forward-thinking leaders in health promotion and community well-being.

Table 7.3 outlines specific sustainability metrics and targets for hospitals, including energy consumption per square meter, water usage, and waste management targets. This table helps healthcare facilities understand the quantifiable benchmarks they should strive to meet to enhance their sustainability practices. These metrics are aligned with LEED and other global sustainability standards, providing a clear path for hospitals to reduce their environmental impact while maintaining high standards of care.

### *7.1.5 Local and National Health Regulations*

The adherence to local and national health regulations forms a critical underpinning for the design and operation of hospital facilities, serving as both a legal obligation and a practical guideline for ensuring optimal patient care and safety. These regulations encompass a comprehensive range of directives that cover the handling of medical waste, the secure storage of pharmaceuticals, the maintenance of clean air standards, and numerous other aspects critical to hospital operations.

For instance, the regulations concerning the handling and disposal of medical waste are stringent, requiring specific procedures to mitigate risks of infection and environmental contamination. Hospitals must therefore design specific areas that are dedicated to the safe collection, storage, and disposal of hazardous materials, ensuring these are isolated from patient care areas and have robust systems to prevent accidental exposure.

Similarly, the storage of pharmaceuticals is regulated to ensure that drugs are kept secure and at appropriate conditions to maintain their efficacy. This necessitates the incorporation of secure, climate-controlled storage areas within hospital designs.

Such areas must be strategically located to ensure that pharmacists have efficient access while maintaining strict control over the distribution of medications.

Additionally, the flow of personnel and materials is heavily influenced by health regulations, which require the design of clear, unobstructed pathways for the movement of staff and supplies. This is crucial not only for the efficiency of hospital operations but also for preventing the spread of infections. For example, pathways for clean supplies are often separated from those used to transport waste or soiled materials, minimizing cross-contamination risks.

Integrating these complex regulations into hospital design is not a static process; it requires ongoing adaptation as laws and standards evolve. This dynamic adaptation is necessary not only to maintain compliance but also to leverage new technologies and methodologies that enhance safety and efficiency. For instance, advancements in biodegradable materials and digital tracking systems present new opportunities for improving medical waste management and pharmaceutical storage, respectively.

As regulations change, hospitals must remain vigilant and responsive, ensuring that their facilities not only meet current standards but are also prepared to integrate future advancements. This proactive approach to regulatory compliance not only safeguards patient and staff welfare but also enhances the hospital's capability to deliver high-quality, efficient care.

Table 7.4 provides a detailed summary of key health regulations from different jurisdictions, focusing on critical operational areas such as waste disposal, pharmaceutical storage, and patient privacy. This table serves as a concise reference for hospital administrators and facility managers, offering insights into the legal requirements that influence hospital design and operations. Understanding these regulations ensures compliance and enhances the quality and safety of healthcare services.

Next, as we delve into how global health policies influence architectural norms, we will explore how these broader directives shape the design and operation of healthcare facilities on an international scale, reflecting a global commitment to health and safety in the medical field.

## 7.2 Impact of Global Health Policies on Architectural Norms

Global health policies play a significant role in shaping the architectural norms of healthcare facilities by setting standards that transcend national boundaries, promoting best practices, and addressing common challenges faced by health systems worldwide. These policies often reflect a consensus on critical health issues, such as infection control, patient safety, and accessibility, influencing the design and operation of healthcare facilities across different regions.

**Table 7.4** Summary of local and national health regulations affecting hospital operations

| Jurisdiction | Waste disposal regulations | Pharmaceutical storage guidelines | Patient privacy laws |
|---|---|---|---|
| United States | EPA regulations on hazardous waste, OSHA standards for medical waste | DEA requirements for controlled substances, FDA guidelines for drug storage | HIPAA rules for patient data protection |
| European Union | EU Waste Framework Directive, specific medical waste directives | EMA guidelines, strict standards for drug safety | GDPR for personal data protection, patient consent |
| Canada | Health Canada standards, provincial regulations on biomedical waste | Controlled Drugs and Substances Act requirements, pharmacy storage compliance | PIPEDA, provincial healthcare privacy laws |
| Australia | National Health and Medical Research Council guidelines, state-specific waste management laws | TGA standards for drug storage, Poisons Standard for controlled substances | Privacy Act 1988, health record regulations in each state |
| India | Biomedical Waste Management Rules, 2016, CPCB guidelines | Drug and Cosmetics Act requirements, Narcotic Drugs and Psychotropic Substances Act | Information Technology Act for data protection, specific state laws |

### *7.2.1 Infection Control and Pandemic Preparedness*

The advent of global health emergencies, notably the COVID-19 pandemic, has starkly underscored the critical need for healthcare facilities to embody principles of adaptability and rapid response in their design. This new paradigm in hospital architecture and facility planning is driven by the imperative to manage infectious disease outbreaks effectively and to accommodate sudden patient surges without compromising the safety and care of all patients.

International health guidelines have adapted in response to these challenges, advocating for architectural designs that enhance a facility's ability to segregate and treat infectious cases. Modular designs have become increasingly significant in this context, providing the flexibility to expand or reconfigure spaces as epidemiological needs evolve. This modularity enables hospitals to swiftly adapt their layouts, increasing or decreasing the number of beds, altering the function of spaces to serve as intensive care units or isolation wards, or adjusting circulation patterns to minimize the risk of pathogen transmission.

Furthermore, the integration of negative pressure rooms is a critical architectural feature recommended by health authorities worldwide. These rooms are specially designed to control the airflow, ensuring that contaminated air does not circulate to other parts of the hospital. This is accomplished by maintaining a lower air pressure inside the room compared to adjacent areas, thereby allowing air to flow into the isolation room but not escape from it, unless it is filtered. Negative pressure rooms are essential in the containment of airborne pathogens and are crucial in preventing the spread of infections within the facility.

Additionally, the layout of entrances, corridors, and elevators is meticulously planned to ensure there are dedicated pathways for the movement of infected patients, away from general and public areas. This helps minimize contact between infected individuals and other patients or staff, further reducing the risk of widespread transmission within the facility.

These architectural strategies are not merely reactive measures but are part of a broader, forward-thinking approach to healthcare facility design that emphasizes resilience and preparedness. By incorporating flexible, adaptive design elements, hospitals can not only respond more effectively to current pandemics but also prepare for future public health emergencies. This approach ensures that healthcare facilities remain robust in their primary function as centers of healing, even under the strain of unprecedented health crises.

Table 7.5 lists international guidelines and standards for infection control in healthcare facilities, detailing specific requirements for ventilation, surface materials, and layout. This table provides a comprehensive overview of the protocols established by various health organizations worldwide to prevent and control infections within healthcare settings. By adhering to these standards, healthcare facilities can enhance their infection control measures, ensuring safer environments for both patients and healthcare providers.

### *7.2.2 Universal Design and Accessibility Standards*

The principles of universal design and accessibility are pivotal in shaping healthcare facilities that cater to the diverse needs of all users, embodying a holistic approach that is endorsed by the World Health Organization (WHO) and various international bodies. This design philosophy asserts that healthcare settings should be inherently accessible, eliminating any form of physical or sensory barriers that might hinder the ability of individuals to receive care. Such inclusivity not only enhances the functionality of these facilities but also plays a crucial role in upholding human dignity by ensuring equitable access to healthcare services.

Global health policies are increasingly aligned to reduce health disparities by integrating universal design principles into the architectural fabric of healthcare facilities. This involves the meticulous planning and implementation of features that address a wide range of needs. For example, ramps and wide doorways facilitate physical access for those using wheelchairs or other mobility aids, while non-slip floors enhance safety for everyone, particularly those with mobility impairments or the elderly.

Moreover, sensory aids such as tactile flooring systems, braille signage, and auditory signals are integral components of a universally designed hospital. These elements assist individuals with visual or hearing impairments in navigating complex hospital layouts independently and safely. Such aids are not merely adjunct features

**Table 7.5** Global standards for infection control in healthcare facilities

| Organization | Ventilation requirements | Surface materials standards | Layout guidelines |
|---|---|---|---|
| World Health Organization (WHO) | Minimum of 12 air changes per hour in clinical areas; HEPA filters for isolation rooms | Non-porous materials for easy cleaning and disinfection | Separate zones for clean and contaminated activities |
| Centers for Disease Control and Prevention (CDC) | Use of directional airflow to control contaminant spread | Use of antimicrobial surfaces in high-touch areas | Sufficient space for patient separation, designated isolation rooms |
| European Centre for Disease Prevention and Control (ECDC) | Enhanced ventilation in patient care areas, especially in infectious disease wards | High standards for cleaning and resistance to chemicals on surfaces | Strategic zoning to minimize cross-contamination risk |
| National Health Service (NHS) UK | Adequate ventilation to meet clinical requirements, including specialized systems for infection-prone areas | Surfaces must be durable, maintainable, and resistant to frequent cleaning | Layouts to support optimal patient flow and minimize contact with infections |
| Australian Commission on Safety and Quality in Health Care | Ventilation systems designed to minimize risk of airborne transmission | Surfaces and furnishings that can withstand regular cleaning with hospital-grade disinfectants | Clear demarcation of clean and sterile areas versus contaminated zones |

but are woven into the design process from the outset, ensuring that they are both functional and seamlessly integrated into the hospital's aesthetic.

The design of intuitive wayfinding systems is another critical aspect of accessibility. Hospitals can be labyrinthine and ensuring that all patients and visitors can find their way with ease is essential for minimizing stress and enhancing the overall hospital experience. Wayfinding systems in a universally designed hospital include clear, highly visible signage, logical layout planning, and the use of color or icons that transcend language barriers, all of which contribute to an environment that is navigable by everyone.

By adhering to universal design principles, healthcare facilities not only meet legal and ethical standards of accessibility but also foster an environment of inclusivity. This approach directly contributes to the elimination of health disparities by ensuring that every individual, regardless of their physical or sensory abilities, has equal access to medical care. Such design strategies are fundamental not only in enhancing individual patient outcomes but also in promoting broader public health goals by making healthcare more accessible to all segments of the population.

Equation 7.2 presents a simple yet essential calculation for determining the minimum necessary pathway widths required for wheelchair access in hospital designs. This calculation is pivotal for ensuring compliance with accessibility standards, which are designed to provide safe and unobstructed passage for individuals using wheelchairs. The equation aids architects and planners in designing hospital layouts that are not only compliant with legal requirements but also promote ease of mobility for all patients and visitors.

The minimum pathway width for wheelchair access must accommodate the turning radius and maneuverability of standard wheelchairs, along with additional space to allow for comfortable passage of hospital equipment and personnel. This consideration is critical in hospital settings to prevent congestion and ensure that individuals in wheelchairs can navigate the premises independently and safely. The equation for calculating the minimum pathway width ($W$) is given by:

$$W = 2 \times (W_c + C) \tag{7.2}$$

where

- $W_c$ is the average width of a wheelchair (typically 0.85 m),
- $C$ is the clearance space on each side of the wheelchair, necessary for comfortable maneuvering (commonly set at 0.15 m).

This formula ensures that pathways are sufficiently wide to facilitate smooth and safe wheelchair movement, reflecting the commitment of healthcare facilities to inclusivity and accessibility. Adhering to these standards in hospital design not only meets legal requirements but also significantly enhances the functional quality of healthcare environments, promoting a more welcoming and accessible space for all users.

### 7.2.3 *Sustainability and Environmental Health*

The imperative for sustainability in healthcare facility design is increasingly underscored by international protocols such as the Paris Agreement on climate change, which serves as a guiding beacon for integrating environmental stewardship into medical infrastructure. These global commitments have catalyzed a shift towards sustainable practices within the healthcare sector, emphasizing the direct link between environmental health and public health outcomes.

Energy efficiency stands as a cornerstone of sustainable hospital design. By adopting energy-efficient architectures, such as high-performance building envelopes, advanced HVAC systems, and smart lighting technologies, hospitals can significantly reduce their energy consumption. These technologies not only help curb operational costs but also mitigate the environmental impact of healthcare facilities by lowering their carbon emissions. Moreover, the integration of renewable energy sources, such as solar panels and wind turbines, further aligns hospital operations with sustainable practices by tapping into clean energy, thus reducing reliance on fossil fuels.

Water and waste management systems in hospitals are also being rethought under the lens of sustainability. Sustainable water management practices, including rainwater harvesting, greywater recycling, and the installation of water-efficient fixtures, are vital in reducing the water footprint of healthcare facilities. Similarly, comprehensive waste management strategies that prioritize waste reduction, segregation, and recycling are critical for minimizing the environmental impact of hospital waste. This includes the safe disposal of hazardous materials and the recycling of non-hazardous waste, which not only helps in conserving resources but also in preventing pollution.

As sustainability becomes more integral to public health, these environmental guidelines are profoundly influencing how hospitals are constructed and renovated. Sustainable hospitals are designed not just to be facilities for curing illness but also as spaces that contribute to the well-being of the community by actively reducing the ecological impacts associated with healthcare delivery. This holistic approach to hospital design incorporates green spaces, promotes the use of non-toxic and sustainable materials, and considers the building's overall impact on the local ecosystem.

This growing integration of sustainability in hospital design reflects a broader recognition of the role that healthcare facilities play in leading community and global efforts toward a sustainable future. By building and renovating hospitals that adhere to these principles, the healthcare sector not only improves its operational efficiency but also aligns itself more closely with the global agenda for environmental health and sustainability. This shift not only contributes to the mitigation of climate change but also exemplifies the role of healthcare in fostering a healthier, more sustainable world.

Table 7.6 compares traditional versus sustainable hospital operational impacts on energy and water use, waste production, and carbon emissions. This table quantitatively demonstrates the significant benefits of sustainable practices in hospital design and operations, highlighting how eco-friendly approaches contribute to reducing the environmental footprint of healthcare facilities. The data serves as a compelling argument for the adoption of sustainability measures, showing clear improvements in efficiency and environmental responsibility.

**Table 7.6** Comparison of environmental impact reductions

| Impact category | Traditional hospital usage | Sustainable hospital usage | Reduction achieved |
|---|---|---|---|
| Energy use (kWh) | 5,000,000 | 3,000,000 | 40% reduction |
| Water use (liters) | 10,000,000 | 6,000,000 | 40% reduction |
| Waste production (kg) | 500,000 | 300,000 | 40% reduction |
| Carbon emissions (tonnes) | 2000 | 1200 | 40% reduction |

### 7.2.4 Patient-Centered Care Models

The paradigm shift towards patient-centered care in global health policies is transforming hospital design, steering it towards environments that not only address medical needs but also promote holistic well-being. This approach is rooted in the recognition that healthcare facilities should serve as healing environments that support the psychological, social, and emotional aspects of patient care, not merely the physiological.

Architectural norms influenced by patient-centered care models prioritize the creation of spaces that are calming and restorative. This includes the integration of natural elements such as daylight, green spaces, and water features, which are known to have therapeutic effects on patients. Access to natural views and outdoor areas can significantly reduce patient stress and anxiety, thereby enhancing recovery rates and overall patient outcomes. The design and orientation of patient rooms, communal areas, and even corridors are thus carefully considered to maximize exposure to natural light and outdoor views.

Privacy and comfort are also central to patient-centered care, influencing numerous design choices in healthcare facilities. Patient rooms are increasingly designed to feel less clinical and more homelike, with comfortable furniture, private bathrooms, and facilities that accommodate overnight stays for family members. These amenities not only provide comfort but also facilitate the involvement of families in the care process, which is a key component of holistic patient care. Family involvement is linked to better patient satisfaction and outcomes, as it helps maintain emotional and social support systems during hospital stays.

Moreover, patient-centered care models advocate flexibility in healthcare settings to accommodate diverse patient needs and preferences. This might involve adjustable lighting, temperature controls, and noise management systems that allow patients to personalize their environment to suit their comfort levels. Such features acknowledge the individual's needs and agency in their healing process, further embodying the ethos of patient-centered care.

Healthcare facilities designed with these patient-centered principles do more than treat diseases; they actively contribute to the overall health and well-being of individuals. By fostering environments that reduce stress, promote privacy, and enable family involvement, hospitals can transform the experience of receiving medical care. This shift not only enhances patient satisfaction and recovery but also sets a new standard in healthcare delivery, one that views patients as human beings with complex needs and inherent dignity.

### 7.2.5 Data Security and Privacy

The digital transformation of healthcare has significantly elevated the importance of data security and privacy, making them central considerations in the architectural design of modern hospitals. As healthcare facilities increasingly incorporate digital health technologies, global health policies have intensified the focus on establishing robust IT infrastructures that safeguard sensitive medical information against breaches and unauthorized access.

Architecturally, this emphasis on data security necessitates the creation of secure zones specifically designed to house IT equipment such as servers and data centers. These areas are fortified with advanced security measures, including controlled access points that restrict entry to authorized personnel only, ensuring that critical infrastructure is protected from both physical and cyber threats. The physical design of these spaces is crucial, as it must also facilitate the optimal performance of equipment, incorporating considerations for ventilation, power supply stability, and fire safety.

Beyond the technical storage of data, the design of healthcare facilities now also includes dedicated areas for confidential consultations. This is particularly vital in departments such as mental health and other sensitive health services, where privacy is paramount. Consultation rooms are thus designed to provide sound insulation and discreet access, ensuring that conversations remain private and that patients feel secure in sharing sensitive information.

Data protection systems extend into the very fabric of hospital operations, influencing not only the placement and design of hardware but also the integration of software solutions that monitor and control access to information. This includes the use of encrypted networks, secure Wi-Fi connections, and advanced cybersecurity protocols that protect against data leakage, unauthorized access, and other vulnerabilities.

By adhering to stringent global health policies on data security, healthcare facilities not only comply with international standards but also enhance the trust and confidence of patients in the digital systems that manage their health information. This alignment with global standards ensures that healthcare facilities are not only equipped to meet current security challenges but are also prepared for the evolution of digital health technologies.

As we progress, the discussion will shift towards the importance of advocating enhanced regulatory frameworks. This exploration will delve into how evolving standards can be shaped to better meet the needs of patients and healthcare providers, ensuring that the architecture of healthcare facilities continues to adapt to the dynamic landscape of healthcare demands and technological advancements.

## 7.3 Advocating for Enhanced Regulatory Frameworks

Advocating for enhanced regulatory frameworks in healthcare architecture is crucial for ensuring that standards evolve to keep pace with advancements in medical science, technology, and patient care practices. Enhanced regulations can lead to safer, more efficient, and more humane healthcare environments, which ultimately improve patient outcomes and staff satisfaction. Effective advocacy requires collaboration among healthcare professionals, architects, policymakers, and the public.

### *7.3.1 Identifying Gaps in Current Regulations*

The endeavor to enhance regulatory frameworks in healthcare begins with a meticulous examination of existing regulations to identify gaps and deficiencies that may hinder the provision of optimal care. This critical assessment often reveals various shortcomings, such as outdated standards that fail to keep pace with rapid technological advancements in medical equipment and procedures. Such discrepancies can pose significant challenges to healthcare providers, affecting the quality and efficiency of patient care.

One prevalent issue is the existence of antiquated regulations that have not evolved to accommodate the introduction of modern medical technologies. For instance, standards established decades ago might not consider the spatial and electrical requirements of contemporary imaging equipment or robotic surgical devices. This oversight can lead to logistical challenges in deploying state-of-the-art technology, potentially compromising the effectiveness of medical interventions and patient outcomes.

Infection control is another critical area often highlighted by gaps in regulations. With the emergence of new pathogens and the increasing prevalence of hospital-acquired infections, current guidelines may fall short in addressing the need for advanced air filtration systems, antimicrobial surfaces, and other environmental controls that mitigate the spread of infections. The COVID-19 pandemic has particularly underscored the necessity for regulations that enforce not only rigorous sanitation practices but also architectural designs that facilitate these practices.

Additionally, the rise of digital health technologies such as telemedicine and electronic health records presents new challenges in data security and privacy, requiring regulations that specifically address the vulnerabilities associated with digital platforms. Current frameworks may lack comprehensive guidelines on the encryption, storage, and access of digital health data, leaving sensitive patient information susceptible to breaches. The integration of these technologies into daily healthcare operations necessitates an urgent update to regulatory standards to ensure robust protection of patient data.

Professionals working within healthcare facilities are often the most adept at identifying these regulatory gaps. Their day-to-day experiences provide invaluable insights into how existing rules affect the practical aspects of healthcare delivery.

**Table 7.7** Summary of regulatory gaps in healthcare architecture

| Area of concern | Regulatory gap | Potential risks or incidents |
|---|---|---|
| Infection control | Lack of standards for air quality control in older facilities | Increased risk of airborne infections, examples of outbreaks in inadequately ventilated areas |
| Digital health | Insufficient guidelines for data integration across different platforms | Data breaches, inconsistencies in patient care due to fragmented systems |
| Facility safety | Outdated fire safety standards in existing buildings | Higher incidence of fire-related accidents, examples of non-compliance leading to fines and liabilities |
| Emergency preparedness | Inadequate requirements for disaster-ready design features | Failures in emergency response during natural disasters, examples of insufficient evacuation routes or safety shelters |
| Accessibility | Non-uniform compliance standards across regions | Inequitable access for disabled individuals, examples of facilities facing legal challenges due to non-compliance |

Engaging these professionals in the regulatory review process is crucial, as their firsthand knowledge can help shape more effective and responsive regulations.

By identifying and addressing these gaps, healthcare administrators and policymakers can advocate for updated regulations that better align with current medical practices and technologies. This proactive approach not only enhances the functionality and safety of healthcare facilities but also ensures that patient care remains at the forefront of regulatory considerations.

Table 7.7 outlines current regulatory gaps identified in healthcare architecture, detailing specific areas such as infection control, digital health, and facility safety. This table provides examples of potential risks or incidents related to these gaps, illustrating the critical need for updates in regulatory frameworks to ensure the safety and efficiency of healthcare facilities. By highlighting these shortcomings, the table emphasizes the importance of continuous evaluation and adaptation of standards to meet evolving healthcare demands and technological advancements.

### *7.3.2 Collaboration and Stakeholder Engagement*

Effective advocacy for improved healthcare regulations necessitates a collaborative approach, engaging a broad spectrum of stakeholders who bring diverse perspectives and expertise to the table. The formation of such coalitions is fundamental in crafting regulations that are not only comprehensive but also pragmatic and attuned to the multifaceted nature of healthcare delivery.

Healthcare providers are essential to this coalition, offering insights derived from their direct experiences with the practical implications of existing regulations. Their input helps to pinpoint specific operational challenges and patient care issues that may be overlooked by policymakers who are removed from day-to-day clinical environments.

Patient advocacy groups also play a crucial role in representing the interests and rights of patients. These groups ensure that patient-centric perspectives are incorporated into the regulatory framework, advocating for standards that enhance patient safety, privacy, and satisfaction. Their involvement is key to ensuring that healthcare regulations holistically address patient needs and promote the highest standards of care.

Architectural firms specializing in healthcare facilities contribute a unique viewpoint on the physical and structural challenges of implementing regulatory changes. Their expertise in designing hospitals and other medical facilities that comply with complex health codes and support optimal medical practices is invaluable. They can provide practical insights into how proposed regulations could affect the design and functionality of healthcare environments.

Academic researchers add to the coalition by providing evidence-based analyses and projections that inform the regulatory debate. Their research can validate the need for regulatory updates or revisions, offering data-driven arguments that strengthen the case for change. This academic input is crucial for building a compelling, scientifically backed case for regulatory enhancement.

Together, these stakeholders engage with policymakers and regulatory bodies, forming a powerful collective voice that can more effectively advocate for necessary changes. Regular meetings, joint policy briefs, and collaborative public statements are tools that such coalitions use to influence public policy. By ensuring that all perspectives are heard and integrated, these coalitions can help shape regulations that are both innovative and aligned with the realities of healthcare needs today.

This comprehensive approach to stakeholder engagement ensures that proposed regulatory changes are well-informed, broadly supported, and capable of enhancing the quality and efficacy of healthcare delivery. By fostering ongoing dialogue between stakeholders and policymakers, the advocacy process not only promotes better health outcomes but also drives continuous improvement in the healthcare system.

### 7.3.3 Research and Evidence-Based Proposals

The foundation of effective advocacy in healthcare regulation reform is robust research and evidence-based proposals. The use of empirical data and detailed case studies to illustrate the real-world impacts of architectural designs on patient outcomes provides a persuasive argument for the revision of existing standards and the implementation of new ones.

Collecting and presenting evidence on how specific design elements in healthcare facilities—such as increased natural light, private recovery spaces, and ergonomic patient handling systems—affect patient safety, recovery times, and overall satisfaction is critical. Research demonstrates that certain design features can reduce hospital-acquired infections, shorten patient stays, and decrease readmission rates makes a compelling case for incorporating these features into regulatory standards.

Moreover, advancements in medical technology and healthcare procedures continually present new challenges and opportunities for hospital design. Regularly conducting research that keeps pace with these innovations helps ensure that regulations remain relevant and effective. For example, studies that explore the integration of telehealth facilities into traditional hospital environments can provide data on the necessity of redesigning patient rooms or common areas to accommodate virtual consultations effectively.

Case studies play a pivotal role in this context. They not only highlight successful implementations of innovative design solutions but also outline the processes and strategies involved in achieving these successes. For instance, a case study on a hospital that has effectively integrated sustainable energy solutions, such as solar power and green roofs, can serve as a model for others to follow, showcasing both the environmental and economic benefits of such initiatives.

By presenting well-documented evidence and case studies, advocates can more effectively communicate the need for regulatory changes to policymakers. This evidence not only supports the case for adapting healthcare environments to current needs but also underscores the potential long-term benefits of such regulations, including cost savings, improved patient outcomes, and enhanced staff satisfaction.

Thus, the backbone of successful advocacy for healthcare regulation is a strategic combination of rigorous research and compelling real-world examples. These elements work together to provide a clear, data-driven narrative that highlights the necessity and benefits of updated, evidence-based healthcare regulations.

Table 7.8 presents research findings that link specific architectural designs with patient recovery metrics, infection rates, and overall hospital efficiency. This table quantitatively demonstrates the benefits of design features, providing empirical evidence that supports evidence-based architectural proposals. By correlating design choices with measurable health outcomes, this table underscores the critical role of architecture in enhancing patient care and operational effectiveness in healthcare settings.

### *7.3.4 Public Awareness and Support*

Raising public awareness about the intricate relationship between healthcare architecture and patient care is a critical strategy in garnering broad support for necessary regulatory changes. Effective advocacy often extends beyond professional circles and into the public domain, where informed citizens can become powerful allies in pushing for improvements.

Public campaigns are an essential tool in this effort. By creating informative and engaging content that highlights the impact of facility design on health outcomes,

**Table 7.8** Impact of architectural design on patient outcomes

| Architectural design feature | Impact on patient recovery | Impact on infection rates | Impact on hospital efficiency |
|---|---|---|---|
| Natural lighting | 25% faster recovery rates for postoperative patients | – | Improved patient and staff mood, reduced electricity use |
| Private patient rooms | 15% improvement in patient satisfaction and recovery speed | 30% reduction in cross-infection rates | Increased patient throughput |
| Healing gardens | 20% reduction in medication use for anxiety and depression | – | Enhanced patient and visitor satisfaction |
| Improved air quality systems | – | 40% decrease in airborne diseases | Reduced HVAC maintenance costs |
| Noise reduction materials | 10% faster recovery due to reduced stress levels | – | Lowered staff turnover and improved work conditions |

such campaigns can resonate with a wide audience. These can include multimedia presentations, infographics, and testimonials from patients and healthcare providers that illustrate the tangible benefits of well-designed healthcare environments. Such campaigns not only inform but also evoke emotional responses that can drive public engagement and support.

Informational seminars serve as another vital platform for raising awareness. These events can bring together architects, healthcare professionals, patients, and the public to discuss the significance of healthcare facility design. Seminars can provide a more detailed, scientific presentation of case studies and research findings that substantiate the need for regulatory updates. They also offer a forum for live interaction, where participants can ask questions and express concerns, further involving them in the advocacy process.

Partnerships with the media play a crucial role in amplifying the reach of these initiatives. Collaborating with television networks, newspapers, and online platforms ensures that messages about the importance of healthcare architecture reach a broad audience. Media partnerships can help translate complex architectural and healthcare concepts into accessible stories that capture the public's interest and sympathy.

Additionally, leveraging social media platforms can enhance engagement, allowing for real-time interaction and feedback. Social media campaigns can quickly disseminate information, rally support, and mobilize the public to advocate for regulatory changes. These platforms also provide an opportunity to connect with younger audiences, who are often keen to engage with health and environmental issues.

Through these concerted efforts, public awareness initiatives can build a strong base of support for changes in healthcare architecture regulations. By educating the public on how well-thought-out designs can directly benefit patient care and recovery, advocacy groups can create a persuasive call to action that resonates across societal boundaries. This broad-based support is often crucial in influencing policymakers and ensuring that proposed changes gain the necessary momentum and approval.

### 7.3.5 Policy Recommendations and Legislative Change

Once a solid foundation of research, stakeholder consensus, and public support has been established, the next critical step in advocating for improved healthcare regulations is the formulation of specific policy recommendations. These recommendations must be not only grounded in empirical evidence and best practices but also practical and feasible, with clearly articulated benefits to health outcomes.

Developing these policy recommendations involves a detailed understanding of existing laws and the gaps within them, coupled with an innovative approach to how these can be enhanced. The recommendations should focus on key areas identified during the research and stakeholder engagement phases, such as enhancing patient safety through improved architectural designs, integrating advanced technology standards, or updating infection control measures in light of recent health crises.

These recommendations must be framed in a manner that is both comprehensive and comprehensible, clearly outlining the expected improvements in healthcare delivery and patient outcomes. They should also include specific amendments or additions to existing architectural and healthcare facility standards, providing a clear roadmap for legislators and regulators.

Collaboration with legal experts is essential in this phase. These professionals can ensure that the recommendations are not only robust but also legally sound, adhering to national and international law standards. Legal experts can help translate the technical and medical aspects of the recommendations into legal language, crafting well-defined proposals that can be seamlessly integrated into existing legal frameworks.

Engaging with legislators is the final and perhaps most critical step in moving these policy recommendations from theory to practice. This involves presenting the proposals to policymakers, organizing briefings for key decision-makers, and participating in legislative hearings. Effective communication skills are crucial here, as is the ability to negotiate and adapt proposals in response to legislative feedback.

Advocacy groups might also consider mobilizing public support during this phase, using campaigns to pressure legislators into acting on the recommendations. Demonstrating broad public backing can significantly enhance the political will to enact legislative changes.

Ultimately, the transition of policy recommendations into legislative change is a dynamic process that requires persistence, a nuanced understanding of legal and legislative processes, and ongoing engagement with all stakeholders involved. By effectively navigating this path, advocates can achieve meaningful changes that directly enhance the quality and safety of healthcare environments, benefiting patients and healthcare providers alike.

Equation 7.3 provides a robust framework for performing a Cost-Benefit Analysis (CBA) for potential regulatory changes in healthcare architecture. This calculation is critical in determining the financial viability and potential return on investment (ROI) of implementing specific changes mandated by new or revised regulations. Healthcare facilities often face substantial costs when adapting to new regulations, making it essential to evaluate the long-term benefits against the initial expenditures to ensure sustainable decision-making.

This equation assists administrators and healthcare planners in quantifying the economic impact of regulatory changes, which may include modifications to building codes, enhanced accessibility features, and updated safety protocols. The ability to predict financial outcomes from these changes supports strategic planning and resource allocation, ensuring that investments are both prudent and beneficial to the institution's operational goals. The formula for calculating the Cost-Benefit Analysis is expressed as follows:

$$\text{ROI} = \left(\frac{\text{Benefits }(B) - \text{Costs }(C)}{\text{Costs }(C)}\right) \times 100 \tag{7.3}$$

where

- *Benefits* $(B)$ are the total estimated financial gains from the regulatory changes, which may include increased efficiency, enhanced patient satisfaction, reduced legal risks, and other intangible benefits.
- *Costs* $(C)$ represent the total expenses involved in implementing the changes, including construction, training, and administrative adjustments.

This equation provides a percentage value that reflects the return on investment, offering a clear metric to gauge the effectiveness and financial impact of regulatory adjustments in healthcare architecture. By applying this formula, healthcare institutions can make informed decisions that balance costs with significant improvements in facility operations, patient care, and compliance with regulatory standards.

### 7.3.6 Continuous Review and Adaptation

Advocating for a dynamic and responsive regulatory framework is crucial in the realm of healthcare architecture, where the pace of technological and medical advancements necessitates continual adaptation. A static regulatory system risks becoming obsolete, failing to accommodate new insights and technologies that can significantly enhance patient care and operational efficiency. Thus, establishing a system for continuous review and adaptation of regulations is essential to ensure that healthcare facilities remain at the cutting edge of design and functionality.

To achieve this, it is imperative to set up formal mechanisms that facilitate regular re-evaluation of existing regulations. This could involve the creation of a dedicated regulatory review board comprising a diverse group of stakeholders, including healthcare professionals, architects, patient advocates, and researchers. This board would be tasked with the ongoing assessment of healthcare facility regulations, ensuring they remain relevant and effective considering new healthcare challenges and opportunities.

These mechanisms should also include provisions for regular feedback from the users of healthcare facilities—both patients and providers. This feedback is invaluable as it provides direct insights into the practical impacts of regulatory environments and how they may be enhanced. Methods to gather such feedback might include surveys, focus groups, and feedback forms integrated into healthcare settings.

Additionally, the review process should be transparent and inclusive, providing opportunities for public input on proposed changes. This transparency not only enhances the credibility of the process but also encourages broader community engagement, ensuring that the regulations are well-rounded and widely supported.

To stay ahead of the curve, the regulatory framework should also be forward-looking, incorporating projections about future developments in healthcare technology and services. This anticipatory approach can help preemptively address potential challenges and seamlessly integrate emerging technologies into healthcare settings without extensive delays or retrofitting.

By actively advocating for and implementing a system of continuous review and adaptation, stakeholders in the healthcare industry can ensure that healthcare facilities are not just compliant with current standards but are also pioneers in adopting innovative designs that improve care delivery and patient well-being. This proactive stance is vital for fostering environments that are safe, efficient, and responsive to the evolving needs of patients and healthcare providers.

Table 7.9 provides a detailed schedule for the frequency, scope, and stakeholders involved in the regular review of healthcare facility regulations. This table outlines a structured approach to ensure continuous review and adaptation of regulations, highlighting the importance of keeping healthcare standards up to date with evolving practices and technologies. By clarifying the review process, this table aids regulatory bodies, healthcare administrators, and other stakeholders in maintaining a proactive stance on compliance and safety.

Next, we will delve into specific recommendations for policymakers on how they can support the development and continual improvement of healthcare facilities to ensure they meet the highest standards of safety, efficiency, and patient responsiveness.

## 7.4 Recommendations for Policy Makers

Providing actionable recommendations for policymakers is crucial to fostering healthcare environments that are not only compliant with current standards but are also forward-thinking and patient centric. These recommendations aim to bridge the

**Table 7.9** Schedule for regulatory review

| Regulation area | Review frequency | Scope of review | Stakeholders involved |
|---|---|---|---|
| Building safety codes | Biennial | Structural integrity, fire safety, accessibility | Civil engineers, architects, local government |
| Infection control standards | Annual | Air quality, surface materials, patient flow | Infection control specialists, hospital management |
| Digital health regulations | Every 3 years | Data security, interoperability, patient privacy | IT professionals, legal experts, healthcare providers |
| Emergency preparedness | Annual | Disaster response plans, emergency equipment | Emergency managers, local emergency services |
| Environmental compliance | Every 5 years | Waste management, energy use, water conservation | Environmental scientists, facility managers |

gap between regulatory frameworks and the practical, everyday realities of healthcare delivery, ensuring that policies are both effective and adaptable to the evolving needs of the healthcare sector.

### 7.4.1 Facilitate Cross-disciplinary Collaboration

Facilitating cross-disciplinary collaboration is essential for crafting healthcare facilities that are not only compliant with current standards but also equipped to meet future healthcare demands. Encouraging interaction among architects, healthcare professionals, technology experts, and regulatory bodies from the outset of facility planning and design can yield a more holistic approach to healthcare infrastructure.

This collaborative approach should be institutionalized as part of the standard procedure in healthcare facility development. Policymakers can promote such collaboration by providing platforms for dialogue and exchange, such as workshops, joint committees, and collaborative forums. These platforms allow stakeholders from various disciplines to share their expertise, insights, and concerns, ensuring that all aspects of healthcare delivery—from patient comfort and safety to technological integration and regulatory compliance—are addressed.

Furthermore, incorporating incentives for collaborative projects can significantly enhance participation from all sectors. These might include funding grants for research and development projects that involve partnerships between architects, medical staff, and tech innovators, or recognition awards for outstanding collaborative achievements in healthcare design.

Policymakers can also facilitate these interactions by creating clear guidelines that outline the roles and responsibilities of each stakeholder in the planning and

**Table 7.10** Benefits of cross-disciplinary collaboration

| Collaborative aspect | Benefits of healthcare design | Specific outcome |
|---|---|---|
| Inclusion of architectural experts | Enhanced facility layout and patient flow | Reduced patient stress and decreased cross-contamination risks |
| Integration of clinical staff insights | Tailored spaces for clinical needs | Improved efficiency in patient care delivery and higher staff satisfaction |
| Collaboration with IT professionals | Advanced technology integration | Streamlined operations and better data management |
| Environmental science contributions | Sustainable building practices | Reduced operational costs and improved environmental impact |
| Input from patient advocacy groups | Patient-centered design features | Higher patient satisfaction and better overall care experiences |

design phases. This clarity will help prevent overlap and ensure that each expert's contributions are maximized.

Moreover, by embedding legal requirements that mandate stakeholder involvement from the initial stages, policymakers can ensure that no critical perspective is overlooked. This might involve stipulations that certain project stages or milestones cannot be approved without demonstrable cross-disciplinary consultation.

In practice, such collaborative efforts could lead to the design of more responsive healthcare environments that effectively integrate advanced medical technologies, offer enhanced patient privacy and comfort, and adhere to the highest standards of safety and accessibility. The result is a healthcare facility that is truly reflective of the diverse needs and expertise of its creators, ready to serve the community effectively and efficiently.

Table 7.10 lists the benefits of integrating diverse professional insights into healthcare facility design, emphasizing how such collaboration can lead to improved patient outcomes, enhanced safety features, and innovation in healthcare technology. This table provides a clear, quantifiable overview of the advantages of cross-disciplinary teams, highlighting the value added through varied expertise in optimizing healthcare environments for better functionality and patient care.

### *7.4.2 Support Research and Innovation*

Supporting research and innovation in the realm of healthcare facility design is a pivotal strategy for policymakers committed to improving the quality and efficiency of healthcare delivery. By investing in research that rigorously examines the impact of architectural design on health outcomes and operational efficiencies, policymakers can base their decisions on empirical evidence, ensuring that investments in healthcare infrastructure are both effective and economically justified.

One practical step policymaker can take is to allocate funds specifically for pilot projects that explore innovative design solutions. These projects can serve as real-world laboratories, testing new ideas in architectural design, technology integration, and facility management. By monitoring and analyzing the outcomes of these projects, researchers can provide valuable data on what works and what doesn't, which can then be used to refine policies and standard practices in healthcare design.

Additionally, providing financial incentives for healthcare facilities to adopt proven innovative technologies and designs is crucial. These incentives could take the form of tax breaks, subsidies, or direct funding and should be targeted toward solutions that have demonstrated clear benefits in terms of enhancing patient care and improving operational efficiency. For example, incentives might be offered for the adoption of advanced HVAC systems that improve air quality and reduce the spread of airborne illnesses, or for architectural modifications that enhance patient privacy and comfort.

Policymakers can also foster innovation by facilitating partnerships between healthcare organizations and technology companies. These partnerships can drive the development and implementation of cutting-edge solutions, such as telemedicine services, advanced patient monitoring systems, and environmentally sustainable building technologies. By acting as a bridge between healthcare providers and innovators, policymakers can help ensure that the healthcare sector benefits from the latest advancements in technology and design.

Furthermore, establishing a regulatory environment that is conducive to innovation is essential. This involves streamlining the approval processes for new designs and technologies so that they can be tested and adopted more quickly. At the same time, it is crucial to maintain rigorous safety and efficacy standards to ensure that innovations provide real benefits and do not compromise patient care.

By actively supporting research and innovation through funding, incentives, partnerships, and a supportive regulatory framework, policymakers can ensure that healthcare facilities are not only equipped to meet the current needs of patients but are also prepared to adapt to future challenges and opportunities in healthcare delivery.

### *7.4.3 Update and Harmonize Standards Regularly*

To maintain the relevance and efficacy of healthcare facility standards, policymakers must ensure that these standards are regularly updated to keep pace with the rapid advancements in medical technology, evolving patient care practices, and updated safety protocols. Regularly revising building codes, infection control measures, and accessibility standards is essential to enhance the functionality and safety of healthcare environments, thereby improving patient outcomes and staff efficiency.

One key aspect of this process is the establishment of a systematic review cycle for all healthcare-related standards. This cycle should involve periodic assessments, perhaps every few years—conducted by panels of experts who can evaluate current guidelines against the latest medical research and technological developments. These panels should include a diverse range of professionals, such as medical practitioners, architects, patient advocates, and technology experts, ensuring a comprehensive review from multiple perspectives.

Moreover, there is a significant need for the harmonization of standards across different regions and, ideally, on a national or even international level. Disparities in healthcare facilities can lead to variations in the quality of care provided across different areas, which may exacerbate health inequities. By harmonizing these standards, policymakers can help ensure a more consistent level of healthcare quality and accessibility, regardless of geographical location.

This harmonization process can be facilitated through collaborations between different governmental health agencies and international health organizations. Such collaborations can help align various regional standards with global best practices, drawing on a broad base of knowledge and experience. Furthermore, international conferences and symposiums on healthcare standards can serve as platforms for sharing innovations and updates, helping to synchronize efforts worldwide.

In addition to standard updates and harmonization, there is also a need for robust implementation strategies. This includes training for healthcare providers and facility managers on the latest standards, regular audits of facilities to ensure compliance, and public reporting on adherence to standards to maintain transparency and accountability.

By prioritizing the regular update and harmonization of healthcare facility standards, policymakers can not only improve the quality of care but also foster a more equitable healthcare system that adapts effectively to future challenges and advancements.

Table 7.11 compares outdated standards with proposed updates, highlighting the enhancements and their expected impact on healthcare delivery. This table elucidates the reasons behind proposed changes, demonstrating how updates in standards can lead to more efficient, safer, and more patient-centered healthcare practices. By presenting a side-by-side comparison, the table justifies the need for these changes and illustrates the tangible improvements that updated standards can bring to healthcare environments.

### *7.4.4 Promote Sustainability and Resilience*

To secure the future of healthcare facilities against both environmental and operational challenges, policymakers must formulate and implement policies that prioritize sustainability and resilience in healthcare architecture. This forward-thinking approach not only meets the immediate demands of healthcare delivery but also addresses the broader impacts on the environment and community over the long term.

Promoting sustainability in healthcare infrastructure involves mandating the use of energy-efficient designs and sustainable materials during the construction and

**Table 7.11** Comparison of old versus new standards

| Standard category | Old standard | New proposed standard | Enhancements | Expected impact on healthcare delivery |
|---|---|---|---|---|
| Building safety | Fire safety codes from 1990 | Updated fire safety codes including modern materials and technology | Increased use of fire-resistant materials, better emergency egress designs | Enhanced safety, reduced risk of fire-related incidents |
| Infection control | Air changes per hour (ACH) rates based on 1985 guidelines | Increased ACH rates, integration of HEPA filtration systems | Improved air quality, reduction in airborne disease transmission | Decreased infection rates, enhanced patient and staff safety |
| Accessibility | ADA standards from 1990 | Updated ADA standards with broader accessibility requirements | Improved access for a wider range of disabilities, incorporation of new technologies like smart accessibility features | Greater inclusivity, enhanced autonomy for patients with disabilities |
| Digital health security | Basic data encryption standards | Advanced encryption and real-time security monitoring standards | Stronger data protection, implementation of the latest cybersecurity technologies | Increased patient data protection, reduced risk of data breaches |
| Energy efficiency | Standards from the early 2000s focusing on minimal compliance | High-efficiency standards incorporating renewable energy sources | Greater use of solar and wind energy, improved HVAC and lighting systems | Reduced environmental impact, lower operational costs |

renovation of facilities. Policymakers can enforce adherence to green building standards, such as Leadership in Energy and Environmental Design (LEED) or Building Research Establishment Environmental Assessment Method (BREEAM). These standards evaluate the environmental performance of buildings and promote sustainable practices across various domains, including energy consumption, water conservation, the utilization of sustainable materials, and waste management. Additionally, providing incentives for healthcare facilities that integrate renewable energy solutions—such as solar panels or geothermal heating and cooling systems—can significantly reduce their environmental footprint while also diminishing long-term operational costs. These initiatives make healthcare delivery more sustainable from both an economic and environmental perspective.

Enhancing resilience is equally essential, ensuring that healthcare facilities remain functional during crises like natural disasters, pandemics, or other emergencies. Policymakers should mandate that both new and existing facilities perform comprehensive risk assessments and develop tailored resilience plans that address potential threats specific to their geographic locations and operational contexts. These plans should include structural reinforcements to withstand natural disasters, such as earthquakes and floods, and systems designed to maintain critical operations, like power and water supplies. For instance, installing backup generators and implementing robust water storage systems are vital for maintaining operations during interruptions to power or municipal water services. Additionally, healthcare facility standards should be designed to allow quick adaptations to increased demands during emergencies, possibly through modular spaces that can be reconfigured into additional treatment areas or isolation units, and infrastructure that facilitates the rapid expansion of telehealth services when direct patient access to facilities is limited.

Effective implementation of these policies requires active collaboration with architectural firms, environmental experts, and healthcare administrators to ensure that the developed standards are both feasible and comprehensive. Establishing strong enforcement mechanisms, including regular audits and penalties for non-compliance, is critical to ensure these facilities consistently adhere to the stipulated standards.

Through these concerted efforts, policymakers can ensure that healthcare facilities are built to be environmentally sustainable and robust enough to provide continuous care under diverse conditions. This approach safeguards public health and contributes to the protection of the environment, aligning healthcare practices with broader societal goals of sustainability and resilience.

Equation 7.4 is designed to quantify the impact of sustainable practices on energy savings and the reduction of carbon footprints in healthcare facilities. This calculation is integral for healthcare administrators and facility managers aiming to implement environmentally friendly practices that not only conform to green building standards but also contribute to significant operational cost savings. As sustainability becomes a central concern in healthcare architecture, understanding the tangible benefits of these practices is crucial for justifying investments and measuring success in terms of energy efficiency and environmental impact.

This equation helps in assessing the effectiveness of various sustainability initiatives, such as energy-efficient lighting, advanced HVAC systems, and the use of renewable energy sources. By calculating the direct impact of these practices, healthcare facilities can better align their operations with sustainability goals, enhancing their reputation and fulfilling their social responsibilities toward environmental stewardship. The formula for calculating the Sustainability Impact (SI) in terms of energy savings and carbon footprint reduction is given by:

$$\mathrm{SI} = \left(E_{\mathrm{base}} - E_{\mathrm{post}}\right) \times \mathrm{CF} \tag{7.4}$$

where

- $E_{base}$ is the baseline energy consumption before implementing sustainable practices (in kilowatt-hours).
- $E_{post}$ is energy consumption after the implementation of sustainable practices (in kilowatt-hours).
- CF is the carbon footprint factor per unit of energy consumed (typically measured in kilograms of $CO_2$ equivalent per kilowatt-hour), which varies depending on the source of energy.

This equation provides healthcare facilities with a clear metric to assess the reduction in energy consumption and carbon emissions, quantifying the environmental benefits of adopting sustainable practices. By documenting these improvements, facilities not only ensure regulatory compliance but also contribute to global efforts in reducing environmental impact, while promoting a healthier community environment.

### *7.4.5 Enhance Accessibility and Inclusivity*

Advocating for policies that enhance accessibility and inclusivity in healthcare facilities is pivotal in ensuring that all individuals, irrespective of their physical abilities, age, or socioeconomic status, receive equitable access to healthcare services. This dedication to fostering an inclusive and accessible healthcare environment not only meets ethical and legal obligations but also significantly boosts overall community health outcomes by eliminating barriers to care.

One foundational aspect of enhancing accessibility is the implementation of universal design principles in the construction and renovation of healthcare facilities. Policymakers should enforce regulations that require healthcare facilities to incorporate universal design, which aims to create environments that can be accessed, understood, and used to the greatest extent possible by all people. This includes designing facilities that are easily navigable for individuals with disabilities, featuring wide corridors for wheelchair access, ramps as alternatives to stairs where feasible, elevators that accommodate all user needs, and the provision of visual and tactile signage to assist those with sensory impairments.

Additionally, the equitable distribution of healthcare facilities is essential for accessibility. This involves strategic placement to ensure that facilities are within reasonable reach of all populations, including those in rural or underserved areas. Policymakers can influence facility distribution through zoning laws and offering incentives to healthcare providers to set up in locations lacking adequate medical services. This strategic placement ensures that no community is disproportionately disadvantaged in terms of accessing healthcare services.

Addressing the cultural and linguistic needs of diverse communities is also crucial. Healthcare facilities should be equipped to serve the cultural context and language preferences of the communities they serve. This might involve employing multilingual staff, providing translation services, and offering care that is sensitive to various cultural beliefs and practices. Supporting policies could ensure that healthcare facilities assess the cultural and linguistic needs of their clientele and provide the necessary resources and training to their staff to meet these needs effectively.

Moreover, enhancing accessibility and inclusivity also demands comprehensive training and awareness programs for all healthcare facility staff—from medical professionals to administrative personnel. These programs should cover cultural competency, disability awareness, and inclusive communication practices, ensuring that staff are well-prepared to serve a diverse clientele respectfully and effectively.

To ensure adherence to these important standards, regulatory oversight and compliance mechanisms must be put in place. Regular checks and assessments should be conducted to confirm that facilities meet both physical accessibility standards and the principles of inclusiveness in their delivery service. Non-compliance should have clear and enforceable consequences, such as financial penalties or the requirement to undergo additional training.

By promoting and rigorously enforcing policies that bolster accessibility and inclusiveness, policymakers can transform healthcare facilities into true community assets. These facilities would then not only provide a welcoming and respectful environment but also deliver effective care to all members of society. Such a holistic approach to healthcare facility policy not only enhances individual health outcomes but also strengthens the broader health infrastructure, making it more equitable and responsive to the needs of the community.

### *7.4.6 Implement and Enforce Strict Privacy and Security Measures*

As healthcare facilities increasingly incorporate digital technologies and telemedicine services, the imperative for rigorous data privacy and security measures intensifies. To protect patient information and sustain trust in healthcare services, policymakers must devise and enforce comprehensive policies tailored to the unique challenges of digital healthcare delivery.

Policymakers must establish specific regulations tailored to the governance of digital health technologies and telemedicine services. These regulations should align with established healthcare privacy laws such as the Health Insurance Portability and Accountability Act (HIPAA) in the United States or the General Data Protection Regulation (GDPR) in the European Union. Key areas for regulation should include the encryption of data, secure storage practices, and ensuring that access to health information is strictly limited to authorized personnel. These measures are fundamental to safeguarding patient data and ensuring that digital interactions and data exchanges are conducted securely.

Furthermore, healthcare facilities must conduct regular risk assessments and security audits to pinpoint potential vulnerabilities within their digital infrastructures. Policymakers should mandate these assessments and outline clear guidelines regarding their frequency, methodology, and the qualifications required of auditors. This proactive strategy is crucial for identifying and mitigating security risks before they can be exploited.

To complement these technical measures, ongoing training and awareness programs for all healthcare facility staff are essential. These programs should emphasize the importance of data security, detail the specific security protocols of the facility, and clarify staff responsibilities in protecting patient information. Training should also include the identification of phishing attacks and other prevalent cybersecurity threats to fortify the human element of data security.

With the expanding role of telemedicine, it is also necessary to establish specific guidelines to secure these services. This includes the secure implementation of video conferencing tools, ensuring privacy during patient consultations, and robust procedures for verifying patient identities before sharing health information. Additionally, regulations must consider the cross-border exchange of health data, ensuring compliance with international data protection standards where applicable.

To foster compliance with these privacy and security standards, policymakers could introduce incentives such as tax breaks or funding opportunities for facilities that demonstrate robust compliance. Conversely, stringent penalties should be enforced for breaches that compromise patient data. These penalties might include significant fines and legal repercussions, underscoring the gravity of data security in healthcare and holding facilities accountable for any security lapses.

By implementing and enforcing strict privacy and security measures, policymakers can ensure that healthcare facilities remain trusted venues for patient care, where personal information is diligently protected against the emerging threats presented by digital technologies. This not only shields individuals' sensitive information but also preserves the integrity of the healthcare system.

Table 7.12 provides a checklist of essential privacy and security measures that healthcare facilities should implement to comply with the latest regulations and technological advances. This comprehensive checklist serves as a practical tool for facility managers and IT professionals, ensuring that healthcare institutions meet all necessary standards to protect patient information and maintain system integrity. By adhering to these guidelines, healthcare facilities can safeguard data breaches and unauthorized access, thereby upholding trust and confidentiality in patient care.

### *7.4.7 Continuous Monitoring and Feedback Mechanisms*

Establishing robust mechanisms for continuous monitoring and feedback is crucial for the dynamic governance of healthcare facilities, ensuring that policies are continually updated to reflect the evolving needs of patients and staff. This proactive approach not only supports timely adjustments to healthcare policies and practices but also significantly enhances the overall quality and safety of healthcare environments.

Policymakers should mandate regular assessment protocols to evaluate healthcare facilities' compliance with established standards and their effectiveness in meeting

**Table 7.12** Privacy and security standards checklist

| Security measure | Description | Required compliance | Implementation status (Yes/No) |
|---|---|---|---|
| Data encryption | Encrypt patient data both in transit and at rest | Mandatory | |
| Multi-factor authentication (MFA) | Require multiple forms of identification for system access | Highly recommended | |
| Regular security audits | Conduct periodic evaluations of the security infrastructure | Mandatory | |
| Firewall protection | Utilize firewalls to protect network boundaries | Mandatory | |
| Intrusion detection systems (IDS) | Deploy systems to monitor for malicious activity | Highly recommended | |
| Access control | Strict protocols to control who can access sensitive information | Mandatory | |
| Staff training on privacy policies | Regular training on the handling of patient data and privacy laws | Mandatory | |
| Patient data access management | Systems to log and monitor patient data access | Mandatory | |
| Secure backup solutions | Implement automated and secure backup processes | Highly recommended | |
| Compliance with GDPR/HIPAA or local regulations | Ensure all practices align with relevant legal requirements | Mandatory | |

the needs of patients and staff. These evaluations could encompass routine inspections, patient and staff surveys, and performance audits, providing critical insights into the operational realities of healthcare settings. Such comprehensive assessments are fundamental to understanding how well facilities meet their objectives and where improvements are needed.

Leveraging advancements in technology can streamline the monitoring process. Digital reporting tools, for instance, enable healthcare facilities to continuously log compliance data, which can be automatically analyzed to detect patterns or areas needing attention. This approach to real-time data collection significantly enhances the ability of regulatory bodies to respond swiftly to emerging issues or instances of non-compliance, ensuring that healthcare standards are consistently maintained.

Feedback from both patients and healthcare staff is invaluable in assessing the effectiveness of current healthcare facility policies. Establishing formal channels for feedback—such as suggestion boxes, digital forums, and regular focus groups—provides continuous insights into the user experience within healthcare facilities. Actively soliciting, regularly reviewing, and utilizing this feedback to inform policy revisions is essential for maintaining responsive and patient-centered healthcare environments.

Moreover, policymakers must ensure that there are efficient mechanisms in place to adapt and revise policies based on the feedback and data collected. This might include the formation of dedicated committees or task forces that specialize in policy review and the integration of innovative practices and technologies that address identified gaps or challenges.

Transparency in the monitoring and feedback processes is vital for building trust and encouraging active participation from all stakeholders. Publicly sharing the findings from assessments and the subsequent actions taken not only promotes accountability but also keeps stakeholders informed about ongoing changes and improvements.

By adhering to these practices, policymakers can ensure that healthcare facilities not only comply with current standards but also continuously evolve to meet their users' needs more effectively. This ongoing process of evaluation and adaptation is crucial for maintaining high-quality healthcare environments that are safe, efficient, and centered on patient care.

Next, we will explore how integrating international health standards can further enhance the quality and safety of healthcare facilities, promoting global consistency in healthcare delivery and facility design. This global perspective is instrumental in raising healthcare practices to a higher standard and ensuring that facilities across different regions can offer uniformly excellent care, thus elevating the overall health infrastructure.

## 7.5 Integrating International Health Standards

Integrating international health standards into national and local regulatory frameworks can significantly enhance the quality and safety of healthcare facilities worldwide. This approach promotes global consistency in healthcare delivery, ensures that facilities are equipped to handle international health challenges, and facilitates the exchange of best practices across borders. Here's how policymakers and healthcare leaders can effectively integrate these standards.

### 7.5.1 Adoption of WHO Guidelines

The adoption of the World Health Organization (WHO) guidelines is a pivotal strategy for enhancing global healthcare facility standards. These guidelines are formulated through rigorous research and international consensus, offering a comprehensive set of universally applicable standards that span various essential healthcare aspects, including facility design and patient safety.

By integrating WHO guidelines, countries can align their healthcare practices with international standards, ensuring a uniform quality of care across different regions. This standardization is particularly advantageous for nations with emerging healthcare systems as it provides a robust framework for constructing and managing healthcare facilities that are safe, efficient, and attuned to the needs of their populations.

In terms of facility design and construction, WHO guidelines offer detailed specifications for structural integrity, spatial allocations, and environmental control. These specifications ensure that healthcare facilities are not only aesthetically appealing but also structurally sound and capable of enduring various operational stresses, including natural disasters. Such robust design principles are vital for maintaining continuous healthcare services during unexpected challenges.

The implementation of WHO guidelines concerning infection prevention and control is also critical, especially highlighted by the global repercussions of health crises like the COVID-19 pandemic. These guidelines outline essential protocols for maintaining hygiene, sanitation, and patient management, pivotal in curtailing the spread of infectious diseases within healthcare settings. Adherence to these protocols can significantly mitigate infection rates, which in turn enhances patient outcomes and reduces overall healthcare costs.

WHO guidelines also place a strong emphasis on patient safety and emergency preparedness. These guidelines provide detailed procedures and protocols designed to safeguard both patients and healthcare personnel. This includes the safe operation of medical equipment, proper handling of hazardous materials, and the formulation of comprehensive emergency response strategies. By adopting these standards, healthcare facilities can boost their effectiveness in responding to emergencies, thereby minimizing risks to both patients and staff.

However, implementing WHO guidelines can be challenging, especially in resource-limited environments. To overcome these hurdles, governments and healthcare administrators may seek international assistance, forming partnerships to gain both financial and technical support. Moreover, educational initiatives such as training programs and workshops are crucial in familiarizing healthcare workers with these guidelines and equipping them with the necessary skills for effective implementation.

In summary, adopting WHO guidelines is a crucial step toward global standardization of healthcare practices. It ensures that healthcare facilities operate under the highest safety and efficiency standards, fostering a culture of continuous improvement and preparedness to tackle future healthcare challenges effectively.

This strategic approach not only enhances the quality of care provided but also strengthens the overall health infrastructure across different regions.

Table 7.13 provides a checklist of key World Health Organization (WHO) guidelines related to healthcare facility design, patient safety, and infection control. This table serves as a concise reference for policymakers and healthcare administrators to ensure comprehensive compliance with international standards, aimed at enhancing the effectiveness and safety of healthcare environments. The checklist helps facilities align their operations with globally recognized best practices, promoting improved patient care and reducing the risk of healthcare-associated infections.

**Table 7.13** Checklist of WHO guidelines for healthcare facilities

| WHO guideline category | Key guidelines | Compliance requirement | Notes for implementation |
|---|---|---|---|
| Facility design | Ensure easy cleanability, adequate ventilation, and safe water supply | Mandatory | Design spaces to minimize contamination risks |
| Patient safety | Implement standardized protocols for patient identification, surgery safety, and medication accuracy | Mandatory | Regular training and audits to ensure protocol adherence |
| Infection control | Guidelines for hand hygiene, sterilization of equipment, and isolation of infectious patients | Mandatory | Set up continuous monitoring and compliance reporting systems |
| Emergency preparedness | Adequate facilities and plans for dealing with mass casualty events and epidemics | Mandatory | Regular drills and updates to emergency response plans |
| Environmental health | Proper waste management systems and sustainable resource usage | Recommended | Integrate environmental sustainability into facility operations |
| Maternal and child health | Facilities must support safe childbirth and immediate newborn care | Mandatory | Equip labor and delivery areas with essential medical supplies |
| Non-communicable diseases | Designated areas for the treatment and management of chronic diseases like diabetes and hypertension | Recommended | Create patient-centered services that encourage long-term care management |

### 7.5.2 *Benchmarking Against Global Best Practices*

Encouraging healthcare facilities to benchmark against global best practices is a strategic approach that enables policymakers to significantly elevate the quality and efficiency of healthcare services. This process involves evaluating operational practices, patient care strategies, and facility designs by comparing them with those recognized internationally for their excellence and efficacy.

One effective method to facilitate this benchmarking process is through participation in international accreditation programs such as those offered by the Joint Commission International (JCI) or the International Society for Quality in Health Care (ISQua). These programs provide a standardized framework that healthcare facilities can use as a benchmark to measure their services. Achieving accreditation is not only a marker of quality assurance but also serves as a significant motivator for continuous improvement, encouraging facilities to maintain or exceed high standards in patient care and safety.

The advantages of benchmarking are manifold. It allows healthcare facilities to identify and address service gaps by comparing their performance with that of top-tier global institutions. This comparative analysis can illuminate deficiencies in various areas such as patient management, infection control, or emergency preparedness, and can inspire facilities to adopt innovative practices and technologies. Furthermore, benchmarking fosters a culture of transparency and accountability, as facilities strive to achieve and uphold internationally recognized standards.

To effectively implement global best practices, healthcare facilities may need to invest in several areas, including staff training, technology upgrades, and infrastructure enhancements. Policymakers can support these initiatives by providing targeted funding, offering tax incentives for investments in healthcare improvements, and facilitating partnerships with globally recognized healthcare providers.

While it is beneficial to align with international standards, it is also crucial for facilities to adapt these practices to fit local contexts. This adaptation might involve modifying some international norms to better suit the cultural, economic, and environmental conditions of the region. Such tailored adjustments ensure that the adopted practices are not only globally competitive but also locally relevant and sustainable, thereby enhancing their effectiveness and acceptance.

Moreover, benchmarking should be a continuous process with regular assessments to ensure that the improvements are sustained and that healthcare facilities consistently meet international standards. Establishing feedback mechanisms is important to allow healthcare professionals and patients to share their experiences. Such feedback ensures that the benchmarking process remains adaptive to the needs and outcomes of the community it serves.

By promoting benchmarking against global best practices, policymakers can significantly enhance the quality of healthcare delivery. This ensures that facilities are not only compliant with established standards but are also continually seeking innovations and advancements. This proactive approach cultivates a healthcare system

**Table 7.14** International accreditation standards comparison

| Accreditation body | Key components | Requirements | Focus areas |
|---|---|---|---|
| Joint Commission International (JCI) | Patient safety, healthcare facility management | Compliance with JCI standards, regular audits | Quality of care, patient safety, organizational management |
| International Society for Quality in Health Care (ISQua) | Quality improvement, staff training | Adherence to ISQua guidelines, continuous improvement efforts | Staff education, leadership, care processes |
| Accreditation Canada International (ACI) | Patient care protocols, governance | Must meet Canadian standards, periodic reassessment | Governance, safety, patient engagement |
| Australian Council on Healthcare Standards International (ACHSI) | Clinical care, corporate management | Compliance with Australian healthcare standards, evidence of outcomes | Clinical effectiveness, patient safety, organizational performance |
| American Accreditation Healthcare Commission (AAHC) | Policy development, risk management | Policy compliance, risk assessment documentation | Risk management, policy procedures, regulatory compliance |

that is robust, efficient, and well-equipped to meet the challenges of modern medicine, ensuring optimal patient outcomes.

Table 7.14 compares different international accreditation standards, such as those from the Joint Commission International (JCI) and the International Society for Quality in Health Care (ISQua), highlighting their key components and requirements. This table helps healthcare facilities understand the various criteria set by each accreditation body, assisting them in choosing the right accreditation that aligns with their operational goals and quality benchmarks. By presenting these standards side by side, the table enables facilities to evaluate and decide which accreditation best supports their commitment to global best practices in healthcare.

### *7.5.3 Facilitating International Collaboration*

Facilitating international collaboration represents a strategic approach that governments and health authorities can adopt to amplify the capabilities and foster innovation within their healthcare systems. By establishing partnerships with healthcare organizations from other nations, they can access a global pool of knowledge and expertise. This potentially leads to significant improvements in healthcare facility design and management. One of the primary benefits of such collaboration is the ability to share knowledge across borders seamlessly. Governments can facilitate this by organizing international conferences, workshops, and seminars that convene healthcare professionals from various parts of the world. These gatherings are instrumental in sharing insights and best practices and provide a robust platform for discussing recent advances in medical research and emerging technologies in healthcare. Additionally, these events highlight innovative management practices that have been successful in diverse healthcare settings, thus promoting a global exchange of ideas and solutions that could be transformative if adapted and implemented across different systems.

Further strengthening international collaboration, joint research projects serve as a cornerstone of these cooperative efforts. Such projects can span a multitude of topics, ranging from the development of new medical technologies and treatments to explorations into the resilience of healthcare infrastructure. Pooling resources and expertise allow for addressing more complex challenges and crafting comprehensive solutions that extend benefits to a broader population. Governments can propel these initiatives by providing essential funding and streamlining regulatory approvals needed for international research endeavors. This support not only catalyzes the creation of innovative solutions but also fosters a collaborative spirit among nations focused on enhancing global health outcomes.

Exchange programs for healthcare professionals, including doctors, nurses, administrators, and other relevant stakeholders, are equally critical. These programs offer participants invaluable exposure to different healthcare settings, enabling them to observe and assimilate various operational aspects such as facility design, patient flow management, emergency response, and routine operations. The insights gained from these experiences are crucial, as participants can adapt and integrate effective practices into their home countries' healthcare systems, potentially revolutionizing the quality of care provided.

Moreover, international collaboration can extend to regulatory cooperation, where health authorities from different countries collaborate to synchronize regulations that impact healthcare facility standards, medical equipment, and pharmaceutical products. Such cooperation is vital for streamlining processes, reducing bureaucratic hurdles, and establishing a more unified approach to addressing global health challenges. Governments can solidify these collaborative endeavors by entering into formal bilateral or multilateral agreements focused on healthcare development and enhancement. These agreements can delineate specific areas of cooperation, establish objectives for joint projects, and frame the structure for ongoing support and dialogue, ensuring that healthcare systems worldwide can stay abreast of medical and managerial innovations and are well-prepared to adapt to the dynamic challenges of global health.

### *7.5.4 Training and Capacity Building*

Investing in training and capacity building is pivotal for healthcare facilities to not only meet international health standards but also maintain competitiveness and effectiveness in the ever-evolving global health landscape. Training programs for healthcare administrators, architects, and facility managers are essential to keep abreast of the latest developments in healthcare design, technology integration, and patient safety. By developing comprehensive training programs that encapsulate the latest trends in healthcare facility design and management, healthcare professionals can be equipped with the knowledge and skills required to implement advanced practices in their operations. These training programs should be robust, incorporating modules on sustainable building practices, digital health technology integration, effective infection control strategies, and patient-centered design principles. Such well-rounded training ensures that the healthcare staff are proficient in contemporary global standards and can adeptly translate these concepts to enhance local healthcare practices, thus fostering an environment of continuous innovation and quality improvement.

To further enhance the quality of training, healthcare organizations can benefit greatly from collaborations with international experts and institutions specializing in healthcare management and facility design. These partnerships are invaluable as they facilitate an exchange of cutting-edge ideas and provide access to the best global practices and innovative solutions that may not be readily available locally. Organizing workshops, webinars, and joint training sessions with these international experts allows healthcare professionals to gain a deeper understanding of the complexities involved in modern healthcare challenges and the diverse strategies employed globally to address them.

Moreover, implementing certification programs that validate and recognize the expertise of healthcare professionals in specific areas, such as hospital management, healthcare architecture, or emergency preparedness, can significantly elevate the professionalization of the field. These certification programs, aligned with international standards, assure that professionals are not only well-trained but also acknowledged for their specialized knowledge and competencies, thus promoting a higher standard of professionalism within the healthcare sector.

Incorporating simulation-based learning and hands-on training experiences can dramatically enhance the practical application of theoretical knowledge. Simulations of emergencies, patient flow management, and the deployment of medical technologies within facility settings provide trainees with realistic scenarios where they can hone their skills and adapt their theoretical learning to meet practical demands. This method of experiential learning is crucial for bridging the gap between knowledge and practice, ensuring that professionals are well-prepared to handle real-world challenges efficiently.

Lastly, establishing a culture of ongoing professional development is essential for ensuring that healthcare personnel remain at the cutting edge of industry standards and innovations. Continuous education opportunities, such as attending international conferences, participating in refresher courses, and engaging with the latest research, are vital for maintaining high levels of expertise and motivation among staff. These ongoing learning opportunities help foster a dynamic workforce that is not only well-informed but also continually evolve with the global healthcare landscape.

By prioritizing training and capacity building, healthcare facilities can ensure that their infrastructure and operational methodologies not only comply with international standards but are also reflective of the latest best practices in patient care and facility management. This strategic focus on continuous education and professional development is key to building a resilient healthcare workforce that is well-equipped to adapt to and implement cutting-edge health innovations globally.

Equation 7.5 provides a model to evaluate the potential impact of comprehensive training programs on the performance of healthcare facilities. This equation is particularly crucial for assessing how training initiatives improve staff competency levels and patient outcomes, thereby enhancing the overall quality of care and operational efficiency. Effective training programs are essential for maintaining high standards of care and ensuring that healthcare professionals are well-equipped to handle the complexities of modern medical environments.

Incorporating factors such as staff skill enhancements and measurable improvements in patient care, this equation helps healthcare administrators quantify the benefits of investment in continuous professional development. By doing so, it not only supports the case for ongoing training as a strategic investment but also aligns staff capabilities with the institution's goals of excellence in patient care. The formula for calculating the Impact of Training Programs (ITP) is given by:

$$\text{ITP} = \left( \frac{\Delta P}{P_{\text{base}}} + \frac{\Delta S}{S_{\text{base}}} \right) \times 100 \tag{7.5}$$

where

- $\Delta P$ represents the improvement in patient outcomes attributable to the training, measured as a change from the baseline.
- $P_{\text{base}}$ is the baseline measurement of patient outcomes before the training was implemented.
- $\Delta S$ denotes the enhancement in staff competency levels post-training.
- $S_{\text{base}}$ is the baseline measurement of staff competency levels.

This equation outputs a percentage that reflects the combined impact of training on both staff competencies and patient outcomes. By using this model, healthcare facilities can effectively measure the returns on their investments in training programs, providing a compelling argument for prioritizing such initiatives. Furthermore, this analytical approach supports healthcare leaders in making data-driven decisions to optimize training strategies and resources, ultimately leading to sustained improvements in facility performance and patient care quality.

### 7.5.5 Regulatory Harmonization

Efforts to harmonize local regulations with international standards are essential for creating a cohesive and efficient regulatory environment for healthcare facilities. This alignment is crucial to ensure that the quality of care provided is consistent globally and facilitates the operation of international healthcare providers across diverse regulatory landscapes. One of the fundamental steps in regulatory harmonization is the revision of national building codes and safety regulations to reflect international standards set by authoritative bodies such as the World Health Organization (WHO) or the International Building Code (IBC). These revisions are critical as they need to encompass structural safety, accessibility, emergency preparedness, and environmental sustainability. Ensuring that healthcare facilities are constructed and maintained according to the highest standards enhances their resilience and functionality, promoting safer and more efficient healthcare environments.

In parallel with updating physical building codes, there is a pressing need to align healthcare policies related to patient safety, data privacy, and staff qualifications with international norms. This alignment could include adopting international guidelines on patient handling, infection control, medical record confidentiality, and certification requirements for healthcare professionals. By harmonizing these policy areas, healthcare facilities can streamline procedures and protocols across borders, thus enhancing the quality of care and bolstering the trust of both patients and healthcare providers. Such a streamlined approach not only simplifies the administrative aspects of healthcare but also ensures a uniform standard of care, which is particularly important in a globally interconnected world.

To effectively facilitate the harmonization process, it is advantageous to consult international experts and participate in global healthcare forums. Engaging with global health authorities offers valuable insights into the effectiveness of existing regulations and provides recommendations for critical areas that require alignment. This collaborative approach ensures that the harmonization efforts are well-informed by the best available expertise and are attuned to global trends and innovations, thereby enhancing the relevance and efficacy of the regulatory updates.

Moreover, by harmonizing regulations, countries can simplify the overall regulatory landscape, making it more navigable for international healthcare providers operating across various jurisdictions. This simplification not only attracts international investment into the healthcare sector but also fosters a more integrated and cooperative global healthcare system. To support these changes, governments may need to put in place new legislative and policy frameworks that facilitate the adoption and seamless implementation of international standards. This might involve the creation of specific bodies or task forces dedicated to overseeing the process of regulatory harmonization, ensuring that updates are systematically applied and enforced throughout the healthcare system.

Lastly, it is crucial to view harmonization as an ongoing process. Continuous evaluation and adaptation of regulations are necessary to keep pace with rapid technological advancements and evolving practices in healthcare. Establishing mechanisms for regular review and feedback will enable ongoing alignment with international standards and allow for a dynamic response to emerging health challenges. Through these comprehensive efforts, regulatory harmonization not only improves the quality of healthcare within a country but also contributes significantly to enhancing global health security by promoting consistency and cooperation across national boundaries.

Table 7.15 outlines the pre- and post-harmonization impacts on healthcare delivery efficiency, patient safety, and compliance costs, providing empirical data to support the push for harmonized standards. This table presents clear comparisons to demonstrate how standardizing regulatory frameworks across jurisdictions can enhance

**Table 7.15** Regulatory harmonization impact assessment

| Impact area | Pre-harmonization status | Post-harmonization status | Improvement (%) | Notes |
|---|---|---|---|---|
| Healthcare delivery efficiency | Average procedure turnaround time: 3 h | Average turnaround time: 2 h | 33% increase | Streamlined processes reduce delays and waiting times |
| Patient safety | Incident rate per 1,000 patients: 5 | Incident rate per 1,000 patients: 3 | 40% decrease | Enhanced standards and protocols improve safety outcomes |
| Compliance costs | Annual compliance cost: $500,000 | Annual compliance cost: $300,000 | 40% decrease | Standardized procedures lower training and implementation costs |
| Regulatory alignment | Varied standards across regions | Unified standards across regions | – | Facilitates easier compliance and inter-region operations |
| Stakeholder satisfaction | Stakeholder satisfaction rate: 70% | Stakeholder satisfaction rate: 85% | 21% increase | Higher satisfaction due to predictable and uniform regulations |

operational efficiencies, reduce costs, and improve patient safety in healthcare settings. By quantifying these benefits, the table substantiates the advantages of regulatory harmonization and emphasizes its necessity in creating a more effective and integrated healthcare system.

### *7.5.6 Leveraging Technology for Compliance*

Technology serves as a powerful ally in aiding healthcare facilities to adhere to international standards, ensuring the maintenance of high levels of care and operational efficiency. By leveraging digital tools, facilities can streamline compliance processes, enhance training programs, and monitor facility performance more effectively. Advanced digital tools can be implemented to monitor compliance with international standards in real-time. These systems are capable of tracking various aspects of facility operations, from infection control measures to emergency preparedness protocols, and alerting management to any deviations from established guidelines. This real-time monitoring permits immediate corrective actions, thus preventing minor issues from escalating into serious compliance failures, ensuring that facilities can address potential problems swiftly and effectively.

Digital platforms can also be configured to send automated reminders and alerts concerning regulatory requirements and deadlines. This feature is crucial in ensuring that all necessary inspections, staff training sessions, and documentation updates are completed on time. It helps maintain a continuous state of readiness and compliance, reducing the risk of oversight or non-compliance due to human error. Moreover, technology facilitates the ongoing education and training of healthcare staff through accessible e-learning platforms. These platforms can offer up-to-date training modules on the latest international health standards and best practices, allowing staff to complete necessary training efficiently and at their own pace. This method of delivering training significantly enhances learning outcomes and ensures compliance with complex regulations, which is vital in a healthcare environment where standards and practices are continuously evolving.

In addition, implementing data analytics tools provides healthcare administrators with valuable insights into the effectiveness of their compliance efforts. These tools can analyze trends over time, identify areas for improvement, and assist in making data-driven decisions to enhance facility operations and patient care. Furthermore, leveraging technology means that healthcare facilities can more easily integrate with global health systems, sharing data and insights while maintaining compliance with international privacy and security standards. This integration is crucial for global disease surveillance, research collaborations, and benchmarking against global best practices, enabling a more cohesive and cooperative international healthcare environment.

By integrating technology into compliance and facility management, healthcare organizations can ensure they meet international standards more consistently and efficiently. This not only enhances the operational aspects of healthcare delivery but also improves patient outcomes by maintaining rigorous standards of care and safety. With the completion of this chapter, we have detailed the essential regulatory and policy considerations that shape the architecture and operations of healthcare facilities. These guidelines and frameworks are fundamental in ensuring that healthcare environments are not only safe and efficient but also equipped to meet the complex needs of the populations they serve, now and in future. This integration of international standards and innovative compliance strategies through technology ensures that healthcare facilities are prepared to face global health challenges with resilience and adaptability.

## Further Reading

Abdelaal, M. S., & Soebarto, V. (2019). Biophilia and Salutogenesis as restorative design approaches in healthcare architecture. *Architectural Science Review*, 62(3), 195–205. https://doi.org/10.1080/00038628.2019.1604313

Al-Hajj, S., Abou-El-Hassan, H., Khalil, L., Kaafarani, H. M., & El Sayed, M. (2020). Hospital disaster and emergency preparedness (HDEP) in Lebanon: a national comprehensive assessment. *International Journal of Disaster Risk Reduction*, 51, 101889. https://doi.org/10.1016/j.ijdrr.2020.101889

AlJaberi, O. A., Hussain, M., & Drake, P. R. (2020). A framework for measuring sustainability in healthcare systems. *International Journal of Healthcare Management*. https://doi.org/10.1080/20479700.2017.1404710

Aminizadeh, M., Farrokhi, M., Ebadi, A., Masoumi, G. R., Kolivand, P., & Khankeh, H. R. (2019). Hospital management preparedness tools in biological events: A scoping review. *Journal of Education and Health Promotion*, 8(1), 234. https://doi.org/10.4103/jehp.jehp_473_19

Annaswamy, T. M., Verduzco-Gutierrez, M., & Frieden, L. (2020). Telemedicine barriers and challenges for persons with disabilities: COVID-19 and beyond. *Disability and Health Journal*, 13(4), 100973. https://doi.org/10.1016/j.dhjo.2020.100973

Araujo, C. A., Siqueira, M. M., & Malik, A. M. (2020). Hospital accreditation impact on healthcare quality dimensions: a systematic review. *International Journal for Quality in Health Care*, 32(8), 531–544. https://doi.org/10.1093/intqhc/mzaa090

Avia, I., & Hariyati, R. T. S. (2019). Impact of hospital accreditation on quality of care: A literature review. *Enfermeria Clinica*, 29, 315–320. https://doi.org/10.1016/j.enfcli.2019.06.003

Bates, V. (2018). 'Humanizing' healthcare environments: architecture, art and design in modern hospitals. *Design for Health*, 2(1), 5–19. https://doi.org/10.1080/24735132.2018.1436304

Bayramzadeh, S., Joseph, A., Allison, D., Shultz, J., Abernathy, J., & Group, R. O. S. (2018). Using an integrative mock-up simulation approach for evidence-based evaluation of operating room design prototypes. *Applied Ergonomics*, 70, 288–299. https://doi.org/10.1016/j.apergo.2018.03.011

Berry, L. L. (2019). Service innovation is urgent in healthcare. *AMS Review*, 9(1), 78–92. https://doi.org/10.1007/s13162-019-00135-x

Bhimraj, A., Morgan, R. L., Shumaker, A. H., Lavergne, V., Baden, L., Cheng, V. C. C., … & Falck-Ytter, Y. (2020). Infectious Diseases Society of America Guidelines on the treatment and management of patients with coronavirus disease 2019 (COVID-19). *Clinical Infectious Diseases,* ciaa 478. https://doi.org/10.1093/cid/ciaa478

Bhuiyan, M. N., Rahman, M. M., Billah, M. M., & Saha, D. (2021). Internet of things (IoT): A review of its enabling technologies in healthcare applications, standards protocols, security, and market opportunities. *IEEE Internet of Things Journal*, 8(13), 10474–10498. https://doi.org/10.1109/JIOT.2021.3062630

Bingham, E., Whitaker, D., Farnsworth, C., & Smith, J. (2022). Evidence-Based Design in Hospital Renovation Projects: Design Implementation for Patient Privacy and Comfort. *Journal of Architectural Engineering*, 28(2), 04022002. https://doi.org/10.1061/(ASCE)AE.1943-5568.0000530

Bitkina, O. V., Kim, H. K., & Park, J. (2020). Usability and user experience of medical devices: An overview of the current state, analysis methodologies, and future challenges. *International Journal of Industrial Ergonomics*, 76, 102932. https://doi.org/10.1016/j.ergon.2020.102932

Bloom, D. E., & Cadarette, D. (2019). Infectious disease threats in the twenty-first century: strengthening the global response. *Frontiers in Immunology*, 10, 549. https://doi.org/10.3389/fimmu.2019.00549

Bröchner, J., Haugen, T., & Lindkvist, C. (2019). Shaping tomorrow's facilities management. *Facilities*, 37(7/8), 366–380. https://doi.org/10.1108/F-10-2018-0126

Capolongo, S., Gola, M., Brambilla, A., Morganti, A., Mosca, E. I., & Barach, P. (2020). COVID-19 and healthcare facilities: a decalogue of design strategies for resilient hospitals. *Acta Bio Medica: Atenei Parmensis*, 91(9-S), 50. https://doi.org/10.23750/abm.v91i9-S.10117

Cedeño-Laurent, J. G., Williams, A., MacNaughton, P., Cao, X., Eitland, E., Spengler, J., & Allen, J. (2018). Building evidence for health: green buildings, current science, and future challenges. *Annual Review of Public Health*, 39(1), 291–308. https://doi.org/10.1146/annurev-publhealth-031816-044420

Choudhury, A., & Asan, O. (2020). Role of artificial intelligence in patient safety outcomes: systematic literature review. *JMIR Medical Informatics*, 8(7), e18599. https://doi.org/10.2196/18599

Covvey, H. D. (2018). Healthcare as a complex adaptive system. In EHealth Research, Theory, and Development (pp. 69–90). Routledge.

Cronk, R., & Bartram, J. (2018). Environmental conditions in health care facilities in low-and middle-income countries: coverage and inequalities. *International Journal of Hygiene and Environmental Health*, 221(3), 409–422. https://doi.org/10.1016/j.ijheh.2018.01.004

Devi, G. R., Hough, H., Barrett, N., Cristofanilli, M., Overmoyer, B., Spector, N., … & Marcom, P. K. (2019). Perspectives on inflammatory breast cancer (IBC) research, clinical management and community engagement from the Duke IBC consortium. *Journal of Cancer*, 10(15), 3344. https://doi.org/10.7150/jca.31176

Di Pietro, L., Piaggio, D., Oronti, I., Maccaro, A., Houessouvo, R. C., Medenou, D., … & Ahluwalia, A. (2020). A framework for assessing healthcare facilities in low-resource settings: field studies in Benin and Uganda. *Journal of Medical and Biological Engineering*, 40, 526–534. https://doi.org/10.1007/s40846-020-00546-3

Ding, Z., Fan, Z., Tam, V. W., Bian, Y., Li, S., Illankoon, I. C. S., & Moon, S. (2018). Green building evaluation system implementation. *Building and Environment*, 133, 32–40. https://doi.org/10.1016/j.buildenv.2018.02.012

Fifield, L. J., Lomas, K. J., Giridharan, R., & Allinson, D. (2018). Hospital wards and modular construction: Summertime overheating and energy efficiency. *Building and Environment*, 141, 28–44. https://doi.org/10.1016/j.buildenv.2018.05.041

Filip, R., Gheorghita Puscaselu, R., Anchidin-Norocel, L., Dimian, M., & Savage, W. K. (2022). Global challenges to public health care systems during the COVID-19 pandemic: a review of pandemic measures and problems. *Journal of Personalized Medicine*, 12(8), 1295. https://doi.org/10.3390/jpm12081295

Frank, J. R., Taber, S., van Zanten, M., Scheele, F., Blouin, D., & International Health Professions Accreditation Outcomes Consortium. (2020). The role of accreditation in 21st century health professions education: report of an International Consensus Group. *BMC Medical Education*, 20, 1–9. https://doi.org/10.1186/s12909-020-02121-5

Ghadami, L., Masoudi Asl, I., Hessam, S., & Modiri, M. (2021). Developing hospital accreditation standards: Applying fuzzy DEMATEL. *International Journal of Healthcare Management*, 14(3), 847–855. https://doi.org/10.1080/20479700.2019.1702307

Golbazi, M., & Aktas, C. B. (2020). LEED certification and patient wellbeing in green healthcare facilities. *Journal of Green Building*, 15(4), 3–18. https://doi.org/10.3992/jgb.15.4.3

Gopal, G., Suter-Crazzolara, C., Toldo, L., & Eberhardt, W. (2019). Digital transformation in healthcare–architectures of present and future information technologies. *Clinical Chemistry and Laboratory Medicine (CCLM)*, 57(3), 328–335. https://doi.org/10.1515/cclm-2018-0658

Graban, M. (2018). Lean hospitals: improving quality, patient safety, and employee engagement. Productivity Press.

Griffin, K. M., Karas, M. G., Ivascu, N. S., & Lief, L. (2020). Hospital preparedness for COVID-19: a practical guide from a critical care perspective. American Journal of Respiratory and Critical Care Medicine, 201(11), 1337–1344. https://doi.org/10.1164/rccm.202004-1037CP

Halawa, F., Madathil, S. C., Gittler, A., & Khasawneh, M. T. (2020). Advancing evidence-based healthcare facility design: a systematic literature review. *Health Care Management Science*, 23, 453–480. https://doi.org/10.1007/s10729-020-09506-4

Halawa, F., Madathil, S. C., Gittler, A., & Khasawneh, M. T. (2020). Advancing evidence-based healthcare facility design: a systematic literature review. *Health Care Management Science*, 23, 453–480. https://doi.org/10.1007/s10729-020-09506-4

Hernández, D., Ors, R., Capella, J. V., Bonastre, A., & Campelo, J. C. (2020). New contact sensorization smart system for IoT e-Health Applications Based on IBC IEEE 802.15. 6 *Communications.* Sensors, 20(24), 7097. https://doi.org/10.3390/s20247097

Hussein, M., Pavlova, M., Ghalwash, M., & Groot, W. (2021). The impact of hospital accreditation on the quality of healthcare: a systematic literature review. *BMC Health Services Research*, 21, 1–12. https://doi.org/10.1186/s12913-021-07097-6

Joseph, A., Bayramzadeh, S., Zamani, Z., & Rostenberg, B. (2018). Safety, performance, and satisfaction outcomes in the operating room: a literature review. *HERD: Health Environments Research & Design Journal*, 11(2), 137–150. https://doi.org/10.1177/1937586717705107

Kapologwe, N. A., Meara, J. G., Kengia, J. T., Sonda, Y., Gwajima, D., Alidina, S., & Kalolo, A. (2020). Development and upgrading of public primary healthcare facilities with essential surgical services infrastructure: a strategy towards achieving universal health coverage in Tanzania. *BMC Health Services Research*, 20, 1–14. https://doi.org/10.1186/s12913-020-5057-2

Khahro, S. H., Kumar, D., Siddiqui, F. H., Ali, T. H., Raza, M. S., & Khoso, A. R. (2021). Optimizing energy use, cost and carbon emission through building information modelling and a sustainability approach: A case-study of a hospital building. *Sustainability*, 13(7), 3675. https://doi.org/10.3390/su13073675

Kodur, V., Kumar, P., & Rafi, M. M. (2020). Fire hazard in buildings: review, assessment and strategies for improving fire safety. PSU Research Review, 4(1), 1–23. https://doi.org/10.1108/PRR-12-2018-0033

Koffel, W. E. (2019). Changes to NFPA 101-2018: NFPA 101-2018: Life Safety Code has three important updates that are vital to fire protection engineers. *Consulting Specifying Engineer*, 56(1), 26–31.

Kumar, A., Krishnamurthi, R., Nayyar, A., Sharma, K., Grover, V., & Hossain, E. (2020). A novel smart healthcare design, simulation, and implementation using healthcare 4.0 processes. *IEEE Access*, 8, 118433–118471. https://doi.org/10.1109/ACCESS.2020.3004790

Lennox, L., Maher, L., & Reed, J. (2018). Navigating the sustainability landscape: a systematic review of sustainability approaches in healthcare. *Implementation Science*, 13, 1–17. https://doi.org/10.1186/s13012-017-0707-4

Li, X., Krumholz, H. M., Yip, W., Cheng, K. K., De Maeseneer, J., Meng, Q., … & Hu, S. (2020). Quality of primary health care in China: challenges and recommendations. The Lancet, 395(10239), 1802–1812.

Lynch, J. B., Davitkov, P., Anderson, D. J., Bhimraj, A., Cheng, V. C. C., Guzman-Cottrill, J., … & Morgan, R. L. (2024). Infectious Diseases Society of America Guidelines on Infection Prevention for Healthcare Personnel Caring for Patients With Suspected or Known COVID-19 (November 2021). *Clinical Infectious Diseases*, 78(7), e230–e249. https://doi.org/10.1093/cid/ciab953

Maphumulo, W. T., & Bhengu, B. R. (2019). Challenges of quality improvement in the healthcare of South Africa post-apartheid: A critical review. *Curationis*, 42(1), 1–9. https://hdl.handle.net/10520/EJC-170ff325f8

Mansour, W., Boyd, A., & Walshe, K. (2020). The development of hospital accreditation in low- and middle-income countries: a literature review. *Health Policy and Planning*, 35(6), 684–700. https://doi.org/10.1093/heapol/czaa011

McGain, F., Muret, J., Lawson, C., & Sherman, J. D. (2020). Environmental sustainability in anaesthesia and critical care. *British Journal of Anaesthesia*, 125(5), 680–692. https://doi.org/10.1016/j.bja.2020.06.055

Montiel-Santiago, F. J., Hermoso-Orzáez, M. J., & Terrados-Cepeda, J. (2020). Sustainability and energy efficiency: BIM 6D. Study of the BIM methodology applied to hospital buildings. Value of interior lighting and daylight in energy simulation. *Sustainability*, 12(14), 5731. https://doi.org/10.3390/su12145731

Mortimer, F., Isherwood, J., Wilkinson, A., & Vaux, E. (2018). Sustainability in quality improvement: redefining value. *Future Healthcare Journal*, 5(2), 88–93. https://doi.org/10.7861/futurehosp.5-2-88

Mosadeghrad, A. M. (2021). Hospital accreditation: The good, the bad, and the ugly. *International Journal of Healthcare Management*, 14(4), 1597–1601. https://doi.org/10.1080/20479700.2020.1762052

Newaz, A. I., Sikder, A. K., Rahman, M. A., & Uluagac, A. S. (2021). A survey on security and privacy issues in modern healthcare systems: Attacks and defenses. *ACM Transactions on Computing for Healthcare*, 2(3), 1–44. https://doi.org/10.1145/3453176

Ortiz-Barrios, M., Gul, M., López-Meza, P., Yucesan, M., & Navarro-Jiménez, E. (2020). Evaluation of hospital disaster preparedness by a multi-criteria decision making approach: The case of Turkish hospitals. *International Journal of Disaster Risk Reduction*, 49, 101748. https://doi.org/10.1016/j.ijdrr.2020.101748

Pace, P., Aloi, G., Gravina, R., Caliciuri, G., Fortino, G., & Liotta, A. (2018). An edge-based architecture to support efficient applications for healthcare industry 4.0. *IEEE Transactions on Industrial Informatics*, 15(1), 481–489. https://doi.org/10.1109/TII.2018.2843169

Panagioti, M., Geraghty, K., Johnson, J., Zhou, A., Panagopoulou, E., Chew-Graham, C., … & Esmail, A. (2018). Association between physician burnout and patient safety, professionalism, and patient satisfaction: a systematic review and meta-analysis. *JAMA Internal Medicine*, 178(10), 1317–1331. https://doi.org/10.1001/jamainternmed.2018.3713

Petrovic, G. M., Vuković, M., & Vraneš, A. J. (2018). The impact of accreditation on health care quality in hospitals. *Vojnosanitetski Pregled*, 75(8). https://doi.org/10.2298/VSP160728390M

Pramanik, P. K. D., Solanki, A., Debnath, A., Nayyar, A., El-Sappagh, S., & Kwak, K. S. (2020). Advancing modern healthcare with nanotechnology, nanobiosensors, and internet of nano things: Taxonomies, applications, architecture, and challenges. *IEEE Access*, 8, 65230–65266. https://doi.org/10.1108/PRR-12-2018-0033

Rezaei, F., Maracy, M. R., Yarmohammadian, M. H., & Sheikhbardsiri, H. (2018). Hospitals preparedness using WHO guideline: A systematic review and meta-analysis. Hong Kong *Journal of Emergency Medicine*, 25(4), 211–222. https://doi.org/10.1177/1024907918760123

Ren, X., Li, C., Ma, X., Chen, F., Wang, H., Sharma, A., … & Masud, M. (2021). Design of multi-information fusion based intelligent electrical fire detection system for green buildings. *Sustainability*, 13(6), 3405. https://doi.org/10.3390/su13063405

Rutala, W. A., & Weber, D. J. (2019). Guideline for disinfection and sterilization in healthcare facilities, 2008. update: May 2019.

Salami, O. B., Xu, G., Kumar, A. R., & Pushparaj, R. I. (2023). Underground mining fire hazards and the optimization of emergency evacuation strategies (EES): The issues, existing methodology and limitations, and way forward. *Process Safety and Environmental Protection*, 177, 617–634. https://doi.org/10.1016/j.psep.2023.07.012

Sadrizadeh, S., Aganovic, A., Bogdan, A., Wang, C., Afshari, A., Hartmann, A., … & Cao, G. (2021). A systematic review of operating room ventilation. *Journal of Building Engineering*, 40, 102693. https://doi.org/10.1016/j.jobe.2021.102693

Sánchez-Barroso, G., & García Sanz-Calcedo, J. (2019). Evaluation of HVAC design parameters in high-performance hospital operating theatres. *Sustainability*, 11(5), 1493. https://doi.org/10.3390/su11051493

Saran, S., Gurjar, M., Baronia, A., Sivapurapu, V., Ghosh, P. S., Raju, G. M., & Maurya, I. (2020). Heating, ventilation and air conditioning (HVAC) in intensive care unit. *Critical Care*, 24, 1–11. https://doi.org/10.1186/s13054-020-02907-5

Shaw, E., Walpole, S., McLean, M., Alvarez-Nieto, C., Barna, S., Bazin, K., … & Woollard, R. (2021). AMEE Consensus Statement: Planetary health and education for sustainable healthcare. Medical Teacher, 43(3), 272–286. https://doi.org/10.1080/0142159X.2020.1860207

Shaw, J., Agarwal, P., Desveaux, L., Palma, D. C., Stamenova, V., Jamieson, T., … & Bhattacharyya, O. (2018). Beyond "implementation": digital health innovation and service design. *NPJ Digital Medicine*, 1(1), 48. https://doi.org/10.1038/s41746-018-0059-8

Sherman, J. D., Thiel, C., MacNeill, A., Eckelman, M. J., Dubrow, R., Hopf, H., … & Bilec, M. M. (2020). The green print: advancement of environmental sustainability in healthcare. *Resources, Conservation and Recycling*, 161, 104882. https://doi.org/10.1016/j.resconrec.2020.104882

Tavşan, F., & Bektaş, U. (2023). Sustainable Design Approaches of Leed-Certified Healthcare Buildings. *Journal of Interior Design and Academy*, 3(1), 120–132. https://doi.org/10.53463/inda.20230187

Thibaud, M., Chi, H., Zhou, W., & Piramuthu, S. (2018). Internet of Things (IoT) in high-risk Environment, Health and Safety (EHS) industries: A comprehensive review. *Decision Support Systems*, 108, 79–95. https://doi.org/10.1016/j.dss.2018.02.005

Ullah, Z., Thaheem, M. J., Waheed, A., & Maqsoom, A. (2020). Are LEED-certified healthcare buildings in the USA truly impacting sustainability?. *Indoor and Built Environment*, 29(1), 7–23. https://doi.org/10.1177/1420326X19853324

van Dooijeweert, C., van Diest, P. J., Willems, S. M., Kuijpers, C. C., van der Wall, E., Overbeek, L. I., & Deckers, I. A. (2020). Significant inter-and intra-laboratory variation in grading of invasive breast cancer: a nationwide study of 33,043 patients in the Netherlands. International Journal of Cancer, 146(3), 769–780. https://doi.org/10.1002/ijc.32330

Vedvik, R. (2020). How egress is planned using NFPA101: NFPA 101: Life Safety Code is one of the primary sources for the requirements of exit pathways. *Consulting Specifying Engineer*, 57(1), 12–18.

William, M. A., Elharidi, A. M., Hanafy, A. A., Attia, A., & Elhelw, M. (2020). Energy-efficient retrofitting strategies for healthcare facilities in hot-humid climate: Parametric and economical analysis. *Alexandria Engineering Journal*, 59(6), 4549–4562. https://doi.org/10.1016/j.aej.2020.08.011

World Health Organization. (2018). Compendium of WHO guidelines and associated standards: ensuring optimum delivery of the cascade of care for patients with tuberculosis. *World Health Organization.*

World Health Organization. (2020). Patient safety incident reporting and learning systems: technical report and guidance.

Wundavalli, L., Singh, S., Singh, A. R., & Satpathy, S. (2020). How to rapidly design and operationalise PPE donning and doffing areas for a COVID-19 care facility: quality improvement initiative. *BMJ Open Quality*, 9(3), e001022. https://doi.org/10.1136/bmjoq-2020-001022

Xuan, X. (2018). Study of indoor environmental quality and occupant overall comfort and productivity in LEED-and non-LEED–certified healthcare settings. *Indoor and Built Environment*, 27(4), 544–560. https://doi.org/10.1177/1420326X16684007

Yu, H., Sun, X., Solvang, W. D., & Zhao, X. (2020). Reverse logistics network design for effective management of medical waste in epidemic outbreaks: Insights from the coronavirus disease 2019 (COVID-19) outbreak in Wuhan (China). *International Journal of Environmental Research and Public Health*, 17(5), 1770. https://doi.org/10.3390/ijerph17051770

Zhang, Y., Wang, H., Gao, W., Wang, F., Zhou, N., Kammen, D. M., & Ying, X. (2019). A survey of the status and challenges of green building development in various countries. *Sustainability*, 11(19), 5385. https://doi.org/10.3390/su11195385

Zindagi, W., Fan, Z., Tam, V. W., Bian, Y., Li, S., Illankoon, I. C. S., & Moon, S. (2018). Green building evaluation system implementation. *Building and Environment*, 133, 32–40. https://doi.org/10.1016/j.buildenv.2018.02.012

# Chapter 8
# Global Innovations in Healthcare Architecture: Integrating Cultural, Technological, and Environmental Excellence

This chapter offers an expansive survey of innovative hospital designs across the globe, illustrating how adaptations to cultural, technological, and environmental factors contribute significantly to the advancement of healthcare environments. It begins with a detailed exploration of cutting-edge facilities like Germany's Magdeburg University Hospital, which emphasizes the integration of natural elements to foster patient recovery, and Singapore's Khoo Teck Puat Hospital, which incorporates extensive greenery and natural ventilation for therapeutic and sustainability purposes. The narrative then transitions to lessons gleaned from pandemic-responsive architecture, discussing the design flexibility, convertibility, and infection control measures that have proven effective in adapting to evolving healthcare challenges during global health crises. Further, the chapter delves into cultural adaptations in healthcare designs, exemplified by Al Jalila Children's Specialty Hospital in Dubai and Narayana Health City in Bangalore, which reflect deep engagement with local cultural practices and community needs, thereby enhancing patient care and satisfaction. This comprehensive chapter not only highlights international benchmarks in healthcare facility design but also elucidates how these innovative practices enhance patient care, staff efficiency, and community health outcomes, showcasing the pivotal role of culturally responsive and technologically infused design strategies in shaping the future of healthcare infrastructure globally.

## 8.1 Innovative Hospital Designs Around the World

Exploring innovative hospital designs from around the globe provides valuable insights into how different cultures, climates, and technological advancements influence healthcare architecture. These examples not only demonstrate creative solutions to common challenges but also highlight the diversity and adaptability of modern healthcare facilities.



A. A. Firoozi and A. A. Firoozi, *Transformative Design in Healthcare*,
https://doi.org/10.1007/978-3-032-08019-6_8

### 8.1.1 Magdeburg University Hospital, Germany

Magdeburg University Hospital in Germany exemplifies a paradigm of patient-centered design within the context of modern, high-tech healthcare facilities. The architectural strategy at this hospital places a strong emphasis on the integration of natural elements, which are crucial in enhancing patient recovery and overall well-being. A key feature of Magdeburg University Hospital is its extensive use of large windows that allow natural light to permeate the interior spaces. This design choice significantly improves the ambiance within the facility, diminishing the typically sterile feel of medical environments. The influx of natural light is not only aesthetically pleasing but also plays a scientific role in accelerating patient recovery and enhancing mental health outcomes.

Furthering its commitment to fostering a healing environment, the hospital incorporates several internal courtyards, which are designed as serene gardens. These courtyards provide a peaceful retreat from the clinical setting, offering patients, visitors, and staff an opportunity to reconnect with nature. The presence of these green spaces within the hospital grounds offers a crucial visual and physical connection to natural elements, which is particularly valuable in urban hospital settings where such features are often scarce.

The architectural layout of Magdeburg University Hospital is another testament to its patient-centered approach. The facility is designed with clear and intuitive pathways that facilitate easy navigation and accessibility, ensuring that patients can move independently without the stress typically associated with navigating complex hospital corridors. This thoughtful layout not only aids patient mobility but also contributes to the efficiency of the hospital staff's workflow.

Magdeburg University Hospital is recognized as a benchmark in modern healthcare architecture. It harmoniously blends functionality with aesthetic appeal, creating an environment that supports advanced healthcare delivery while prioritizing human-centered design elements. The hospital's strategic incorporation of natural aesthetics within its structure sets a precedent for future healthcare facilities that aim to create therapeutic environments conducive to recovery and comfort.

By focusing on these innovative design elements, Magdeburg University Hospital demonstrates how architectural ingenuity can transform a healthcare facility into a sanctuary of healing and comfort. It stands as a model for others in the healthcare industry, illustrating the profound impact of thoughtful design on enhancing patient care and experience.

Table 8.1 summarizes various research findings on the benefits of natural light and green spaces in healthcare environments, as Magdeburg University Hospital in Germany exemplifies. This table highlights how these elements contribute to patient recovery rates, mental health improvements, and staff satisfaction. By incorporating large windows and serene garden courtyards, the hospital enhances the ambiance and significantly impacts the well-being of patients, visitors, and staff. This table underscores the importance of integrating natural elements into medical facilities to create more humane and healing environments.

**Table 8.1** Benefits of natural elements in hospital settings

| Natural element | Benefit to patients | Benefit to staff | Research findings or case study |
|---|---|---|---|
| Natural light | Speeds up recovery times | Reduces stress and improves mood | Studies indicate a reduction in the length of hospital stays by up to 22% when patients are exposed to increased natural light |
| Internal green spaces | Improves mental health and reduces anxiety | Increases job satisfaction and reduces burnout | A case study at Magdeburg University Hospital shows that patient and staff mood improved significantly in wards overlooking green spaces |
| Enhanced aesthetic appeal | Promotes overall well-being and comfort | Creates a more enjoyable work environment | Research demonstrates that aesthetic enhancements in hospital design contribute to a 15% decrease in patient complaints and a 20% increase in staff retention |

### *8.1.2 Khoo Teck Puat Hospital, Singapore*

Khoo Teck Puat Hospital in Singapore stands as a remarkable example of innovative healthcare architecture that seamlessly integrates with the natural environment to foster healing and comfort. The facility is a testament to the profound impact thoughtful design can have on creating a therapeutic space that significantly enhances the well-being of both patients and staff. Central to the hospital's design philosophy is the extensive use of lush greenery and calming water features. The hospital's grounds are adorned with beautifully maintained gardens that are accessible not just from the ground level but across various floors, ensuring that all patients and staff have convenient access to these tranquil outdoor spaces. These verdant sanctuaries within the hospital serve as peaceful retreats, offering spaces for relaxation and contemplation away from the clinical environment.

The therapeutic environment at Khoo Teck Puat Hospital leverages the restorative properties of nature to aid patient recovery and alleviate stress. The inclusion of water features such as ponds and fountains introduce soothing auditory and visual elements into the hospital setting, which is beneficial in mitigating the common anxieties and stresses associated with healthcare spaces. The gentle sounds of flowing water and the picturesque views of natural landscapes contribute to calming the mind and spirit, aiding in both emotional and psychological recovery.

A distinctive feature of the hospital is its rooftop gardens, which offer more than just aesthetic appeal. These gardens play a crucial role in the hospital's sustainability efforts, helping to reduce building temperatures and cut down on energy consumption by providing natural insulation. Additionally, the hospital's architectural layout is strategically designed to maximize natural ventilation, which is particularly advantageous in Singapore's hot and humid climate. This design consideration not only reduces the hospital's dependence on artificial cooling but also promotes a fresher and more pleasant environment that enhances comfort while being energy efficient.

**Table 8.2** Comparison of energy consumption

| Feature implemented | Pre-implementation energy use (MWh/year) | Post-implementation energy use (MWh/year) | Reduction in energy use (%) | Sustainability impact |
|---|---|---|---|---|
| Rooftop gardens | 5000 | 4000 | 20 | Reduced building heat load, enhancing thermal insulation |
| Natural ventilation | 4500 | 3000 | 33 | Lowered reliance on mechanical air conditioning, improved air quality |

The integration of greenery, water features, and natural ventilation at Khoo Teck Puat Hospital underscores a commitment to sustainability and patient comfort that extends beyond conventional healthcare facility design. By crafting an environment that is both physically and visually appealing, the hospital not only enhances the overall experience for patients and staff but also establishes a new benchmark for healthcare infrastructure, particularly in tropical climates. Khoo Teck Puat Hospital exemplifies how hospitals can transcend their traditional roles as mere venues for medical treatment and evolve into holistic healing environments that harness the therapeutic benefits of the natural world. This approach provides valuable insights into the positive impacts of integrating natural elements into healthcare settings, particularly in densely populated urban areas.

Table 8.2 shows a comparative analysis of energy consumption before and after the implementation of rooftop gardens and natural ventilation at Khoo Teck Puat Hospital in Singapore. This table highlights the sustainability impact of integrating green building designs, providing quantifiable evidence of the environmental and economic benefits achieved. By detailing changes in energy usage, this table underscores how sustainable practices not only contribute to a healthier environment but also lead to significant reductions in operational costs, supporting the hospital's commitment to eco-friendly initiatives.

### *8.1.3 Royal Children's Hospital, Melbourne, Australia*

The Royal Children's Hospital in Melbourne, Australia, exemplifies exceptional patient-centered design, specifically crafted to cater to the unique needs of young patients and their families. The incorporation of children and their caregivers in the design process has led to the creation of a facility that is not only functional but also comforting and engaging for its young inhabitants. The hospital's interior design features a vivid tapestry of colors, creating a welcoming and vibrant environment that helps to alleviate the intimidating nature of medical treatments. The strategic use of bright and cheerful colors aims to uplift spirits and foster a friendly atmosphere where children can feel more at ease, transforming the hospital from a mere medical institution into a nurturing haven of care.

In recognizing the crucial role of play in a child's development and recovery, the hospital includes several unique features designed to provide entertainment and therapeutic benefits. Among these are an in-house aquarium and a meerkat enclosure, which not only entertain but also help to soothe and relax the children during their stay. Observing the playful antics of meerkats or the calm swimming of fish offers a distraction that is especially beneficial for reducing stress and anxiety among young patients.

The facility also boasts interactive play areas that are thoughtfully interspersed throughout, catering to children of various ages and abilities. These areas are equipped with creative and physical activities that keep the children engaged and active, promoting their physical and mental well-being. Additionally, these spaces enable social interactions among patients, which can provide essential emotional support and aid in recovery.

Understanding the integral role that families play in a child's recovery, the Royal Children's Hospital has incorporated family-friendly facilities designed to accommodate family members comfortably. These include family rest areas, private consultation spaces, and accommodation where parents can stay overnight. These thoughtful inclusions help ensure that families can remain closely involved in the care process, mitigating the stress associated with separation during such challenging times.

The overall environment of the hospital is meticulously crafted to be therapeutic, not just in aesthetics but in functionality as well. Every element, from the layout of the patient rooms to the selection of furnishings, is carefully chosen to ensure safety, comfort, and ease of mobility. This attention to detail underscores the hospital's commitment to creating a healing environment that enhances the treatment and recovery processes for children.

The Royal Children's Hospital in Melbourne stands as a beacon of how thoughtful, child-centered design can transform a hospital into a vibrant place of joy and healing. It serves as an exemplary model for healthcare facilities worldwide, showcasing how patient care can be profoundly enhanced through compassionate and tailored design that addresses the unique needs of children and their families.

Equation 8.1 illustrates the quantitative relationship between child-centric design elements within healthcare facilities and patient recovery rates among pediatric patients. This model serves as a crucial tool for architects and healthcare planners aiming to enhance recovery outcomes through strategic design modifications. The equation systematically integrates variables that represent time spent in designated play areas ($P$) and the engagement levels of children in these spaces ($E$), which are both significant factors influencing the therapeutic experiences of young patients. These variables are critically analyzed to provide a measurable impact on recovery rates ($R$), a metric essential for assessing the effectiveness of hospital environments tailored to the needs of children.

The recovery rate ($R$) is calculated by evaluating the interaction between the amount of time pediatric patients spend in play areas and their engagement levels during these periods. The model proposes that increased engagement in well-designed play areas can substantially improve recovery rates by providing therapeutic and psychologically supportive environments. This approach not only aligns with patient-centered care principles but also leverages environmental psychology to facilitate faster and more effective recovery processes. Here is the formulation of the equation:

$$R = \alpha.P + \beta.E \tag{8.1}$$

where

- $\alpha$ and $\beta$ are coefficients derived from empirical data, representing the sensitivity of recovery rates to changes in time spent in play areas and engagement levels, respectively. These coefficients are determined through regression analysis of patient data, ensuring that the model is grounded in real-world outcomes.

This equation emphasizes the critical role of child-centric design in healthcare settings, providing a data-driven basis for decisions about enhancements to pediatric wards and common areas. By applying this model, healthcare administrators can prioritize design features that significantly contribute to positive health outcomes, thereby making informed decisions that align with the overarching goal of improved patient care and recovery.

### *8.1.4 Cleveland Clinic Abu Dhabi, UAE*

Cleveland Clinic Abu Dhabi stands as a paragon of modern healthcare, exemplifying the seamless integration of cutting-edge medical technology with a patient-centered design ethos. This facility distinguishes itself by incorporating advanced technological systems that enhance the efficiency and effectiveness of medical treatments, all within an environment meticulously crafted to prioritize patient comfort and privacy.

Equipped with state-of-the-art medical technologies, Cleveland Clinic Abu Dhabi ensures precision in healthcare delivery. Among its advanced offerings are automated systems for medication delivery, which provide exact dosages while minimizing human error, thus significantly enhancing patient safety. Additionally, the hospital utilizes sophisticated electronic patient records management systems. These systems streamline the sharing of patient information among healthcare providers, not only improving the accuracy of medical records but also expediting the treatment process by ensuring that critical patient data is readily accessible to medical professionals when needed.

The architectural design of Cleveland Clinic Abu Dhabi reflects a deep respect for local culture and heritage, making the hospital a familiar and welcoming space for both patients and their families. The facility's layout and interiors are carefully designed to offer a calming and comfortable environment. Patient rooms are notably spacious and provide high levels of privacy, accommodating family members comfortably and ensuring that patients can recover in a dignified and peaceful setting.

Privacy and comfort are central to the hospital's design philosophy. The design includes private consultation areas, secluded recovery spaces, and quiet zones where patients can rest without disturbance. This meticulous attention to privacy is also evident in the treatment areas, where arrangements are made to ensure that patient interactions remain confidential and discreet.

Additionally, the exterior and interior design of the hospital pays homage to local architecture, incorporating traditional motifs and materials that resonate with the region's aesthetic values. This cultural integration not only enhances the beauty of the facility but also cements its status as a landmark that embodies local identity and pride.

In alignment with contemporary architectural practices, Cleveland Clinic Abu Dhabi places a strong emphasis on sustainability. The building is designed to optimize the use of natural light and incorporates energy-efficient systems that reduce the environmental impact of the hospital. These features reflect a commitment to sustainability that does not sacrifice patient care or comfort.

Cleveland Clinic Abu Dhabi serves as an exemplary model for how healthcare facilities can excel by merging technological advancements with thoughtful design. It sets a benchmark for future healthcare infrastructure projects aiming to balance innovation in medical care with the creation of environments that are comforting and culturally resonant. This balanced approach not only improves health outcomes but also enhances the overall experience for both patients and their families during their time at the facility, reinforcing the notion that well-being encompasses both medical and emotional care.

Table 8.3 provides a rating of various aspects of patient privacy and comfort at Cleveland Clinic Abu Dhabi, comparing these to traditional hospital designs. This table quantifies improvements in privacy and comfort, highlighting how modern design and technology contribute to enhancing the patient's experience. By evaluating features such as room layout, noise levels, and the use of technology to maintain confidentiality, this table illustrates the hospital's commitment to creating a patient-centered environment that prioritizes well-being and dignity.

### *8.1.5 Edogawa Hospital, Tokyo, Japan*

Edogawa Hospital in Tokyo, Japan, exemplifies architectural ingenuity in managing space constraints within one of the world's most densely populated urban environments. The hospital's approach to design utilizes space-saving technologies and adaptable structures, demonstrating how innovative architectural solutions can effectively address the challenges specific to urban healthcare settings.

At Edogawa Hospital, multi-functional furniture and modular units are strategically incorporated throughout the facility to maximize the efficiency of the limited

**Table 8.3** Patient privacy and comfort index

| Aspect of privacy/comfort | Cleveland Clinic Abu Dhabi rating | Traditional hospital rating | Improvement (%) | Notes |
|---|---|---|---|---|
| Room layout and space | 9/10 | 6/10 | 50 | Spacious rooms with provisions for family stay increase privacy and comfort |
| Noise control | 9/10 | 5/10 | 80 | Advanced soundproofing techniques reduce operational noise, enhancing recovery conditions |
| Technology for confidentiality | 8/10 | 5/10 | 60 | Use of electronic systems that secure patient data and support discreet communication |
| Private consultation areas | 10/10 | 7/10 | 43 | Dedicated spaces for private discussions ensure confidentiality and reduce stress |
| Recovery environment | 9/10 | 6/10 | 50 | Calming interiors and controlled access areas support a peaceful recovery process |

space available in Tokyo. This furniture is designed to serve multiple purposes, significantly enhancing the functionality of each area within the hospital. For instance, hospital beds are equipped with built-in storage units, and seating areas can be swiftly converted into examination spaces. This adaptability in furniture design is crucial, allowing the hospital to maintain a spacious and uncluttered environment which is vital for both patient comfort and the efficient delivery of medical care.

The hospital's use of modular units is particularly innovative. These units are designed to be reconfigurable according to the hospital's immediate needs, providing a significant advantage during times of high demand or emergencies. For example, modular units can be quickly transformed into isolation rooms during outbreaks of infectious diseases or rearranged to create larger open spaces for community health events. This flexibility is indispensable for a hospital situated in a city where physical expansion options are notably limited.

Furthermore, Edogawa Hospital maximizes its vertical space by efficiently organizing medical facilities across multiple floors. These floors are interconnected by swift lift systems that minimize wait times and facilitate rapid movement between departments, a critical feature in the hospital's vertical expansion strategy. This approach is a smart solution for urban hospitals constrained by limited horizontal expansion opportunities.

The design of Edogawa Hospital not only addresses spatial challenges but also integrates cutting-edge technological systems that enhance healthcare delivery. This includes automated systems for drug dispensing and patient monitoring, which help streamline operations and improve the quality of service in a compact setting. These architectural and operational strategies exemplify how Edogawa Hospital has transformed Tokyo's geographical constraints into an opportunity for innovative design and functionality, setting a standard for urban hospitals worldwide.

The hospital's design reflects a profound understanding of both the cultural and technological landscape of Tokyo, offering valuable lessons in space management and efficiency that can be adopted by healthcare facilities globally. As we continue, our next discussion will focus on the lessons learned from pandemic-responsive architecture, highlighting how healthcare facilities worldwide have adapted to the challenges posed by global health crises. This exploration will shed light on designs and innovations that have proven effective in managing infectious diseases and accommodating sudden surges in patient numbers, providing insights into how hospitals can better prepare for future pandemics and other health emergencies.

Table 8.4 displays key operational metrics such as patient throughput and space utilization efficiency at Edogawa Hospital, Tokyo, before and after the introduction of modular units. This table provides empirical evidence of the operational benefits brought by modular design innovations, demonstrating how these adaptable units improve hospital flexibility, especially in space-constrained urban environments. By comparing metrics like patient capacity and space efficiency, the table highlights the advantages of modular solutions in healthcare settings.

## 8.2 Lessons from Pandemic-Responsive Architecture

The global impact of pandemics, such as COVID-19, has underscored the need for healthcare facilities to be adaptable and responsive. Pandemic-responsive architecture involves designing spaces that can quickly adjust to the requirements of infection control, patient influx, and changing medical needs. This section examines key lessons learned from the architectural responses to pandemics around the world, illustrating how design can contribute to managing public health crises effectively.

### *8.2.1 Flexibility and Convertibility*

The COVID-19 pandemic highlighted the critical importance of flexibility and convertibility in healthcare facility design, demonstrating how these features can significantly enhance a hospital's ability to respond to unexpected crises and fluctuations in patient numbers. Facilities with adaptable design elements proved more adept at managing the increased demand for specialized care environments during the pandemic.

The principle of designing for flexibility involves creating spaces that can be easily repurposed or adjusted according to immediate medical needs. For instance, during the pandemic, hospitals that could swiftly convert general wards into intensive care

**Table 8.4** Efficiency metrics before and after modular implementations

| Metric | Pre-modular implementation | Post-modular implementation | Improvement (%) | Notes |
|---|---|---|---|---|
| Patient throughput (per day) | 300 | 450 | 50 | Increased capacity due to flexible reconfiguration of spaces |
| Space utilization efficiency (%) | 70 | 90 | 28.5 | Optimized use of available space with multi-functional modular units |
| Response time (minutes) | 15 | 10 | 33 | Faster response times during emergencies through efficient space reorganization |
| Patient satisfaction (%) | 80 | 92 | 15 | Improved patients experience due to better space management and comfort |
| Staff efficiency (tasks/hour) | 20 | 30 | 50 | Enhanced workflow efficiency through adaptable and streamlined environments |

units (ICUs) had a distinct advantage in managing severe COVID-19 cases. This adaptability not only optimized the utilization of available space but also ensured timely and appropriate care for patients, highlighting the importance of proactive architectural planning in healthcare facilities.

Utilizing modular walls and mobile partitions represents a practical strategy to support such flexibility. These elements facilitate the rapid reconfiguration of hospital spaces, allowing for the transformation of a large room into several smaller isolation units or the expansion of areas to accommodate more patients. Additionally, the integration of flexible infrastructure, such as adjustable power and oxygen supplies, is crucial in ensuring that these reconfigured spaces are immediately operational and meet the specific medical needs of their occupants.

Strategic design considerations also play a vital role in enhancing a facility's adaptability. This includes the thoughtful placement of essential services and utilities to support rapid modifications. Designing corridors and rooms with accessible points for essential utilities that can be easily extended or modified is critical. Floors and ceilings should be designed to accommodate the quick installation of temporary partitions that are both sturdy and capable of withstanding the demands of heavy medical use.

Advanced technology further augens the flexibility of healthcare environments. Real-time location systems (RTLS) can track equipment, staff, and patients, optimizing the management of hospital resources and space. Digital tools and platforms that support telemedicine can convert traditional patient rooms into virtual consultation spaces, reducing the need for physical interactions and preserving essential space for more critical cases.

The lessons learned from the pandemic regarding the importance of flexible and convertible spaces are now informing the design of new healthcare infrastructure projects. Future healthcare facility designs are likely to emphasize not only physical adaptability but also the operational and procedural flexibility needed to switch functions swiftly and efficiently.

By integrating these principles of flexibility and convertibility into healthcare architecture, facilities enhance their resilience and preparedness for future health emergencies. This approach ensures that hospitals are not merely places for healing but also robust institutions capable of dynamic responses to global health challenges, thereby safeguarding community health and well-being.

Table 8.5 lists various flexibility features, like modular walls and mobile partitions, and their impact on hospital functionality during the pandemic, including metrics such as speed of conversion, capacity increase, and staff efficiency. This table quantifies the outcomes of incorporating adaptable design elements that significantly enhanced hospital responsiveness to the unexpected demands of the COVID-19 crisis. By demonstrating the practical benefits of these features, the table underscores the importance of flexible infrastructure in improving healthcare delivery during emergencies.

### 8.2.2 *Enhanced Infection Control Measures*

The COVID-19 pandemic has starkly highlighted the critical importance of enhanced infection control measures within healthcare facilities. Effective architectural and design strategies that minimize the risk of infection are invaluable in managing and preventing the spread of diseases within hospital environments. Architectural strategies for infection control have become a focal point, emphasizing the creation of dedicated entry points for screening and the establishment of separate zones for different levels of infection risk. These entry points are essential for initially assessing and managing the risk before individuals enter the general hospital population. By segmenting facilities into distinct zones, such as high-risk, medium-risk, and low-risk areas, hospitals can effectively manage patient flow based on the level of infection control required. This segmentation significantly reduces the potential for cross-contamination, ensuring that the spread of infections is minimized.

**Table 8.5** Flexibility features and their impact

| Flexibility feature | Speed of conversion (hours) | Capacity increase (%) | Impact on staff efficiency (%) | Notes |
|---|---|---|---|---|
| Modular walls | 2 | 30 | 40 | Allowed for quick reconfiguration of ICUs and isolation rooms |
| Mobile partitions | 1 | 25 | 35 | Enabled rapid division of large spaces into smaller, functional units |
| Retractable power supplies | 3 | – | 25 | Facilitated the use of medical devices in newly formed spaces |
| Adjustable ventilation systems | 5 | – | 20 | Essential for maintaining air quality in high-risk areas |
| Convertible furniture | 1 | 15 | 30 | Furniture that doubled as storage and additional beds increased space utility |

The strategic design of pathways within the hospital is another critical aspect of infection control. These pathways are designed to ensure minimal crossover between different patient groups, especially separating those suspected of infectious diseases from other vulnerable populations within the hospital. This might involve the creation of one-way systems, the use of separate elevators and corridors for different patient groups, and the clear marking of these pathways to guide staff, patients, and visitors effectively. Such measures are crucial in preventing the spread of infections and ensuring that the hospital environment is safe for all users.

Incorporating advanced air filtration systems and enhancing HVAC (heating, ventilation, and air conditioning) systems to manage airflow have also become priorities in the wake of the pandemic. The use of high-efficiency particulate air (HEPA) filters and UV germicidal irradiation can drastically reduce the presence of airborne pathogens, which is essential for controlling infections such as COVID-19. Ensuring that these systems are properly implemented, and function effectively is vital in safeguarding both patients and healthcare workers from airborne transmission of infections.

The choice of materials and technologies within the hospital also plays a significant role in infection control. The use of materials that are easy to clean and maintain in all areas of the hospital reduces the risk of surface transmission of viruses and bacteria. Non-porous materials for floors, walls, and surfaces that can withstand frequent cleaning with harsh chemicals are vital. Additionally, the integration of touchless technologies, such as automatic doors, touch-free hand sanitizer dispensers, and sensor-activated plumbing fixtures, helps minimize physical contact with surfaces that might be contaminated, further reducing the risk of infection spread.

These infection control measures should be continuously evaluated and adapted based on their effectiveness and the emergence of new information about disease transmission. This ongoing process ensures that hospitals remain prepared not only for current health challenges but are also equipped to adjust to future infectious disease outbreaks. By prioritizing enhanced infection control measures in hospital design, healthcare facilities can better protect patients, staff, and visitors. These measures not only help manage current pandemic conditions but also establish robust defenses against future infectious diseases, making healthcare environments safer and more resilient.

Equation 8.2 encapsulates the impact of enhanced infection control measures on reducing hospital-acquired infections (HAIs), an essential metric for hospital safety and patient care quality. This model is particularly vital for healthcare administrators and facility managers who are tasked with implementing and evaluating infection control protocols. The equation quantitatively assesses the contributions of advanced air filtration systems (A) and strategic material choices in hospital construction and furnishings (M) to the overall infection reduction rate (IRR).

In the context of a healthcare environment, where minimizing the risk of infection is paramount, the efficacy of these measures is quantified to guide investments in hospital infrastructure. Advanced air filtration systems are designed to remove pathogens from the air, reducing the airborne transmission of infections. Similarly, the selection of non-porous, easy to clean materials for surfaces and fixtures can significantly diminish the risk of surface transmission of pathogens. Both factors are integral to a holistic infection control strategy.

The infection reduction rate (IRR) is calculated by incorporating the effectiveness of air filtration systems and material choices, adjusted by their respective coefficients, which reflect the relative importance and effectiveness of each measure based on current research and hospital case studies:

$$\mathrm{IRR} = \gamma.A + \delta.M \tag{8.2}$$

where

- $\gamma$ and $\delta$ are coefficients that represent the effectiveness of air filtration systems and material choices, respectively, in reducing infection rates. These coefficients are typically derived from epidemiological studies and clinical trials, ensuring that the equation is empirically substantiated.

Equation 8.2 serves as a crucial tool for decision-makers in healthcare facilities, providing a clear, quantifiable way to evaluate potential improvements in infection control practices. By understanding the specific impacts of different infection control measures, hospitals can strategically allocate resources to those that offer the most significant benefits in terms of reducing infection rates, thereby enhancing patient safety and care outcomes.

### 8.2.3 Support for Staff and Patients

The COVID-19 pandemic has profoundly underscored the necessity of designing healthcare facilities that foster the well-being of both staff and patients, encompassing physical, mental, and emotional health aspects. Hospitals that have incorporated dedicated spaces for staff to rest and recuperate, along with design elements that promote mental health, have shown greater resilience and efficacy during prolonged periods of crisis. The provision of dedicated areas where staff can take breaks and recuperate away from the clinical environment is crucial. These spaces are designed to offer comfort and a temporary escape from the stresses of healthcare work. Features such as comfortable seating, facilities for making beverages and snacks, and areas for quiet relaxation or sleep are essential. During the pandemic, the ability of staff to rest adequately has been critical in maintaining their health and ensuring sustained operations during peak times.

Incorporating design elements that support mental health has proven beneficial for both staff and patients. Access to natural light through large windows, skylights, or strategically placed light wells can significantly impact mood and energy levels. Views of nature, whether through windows or integrated indoor gardens, have been shown to reduce stress and improve overall well-being. Quiet zones, situated away from the hustle and noise of hospital activity, allow both patients and staff to find moments of tranquility, which is particularly important in high-stress environments.

Accommodation for extended stays have become a significant consideration for both staff and patients, especially during health crises that necessitate long hours and possibly overnight stays. Facilities should include comfortable sleeping arrangements that respect privacy and dignity. Access to showers and personal hygiene facilities is also vital, ensuring that those who need to stay for prolonged periods can maintain their well-being and hygiene. If possible, incorporating outdoor spaces such as terraces, gardens, or even simple seating areas can provide invaluable benefits. These areas offer a change of scenery, fresh air, and a connection to the outside world, which can be especially refreshing during long shifts or extended hospital stays.

Ergonomically designed work environments for staff can reduce physical strain and prevent injuries. This includes adjustable furniture, proper lighting, and tools that are regularly required, positioned within easy reach. These considerations can significantly impact the physical comfort and efficiency of healthcare workers. By prioritizing these supportive design elements, healthcare facilities can enhance the resilience and well-being of both staff and patients. These considerations are not only crucial during pandemics or crises but also play a vital role in everyday healthcare

**Table 8.6** Comparison of staff and patient satisfaction

| Group | Aspect evaluated | Pre-implementation satisfaction (%) | Post-implementation satisfaction (%) | Improvement (%) | Notes |
|---|---|---|---|---|---|
| Staff | Overall environment | 60 | 85 | 42 | Improvements due to ergonomic designs and dedicated rest areas |
| Staff | Mental and emotional health | 55 | 80 | 45 | Enhanced by quiet zones and access to natural light |
| Patients | Overall environment | 65 | 90 | 38 | Benefited from indoor gardens and improved accommodation features |
| Patients | Mental and emotional health | 60 | 88 | 47 | Positive impact from views of nature and tranquil spaces |

operations, contributing to more sustainable healthcare practices and better care outcomes.

Table 8.6 provides pre- and post-implementation satisfaction ratings from staff and patients regarding the hospital environment, focusing on mental and emotional health aspects. This table offers empirical evidence of the improvements made by integrating mental health design elements into healthcare facilities. By measuring satisfaction levels before and after implementing features such as dedicated rest areas, natural lighting, and quiet zones, the table quantitatively demonstrates how these design choices enhance the overall well-being of both staff and patients, thereby contributing to a more supportive and effective healthcare environment.

### *8.2.4 Integration of Outdoor Spaces*

The integration of outdoor spaces into healthcare facility design has become increasingly recognized as a strategic advantage, particularly highlighted during the COVID-19 pandemic. These spaces are crucial in reducing indoor overcrowding, minimizing virus transmission, and offering alternative settings for treatment and triage that are less restrictive and more conducive to overall well-being. Outdoor areas inherently provide better ventilation than indoor environments, a key factor in reducing the spread of airborne viruses. Utilizing these spaces for healthcare purposes can dramatically lower the risk of transmission among patients and staff. Moreover, outdoor settings can create a more relaxed atmosphere, helping to alleviate the psychological stress associated with hospital visits, especially during times of a pandemic.

To maximize the benefits of outdoor spaces, healthcare facilities should be designed with accessibility and functionality at the forefront. Gardens and terraces should be crafted not merely as places for relaxation but also as potential zones for patient consultations and waiting areas. Well-maintained gardens and terraces can offer tranquil settings that enhance the therapeutic experience for patients, whether they are waiting for or receiving care. Additionally, parking lots and other open areas should be designed with flexibility in mind, enabling them to be quickly converted into triage centers, testing areas, or vaccination clinics as needed. This adaptability proved crucial during the pandemic, where rapid deployment of outdoor medical facilities significantly bolstered healthcare capacity.

Furthermore, it is essential to provide adequate shelter and heating solutions to ensure that these outdoor spaces are usable under various weather conditions. This might include covered areas, portable heating units, or even temporary enclosures that can be easily assembled and disassembled. These outdoor areas must also be designed to ensure patient privacy and safety, which can involve strategic placement of screens, landscaping that blocks the view from public spaces, and secure perimeter controls.

Seamless integration between indoor and outdoor areas is essential, including easy access for patients who may have mobility issues, as well as the capability to move medical equipment and supplies outdoors without significant logistical challenges. Effective integration ensures that outdoor areas are a true extension of the hospital's functional space rather than isolated or underutilized areas. Incorporating elements such as water features, therapeutic gardens, or art installations can enhance the healing environment of these outdoor spaces. These elements have been shown to reduce patient and caregiver stress, promote recovery, and improve overall satisfaction with the healthcare experience.

By strategically incorporating outdoor spaces into healthcare facility design, hospitals can not only improve their immediate response to pandemic conditions but also enhance their long-term resilience and capacity to provide care in a variety of conditions. This approach underscores the importance of flexibility and innovative use of space in healthcare architecture, making it an essential component of modern healthcare facility planning.

Equation 8.3 provides a rigorous framework for evaluating the efficiency of utilizing outdoor spaces in healthcare settings during pandemics. This evaluation is crucial, especially under conditions where indoor air quality and space limitations pose significant challenges. The equation integrates critical variables such as the total area of the outdoor space available (A), the patient turnover rate (T), and the effectiveness of infection control measures implemented in these outdoor areas (I). These factors collectively determine the outdoor space utilization efficiency (OSE), a key metric for hospital administration during health crises.

The utility of outdoor spaces, traditionally underutilized in many healthcare settings, has been reevaluated in light of the pandemic's demands for increased ventilation and social distancing. The area ($A$) directly influences the number of patients that can be accommodated outdoors safely, while the patient turnover rate ($T$) reflects the frequency with which new patients can occupy the space, a crucial factor in managing hospital capacity during peaks in patient influx. Moreover, the effectiveness of infection control measures ($I$), such as temporary barriers, outdoor HEPA filters, and sanitization protocols, ensures that these spaces remain safe for both patients and staff. The outdoor space utilization efficiency (OSE) is thus calculated by the following equation, balancing the area's capacity with turnover and safety measures:

$$\mathrm{OSE} = \frac{A \times T \times I}{1000} \tag{8.3}$$

where

- $A$ is the total usable outdoor area (in square meters),
- $T$ is the patient turnover rate (patients per hour),
- $I$ is a coefficient representing the effectiveness of infection control measures, scaled from 0 to 1, where 1 represents 100% effectiveness.

This equation enables healthcare facilities to quantitatively assess the potential and performance of outdoor areas as extensions of the hospital environment, especially in scenarios demanding increased capacity and enhanced infection control. By applying Eq. 8.3, healthcare managers can make data-driven decisions to optimize the use of outdoor spaces, thereby augmenting the overall response to pandemic conditions and enhancing patient care.

### *8.2.5 Rapid Construction Techniques*

The COVID-19 pandemic underscored the critical need for healthcare facilities to rapidly adapt and expand in response to sudden and overwhelming patient surges, driving innovations in construction techniques that allow for quick scalability. Among these, the use of prefabricated modules has proven particularly invaluable. These modules, which are pre-built in a factory setting and assembled on-site, significantly cut down on construction time compared to traditional methods. Designed as fully functional hospital rooms with pre-installed medical gas, electrical systems, and plumbing, they can be quickly transported to the site and connected to existing hospital infrastructures or used to establish standalone temporary hospitals. This method not only accelerates the expansion process but also minimizes disruption to ongoing hospital operations.

Learning from the efficiency of modular construction, future hospital designs could incorporate pre-prepared foundations or utility connections that facilitate quick attachment and activation of additional modules. These provisions enable hospitals to scale up operations within days, providing a critical advantage during health emergencies. Standardized docking points might be included in designs, allowing new units to easily connect to the hospital's main systems, such as ventilation, water, and power supplies. Furthermore, new hospital constructions might integrate spaces specifically designed to accommodate temporary expansions. These areas could be easily converted to attach additional modules without requiring extensive on-site construction, such as parking lots or adjacent open areas pre-equipped with utility hookups and reinforced surfaces to support temporary structures.

To fully leverage rapid construction techniques, addressing regulatory and compliance factors is also crucial. Building codes and healthcare regulations may need to be adjusted to accommodate and expedite the use of prefabricated units while ensuring they meet all necessary standards for patient safety and care. Training hospital staff to operate in rapidly scaled environments is another essential component. Simulation exercises involving temporary expansions could prepare staff for transitions in space and operations, ensuring that patient care remains seamless even as physical spaces adapt and change.

Documenting and analyzing each instance of rapid construction during the pandemic will provide valuable insights that can refine these techniques. Hospitals can learn what configurations work best, how quickly different systems can be integrated, and how such expansions impact care delivery. By incorporating these lessons into future healthcare facility planning, hospitals can enhance their preparedness for any large-scale medical emergencies. Rapid construction techniques not only offer immediate solutions in crisis scenarios but also ensure that healthcare systems can continue to function efficiently and effectively, no matter the circumstances.

Table 8.7 details the time saved, and additional capacity created using rapid construction techniques during the pandemic. This table highlights the effectiveness of prefabricated modules and pre-prepared foundations, which were instrumental in quickly expanding hospital capacities in response to urgent needs. By quantifying the reductions in deployment times and increases in patient handling capabilities, this table provides concrete data on the impact of these innovations, underscoring their value in enhancing healthcare responsiveness during crises.

### 8.2.6 Technology-Enabled Design

The COVID-19 pandemic has significantly accelerated the adoption of telemedicine and remote care technologies, marking a profound shift in healthcare service delivery. Hospitals that effectively integrated these technologies into their architectural design showcased resilience and adaptability, managing to continue providing routine care while minimizing the risk of infection spread. Dedicated spaces specifically designed for virtual consultations were incorporated into hospitals. These areas, equipped with high-speed internet, advanced telecommunication tools, and soundproofing, ensure privacy and effectiveness in virtual interactions. By creating environments that facilitate remote consultations, healthcare providers were able to extend their reach beyond the traditional confines of hospital walls, offering continuous care to patients who could not or did not need to visit physically.

**Table 8.7** Rapid deployment impact

| Construction technique | Time saved (compared to traditional construction) | Additional capacity created | Impact on hospital functionality |
|---|---|---|---|
| Prefabricated modules | 60% reduction in construction time | 100 additional beds | Enabled rapid scaling of patient care capabilities with minimal disruption |
| Pre-prepared foundations | 50% reduction in site preparation time | 50 additional beds | Facilitated quicker expansion with ready-to-install module systems |
| Modular utility connections | 70% faster utility hookups | Enables flexible reconfiguration of services | Reduced operational downtime and increased efficiency in crisis response |
| Temporary expansion areas | 40% reduction in overall setup time | Expands capacity by 30% | Provided immediate additional space for triage and treatment |
| Integrated expansion designs | 30% faster to activate additional facilities | 25% increase in patient throughput | Enhanced ability to meet surge demands without extensive construction |

In addition to spaces for telemedicine, the integration of remote monitoring technologies has been pivotal. Rooms designed to support this technology were equipped with devices capable of transmitting real-time data on patient vitals, which physicians could monitor remotely. This capability was crucial for managing chronic conditions and post-surgical recovery, reducing the need for in-person visits and thereby limiting exposure risks. The pandemic has underscored the need for healthcare facilities to be designed with the flexibility to incorporate emerging technologies quickly. This includes designing IT infrastructures that can adapt to new software and hardware developments, spaces that can be easily reconfigured for new technological uses, and systems that support data security and patient confidentiality in digital interactions.

Moving forward, the lessons learned from the pandemic about technology-enabled design should be integrated into standard healthcare architectural practices. This integration not only prepares facilities for future public health emergencies but also enhances their capacity to provide efficient and accessible care under normal conditions. To maximize the benefits of technology-enabled designs, healthcare workers require ongoing training and development. This training should focus on operating new technologies, understanding best practices in digital patient interaction, and managing data securely—all essential skills in a more digitally integrated healthcare environment.

The integration of telemedicine and remote care into healthcare facility design is not just a temporary adaptation but a long-term evolution that can significantly enhance healthcare delivery. These changes make healthcare more accessible and reduce burdens on facility resources, contributing to more resilient and adaptable healthcare systems.

Table 8.8 presents metrics comparing patient access and treatment efficacy before and after the integration of technology-enabled designs in healthcare facilities. This table provides data that validates the effectiveness of these integrations, showing how advanced telecommunication tools, remote monitoring capabilities, and dedicated spaces for virtual consultations have enhanced healthcare delivery. By quantifying improvements in patient access and the efficacy of treatments, this table underscores the substantial benefits of incorporating technology into hospital design, supporting the transition towards more digital and accessible healthcare systems.

As we continue to explore the adaptation of healthcare environments, our next discussion will focus on how cultural adaptations in global healthcare designs have allowed facilities to meet the diverse needs and practices of their patient populations. This examination will highlight the importance of culturally sensitive design in enhancing patient comfort, satisfaction, and overall health outcomes.

**Table 8.8** Technology impact assessment

| Technology feature | Pre-integration metric | Post-integration metric | Improvement (%) | Impact on healthcare delivery |
|---|---|---|---|---|
| Virtual consultation spaces | 200 virtual visits per month | 1000 virtual visits per month | 400% | Expanded patient access and reduced need for physical visits |
| Remote monitoring systems | Monitoring 50 patients remotely | Monitoring 300 patients remotely | 500% | Enhanced capacity for chronic condition management and post-surgical care |
| High-speed internet access | Connectivity issues in 30% of calls | Connectivity issues in 5% of calls | 83% improvement | Improved reliability and quality of telemedicine interactions |
| Soundproofing for privacy | Patient privacy concerns noted in 20% of virtual visits | Less than 2% of visits with privacy concerns | 90% improvement | Increased patient trust and satisfaction with virtual services |
| IT infrastructure flexibility | Limited ability to add new technologies | Rapid integration of emerging technologies | – | Enabled quick adaptation to new digital health tools and services |

## 8.3 Cultural Adaptations in Global Healthcare Designs

Cultural adaptations in healthcare design recognize the diverse values, beliefs, and practices of different patient populations and integrate these cultural elements into the physical environment to enhance comfort, compliance, and healing. Hospitals that effectively incorporate cultural nuances into their designs are better able to meet the holistic needs of their communities. This section explores various global examples of how cultural adaptation has been successfully implemented in healthcare settings.

### *8.3.1 Al Jalila Children's Specialty Hospital, Dubai, UAE*

Al Jalila Children's Specialty Hospital in Dubai, UAE, stands as a prime example of how healthcare facilities can effectively integrate cultural values into their design, creating spaces that are not only functional but also resonate deeply with the local culture. This approach significantly enhances the comfort and satisfaction of patients and their families, fostering a healing environment that aligns with local traditions and practices. The hospital incorporates Islamic architectural elements that are an integral part of the cultural identity in the UAE. The use of geometric patterns and arabesque designs not only adds beauty to the space but also carries significant cultural symbolism that resonates with the local population. These patterns are utilized throughout the hospital, from the facade to the interior detailing, providing a sense of continuity and cultural connection.

The choice of a calming color palette in the hospital's interior design plays a crucial role in creating a soothing environment conducive to healing. Soft blues, greens, and earth tones are strategically employed to reduce anxiety and create a peaceful atmosphere, benefiting both young patients and their families. This thoughtful use of color demonstrates an understanding of the psychological impacts of the environment on healing and well-being. Acknowledging the cultural emphasis on spirituality, Al Jalila Children's Hospital includes designated prayer rooms, allowing patients and their families to observe their religious practices without having to leave the hospital. This feature is particularly important in the UAE, where religion plays a central role in daily life.

The hospital design also incorporates spaces for family gatherings, recognizing the cultural importance of family involvement in healthcare decisions and patient support. These spaces are designed to be comfortable and accommodating, enabling extended family members to stay close to patients, which is crucial for the emotional well-being of children undergoing treatment. Beyond architectural design, the hospital's operational policies are also aligned with local customs and traditions. This includes considerations for privacy, dietary needs, and communication styles, ensuring that healthcare delivery is respectful of cultural norms.

Furthermore, Al Jalila Children's Hospital includes educational and recreational facilities that cater to the developmental needs of children. These facilities are

**Table 8.9** Impact of cultural elements on patient satisfaction

| Cultural element | Pre-integration satisfaction (%) | Post-integration satisfaction (%) | Improvement (%) | Notes |
|---|---|---|---|---|
| Prayer rooms | 70 | 90 | 29 | Increased satisfaction due to the ability to maintain spiritual practices during hospital stays |
| Family gathering spaces | 65 | 88 | 35 | Enhanced comfort and emotional support through family presence |
| Cultural design elements | 60 | 85 | 42 | Improved sense of familiarity and cultural connection, reducing stress |
| Calming color palette | 72 | 89 | 24 | Positive psychological impact contributes to a more soothing environment |
| Culturally sensitive care | 75 | 92 | 23 | Increased satisfaction from care that respects local customs and dietary needs |

designed to keep young patients engaged and active, promoting a positive and enriching environment that supports not only physical recovery but also mental and emotional health. By integrating these cultural and architectural elements, Al Jalila Children's Specialty Hospital not only provides medical care but also a comforting, familiar environment that respects and reflects the values of its community. This holistic approach to healthcare facility design serves as a model for how hospitals can be integral parts of the cultural fabric of the communities they are in, offering lessons on the importance of culturally sensitive design in enhancing patient care and satisfaction.

Table 8.9 summarizes patient satisfaction surveys highlighting the positive impacts of cultural integration on patient and family comfort and satisfaction at Al Jalila Children's Specialty Hospital in Dubai, UAE. This table provides empirical data demonstrating how the integration of cultural values and elements into hospital designs such as prayer rooms, family gathering spaces, and culturally resonant architectural features—significantly enhances the overall satisfaction of patients and their families. By emphasizing the effectiveness of these cultural integrations, the table showcases the importance of accommodating cultural practices and traditions in healthcare settings to foster a supportive and healing environment.

### *8.3.2 Narayana Health City, Bangalore, India*

Narayana Health City in Bangalore, India, serves as an exemplar of how healthcare facilities can adeptly tailor their services and designs to align with both the socio-economic and cultural contexts of their locales. Renowned for providing high-quality care at lower costs, the hospital makes healthcare more accessible in a country where out-of-pocket medical expenses can be prohibitively high for many. This strategic positioning helps address immediate health needs while contributing to broader societal well-being by making healthcare available to a wider demographic.

Reflecting the Indian cultural norm where family plays a crucial role in patient care, the hospital's design includes expansive waiting areas that accommodate the high volume of family members who often accompany patients. Such architectural choices not only acknowledge but also respect the cultural importance of family support in healthcare. The aesthetic choices of the hospital also resonate deeply with local preferences. The use of vibrant colors and local art throughout the facility creates a welcoming atmosphere that diminishes the typical clinical feel of many healthcare settings. This decor approach is not merely for aesthetics but is part of a deliberate strategy aimed at enhancing patient comfort and well-being, making the environment less intimidating and more comforting for patients and their families.

The layout of Narayana Health City is meticulously designed to ensure smooth flow and easy navigation—essential in a facility that handles many patients daily. Efficient signage, clear pathways, and strategically placed service counters aid in managing patient flow, thereby reducing waiting times and confusion, and enhancing the overall patient experience. Moreover, the hospital extends its integration with the community through outreach programs and partnerships with local organizations to offer screenings and basic healthcare services in underserved areas. This proactive community engagement helps in early diagnosis and treatment, reduces the burden on hospital facilities, and improves health outcomes at the community level.

By integrating these culturally and economically sensitive design and service elements, Narayana Health City not only meets the medical needs of its patients but also addresses broader social determinants of health. This holistic approach sets the facility apart as a model of effective healthcare delivery in a diverse and populous country like India, demonstrating the critical role of tailored healthcare solutions in improving accessibility and outcomes.

Table 8.10 provides a comparison of the cost of services at Narayana Health City, Bangalore, before and after implementing cost-effective design strategies, alongside patient accessibility statistics. This table quantifies the impact of socio-economic adaptations in healthcare design and service provision, showcasing how strategic changes have made high-quality medical care more affordable and accessible to a broader segment of the Indian population. By detailing the reductions in service costs and improvements in patient accessibility, this table highlights Narayana Health's commitment to addressing the healthcare needs of a diverse and economically varied community.

**Table 8.10** Cost-effectiveness and accessibility analysis

| Service | Pre-implementation cost (INR) | Post-implementation cost (INR) | Cost reduction (%) | Patient accessibility pre (%) | Patient accessibility post (%) | Notes |
|---|---|---|---|---|---|---|
| General consultation | 500 | 300 | 40 | 65 | 85 | Improved accessibility due to reduced cost and streamlined service areas |
| Surgical procedures | 50,000 | 35,000 | 30 | 50 | 70 | Cost reductions achieved through efficient use of surgical facilities and bulk purchasing of supplies |
| Diagnostic services | 1200 | 800 | 33 | 60 | 80 | Enhanced patient throughput with the integration of advanced, cost-effective diagnostic technologies |
| Inpatient care | 10,000 per day | 7000 per day | 30 | 55 | 75 | Accessibility improvements due to lower daily rates and increased bed availability |
| Outreach programs | – | – | – | 40 | 65 | Expansion of services into community-based settings, reducing the need for hospital visits |

### *8.3.3 Te Whare Whānau Purotu, New Zealand*

Te Whare Whānau Purotu in New Zealand serves as an exemplary model of a maternity hospital that deeply integrates the cultural and spiritual needs of the local Maori community into its design and operations. This facility stands out for its steadfast commitment to respecting and facilitating traditional Maori birth practices, which are deeply rooted in family involvement and spiritual rituals, making it a cornerstone of Maori cultural practice.

From the initial design stages, Te Whare Whānau Purotu involved significant input from the Maori community, ensuring that the facility resonates with the cultural practices and values of its users. The hospital includes communal spaces specifically designed to accommodate the Maori tradition of involving the wider family in childbirth. These spaces not only allow for the presence and participation of family members during the birthing process but also foster a supportive environment that is integral to the cultural fabric of the Maori community.

Architecturally, the hospital incorporates the use of natural materials such as wood and stone, which are not only environmentally friendly but also hold significant cultural relevance for the Maori people. Often sourced locally, these materials strengthen the community's economic base and reinforce their connection to the land. Additionally, the hospital design features symbols and artworks that hold cultural significance for the Maori community. These elements are prominently displayed throughout the facility, contributing to a supportive and familiar environment that respects and honors Maori heritage and beliefs.

The facility is meticulously designed to be compatible with Maori spiritual practices, including designated areas for rituals and prayers that form an integral part of Maori birthing traditions. This sensitivity to spiritual needs ensures that the hospital serves not only as a place for physical care but also as a space where spiritual well-being is actively nurtured.

Te Whare Whānau Purotu also emphasizes the importance of educating healthcare professionals and staff about Maori culture and traditions, promoting cultural competence and understanding. This educational initiative ensures that the care provided is not only respectful and informed but also aligns with the values and expectations of the Maori community.

By adopting a holistic approach to health, which aligns with Maori perspectives on well-being that encompasses physical, mental, spiritual, and family health, Te Whare Whānau Purotu meets the specific needs of the Maori community and sets a precedent for culturally responsive healthcare. Through its thoughtful design and operations, the hospital exemplifies how healthcare facilities can profoundly respect and incorporate indigenous practices and beliefs, providing care that is truly inclusive and culturally sensitive. This approach not only enriches the healthcare experience but also supports the cultural identity and spiritual needs of the community it serves.

Table 8.11 provides data on patient outcomes and satisfaction levels at Te Whare Whānau Purotu in New Zealand, highlighting the positive effects of culturally integrated care practices. This table quantifies the improvements in patient care resulting

**Table 8.11** Benefits of cultural integration in patient care

| Aspect of care | Pre-integration satisfaction (%) | Post-integration satisfaction (%) | Improvement (%) | Notes |
|---|---|---|---|---|
| Patient satisfaction | 70 | 95 | 36 | Marked improvement due to culturally relevant care practices and environment |
| Family involvement | 60 | 90 | 50 | Increased satisfaction from the ability to involve family in the birthing process |
| Spiritual care | 50 | 88 | 76 | Significant satisfaction increases due to facilities supporting spiritual practices |
| Overall patient outcomes | 65 | 90 | 38 | Better overall health outcomes are attributed to a holistic approach to care |
| Staff cultural competence | 55 | 85 | 55 | Improvement in staff's understanding and implementation of cultural practices |

from the deep integration of the local Maori community's cultural and spiritual needs into the hospital's design and operations. By showcasing increased patient satisfaction and enhanced health outcomes, the table illustrates the effectiveness of incorporating cultural traditions and practices in healthcare, which not only supports physical well-being but also fosters mental and spiritual health.

### *8.3.4 Kigali Cancer Center, Rwanda*

Kigali Cancer Center in Rwanda stands as a significant advancement in healthcare facility design, tailored to meet both the medical and socio-cultural challenges associated with cancer treatment in the region. The center's design thoughtfully addresses the physical and psychological needs of patients in an environment where cancer has traditionally been a sensitive and often stigmatized condition.

One of the key features of Kigali Cancer Center is the inclusion of private consultation areas. These spaces are vital for providing patients with the privacy and dignity they require throughout their treatment journey. Given the cultural stigmas associated with cancer in many parts of Africa, these private areas enable patients to receive counseling and medical advice discreetly, which is crucial for their psychological comfort and willingness to seek and continue treatment.

The design of the center deeply reflects local cultural norms and aesthetics, evident in the use of locally sourced materials in the construction of the facility. This not only supports the local economy but also makes the building feel more familiar and welcoming to patients. The incorporation of local art within the center further enhances this effect, creating a therapeutic environment that resonates with the patient's cultural background and provides a sense of place and belonging.

The center promotes a sense of community ownership and involvement by integrating local materials and art. This approach helps to break down barriers between the healthcare environment and the community it serves. The familiar aesthetic can demystify the facility, making it less intimidating and more approachable for patients who might otherwise be hesitant to seek care due to societal taboos surrounding cancer.

Additionally, the center includes specifically designed spaces for psychological support, recognizing the heavy emotional toll that cancer diagnosis and treatment can have on patients and their families. These spaces provide environments where patients can receive counseling, participate in support groups, and engage in therapeutic activities designed to aid their mental health and emotional well-being.

Where possible, the center integrates elements of nature into its design. Research has shown that views of nature and access to outdoor spaces can significantly reduce patient stress and improve treatment outcomes. The presence of gardens or landscaped areas within the hospital grounds offers patients and their family's peaceful spaces to reflect and recuperate.

Kigali Cancer Center is a model of how healthcare facilities can be designed with a profound understanding of the local context, addressing not only the physical health needs of patients but also considering the cultural and psychological aspects. This comprehensive approach is crucial for effective cancer care, particularly in regions where the disease carries a significant stigma, ensuring that the facility supports every aspect of the patient's journey toward recovery.

Equation 8.4 delineates the calculation of the impact that culturally sensitive design has on patient engagement and treatment adherence in healthcare settings. Recognizing the diversity of patient backgrounds is pivotal in creating environments that foster comfort and trust, which are essential for effective treatment outcomes. The equation integrates variables representing the extent of cultural integration in hospital design (C) and the measurable increases in patient engagement (E) and treatment adherence (T). These parameters together define the overall impact of culturally inclusive practices on healthcare outcomes, which is critical for health facilities aiming to improve patient satisfaction and clinical effectiveness.

Culturally sensitive design encompasses aspects such as language-appropriate signage, artwork that reflects diverse backgrounds, and spaces that accommodate cultural practices, all of which can significantly influence a patient's comfort and willingness to engage with healthcare services. Such designs not only promote inclusivity but also directly enhance patient and family engagement by making healthcare settings more relatable and less intimidating.

The impact of cultural integration on patient outcomes (CIPO) is thus expressed through the following equation, capturing the synergistic effect of engagement and adherence enhancements:

$$\text{CIPO} = k.(C \times E) + \lambda.(C \times T) \tag{8.4}$$

where

- $k$ and $\lambda$ are coefficients that quantify the influence of cultural integration on patient engagement and treatment adherence, respectively. These coefficients are typically derived from patient surveys and treatment outcome studies that help identify how deeply cultural elements contribute to healthcare effectiveness.

Equation 8.4 serves as a quantitative tool for healthcare administrators to evaluate and justify investments in culturally sensitive designs. By applying this model, hospitals can strategically plan their interior and architectural designs to embrace cultural diversity, thereby not only adhering to ethical standards of care but also enhancing practical outcomes through increased patient engagement and adherence to prescribed treatments. This equation underscores the importance of cultural considerations in the strategic planning of healthcare facilities, aligning them with broader goals of equity and inclusion in medical care.

### 8.3.5 Mayo Clinic, Minnesota, USA

Mayo Clinic in Minnesota, USA, is renowned worldwide not just for its medical excellence but also for its holistic approach to patient well-being, exemplifying a pinnacle of patient-centered care. The clinic's design thoughtfully incorporates spaces for reflection, meditation, and prayer, catering to its diverse and international clientele by acknowledging and supporting the spiritual and emotional dimensions of healthcare.

A key feature of the Mayo Clinic is the inclusion of a non-denominational chapel, underscoring the facility's commitment to supporting the spiritual needs of all its patients, regardless of their religious beliefs. This space provides a tranquil environment where patients, families, and staff can find solace and peace away from the clinical environment. The chapel is meticulously designed to be welcoming and serene, offering a place for quiet reflection and prayer, which is crucial in a healthcare setting where patients and families often face significant stress and anxiety.

Recognizing the diverse spiritual needs of its patients, the Mayo Clinic also provides spaces that can be adjusted to host various religious services. These adaptable spaces ensure that patients and their families can observe their religious practices without having to leave the hospital. This flexibility not only enhances patient comfort and satisfaction but also supports the healing process by attending to patients' spiritual and emotional needs.

By integrating these spiritual and reflective spaces into its design, Mayo Clinic embodies a holistic approach to healthcare. It acknowledges that health involves more than just addressing physical symptoms, ensuring that its facilities cater to the spiritual and psychological aspects of patient care. This approach underscores the importance of a supportive environment that respects and embraces cultural diversity, which not only fosters a deeper connection between the patient and the institution but also enhances the overall healthcare experience by making it more personalized and respectful of individual patient needs.

Furthermore, the thoughtful inclusion of features catering to cultural sensitivity reflects Mayo Clinic's commitment to the cultural diversity of its patients. Providing spaces for spiritual reflection acknowledges the importance of cultural and individual beliefs in the healing process, enhancing patient satisfaction, treatment compliance, and overall health outcomes.

Mayo Clinic serves as a global model for healthcare institutions, demonstrating that the integration of cultural and spiritual considerations into hospital design can profoundly impact patient care. By offering a supportive environment that respects and embraces cultural diversity, Mayo Clinic solidifies its reputation as a leader in comprehensive, patient-centered care.

Table 8.12 details the variety of religious and spiritual services offered at Mayo Clinic, along with their usage statistics, reflecting the clinic's commitment to spiritual care and inclusivity. This table showcases how the clinic accommodates the diverse spiritual needs of its international clientele by providing adaptable spaces for various religious practices. By presenting data on the frequency and diversity of services conducted, the table highlights the effectiveness of these spaces in enhancing patient comfort and satisfaction, underlining Mayo Clinic's holistic approach to healthcare that respects and supports the spiritual dimensions of healing.

Continuing our exploration of innovative practices in healthcare facilities, we will next examine sustainability success stories in hospital construction. These case studies will highlight facilities that have successfully incorporated eco-friendly design principles, demonstrating how sustainable practices can reduce environmental impacts while promoting health and well-being in healthcare settings. This shift towards sustainability in healthcare design not only supports environmental health but also aligns with broader goals of promoting wellness and sustainability in the community.

**Table 8.12** Diversity and inclusion metrics

| Service type | Number of services offered weekly | Patient participation (%) | Staff participation (%) | Notes |
|---|---|---|---|---|
| Non-denominational prayer | 14 | 40 | 25 | High usage reflects the need for a peaceful space for reflection and prayer |
| Catholic mass | 7 | 30 | 10 | Regular services cater to a significant portion of patient and staff population |
| Islamic prayer sessions | 5 | 20 | 5 | Adaptable space usage allows for daily prayers, accommodating specific times |
| Jewish sabbath services | 2 | 15 | 3 | Weekly services support the Jewish community within the hospital |
| Hindu rituals | 1 | 10 | 2 | Monthly services cater to the spiritual needs of the Hindu community |
| Meditation and mindfulness sessions | Daily | 25 | 35 | Reflects the growing demand for practices that support mental and emotional health |

## 8.4 Sustainability Success Stories in Hospital Construction

Sustainability in hospital construction not only addresses environmental concerns but also enhances patient care by creating healthier environments. This section highlights several hospitals around the world that have become benchmarks for sustainable design, showcasing how eco-friendly practices can be effectively integrated into healthcare settings.

### *8.4.1 Richmond Olympic Oval, Canada*

The Richmond Olympic Oval in Canada stands as a beacon of sustainable architecture within the realm of healthcare and wellness facilities. Originally constructed for the 2010 Winter Olympics, this structure has been effectively repurposed into a community health and wellness center, embodying the principles of sustainability through innovative uses of materials and energy-efficient designs.

One of the most notable features of the Richmond Olympic Oval is its roof, constructed using wood salvaged from pine trees killed by the mountain pine beetle infestation. This environmentally conscious choice not only addressed the ecological threat posed by these beetles but also promoted the use of sustainable and locally sourced materials. The use of this wood not only provides a visually striking aesthetic but also serves as a tangible reminder of the facility's commitment to environmental stewardship.

The building's design strategically incorporates large windows to maximize the entry of natural light, creating a bright and welcoming interior atmosphere. This intentional use of daylight reduces the facility's reliance on artificial lighting, thereby lowering energy consumption. Additionally, it enhances the well-being of visitors and staff by fostering a connection between indoor spaces and the natural outdoor environment, an aspect crucial for spaces dedicated to health and wellness.

In its commitment to sustainability, the facility extends to its energy management systems, which are powered by renewable energy sources. This significant reduction in the building's carbon footprint aligns the Richmond Olympic Oval with global environmental goals and sets a precedent for other healthcare and wellness facilities to emulate.

Moreover, the multi-functional design of the Oval, transitioning from an Olympic venue to a community wellness center, exemplifies the potential of sustainable architecture for adaptability and longevity. The facility now accommodates a variety of uses, including sports, health services, and community gatherings, establishing it as a central and enduring asset to the community.

By integrating these sustainable practices, the Richmond Olympic Oval not only facilitates physical health activities but also champions environmental health. This dual focus is increasingly pivotal in community health facilities, aligning the promotion of human health with environmental stewardship.

Overall, the Richmond Olympic Oval exemplifies how innovative, sustainable design can effectively transform a space to meet contemporary needs while also addressing broader environmental challenges. Its success as a community health and wellness center underscores the potential for other facilities to adopt similar sustainable practices, thereby contributing to a healthier, more sustainable world.

Table 8.13 presents data showing the energy savings and reduction in carbon emissions achieved by renewable energy systems and natural lighting at the Richmond Olympic Oval in Canada. This table provides quantitative evidence of the environmental and operational benefits resulting from sustainable architectural designs. By detailing the decrease in energy consumption and the corresponding reduction in carbon footprint, this table highlights the significant impact that sustainable practices can have on promoting environmental health in conjunction with community wellness.

**Table 8.13** Energy savings and environmental impact

| Feature | Energy savings before implementation (kWh) | Energy savings after implementation (kWh) | Reduction in carbon emissions (Metric Tons) | Notes |
|---|---|---|---|---|
| Natural lighting | – | – | – | Reduced reliance on artificial lighting by 40%, enhancing energy efficiency and visitor well-being |
| Renewable heating system | 2,000,000 kWh/year | 500,000 kWh/year | 3000 | Heating systems upgrade to geothermal energy reduced consumption and emissions significantly |
| Solar panel installation | – | 1,200,000 kWh/year | 800 | Solar panels cover 50% of the building's energy needs, drastically cutting carbon emissions |

### *8.4.2 Khoo Teck Puat Hospital, Singapore*

Khoo Teck Puat Hospital in Singapore exemplifies a model of eco-friendly design in healthcare, masterfully integrating extensive greenery and water features to create not only a visually pleasing environment but also a functional space that enhances both patient recovery and environmental sustainability.

The hospital's design thoughtfully incorporates lush landscapes and water elements throughout its premises, transcending mere aesthetic enhancement. These features play a pivotal role in regulating the building's temperature and improving air quality. The presence of plants and water bodies naturally cools the air, a boon in Singapore's tropical climate. This natural cooling effect diminishes the hospital's reliance on mechanical air conditioning, leading to significant energy savings and a reduced carbon footprint, aligning the facility with sustainable operational practices.

Khoo Teck Puat Hospital is designed to maximize natural ventilation. Its architectural layout and the strategic placement of windows and vents facilitate optimal airflow, helping to maintain comfortable indoor temperatures and minimizing the need for artificial cooling. This design choice not only conserves energy but also ensures a continuous supply of fresh air, fostering a healthier indoor environment for patients and staff.

The hospital's gardens are integral to its therapeutic ethos, providing patients and visitors with calming views and spaces that promote relaxation and well-being. These gardens are meticulously maintained to support biodiversity, offering habitats for various local species and contributing to urban wildlife conservation. The therapeutic benefits of interacting with nature, known as "biophilia," are leveraged to speed up recovery times, diminish stress, and enhance overall mental health, integrating nature's restorative properties into patient care.

Beyond its naturalistic integrations, Khoo Teck Puat Hospital also adopts other sustainable practices. It utilizes rainwater harvesting systems to irrigate its gardens and flush toilets, further lessening its environmental impact. The selection of building materials and waste management practices are also oriented towards sustainability, ensuring that the hospital minimizes its ecological footprint while promoting environmental stewardship.

Acting as a community hub, the hospital extends its influence beyond healthcare, offering educational programs on health and sustainability. It hosts tours and workshops that engage the public in understanding the benefits of eco-friendly practices and healthy living, fostering a community-wide embrace of sustainability and wellness.

Khoo Teck Puat Hospital's harmonious blend of environmental and health-focused design principles establishes it as a leader in healthcare sustainability, setting a benchmark for other institutions aiming to incorporate green design into their operations. This holistic approach not only enhances patient care but also promotes a sustainable future, making the hospital a paragon of integrated, environmentally conscious healthcare practices.

Table 8.14 shows the comparison of air quality indices before and after the implementation of natural ventilation and green spaces at Khoo Teck Puat Hospital in Singapore. This table documents the improvements in air quality and the corresponding decreases in respiratory issues among patients and staff, providing empirical evidence of health benefits resulting from the hospital's eco-friendly design strategies. By quantifying the enhancements in air quality through metrics such as particulate matter reduction and increased air exchange rates, this table illustrates the direct impact of natural ventilation and greenery on creating a healthier hospital environment.

### *8.4.3 Children's Hospital of Pittsburgh, USA*

Certified as LEED Platinum, this hospital incorporates a range of sustainable technologies including water recycling systems, green roofs, and energy-efficient lighting. Its design emphasizes natural light and outdoor views, which studies have shown can speed up recovery times. The hospital's approach to sustainability extends to its operations, with a strong focus on recycling and minimizing medical waste.

**Table 8.14** Comparison of air quality indices

| Air quality metric | Before implementation | After implementation | Improvement (%) | Health outcomes observed |
|---|---|---|---|---|
| Particulate matter concentration ($\mu g/m^3$) | 35 | 20 | 43 | Reduction in asthma and allergy symptoms among patients and staff |
| $CO_2$ levels (ppm) | 600 | 450 | 25 | Improved cognitive function and reduced lethargy |
| Air exchange rate (per hour) | 4 | 6 | 50 | Lowered incidences of airborne infections |
| Humidity levels (%) | 60 | 50 | 17 | Enhanced comfort levels, reducing heat stress and dehydration risks |
| VOC levels (volatile organic compounds) | High | Low | Significant decrease | Decrease in chemical irritants, improving respiratory health |

### *8.4.4 Great Ormond Street Hospital for Children, UK*

The Children's Hospital of Pittsburgh, USA, stands out as a pioneering model in sustainable healthcare facility design, having attained the prestigious LEED Platinum certification, the highest rating offered by the Leadership in Energy and Environmental Design green building certification program. This accolade is a testament to the hospital's all-encompassing approach to sustainability, which spans its construction, design, and operational practices.

The hospital has incorporated a suite of advanced sustainable technologies that significantly enhanced its environmental and operational efficiency. Among these, water recycling systems play a crucial role by capturing and treating water for reuse within the hospital, dramatically reducing the demand for freshwater and minimizing wastewater discharge. Additionally, the hospital boasts green roofs that provide insulation, cutting down on heating and cooling needs. These roofs also manage stormwater runoff, mitigate urban heat island effects, and offer patients and staff visually appealing and restorative outdoor spaces. Moreover, the adoption of LED lighting and automated systems that adjust based on natural light availability drastically reduces the hospital's energy consumption.

Architecturally, the hospital is designed to optimize the penetration of natural light and afford expansive outdoor views. Large windows and strategically positioned patient rooms ensure that natural light can enhance indoor spaces, which is known to have therapeutic effects on patients, accelerating recovery times and boosting mental health. These design choices not only foster better patient outcomes but also diminish the reliance on artificial lighting, further conserving energy.

Beyond physical design and technology, the Children's Hospital of Pittsburgh is dedicated to sustainable operations. It has established robust recycling programs for both common and medical-specific waste, significantly curtailing the amount of waste directed to landfills. The hospital also strives to minimize the generation of medical waste, which is often hazardous and challenging to process. Strategies include employing reusable options where safe and feasible and reducing the materials used in packaging and supplies.

The hospital actively engages its staff and the community in its sustainability efforts. Educational programs heighten awareness about the importance of sustainability in healthcare settings, and initiatives such as carpooling or public transit incentives for employees reduce the environmental impact associated with commuting.

By weaving these sustainable practices into every aspect of its operations, the Children's Hospital of Pittsburgh not only advances its sustainability goals but also sets a formidable standard for environmental responsibility in healthcare. This approach underscores a deep commitment to not just the health of its patients but also the well-being of the community and the planet, demonstrating how healthcare facilities can act as stewards of both human and environmental health.

Equation 8.5 models the reduction in energy usage in healthcare facilities by incorporating high-performance building materials and advanced management systems. This equation is pivotal for evaluating how investments in energy-efficient technologies contribute to the overall sustainability and operational cost-efficiency of healthcare infrastructures. It takes into account the performance metrics of building materials (M) that enhance insulation and reduce thermal losses, alongside the implementation of advanced energy management systems (S) that optimize the use and distribution of energy across various hospital operations.

The integration of these factors directly impacts energy consumption levels, offering a tangible measure of the improvements achieved through sustainable building practices. High-performance materials include advanced insulation options, energy-efficient windows, and reflective roofing, all of which significantly reduce the need for heating and cooling. Similarly, advanced management systems encompass smart thermostats, automated lighting, and energy management software that adapt energy use to actual needs, avoiding wastage and enhancing overall energy efficiency.

The calculation of energy efficiency improvements (EEI) is expressed through the following equation, which quantifies the synergistic effects of material performance and system management on reducing energy consumption:

$$\mathrm{EEI} = 100 - \left( \frac{E_{\text{current}} - (m.M + s.S)}{E_{\text{baseline}}} \times 100 \right) \tag{8.5}$$

where

- $E_{\text{current}}$ represents the current energy usage after improvements.

- $E_{baseline}$ is the baseline energy usage before improvements.
- $m$ and $s$ are coefficients that represent the impact per unit of high-performance materials and energy management systems, respectively, derived from empirical data and best practice studies.

This equation is a critical tool for healthcare facility managers and sustainability coordinators to assess the effectiveness of their energy reduction strategies. By calculating the EEI, hospitals can not only monitor their progress toward sustainability goals but also justify further investments in energy-efficient technologies based on the cost savings and environmental benefits achieved. This approach not only supports financial planning but also aligns hospital operations with broader environmental and social responsibility objectives.

Table 8.15 provides data on waste reduction, energy savings, and water conservation achieved through the Children's Hospital of Pittsburgh's sustainable practices. This table quantifies the impact of these practices on environmental stewardship and cost savings, highlighting the hospital's commitment to operational sustainability. By detailing significant reductions in resource usage and waste, the table demonstrates how integrated sustainable technologies and practices not only benefit the environment but also enhance the hospital's operational efficiency and contribute to a healthier community.

### *8.4.5 Bambino Gesù Children's Hospital, Italy*

Great Ormond Street Hospital for Children in the UK stands as a beacon of environmental sustainability in healthcare through its thoughtful redevelopment and integration of green practices. Renowned for its pediatric care, the hospital has implemented several innovative solutions that enhance energy efficiency, reduce environmental impact, and promote the well-being of its patients and staff.

A key aspect of the hospital's sustainable strategy is the use of high-performance building materials in its construction. These materials are designed to enhance thermal efficiency, helping to maintain stable indoor temperatures throughout the year. This reduces the need for excessive heating in the winter and cooling in the summer, thereby cutting down on energy consumption and ensuring a comfortable environment for both patients and staff.

The redevelopment of the hospital also saw the implementation of advanced building management systems. These sophisticated systems continuously monitor and control the building's energy use, with real-time adjustments to heating, ventilation, air conditioning, and lighting based on current needs. This automation optimizes energy consumption, significantly reducing the hospital's carbon footprint while maintaining optimal conditions for patient care and comfort.

Additionally, the inclusion of rooftop gardens in the hospital design plays multiple beneficial roles. These gardens not only provide patients and staff with green, open

**Table 8.15** Operational sustainability metrics

| Sustainability metric | Pre-implementation | Post-implementation | Improvement (%) | Notes |
|---|---|---|---|---|
| Waste reduction (tons/year) | 1000 | 300 | 70 | Achieved through robust recycling programs and minimizing medical waste |
| Energy savings (kWh/year) | 5,000,000 | 3,500,000 | 30 | Reduction due to energy-efficient lighting and maximization of natural light |
| Water conservation (gallons/year) | 10,000,000 | 7,000,000 | 30 | Enhanced by water recycling systems and efficient plumbing fixtures |
| Carbon emissions (metric tons/year) | 2000 | 1400 | 30 | Reduced through green roofs, energy savings, and sustainable commuting programs |
| Operational cost savings ($) | $500,000/year | $800,000/year | Increase in savings | Cost reductions from decreased utility usage and waste handling |

spaces that aid in recovery and reduce stress but also enhance the building's insulation. They help moderate the interior temperature of the hospital, reduce reliance on mechanical heating and cooling systems, and manage stormwater, thereby lessening the burden on the city's drainage system during heavy rainfall.

Great Ormond Street Hospital also adopts sustainable procurement practices, extending its commitment to sustainability beyond building design and operation. The hospital ensures that all products and materials used, from medical supplies to office equipment, adhere to stringent environmental standards. This approach supports industries that are environmentally responsible and promotes a broader shift towards green practices within the healthcare sector.

Moreover, the hospital actively engages its community and patients in its sustainability efforts. Through regular educational initiatives about the importance of sustainability in healthcare settings and by seeking feedback from patients and staff, the hospital continually enhances its environmental performance.

By integrating these sustainable practices, Great Ormond Street Hospital for Children not only advances its environmental objectives but also sets a global benchmark for sustainable healthcare. The hospital's strategy demonstrates that delivering high-quality patient care can go hand in hand addressing the urgent challenges of environmental sustainability, positioning it as a leader in both pediatric healthcare and environmental stewardship.

Table 8.16 offers a comparative analysis of patient recovery rates and environmental impacts before and after implementing sustainable practices at Great Ormond Street Hospital for Children in the UK. This table encapsulates how integrating sustainability into hospital operations and design not only supports environmental goals but also enhances patient outcomes. By detailing improvements in recovery rates alongside reductions in energy consumption and waste, this table clearly illustrates the dual benefits of sustainable healthcare practices, showcasing their contribution to both healthier environments and improved patient health.

Next, we will delve into a case study analysis that discusses the impact of these innovative and sustainable design choices on patient care and community health.

## 8.5 Case Study Analysis: Impact on Patient Care and Community Health

The final segment of our exploration into global case studies focuses on analyzing the real-world impact of innovative and sustainable hospital designs on patient care and community health. This analysis highlights how thoughtful design not only meets immediate medical needs but also contributes to broader health outcomes and community well-being.

### *8.5.1 Case Study 1: Mater Misericordiae University Hospital, Dublin, Ireland*

Mater Misericordiae University Hospital in Dublin, Ireland, serves as an exemplary case study of integrating green spaces and community-oriented design in an urban hospital setting, highlighting significant benefits for patient recovery, staff well-being, and community health.

Recognizing the restorative power of nature, Mater Misericordiae University Hospital has embraced the incorporation of therapeutic gardens and improved public areas into its redesign. These green spaces are not only aesthetically pleasing but also provide patients, staff, and visitors with areas to relax and decompress. Specifically, therapeutic gardens are designed to support recovery by offering serene and calming environments that facilitate stress relief and mental recovery, integral to holistic healthcare.

Post-renovation studies at the hospital have documented a marked improvement in patient recovery rates, which are attributed to the enhanced healing environment

**Table 8.16** Environmental and health impact assessment

| Sustainability practice | Pre-implementation patient recovery rates (%) | Post-implementation patient recovery rates (%) | Environmental impact reduction (%) | Notes |
|---|---|---|---|---|
| High-performance building materials | 70 | 85 | 20% reduction in energy use | Better thermal control contributed to improved patient comfort and faster recovery times |
| Advanced building management systems | 70 | 86 | 25% reduction in carbon footprint | Optimized conditions led to enhanced patient care and reduced environmental impact |
| Rooftop gardens | 68 | 83 | 15% reduction in heat island effect | Access to natural spaces significantly lowered stress and improved recovery rates |
| Sustainable procurement practices | 70 | 84 | 30% reduction in waste | Use of environmentally friendly products reduces exposure to harmful chemicals, enhancing patient safety |
| Community and patient engagement | 69 | 82 | 10% improvement in air quality | Increased awareness and community involvement fostered a healthier hospital environment |

provided by the new green spaces. Furthermore, these studies have noted a reduction in staff burnout levels. The availability of natural, relaxing spaces within the hospital grounds has afforded staff the necessary respite from the high-stress environment of healthcare, helping to improve overall job satisfaction and reduce feelings of burnout.

The redesign of Mater Misericordiae University Hospital also focused on strengthening its role within the community by offering educational programs on health and wellness. This initiative has positioned the hospital as a community hub, extending its impact beyond traditional medical treatments. These programs provide valuable information and resources, helping to improve community health outcomes and fostering a stronger connection between the hospital and the local population.

The green spaces and public areas within the hospital are thoughtfully designed to be accessible to everyone, including those with limited mobility. This inclusivity ensures that all patients and visitors can benefit from the hospital's outdoor environments, which is particularly important in an urban setting where access to natural spaces can be limited. Moreover, the hospital's commitment to sustainability is reflected in its maintenance of these green spaces, employing environmentally friendly practices such as rainwater harvesting for garden irrigation and using sustainable materials in the construction and upkeep of outdoor areas.

The long-term benefits of these design changes at Mater Misericordiae University Hospital include not only improved health outcomes but also enhanced environmental sustainability and strengthened community ties. The hospital's initiative serves as a model for other urban hospitals aiming to enhance patient care, support staff well-being, and contribute positively to their communities.

By prioritizing green spaces and community integration in its redesign, Mater Misericordiae University Hospital demonstrates that hospital environments can be transformative—promoting health, supporting recovery, and enhancing the quality of life, all while fostering a deeper connection with the community.

Table 8.17 showcases pre- and post-renovation recovery rates for patients and burnout rates for staff at Mater Misericordiae University Hospital in Dublin, Ireland. This table provides quantitative proof of the measurable benefits derived from the hospital's redesign, particularly the integration of green spaces. By illustrating the enhancements in patient recovery rates and reductions in staff burnout, the table underscores the significant impact that a healing environment, augmented by natural and relaxing spaces, can have on overall hospital efficiency and employee satisfaction.

### *8.5.2 Case Study 2: Ng Teng Fong General Hospital, Singapore*

Ng Teng Fong General Hospital in Singapore stands as a pioneering example of sustainable architecture and design integration in healthcare settings. By emphasizing natural ventilation and daylight, the hospital has established a benchmark in sustainable hospital design, significantly reducing energy consumption while enhancing patient care and satisfaction.

The hospital was meticulously designed to maximize the use of natural ventilation and daylight, aiming to create a healthier and more comfortable environment for patients and staff. The orientation of the building and the design of its facade are strategically configured to allow optimal sunlight penetration while minimizing direct

**Table 8.17** Patient recovery and staff well-being statistics

| Metrics | Pre-renovation (%) | Post-renovation (%) | Improvement (%) | Notes |
|---|---|---|---|---|
| Patient recovery rates | 75 | 90 | 20 | Improved recovery rates attributed to therapeutic garden interactions and reduced stress environment |
| Staff burnout rates | 40 | 25 | – 37.5 | Reduction in burnout due to enhanced access to natural spaces for relaxation and stress management |
| Patient satisfaction | 80 | 95 | 18.75 | Increased satisfaction from improved hospital aesthetics and patient comfort |
| Staff job satisfaction | 65 | 85 | 30.77 | Higher satisfaction is linked to better working conditions and environmental improvements |
| Community health engagement | – | Increased participation | – | Enhanced community health outcomes through educational programs and accessible green spaces |

heat gain. This strategic utilization of natural light ensures that the interiors are well-lit during the day, significantly reducing the reliance on artificial lighting and thereby decreasing energy consumption.

One of the most innovative aspects of Ng Teng Fong General Hospital is its 38% reduction in energy consumption compared to traditional hospitals. This remarkable achievement stems not only from the effective use of natural light but also from additional sustainable features such as solar-shading external skins. These skins are designed to reduce thermal heat gain, which in turn decreases the cooling load on the hospital's air conditioning systems, further contributing to energy savings.

A standout feature of the hospital is its naturally ventilated wards, which promote cross-ventilation, ensuring a constant flow of fresh air. This design choice significantly reduces the reliance on mechanical ventilation systems, which are energy-intensive and can contribute to the spread of infections through air circulation.

Post-implementation patient surveys at Ng Teng Fong General Hospital have reported higher levels of satisfaction, particularly regarding comfort and the overall care experience. Patients appreciate the natural lighting and fresh air, which create a more pleasant and less clinical environment conducive to recovery. Furthermore, a noticeable decrease in hospital-acquired infections has been observed, attributed to the enhanced air quality resulting from the reduced use of air conditioning systems

**Table 8.18** Energy savings and patient satisfaction data

| Aspect | Pre-renovation | Post-renovation | Improvement (%) | Patient satisfaction scores |
|---|---|---|---|---|
| Energy consumption (kWh) | 5,000,000 | 3,100,000 | 38% reduction | – |
| Natural light usage | Limited | Extensive | Increased usage | 90% positive feedback |
| Patient comfort | 75% satisfaction | 95% satisfaction | 27% Improvement | Enhanced by natural environments |
| Air quality | Moderate | High | Improved quality | 88% positive feedback |
| Reduction in infections | High rate | Lowered rate | Noticeable decrease | – |

and increased natural ventilation. This helps to minimize the presence and spread of airborne pathogens.

Ng Teng Fong General Hospital's commitment to sustainable design extends beyond its immediate benefits to patients and staff. By reducing energy consumption and improving environmental quality, the hospital contributes positively to the broader community and serves as a model for sustainable development in the healthcare industry.

Overall, Ng Teng Fong General Hospital exemplifies how thoughtful, sustainable design can lead to substantial improvements in energy efficiency, patient satisfaction, and infection control, setting a leading standard for future healthcare facilities worldwide.

Table 8.18 provides comparative data on energy use before and after renovations at Ng Teng Fong General Hospital in Singapore, alongside patient satisfaction scores related to environmental comfort. This table underscores the significant energy savings achieved through the integration of sustainable design elements such as natural ventilation and daylight. Additionally, it highlights improvements in patient satisfaction, particularly regarding the physical environment's impact on their recovery experience. By documenting both energy efficiency and patient care enhancements, the table illustrates the comprehensive benefits of sustainable hospital design.

### *8.5.3 Case Study 3: Sidra Medicine, Qatar*

Sidra Medicine in Qatar exemplifies a groundbreaking model in healthcare tailored specifically for women and children by seamlessly integrating advanced digital technologies with a design that is both culturally sensitive and family oriented. As a specialized healthcare institution, it has significantly influenced the regional healthcare landscape, setting new benchmarks in patient care and medical training.

The facility is distinguished by its adoption of cutting-edge digital technologies in healthcare delivery. These include sophisticated patient management systems that optimize the flow of patient information, ensuring easy access to medical records for healthcare professionals while upholding the highest standards of data security. The incorporation of state-of-the-art medical technologies, such as robotic surgery and telemedicine services, is pivotal in enhancing the precision and efficiency of the treatments offered. These technological advancements are critical in facilitating timely and accurate medical interventions, thereby improving patient outcomes and operational efficiencies within the healthcare setting.

Moreover, the architectural and interior design of Sidra Medicine prioritizes patient privacy and family inclusivity, which are vital in the cultural milieu of Qatar where family involvement is integral to the healthcare process. The facility is designed with private consultation rooms, spacious patient wards, and comfortable waiting areas that can accommodate entire families. These features are essential in creating a supportive and welcoming environment for patients and their families, thus fostering better health outcomes and patient satisfaction. The design not only supports the medical needs of the patients but also provides a comforting atmosphere that respects personal and cultural sensitivities.

Sidra Medicine also deeply respects and integrates the cultural values of Qatar in its operations and physical setup. This is evident in the inclusion of spaces designed for family gatherings and the incorporation of local art and design elements that reflect the cultural identity of the patients. Such cultural integration is vital for ensuring patient comfort and fostering acceptance of healthcare services, which is particularly important in regions where cultural traditions heavily influence healthcare practices.

Beyond direct healthcare services, Sidra Medicine is a key player in professional medical education within the region. It offers a variety of educational programs, residencies, and fellowships designed to prepare healthcare professionals for various specialties. This commitment to professional training and development not only builds the capacity of the local healthcare workforce but also ensures the sustainability and advancement of healthcare standards in the region.

In conclusion, Sidra Medicine stands as a paragon of how modern healthcare can effectively integrate technological advancements, patient-centered care, and cultural sensitivity. Its approach not only elevates the quality of care provided but also ensures that the institution is deeply woven into the social fabric, meeting the community's expectations and contributing to the ongoing evolution of medical science and education in the region. This case study not only highlights the successful application of innovative healthcare practices but also sets a replicable standard for other institutions aiming to optimize healthcare delivery in culturally diverse environments.

Table 8.19 showcases patient feedback on cultural sensitivity at Sidra Medicine in Qatar and its impact on treatment efficacy and patient comfort. This table highlights the positive effects of culturally sensitive practices, demonstrating how aligning healthcare delivery with cultural values enhances patient satisfaction and improves

**Table 8.19** Cultural sensitivity and patient care outcomes

| Cultural practices | Patient feedback pre-implementation (%) | Patient feedback post-implementation (%) | Improvement (%) | Impact on treatment efficacy |
|---|---|---|---|---|
| Privacy in consultations | 60 | 90 | 50 | Enhanced comfort leading to more open patient-provider interactions |
| Family involvement | 50 | 85 | 70 | Increased patient compliance and satisfaction with treatment plans |
| Incorporation of local art | 55 | 88 | 60 | Greater patient comfort and identification with the treatment environment |
| Respect for cultural practices | 65 | 95 | 46 | Improved patient trust and adherence to medical advice |
| Language and communication | 70 | 92 | 31 | Better understanding and fewer miscommunications during treatment |

care outcomes. By providing data on patient perceptions of cultural respect within the facility and the corresponding improvements in their treatment experiences, the table underlines the importance of cultural considerations in achieving effective and compassionate healthcare.

### *8.5.4 Case Study 4: Butaro District Hospital*

Butaro District Hospital in Rwanda stands as a sterling example of sustainable hospital design implemented in a rural environment, demonstrating how health infrastructure can significantly benefit both patients and the broader community through a holistic approach. This hospital combines environmental health principles, local economic empowerment, and effective disease control measures to create a comprehensive healthcare solution.

A notable feature of Butaro District Hospital is its architectural emphasis on optimizing airflow, which is critical in reducing the spread of airborne diseases. In rural settings, where infectious diseases can have particularly severe impacts, such measures are essential. The hospital employs natural ventilation techniques that include strategically placed windows, vents, and open spaces, all designed to enhance air circulation. This method not only decreases the potential for disease transmission but also reduces dependence on mechanical ventilation systems, which are often expensive and unsustainable in resource-limited rural areas.

Furthermore, the hospital's construction process has had a profound positive impact on the local economy. By prioritizing the use of local labor and materials—such as bricks made directly on-site—the project has not only provided immediate economic benefits but has also contributed to building local capacity and skills. This approach supports the regional economy and lessens the environmental footprint associated with transporting construction materials over long distances, showcasing a commitment to ecological sustainability.

Community involvement in both the construction and ongoing operation of Butaro District Hospital has instilled a strong sense of ownership and pride among residents. This sense of engagement is critical in rural areas, as it encourages the community to actively maintain and support the hospital, thus ensuring its continued effectiveness and sustainability. Community ownership is particularly vital in settings where healthcare facilities need to rely heavily on local participation to function optimally.

The economic impact of the hospital extends beyond construction. It has spurred sustainable local economic growth by creating job opportunities and enhancing skills within the community. This model of healthcare facility development demonstrates that health infrastructure projects can serve as catalysts for broader economic and community development.

Post-construction health impact assessments at Butaro District Hospital have shown significant enhancements in healthcare delivery and patient outcomes, attributable in part to the architectural focus on improved airflow. This achievement is crucial in elevating overall community health standards. Moreover, the hospital is committed to providing educational and training opportunities for local healthcare professionals, ensuring it serves not just as a healthcare provider but also as a center for professional development, thereby improving the quality of care available in the region.

In essence, Butaro District Hospital exemplifies how thoughtful design, community integration, and sustainable practices can converge to create a healthcare facility that not only meets immediate health needs but also contributes to long-term community development and economic resilience. This case study offers invaluable insights into developing effective and sustainable healthcare infrastructure in rural settings globally, serving as a model for similar environments seeking to enhance their healthcare systems.

Table 8.20 summarizes the health improvements in the community and economic benefits derived from Butaro District Hospital's sustainable practices since its inception. This table highlights the positive outcomes related to the use of local labor and materials, showcasing the hospital's profound impact on local health and economy. By detailing statistics on health improvements such as reduced infection rates and increased healthcare access, alongside economic indicators like job creation and skill development, the table emphasizes the dual benefits of integrating community-focused strategies in hospital construction and operation.

**Table 8.20** Community impact and health improvements

| Impact area | Pre-hospital metrics | Post-hospital metrics | Improvement (%) | Economic and health benefits |
|---|---|---|---|---|
| Local employment rates | 20% employment | 40% employment | 100% increase | Doubling of local employment rates due to construction and operational jobs |
| Infection rates in community | High | Reduced by 50% | 50% reduction | Major health improvements due to better disease management and prevention |
| Healthcare access | 30% access | 80% access | 167% increase | Significant increase in community access to healthcare services |
| Economic growth | Low | Moderate | Substantial growth | Increased economic activity due to hospital-related spending and services |
| Professional training | Minimal | Extensive | Large-scale development | Development of local healthcare professionals and community education |

### 8.5.5 *Case Study 5: Dell Children's Medical Center of Central Texas, USA*

Dell Children's Medical Center of Central Texas is a leading example of sustainable healthcare design, achieving the distinction of being the world's first hospital to earn LEED Platinum certification. This accolade highlights the center's commitment to environmental sustainability intertwined with patient-centered care, particularly through its innovative use of biophilic design principles that integrate natural elements into the healing environment.

The core of the hospital's design philosophy is the deep integration of biophilic elements, which are instrumental in enhancing the healing process by fostering connections with nature. The hospital incorporates features such as a healing garden and strategically placed windows that provide over 80% of occupied spaces with views of nature or gardens. This pervasive presence of natural elements within the hospital setting plays a crucial role in reducing stress, enhancing mental health, and accelerating recovery times for patients. The intentional inclusion of natural views and elements in the healing environment is based on the understanding that human well-being is deeply connected to our interactions with the natural world.

In addition to its biophilic design, Dell Children's Medical Center incorporates sustainable architectural features that emphasize energy efficiency, the use of environmentally friendly materials, and innovative water conservation techniques. These elements significantly mitigate the hospital's environmental impact, underscoring a dual commitment to planetary health and patient care. The sustainable practices adopted by the hospital not only reduce its carbon footprint but also serve as a testament to the potential for healthcare facilities to contribute positively to environmental stewardship.

Post-occupancy evaluations of Dell Children's Medical Center have demonstrated tangible benefits from its design and architectural choices. Patients have reported increased satisfaction due to the tranquil, nature-infused environment, and there has been a notable decrease in recovery times and medication usage. These outcomes provide empirical support for the concept that the physical environment can significantly influence the efficacy of medical treatments and patient recovery experiences.

The advantages of the hospital's design also extend to its staff. The inclusion of natural light and accessible green spaces within the hospital has led to enhanced job satisfaction, reduced turnover rates, and increased fulfillment among medical professionals. This improved staff morale not only bolsters the work environment but also enhances the quality of care provided to patients, fostering a cycle of well-being and operational efficiency.

Beyond its immediate impact on patients and staff, Dell Children's Medical Center also serves as a community exemplary in environmental stewardship and holistic health practices. The hospital's efforts frequently reach into the community through educational programs and partnerships that advocate health, wellness, and sustainable living.

In summary, Dell Children's Medical Center is a paradigm of how hospital architecture can extend beyond traditional healthcare delivery to embrace broader community health goals and environmental sustainability. The success of its biophilic design offers a compelling blueprint for other healthcare institutions seeking to refine therapeutic environments and adopt a holistic health approach that benefits patients, staff, and the larger community. These insights illustrate that innovative architectural solutions can address immediate healthcare needs while cultivating resilient, healthy communities, thereby enhancing overall quality of life and well-being.

Equation 8.6 quantifies the energy savings achieved through the integration of biophilic elements into healthcare facility design, specifically focusing on natural insulation and light optimization. Biophilic design, which incorporates natural materials and elements into the built environment, has been recognized for enhancing indoor environmental quality and reducing reliance on mechanical heating and cooling systems. This equation evaluates the efficiency gains from using natural materials that provide insulation and designing spaces that maximize natural light, thereby reducing the need for artificial lighting and energy-intensive HVAC systems.

The integration of biophilic design elements is proven to enhance patient and staff well-being while simultaneously offering sustainable energy solutions. Natural insulation involves the use of materials like wood, cork, and bamboo, which have excellent thermal properties, reducing the energy required for heating and cooling. Light optimization includes architectural choices that enhance sunlight penetration, which can significantly decrease the use of electric lighting during daylight hours.

The calculation of energy efficiency from biophilic elements (EEBE) is detailed in the following equation, providing a formula to assess the direct energy savings attributable to these natural design strategies:

$$\text{EEBE} = \eta.I + \theta.L \tag{8.6}$$

where

- $\eta$ and $\theta$ are coefficients that quantify the energy savings per unit area from natural insulation ($I$) and light optimization ($L$), respectively. These coefficients are derived from controlled studies and real-world data that compare energy usage before and after the implementation of biophilic designs.
- $I$ represents the extent of natural insulation used (measured as the area covered or the insulation rating).
- $L$ quantifies the degree of light optimization achieved (measured in terms of reduced lighting energy consumption or increased daylighting area).

Equation 8.6 offers healthcare facility designers and environmental engineers a powerful tool to measure and optimize the energy performance of hospitals and clinics. By applying this equation, facilities can not only ensure a greener, more sustainable approach to healthcare infrastructure but also create environments that promote health and healing through natural connections, ultimately contributing to the facility's operational efficiency and the well-being of its occupants. This strategic application of biophilic principles showcases the tangible benefits of sustainable design in reducing energy costs and enhancing the therapeutic environment.

## 8.6 Case Studies on Failures and Lessons Learned

Examining case studies that detail failures in healthcare facility design and operation is crucial for understanding potential pitfalls and developing strategies to avoid them in the future. This subsection discusses significant healthcare design failures, the lessons learned from these incidents, and how these insights can guide future architectural and operational planning.

### 8.6.1 Case Study 1: Hospital Overcrowding and Inadequate Emergency Facilities

The instance of severe overcrowding at a major city hospital during a flu epidemic vividly highlights the critical need for flexibility and adaptability in hospital design. This scenario, characterized by prolonged wait times and compromised emergency care, starkly emphasizes the limitations of traditional hospital layouts when faced with sudden surges in patient volume. The situation unfolded as the hospital grappled with an overwhelming influx of patients, far exceeding its capacity, which not only strained its resources but also escalated the risk of cross-infection among patients and staff due to prolonged proximity and wait times.

The primary lesson derived from this ordeal underscores the importance of integrating flexibility into hospital design to effectively manage unexpected increases in patient numbers, which may arise during epidemics, natural disasters, or other large-scale emergencies. The incident demonstrated that traditional static hospital designs are often ill-equipped to handle such spikes, highlighting a glaring need for adaptable spaces capable of swiftly transitioning to accommodate varying functional needs.

In response to this crisis, the hospital initiated a strategic redesign incorporating modular components and multi-use spaces into its infrastructure to enhance its responsiveness to similar future events. Modular components, such as movable walls, temporary barriers, and mobile furniture, allow for rapid reconfiguration of hospital layouts. This flexibility facilitates the quick expansion of treatment areas or the conversion of non-clinical spaces into care zones as needed. Additionally, the redesign included multi-use spaces—areas like conference rooms, cafeterias, and lobbies—designed with dual functionality to serve as patient care areas during crises. This approach is vital for maximizing spatial efficiency and adaptability during peak demand periods.

Furthermore, the hospital revamped its triage and emergency protocols to better manage patient flow during high-volume periods. This revision involved leveraging technology to monitor bed availability and patient status, thereby enabling swifter decision-making and more efficient resource allocation. Such improvements in operational procedures are crucial for maintaining care quality during emergencies.

This case study not only prompted significant changes within the affected hospital but also served as a critical wake-up call to other healthcare facilities globally. It has spurred hospitals worldwide to reconsider how flexible design principles can be incorporated into their operations, preparing them to more adeptly handle similar challenges in the future. Through strategic architectural and operational enhancements, hospitals can significantly improve their preparedness for emergencies, thereby ensuring they maintain high standards of care even under the most demanding circumstances. This experience profoundly illustrates the essential role that flexibility plays in the architectural and operational planning of healthcare settings, serving as a valuable lesson in enhancing hospital resilience and response capability.

**Table 8.21** Impact of overcrowding

| Metrics | Before epidemic | During epidemic | Change (%) | Impact on patient care |
|---|---|---|---|---|
| Average wait times (h) | 2 | 8 | 300% increase | Substantial increase leading to delayed treatment and patient dissatisfaction |
| Treatment delays (h) | 1 | 5 | 400% increase | Critical delays in emergencies and routine care impacting patient outcomes |
| Adverse outcomes (incidents) | 10 | 50 | 400% increase | Significant rise in incidents due to delayed treatment and inadequate patient management |
| Bed occupancy rate (%) | 85 | 120 | 41% increase | Overcapacity leading to strained resources and increased cross-infection risk |
| Staff stress levels (scale 1–10) | 5 | 9 | 80% increase | High levels of stress among staff affecting performance and care quality |

Table 8.21 provides quantitative evidence of the challenges faced due to overcrowding at a major city hospital during a flu epidemic. This table highlights critical metrics such as patient wait times, treatment delays, and adverse outcomes, emphasizing the scale and impact of overcrowding on healthcare delivery. By showcasing specific data before and during the epidemic, the table underlines the urgent need for hospital designs that incorporate flexibility and adaptability to effectively manage sudden increases in patient volume and ensure timely, efficient care.

### 8.6.2 Case Study 2: Infection Outbreak Due to Poor Ventilation Design

The serious infection outbreak in a newly constructed hospital wing, attributed to deficiencies in its ventilation system, underscores the indispensable role of HVAC (heating, ventilation, and air conditioning) systems in controlling infections within healthcare environments. Shortly after this wing opened, numerous patients and staff were affected, prompting an investigation that traced the issue back to HVAC design. This design flaw led to the spread of airborne pathogens across multiple areas of the facility due to inadequate air filtration, improper airflow patterns, and insufficient pressure differentials between spaces—key elements in containing airborne contaminants.

This incident brought to light the critical importance of meticulously designing HVAC systems to effectively manage air quality and control infections. Essential aspects of such a design include proper management of airflow to prevent contaminants from moving from infected to clean areas, the installation of high-efficiency particulate air (HEPA) filters to trap and prevent the circulation of airborne pathogens, and the maintenance of appropriate pressure differentials. Isolation rooms, for instance, should maintain negative pressure relative to adjacent areas to ensure that pathogens remain contained.

In response to the outbreak, the hospital initiated several corrective measures to enhance its HVAC system. These included retrofitting the system with HEPA filters to improve the filtration quality significantly, thereby mitigating the risk of spread infection. Additionally, the airflow patterns within the hospital were thoroughly analyzed and redesigned to ensure that air would flow from less contaminated to more contaminated areas, a reversal of the previous problematic airflow. Modifications were also made to establish and maintain proper pressure differentials between adjacent spaces, with particular attention to isolation rooms and other critical care areas, to enhance the containment of pathogens.

The ramifications of this case extend well beyond the individual hospital, affecting hospital design and operations globally. It highlights the necessity for healthcare facilities to integrate advanced HVAC technologies and design principles that prioritize infection control, especially crucial in the design of new hospital wings or during significant renovations. The incident has prompted hospitals to recognize the need for continuous evaluation and updating of their HVAC systems to adapt to emerging diseases that can be transmitted via airborne particles. Regular audits and staying abreast of the latest research in infection control are vital strategies to ensure that healthcare environments remain safe for both patients and staff.

By drawing lessons from this unfortunate event and implementing informed improvements, healthcare facilities can significantly bolster their capabilities to manage and prevent the spread of infections. Such proactive measures are essential in safeguarding the health and well-being of everyone within the hospital environment, demonstrating the profound impact of well-thought-out HVAC system design and maintenance in healthcare settings.

Table 8.22 compares infection rates before and after the redesign of the HVAC systems at a hospital wing where a severe infection outbreak highlighted deficiencies in the original system. This table quantifies the effectiveness of the updates, which included the integration of HEPA filters, the redesign of airflow patterns, and the establishment of proper pressure differentials. By documenting the infection rates before and following the HVAC system overhaul, the table demonstrates significant improvements in air quality and infection control, substantiating the critical impact of well-designed HVAC systems in healthcare settings.

### *8.6.3 Case Study 3: Technology System Failure Impacting Patient Data*

The incident of an electronic health record (EHR) system crash at a hospital serves as a stark reminder of the critical importance of robust IT infrastructure in modern

**Table 8.22** Infection rates before and after HVAC redesign

| HVAC design elements | Infection rates before redesign | Infection rates after redesign | Percentage reduction | Notes |
|---|---|---|---|---|
| Overall infection rates | 25 cases per 1000 patient days | 8 cases per 1000 patient days | 68% reduction | Significant decrease in overall infection rates |
| Airborne infection rates | 15 cases per 1000 patient days | 3 cases per 1000 patient days | 80% reduction | Marked improvement in controlling airborne infections |
| HEPA filter implementation | Not applicable | Implemented | – | HEPA filters contributed to improved air quality |
| Airflow pattern redesign | Inefficient | Optimized | – | Redirection of airflow to prevent cross-contamination |
| Pressure differential maintenance | Inadequate | Properly maintained | – | Ensured containment of airborne pathogens in isolation areas |

healthcare settings. This system failure, attributed to inadequate backup and redundancy systems, resulted in significant data loss, disrupting hospital operations, delaying patient care, and compromising the integrity of vital data. The event exposed several vulnerabilities in the hospital's IT infrastructure that demanded immediate rectification to prevent future occurrences.

The key lessons drawn from this incident underscore the essential need for comprehensive data backup solutions and effective redundancy systems within healthcare facilities. First, it highlighted the necessity of robust backup solutions that encompass both on-site and off-site storage to ensure data protection against local failures. Second, the importance of redundancy systems became apparent, emphasizing that no single point of failure should have the capacity to disrupt the entire IT infrastructure. These systems are crucial in maintaining operational continuity even when individual components fail. Third, regular testing and updating of IT systems were identified as vital practices. These measures are crucial for identifying and mitigating potential vulnerabilities before they can precipitate system-wide failures.

In response to the EHR system crash, the hospital implemented several measures to fortify its IT infrastructure. A tiered data backup strategy was adopted, involving both on-site and off-site data storage. This layered approach guarantees that data recovery can be swiftly executed from multiple sources should one layer fail. Additionally, the hospital invested in upgrading its IT systems to include fault-tolerant systems. These systems are specifically designed to operate reliably and continuously, even when certain components fail, thereby reducing the risk of a total system shutdown and subsequent data loss.

Moreover, the hospital enhanced its IT security measures to shield against cyber threats, which pose a significant risk of data breaches or loss. These enhanced security protocols include the deployment of stronger firewalls, advanced intrusion detection systems, and the conduct of regular security audits to ensure the integrity and safety of the hospital's digital infrastructure.

This case study serves as a crucial reminder for all healthcare organizations about the importance of investing in strong IT infrastructure. As the reliance on digital records and systems grows, hospitals must adhere to best practices in IT management. These practices include implementing robust backup systems, ensuring fault tolerance, and maintaining stringent cybersecurity measures to protect patient information and guarantee the uninterrupted operation of healthcare services.

By learning from this incident and proactively implementing comprehensive IT solutions, hospitals can significantly enhance their resilience against future failures. This proactive approach is essential for ensuring the reliability of critical healthcare operations and safeguarding the privacy and security of patient data, ultimately contributing to the overall efficacy and safety of healthcare delivery.

Table 8.23 presents data detailing the operational impact of a significant IT failure at a hospital, where the electronic health record (EHR) system crashed due to inadequate backup systems. The table includes metrics such as the duration of system downtime, the number of affected patients, and recovery times, emphasizing the severe consequences of IT failures on hospital operations. By quantifying the extent of the disruption—spanning from delayed patient care to prolonged recovery of services, this table underscores the necessity for robust IT infrastructure, effective data backup solutions, and redundancy systems in healthcare settings to maintain uninterrupted operations and safeguard patient data.

### *8.6.4 Case Study 4: Structural Failure in a New Hospital Building*

The structural failure in a new hospital building shortly after its completion serves as a stark reminder of the crucial importance of quality control and the use of certified materials in construction, especially in facilities as critical as hospitals. This incident, which resulted in significant structural damage due to substandard construction materials, not only compromised the structural integrity of the building but also necessitated a temporary closure of the facility. This closure led to extensive repairs and significant disruptions in hospital services, underscoring the severity of the issue.

Key lessons emerged from this troubling incident, highlighting essential aspects of construction that must be rigorously managed to prevent similar occurrences in the future. First, the incident demonstrated the absolute necessity for stringent quality control measures throughout the construction process. Such measures are crucial to ensure that all materials and practices adhere to the highest standards required for hospital infrastructure. Second, the importance of using only certified, high-quality materials was made abundantly clear. These materials guarantee the safety and longevity of the building, which is particularly vital in a hospital setting where

**Table 8.23** Impact of IT failures on hospital operations

| Metric | Value before IT failure | Value during IT failure | Impact |
|---|---|---|---|
| System downtime | 0 h | 72 h | Prolonged interruption in patient care and access to medical records |
| Affected patients | 0 | 500 + | Significant number of patients experienced delays in treatment and access to information |
| Recovery time | Not applicable | 120 h | Extensive time required to restore systems and data integrity |
| Data loss instances | 0 | Multiple instances | Critical loss of patient data affecting ongoing care and records |
| Emergency procedures activated | Not applicable | Yes | Activation of emergency protocols to mitigate the impact of the outage |

failures can result in serious risks to patient and staff safety. Lastly, the meeting of contractors and suppliers was identified as a critical step. It is essential to ensure that all parties involved in the construction process have a proven track record of compliance and quality.

In response to the structural failure, the hospital implemented several corrective measures to strengthen its construction protocols and prevent future incidents. These included requiring comprehensive compliance certificates from all contractors and suppliers involved in building processes. These certifications must come from recognized standards organizations and demonstrate adherence to local building codes. Additionally, the hospital introduced regular audits during the construction phase, conducted by independent inspectors to ensure strict adherence to all specifications and quality standards.

Furthermore, the hospital enhanced its project management practices to include more rigorous oversight of the entire construction process. This involved closer monitoring of material procurement, storage, and usage, as well as ensuring that all construction staff were adequately trained and qualified for their roles. By strengthening these areas, the hospital aimed to enhance the overall integrity and safety of its infrastructure.

This case study underscores the broader implications for healthcare infrastructure development, emphasizing the need for robust oversight and quality assurance at all stages of hospital construction. Ensuring the structural integrity of healthcare facilities is crucial, not just for regulatory compliance but also for the safety of patients, staff, and the broader community. By learning from this incident and instituting stringent control measures, healthcare facilities can better protect against similar failures in the future, ensuring that new buildings are safe, reliable, and constructed to endure. This approach not only protects the hospital's investment in its infrastructure but also upholds its fundamental duty to provide a safe environment for healthcare delivery.

Table 8.24 offers a comprehensive checklist developed in response to a structural failure incident in a new hospital building, designed to ensure the quality and compliance of construction materials and practices. This checklist is a vital tool for project managers, contractors, and quality assurance teams to systematically verify the adherence to safety standards and durability requirements. It covers critical areas such as material certifications, contractor vetting, compliance with local building codes, and regular audit protocols. By providing a structured approach to quality control, this checklist aims to prevent future incidents by ensuring all aspects of construction meet the highest standards required for hospital facilities.

**Table 8.24** Compliance and quality control checklist

| Checklist item | Description | Compliance requirement | Verification method |
|---|---|---|---|
| Material certification | All materials must have up-to-date certifications | Mandatory | Review of documentation |
| Contractor and supplier vetting | Contractors and suppliers must have verified quality histories | Mandatory | Background checks, reference reviews |
| Adherence to building codes | Construction must comply with local and international codes | Mandatory | Inspection reports |
| Installation audits | Regular audits to inspect the installation processes | Periodic | Scheduled audits by independent inspectors |
| Training and qualifications of personnel | All personnel must have relevant qualifications and training | Mandatory | Certificates and training logs |
| Documentation and reporting | Complete documentation of all construction stages | Mandatory | Review of project logs and reports |
| Emergency preparedness | Systems and structures must accommodate emergency protocols | Mandatory | Emergency response drills and system tests |
| Sustainability standards | Compliance with environmental and sustainability standards | Recommended | Sustainability audit and certification review |

### 8.6.5 Case Study 5: Inadequate Space for Evacuation During a Fire

The incident of a fire at a hospital, which resulted in a chaotic and hazardous evacuation, highlights the critical importance of integrating robust emergency and evacuation protocols into hospital design. The fire revealed significant deficiencies in the hospital's emergency preparedness, especially concerning the physical layout of the building. Narrow corridors and an insufficient number of emergencies exits significantly impeded the evacuation process, creating perilous bottlenecks and delaying the safe exit of patients and staff. This incident underscored severe safety risks and emphasized the need for immediate actions to enhance emergency response measures and prevent future tragedies.

Several crucial lessons emerged from this incident, reinforcing the importance of proactive emergency planning in hospital design. First, emergency and evacuation protocols must be integral elements of hospital architecture, prioritized during the design phase rather than as afterthoughts. This involves ensuring that evacuation paths are ample and accessible to all patients, including those with limited mobility. Second, evacuation routes must be marked and consistently free from obstructions to facilitate unhindered egress. Regular checks and maintenance are necessary to always keep these paths clear and functional. Lastly, the adequacy of emergency exits is critical; there must be enough exits, appropriately located and sized, to accommodate the rapid evacuation of all individuals present during an emergency.

In response to the challenges posed by the fire incident, the hospital undertook several significant measures to revamp its emergency preparedness. These included a complete redesign of the evacuation routes to optimize flow and eliminate bottlenecks. The redesign involved widening corridors and strategically increasing the number of emergencies exits to facilitate quicker and safer egress for everyone in the facility. Furthermore, the hospital instituted mandatory regular fire drills and comprehensive safety training for all staff members. These drills are designed to familiarize staff with emergency procedures, reduce panic, and enhance the efficiency of evacuation processes. Additionally, the hospital upgraded its fire safety infrastructure by installing advanced fire detection and suppression systems throughout the facility, which are regularly tested and maintained to ensure optimal functionality.

This case study serves as a poignant reminder to all healthcare facilities of the paramount importance of emergency preparedness. By learning from this incident and integrating robust safety protocols into their operational and architectural strategies, hospitals can significantly enhance overall safety. Such proactive measures ensure swift and orderly responses to emergencies, safeguarding both patients and staff.

Collectively, these case studies underscore that while failures in healthcare facilities can have severe consequences, they also provide vital learning opportunities. By analyzing these incidents and implementing corrective actions, healthcare facilities can enhance their safety protocols, operational efficiency, and resilience. This

**Table 8.25** Emergency response times and outcomes

| Metric | Before redesign | After redesign | Improvement (%) |
|---|---|---|---|
| Average evacuation time (minutes) | 11.8 | 7.2 | 39.0 |
| Number of injuries reported | 7 | 2 | 71.4 |
| Staff satisfaction rate (%) | 63 | 88 | 39.7 |
| Patient satisfaction rate (%) | 58 | 85 | 46.6 |

proactive approach is crucial in fostering a safer environment for both patients and staff, ultimately contributing to better healthcare outcomes.

Table 8.25 provides a detailed comparison of emergency response times and the outcomes for patients and staff during evacuations before and after significant redesign efforts at a hospital. This data clearly illustrates the improved efficiency and safety of evacuation processes as a direct result of enhancements made to the facility's emergency routes and safety protocols. Key metrics included in the table are the reduction in average evacuation time, a decrease in reported injuries during evacuations, and increased satisfaction rates among staff and patients regarding the hospital's emergency response capabilities.

## Further Reading

Aalhashem, N. A., Al-Kaissi, K. A., Khaled, T. T., Al-Sharify, T. A., Al-Sharify, Z. T., Naser, Z. A., & Al-Hamd, R. K. S. (2023, July). The future of an intelligent and responsive architecture design after the COVID-19 epidemic. In *AIP Conference Proceedings* (Vol. 2787, No. 1). AIP Publishing. https://doi.org/10.1063/5.0148420

Abdelaal, M. S., & Soebarto, V. I. (2018). The death of modern hospitals: Towards a comprehensive approach for restorative healthcare environmental design.

Abdelaal, M. S., & Soebarto, V. I. (2019). Biophilia and Salutogenesis as restorative design approaches in healthcare architecture. *Architectural science review*, 62(3), 195–205. https://doi.org/10.1080/00038628.2019.1604313

Abou-Alfa, G. K., Norton, L., & Global Oncology Medical Diplomacy Working Group. (2023). Global Oncology Medical Diplomacy Working Group Inaugural Meeting: Defining worldwide barriers to germline genomics in cancer prevention and management. *Annals of Global Health*, 89(1). https://doi.org/10.5334/aogh.3967

Abou-Nassar, E. M., Iliyasu, A. M., El-Kafrawy, P. M., Song, O. Y., Bashir, A. K., & Abd El-Latif, A. A. (2020). DITrust chain: towards blockchain-based trust models for sustainable healthcare IoT systems. *IEEE access*, 8, 111223–111238. https://doi.org/10.1109/ACCESS.2020.2999468

Aerts, R., Honnay, O., & Van Nieuwenhuyse, A. (2018). Biodiversity and human health: mechanisms and evidence of the positive health effects of diversity in nature and green spaces. *British medical bulletin*, 127(1), 5–22. https://doi.org/10.1093/bmb/ldy021

Alazab, M., Khan, L. U., Koppu, S., Ramu, S. P., Iyapparaja, M., Boobalan, P., … & Aljuhani, A. (2022). Digital twins for healthcare 4.0—Recent advances, architecture, and open challenges. *IEEE Consumer Electronics Magazine*, 12(6), 29–37. https://doi.org/10.1109/MCE.2022.3208986

Alsawaf, E. S., & Albadry, A. M. (2022). Principles for the sustainable design of hospital buildings. *International Journal of Sustainable Development and Planning*, 17(6), 1797–1808. https://doi.org/10.18280/ijsdp.170614

Anåker, A., von Koch, L., Heylighen, A., & Elf, M. (2019). "It's lonely": Patients' experiences of the physical environment at a newly built stroke unit. *HERD: Health Environments Research & Design Journal*, 12(3), 141–152. https://doi.org/10.1177/1937586718806696

Arslan, A., Golgeci, I., Khan, Z., Al-Tabbaa, O., & Hurmelinna-Laukkanen, P. (2021). Adaptive learning in cross-sector collaboration during global emergency: conceptual insights in the context of COVID-19 pandemic. *Multinational Business Review*, 29(1), 21–42. https://doi.org/10.1108/MBR-07-2020-0153

Bharathi, R., Abirami, T., Dhanasekaran, S., Gupta, D., Khanna, A., Elhoseny, M., & Shankar, K. (2020). Energy efficient clustering with disease diagnosis model for IoT based sustainable healthcare systems. *Sustainable Computing: Informatics and Systems*, 28, 100453. https://doi.org/10.1016/j.suscom.2020.100453

Bisoi, A. K., & Mankar, D. D. (2024). Green Hospitals and Healthcare Facilities with Healing Architecture in India. Indian Practitioner, 77(7).

Bokhour, B. G., Fix, G. M., Mueller, N. M., Barker, A. M., Lavela, S. L., Hill, J. N., … & Lukas, C. V. (2018). How can healthcare organizations implement patient-centered care? Examining a large-scale cultural transformation. *BMC health services research*, 18, 1–11. https://doi.org/10.1186/s12913-018-2949-5

Brambilla, A., Sun, T. Z., Elshazly, W., Ghazy, A., Barach, P., Lindahl, G., & Capolongo, S. (2021). Flexibility during the COVID-19 pandemic response: Healthcare facility assessment tools for resilient evaluation. *International Journal of Environmental Research and Public Health*, 18(21), 11478. https://doi.org/10.3390/ijerph182111478

Callegari, B., & Feder, C. (2022). A literature review of pandemics and development: The long-term perspective. *Economics of Disasters and Climate Change*, 6(1), 183–212. https://doi.org/10.1007/s41885-022-00106-w

Capolongo, S., Gola, M., Brambilla, A., Morganti, A., Mosca, E. I., & Barach, P. (2020). COVID-19 and healthcare facilities: a decalogue of design strategies for resilient hospitals. *Acta Bio Medica: Atenei Parmensis*, 91(9-S), 50. https://doi.org/10.23750/abm.v91i9-S.10117

Chen, L. K., Yuan, R. P., Ji, X. J., Lu, X. Y., Xiao, J., Tao, J. B., … & Jiang, L. Z. (2021). Modular composite building in urgent emergency engineering projects: A case study of accelerated design and construction of Wuhan Thunder God Mountain/Leishenshan hospital to COVID-19 pandemic. *Automation in Construction*, 124, 103555. https://doi.org/10.1016/j.autcon.2021.103555

Chen, L., Zhai, C., Wang, L., Hu, X., & Huang, X. (2022). Modular structure construction progress scenario: a case study of an emergency hospital to address the COVID-19 PANDEMIC. *Sustainability*, 14(18), 11243. https://doi.org/10.3390/su141811243

Chanchaichujit, J., Tan, A., Meng, F., & Eaimkhong, S. (2019). Healthcare 4.0. Springer Nature, Singapore. 10, 978–981. https://doi.org/10.1007/978-981-13-8114-0

Chersich, M. F., & Wright, C. Y. (2019). Climate change adaptation in South Africa: a case study on the role of the health sector. *Globalization and health*, 15, 1–16. https://doi.org/10.1186/s12992-019-0466-x

Curry, L. A., Brault, M. A., Linnander, E. L., McNatt, Z., Brewster, A. L., Cherlin, E., … & Bradley, E. H. (2018). Influencing organisational culture to improve hospital performance in care of patients with acute myocardial infarction: a mixed-methods intervention study. *BMJ quality & safety*, 27(3), 207–217. https://doi.org/10.1136/bmjqs-2017-006989

de Oliveira, K. B., dos Santos, E. F., Neto, A. F., de Mello Santos, V. H., & de Oliveira, O. J. (2021). Guidelines for efficient and sustainable energy management in hospital buildings. *Journal of Cleaner Production*, 329, 129644. https://doi.org/10.1016/j.jclepro.2021.129644

Delmonico, D. V. D. G., Santos, H. H. D., Pinheiro, M. A., de Castro, R., & de Souza, R. M. (2018). Waste management barriers in developing country hospitals: Case study and AHP analysis. *Waste Management & Research*, 36(1), 48–58. https://doi.org/10.1177/0734242X17739972

Dehury, R. K., Dehury, P., Sripathi, N., Acharyulu, G. V. R. K., Behera, M. R., & Neeragatti, S. (2023). Health sector development in India: an account from Bhore Committee 1946 to National Health Policy 2017. *Journal of Development Policy and Practice*, 8(2), 209–242. https://doi.org/10.1177/24551333231163666

Dhagarra, D., Goswami, M., & Kumar, G. (2020). Impact of trust and privacy concerns on technology acceptance in healthcare: an Indian perspective. *International journal of medical informatics*, 141, 104164. https://doi.org/10.1016/j.ijmedinf.2020.104164

Dion, H., Evans, M., & Farrell, P. (2023). Hospitals management transformative initiatives; towards energy efficiency and environmental sustainability in healthcare facilities. *Journal of Engineering, Design and Technology*, 21(2), 552–584. https://doi.org/10.1108/JEDT-04-2022-0200

DuBose, J., MacAllister, L., Hadi, K., & Sakallaris, B. (2018). Exploring the concept of healing spaces. *HERD: Health Environments Research & Design Journal*, 11(1), 43–56. https://doi.org/10.1177/1937586716680567

Ebekozien, A., Ayo-Odifiri, S. O., Nwaole, A. N. C., Ibeabuchi, A. L., & Uwadia, F. E. (2022). Barriers in Nigeria's public hospital green buildings implementation initiatives. *Journal of Facilities Management*, 20(4), 586–605. https://doi.org/10.1108/JFM-01-2021-0009

El Messeidy, R. (2019). Application of biophilic patterns in health care environments to enhance healing. *Engineering Research Journal*, 163, 130–143. https://doi.org/10.21608/erj.2019.122518

Fan, K., Chan, E. H., & Chau, C. K. (2018). Costs and benefits of implementing green building economic incentives: Case study of a gross floor area concession scheme in Hong Kong. *Sustainability*, 10(8), 2814. https://doi.org/10.3390/su10082814

Firouzi, F., Rahmani, A. M., Mankodiya, K., Badaroglu, M., Merrett, G. V., Wong, P., & Farahani, B. (2018). Internet-of-Things and big data for smarter healthcare: From device to architecture, applications and analytics. *Future Generation Computer Systems*, 78, 583–586. https://doi.org/10.1016/j.future.2017.09.016

Fu, S. H., Gasparrini, A., Rodriguez, P. S., & Jha, P. (2018). Mortality attributable to hot and cold ambient temperatures in India: a nationally representative case-crossover study. *PLoS medicine*, 15(7), e1002619. https://doi.org/10.1371/journal.pmed.1002619

Gupta, B., & Khanna, T. (2019). A recombination-based internationalization model: Evidence from Narayana health's journey from India to the Cayman Islands. *Organization Science*, 30(2), 405–425. https://doi.org/10.1287/orsc.2018.1241

Gupta, D., Bharathi, R., Elhoseny, M., & Shankar, K. (2020). Energy efficient clustering with disease diagnosis model for IoT based sustainable healthcare systems. *Sustainable Computing: Informatics and Systems*, 28, 100453. https://doi.org/10.1016/j.suscom.2020.100453

Hall, D. A., Zaragoza Domingo, S., Hamdache, L. Z., Manchaiah, V., Thammaiah, S., Evans, C., ... & International Collegium of Rehabilitative Audiology and TINnitus Research NETwork. (2018). A good practice guide for translating and adapting hearing-related questionnaires for different languages and cultures. *International Journal of Audiology*, 57(3), 161–175. https://doi.org/10.1080/14992027.2017.1393565

Haghi, M., Neubert, S., Geissler, A., Fleischer, H., Stoll, N., Stoll, R., & Thurow, K. (2020). A flexible and pervasive IoT-based healthcare platform for physiological and environmental parameters monitoring. *IEEE Internet of Things Journal*, 7(6), 5628–5647. https://doi.org/10.1109/JIOT.2020.2980432

Handtke, O., Schilgen, B., & Mösko, M. (2019). Culturally competent healthcare–A scoping review of strategies implemented in healthcare organizations and a model of culturally competent healthcare provision. *PloS one*, 14(7), e0219971. https://doi.org/10.1371/journal.pone.0219971

Hong, H., & Gasparatos, A. (2020). Eco-industrial parks in China: Key institutional aspects, sustainability impacts, and implementation challenges. *Journal of Cleaner Production*, 274, 122853. https://doi.org/10.1016/j.jclepro.2020.122853

Holeman, I., & Kane, D. (2020). Human-centered design for global health equity. *Information technology for development*, 26(3), 477–505. https://doi.org/10.1080/02681102.2019.1667289

Jennings, V., & Bamkole, O. (2019). The relationship between social cohesion and urban green space: An avenue for health promotion. *International journal of environmental research and public health*, 16(3), 452. https://doi.org/10.3390/ijerph16030452

Jiang, B., Song, Y., Li, H. X., Lau, S. S. Y., & Lei, Q. (2020). Incorporating biophilic criteria into green building rating tools: Case study of Green Mark and LEED. *Environmental Impact Assessment Review*, 82, 106380. https://doi.org/10.1016/j.eiar.2020.106380

Jogia, J., Sharif, A. H., Nawaz, F. A., Khan, A. R., Alawami, R. H., Aljanahi, M. A., & Sultan, M. A. (2022). Comorbidities associated with attention-deficit/hyperactivity disorder in children and adolescents at a tertiary care setting. *Global Pediatric Health*, 9, 2333794X221076607. https://doi.org/10.1177/2333794X221076607

Kaluzny, A. D., Veney, J. E., & Gentry, J. T. (2018). Innovation of health services: A comparative study of hospitals and health departments. In *Organizational Innovation* (pp. 83–114). Routledge.

Khalid, W., Zaki, S. A., Rijal, H. B., & Yakub, F. (2019). Investigation of comfort temperature and thermal adaptation for patients and visitors in Malaysian hospitals. *Energy and Buildings*, 183, 484–499. https://doi.org/10.1016/j.enbuild.2018.11.019

Kondo, M. C., Fluehr, J. M., McKeon, T., & Branas, C. C. (2018). Urban green space and its impact on human health. *International journal of environmental research and public health*, 15(3), 445. https://doi.org/10.3390/ijerph15030445

Kumar, P., Druckman, A., Gallagher, J., Gatersleben, B., Allison, S., Eisenman, T. S., ... & Morawska, L. (2019). The nexus between air pollution, green infrastructure and human health. *Environment international*, 133, 105181. https://doi.org/10.1016/j.envint.2019.105181

Lai, J., Maher, L., Zhou, C., Zhou, Y., Li, C., Fu, J., ... & Wu, Y. (2024). Psychometric testing of the Chinese National Health Service Sustainability Model as an instrument to assess innovation in Chinese nursing settings. *International Journal of Nursing Practice*, 30(3), e13214. https://doi.org/10.1111/ijn.13214

Lahariya, C. (2020). Health & wellness centers to strengthen primary health care in India: concept, progress and ways forward. *The Indian Journal of Pediatrics*, 87(11), 916–929. https://doi.org/10.1007/s12098-020-03359-z

Li, Y., Cao, L., Han, Y., & Wei, J. (2020). Development of a conceptual benchmarking framework for healthcare facilities management: Case study of Shanghai municipal hospitals. *Journal of Construction Engineering and Management*, 146(1), 05019016. https://doi.org/10.1061/(ASCE)CO.1943-7862.0001731

Lovreglio, R., Gonzalez, V., Feng, Z., Amor, R., Spearpoint, M., Thomas, J., ... & Sacks, R. (2018). Prototyping virtual reality serious games for building earthquake preparedness: The Auckland City Hospital case study. *Advanced engineering informatics*, 38, 670–682. https://doi.org/10.1016/j.aei.2018.08.018

Lepore, L., Metallo, C., Schiavone, F., & Landriani, L. (2018). Cultural orientations and information systems success in public and private hospitals: preliminary evidences from Italy. *BMC health services research*, 18, 1–13. https://doi.org/10.1186/s12913-018-3349-6

Mahfouz, N. A., Kizhakkedath, P., Ibrahim, A., El Naofal, M., Ramaswamy, S., Harilal, D., ... & Abou Tayoun, A. (2020). Utility of clinical exome sequencing in a complex Emirati pediatric cohort. *Computational and structural biotechnology journal*, 18, 1020–1027. https://doi.org/10.1016/j.csbj.2020.04.013

Marques, G., Pitarma, R., M. Garcia, N., & Pombo, N. (2019). Internet of things architectures, technologies, applications, challenges, and future directions for enhanced living environments and healthcare systems: a review. *Electronics*, 8(10), 1081. https://doi.org/10.3390/electronics8101081

Mazumdar-Shaw, K. (2018). Leveraging affordable innovation to tackle India's healthcare challenge. IIMB Management Review, 30(1), 37–50. https://doi.org/10.1016/j.iimb.2017.11.003

Mehta, P. (2018). Framework of Indian healthcare system and its challenges: an insight. In *Health Economics and Healthcare Reform: Breakthroughs in Research and Practice* (pp. 405–429). IGI Global. https://doi.org/10.4018/978-1-5225-3168-5.ch023

Ndayishimiye, C., Sowada, C., Dyjach, P., Stasiak, A., Middleton, J., Lopes, H., & Dubas-Jakóbczyk, K. (2022). Associations between the COVID-19 pandemic and hospital infrastructure adaptation and planning—a scoping review. *International journal of environmental research and public health*, 19(13), 8195. https://doi.org/10.3390/ijerph19138195

Nematchoua, M. K., Yvon, A., Kalameu, O., Asadi, S., Choudhary, R., & Reiter, S. (2019). Impact of climate change on demands for heating and cooling energy in hospitals: An in-depth case study of six islands located in the Indian Ocean region. *Sustainable Cities and Society*, 44, 629–645. https://doi.org/10.1016/j.scs.2018.10.031

Naresh, V. S., Pericherla, S. S., Murty, P. S. R., & Reddi, S. (2020). Internet of Things in Healthcare: Architecture, Applications, Challenges, and Solutions. *Computer Systems Science & Engineering*, 35(6).

Alarcón, A. C., Rangel, A. M., ... & Hickie, I. B. (2020). Involving end users in adapting a Spanish version of a web-based mental health clinic for young people in Colombia: exploratory study using participatory design methodologies. *JMIR Mental Health*, 7(2), e15914. https://doi.org/10.2196/15914

Ospina-Pinillos, L., Davenport, T., Mendoza Diaz, A., Navarro-Mancilla, A., Scott, E. M., & Hickie, I. B. (2019). Using participatory design methodologies to co-design and culturally adapt the Spanish version of the mental health eClinic: qualitative study. *Journal of medical Internet research*, 21(8), e14127. https://doi.org/10.2196/14127

Pace, P., Aloi, G., Gravina, R., Caliciuri, G., Fortino, G., & Liotta, A. (2018). An edge-based architecture to support efficient applications for healthcare industry 4.0. *IEEE Transactions on Industrial Informatics*, 15(1), 481–489. https://doi.org/10.1109/TII.2018.2843169

Putrino, D., Ripp, J., Herrera, J. E., Cortes, M., Kellner, C., Rizk, D., & Dams-O'Connor, K. (2020). Multisensory, nature-inspired recharge rooms yield short-term reductions in perceived stress among frontline healthcare workers. *Frontiers in Psychology*, 11, 560833. https://doi.org/10.3389/fpsyg.2020.560833

Rabin, B. A., McCreight, M., Battaglia, C., Ayele, R., Burke, R. E., Hess, P. L., ... & Glasgow, R. E. (2018). Systematic, multimethod assessment of adaptations across four diverse health systems interventions. *Frontiers in public health*, 6, 102. https://doi.org/10.3389/fpubh.2018.00102

Rasheed, H. (2024). Therapeutic Architecture Illumination And Environments Within Healing Space. https://doi.org/10.20944/preprints202402.1305.v1

Rathod, S., Gega, L., Degnan, A., Pikard, J., Khan, T., Husain, N., ... & Naeem, F. (2018). The current status of culturally adapted mental health interventions: a practice-focused review of meta-analyses. *Neuropsychiatric disease and treatment*, 165–178.

Reyes-Riveros, R., Altamirano, A., De La Barrera, F., Rozas-Vásquez, D., Vieli, L., & Meli, P. (2021). Linking public urban green spaces and human well-being: A systematic review. *Urban forestry & urban greening*, 61, 127105. https://doi.org/10.1016/j.ufug.2021.127105

Ranganathan, P., Sengar, M., Chinnaswamy, G., Agrawal, G., Arumugham, R., Bhatt, R., ... & Pramesh, C. S. (2021). Impact of COVID-19 on cancer care in India: a cohort study. *The Lancet Oncology*, 22(7), 970–976.

Saravanan, V. S. (2018). Contestation and negotiation of urban health in India: A situated political approach. *World Development*, 104, 375–387. https://doi.org/10.1016/j.worlddev.2017.12.003

Scotti, F., Pierri, F., Bonaccorsi, G., & Flori, A. (2022). Responsiveness of open innovation to COVID-19 pandemic: The case of data for good. *PLoS One*, 17(4), e0267100. https://doi.org/10.1371/journal.pone.0267100

Setola, N., Naldi, E., Arnetoli, M. V., Marzi, L., & Bologna, R. (2022). Hospital responses to COVID-19: evidence from case studies to support future healthcare design research. *Facilities*, 40(1/2), 131–145. https://doi.org/10.1108/F-03-2021-0023

Sharifi, A., Khavarian-Garmsir, A. R., & Kummitha, R. K. R. (2021). Contributions of smart city solutions and technologies to resilience against the COVID-19 pandemic: A literature review. *Sustainability*, 13(14), 8018. https://doi.org/10.3390/su13148018

Sharma, A., Borah, S. B., & Moses, A. C. (2021). Responses to COVID-19: The role of governance, healthcare infrastructure, and learning from past pandemics. *Journal of business research*, 122, 597–607. https://doi.org/10.1016/j.jbusres.2020.09.011

Shen, W., Tang, W., Siripanan, A., Lei, Z., Duffield, C. F., & Hui, F. K. P. (2018). Understanding the green technical capabilities and barriers to green buildings in developing countries: A case study of Thailand. *Sustainability*, 10(10), 3585. https://doi.org/10.3390/su10103585

Shepherd, S. M., Willis-Esqueda, C., Newton, D., Sivasubramaniam, D., & Paradies, Y. (2019). The challenge of cultural competence in the workplace: perspectives of healthcare providers. *BMC health services research*, 19, 1–11. https://doi.org/10.1186/s12913-019-3959-7

Singh, P., Dixit, V., & Kaur, J. (2019). Green healthcare for smart cities. In *Green and Smart Technologies for Smart Cities* (pp. 91–130). CRC Press.

Slewa-Younan, S., Nguyen, T. P., Al-Yateem, N., Rossiter, R. C., & Robb, W. (2020). Causes and risk factors for common mental illnesses: the beliefs of paediatric hospital staff in the United Arab Emirates. *International Journal of Mental Health Systems*, 14, 1–16. https://doi.org/10.1186/s13033-020-00367-6

Sowada, C., Dyjach, P., Ndayishimiye, C., Stasiak, A., Middleton, J., Lopes, H., & Dubas-Jakóbczyk, K. (2022). Associations between the COVID-19 pandemic and hospital infrastructure adaptation and planning—a scoping review. *International journal of environmental research and public health*, 19(13), 8195. https://doi.org/10.3390/ijerph19138195

Sudhakar, K., Winderl, M., & Priya, S. S. (2019). Net-zero building designs in hot and humid climates: A state-of-art. *Case Studies in Thermal Engineering*, 13, 100400. https://doi.org/10.1016/j.csite.2019.100400

Sulis, E., Amantea, I. A., Aldinucci, M., Boella, G., Marinello, R., Grosso, M., ... & Ambrosini, S. (2024). An ambient assisted living architecture for hospital at home coupled with a process-oriented perspective. *Journal of Ambient Intelligence and Humanized Computing*, 15(5), 2737–2755. https://doi.org/10.1007/s12652-022-04388-6

Teo, K. W., Hu, Y., Chew, K. T., Pek, W. Y., Chua, H. C., Matchar, D. B., & Ng, Y. F. (2023). Health System Transformation Playbook and Unified Care Model: an integrated design, systems & complexity thinking approach to health system transformation. *Frontiers in health services*, 3, 1157038. https://doi.org/10.3389/frhs.2023.1157038

Tsai, T. I., Luck, L., Jefferies, D., & Wilkes, L. (2024). Challenges in adapting a survey: ensuring cross-cultural equivalence. *Nurse researcher*, 32(2). https://doi.org/10.7748/nr.2018.e1581

Ugail, H., Aggarwal, R., Iglesias, A., Howard, N., Campuzano, A., Suárez, P., ... & Muhammad, K. (2021). Social distancing enhanced automated optimal design of physical spaces in the wake of the COVID-19 pandemic. *Sustainable Cities and Society*, 68, 102791. https://doi.org/10.1016/j.scs.2021.102791

Ulrich, R. S., Cordoza, M., Gardiner, S. K., Manulik, B. J., Fitzpatrick, P. S., Hazen, T. M., & Perkins, R. S. (2020). ICU patient family stress recovery during breaks in a hospital garden and indoor environments. *HERD: Health environments research & design journal*, 13(2), 83–102. https://doi.org/10.1177/1937586719867157

Uslu, B. Ç., Okay, E., & Dursun, E. (2020). Analysis of factors affecting IoT-based smart hospital design. *Journal of Cloud Computing*, 9(1), 67. https://doi.org/10.1186/s13677-020-00215-5

Verderber, S., Gray, S., Suresh-Kumar, S., Kercz, D., & Parshuram, C. (2021). Intensive care unit built environments: a comprehensive literature review (2005–2020). *HERD: Health Environments Research & Design Journal*, 14(4), 368–415. https://doi.org/10.1177/19375867211009273

Viryasiri, T., Laovisutthichai, V., Sangnin, K., Dhanakoses, K., Roopkaew, P., & Viryasiri, P. (2024). Hospital design principles implementation: Reflections from practitioners in Thailand. *Journal of Asian Architecture and Building Engineering*, 23(1), 43–56. https://doi.org/10.1080/13467581.2023.2215842

Wang, K., Ke, Y., & Sankaran, S. (2019). Public-private partnerships in non-profit hospitals: Case study of China. *The International journal of health planning and management*, 34(4), e1862–e1898. https://doi.org/10.1002/hpm.2891

Zhang, Y., Tzortzopoulos, P., & Kagioglou, M. (2019). Healing built-environment effects on health outcomes: Environment–occupant–health framework. *Building research & information*, 47(6), 747–766. https://doi.org/10.1080/09613218.2017.1411130

Zhao, J., Liu, Y., & Zhou, P. (2018). Framing a sustainable architecture for data analytics systems: An exploratory study. *IEEE Access*, 6, 61600–61613. https://doi.org/10.1109/ACCESS.2018.2872673

# Chapter 9
# Navigating the Complex Landscape of Healthcare Innovation: Strategies for Overcoming Design and Implementation Challenges

This chapter delves into the myriad challenges healthcare facilities face when implementing advanced designs and offers strategic recommendations to navigate these hurdles effectively. It begins by addressing the financial constraints that often impede the adoption of innovative technologies and designs, suggesting tactics such as external funding, phased implementations, and forming strategic partnerships to alleviate financial burdens. The chapter then explores regulatory hurdles, emphasizing the importance of early engagement with regulatory bodies and advocating for flexible regulations that accommodate technological advancements without compromising safety or quality. The discussion extends to overcoming resistance to change within healthcare organizations by fostering a culture that values continuous education, stakeholder involvement, and clear communication. Technical challenges are also scrutinized, with solutions involving ongoing training, careful technology selection, and robust vendor partnerships to ensure seamless integration of new systems. Furthermore, the chapter discusses cultural barriers in global healthcare settings, recommending strategies like community engagement and the inclusion of culturally relevant design elements to enhance patient care and facility acceptance. Collectively, these insights provide a comprehensive roadmap for healthcare facilities striving to incorporate advanced architectural and technological innovations while navigating the financial, regulatory, cultural, and technical landscapes that shape modern healthcare environments.

## 9.1 Overcoming Barriers to Implementing Advanced Designs

Implementing advanced designs in healthcare facilities often involves navigating a range of challenges that can hinder progress and innovation. These barriers can be financial, regulatory, or stem from a lack of awareness about the benefits of modern



A. A. Firoozi and A. A. Firoozi, *Transformative Design in Healthcare*,
https://doi.org/10.1007/978-3-032-08019-6_9

design practices. Understanding these challenges is the first step toward developing effective strategies to overcome them.

### 9.1.1 Financial Constraints

Financial constraints pose a formidable challenge for healthcare facilities aspiring to implement advanced designs and technologies. The substantial costs associated with these innovations, including the procurement of cutting-edge technology and sustainable materials, often discourage the necessary enhancements that could significantly improve service delivery and patient care. Despite these financial hurdles, healthcare organizations can adopt various strategic measures to secure the needed investments for upgrading their infrastructure.

A primary strategy for overcoming financial barriers is to explore external funding opportunities that can provide substantial support for infrastructure enhancements. Healthcare facilities can access diverse funding sources to facilitate their projects. These include grants from government bodies, non-profit organizations, and international health agencies, which are often earmarked for projects that promote healthcare innovation and offer substantial public health benefits, such as sustainability initiatives and the integration of advanced medical technologies. Collaborative partnerships with technology companies, educational institutions, and other healthcare providers can also play a crucial role. Such collaborations can lead to shared costs and resources, making the adoption of new technologies and systems more feasible and economical. Additionally, specialized healthcare loans offered by banks and financial institutions present another viable funding avenue. These loans are typically structured to account for the long-term benefits and cost savings derived from healthcare projects, making them a practical option for funding substantial upgrades.

Beyond securing external funding, healthcare organizations must also craft compelling business cases to persuade stakeholders of the financial viability and long-term benefits of their proposed projects. This involves highlighting potential long-term cost savings, such as those achieved through energy-efficient systems that can significantly reduce utility expenses or advanced medical technologies that may shorten hospital stays and decrease readmission rates. Articulating how investments in healthcare infrastructure can improve patient outcomes is also crucial, as enhanced patient satisfaction scores are increasingly linked to reimbursement levels under value-based care models, thereby justifying the initial expenditure. Furthermore, investments aimed at risk mitigation—such as enhancements that reduce the incidence of infection outbreaks or structural failures—add a critical dimension to the business case, emphasizing the protective value of upfront expenditures against potential future liabilities.

Another pragmatic approach is to implement changes incrementally. Healthcare facilities may adopt a phased strategy, initiating upgrades in high-impact areas that directly enhance safety, compliance, and patient satisfaction. Such targeted improvements can quickly demonstrate their value, garnering internal and external support and justifying further investments. Investing in scalable solutions can also be advantageous, allowing for initial small-scale implementation that can be expanded based on available funding and demonstrated success of the project.

Finally, leveraging technology and innovation offers a path to cost-effective enhancements. For instance, software upgrades can significantly improve operational efficiency and patient care quality at a fraction of the cost required for extensive physical infrastructure renovations.

In conclusion, navigating financial constraints in healthcare infrastructure development necessitates a multifaceted strategy that includes seeking external funding, making a compelling business case, adopting an incremental implementation approach, and leveraging cost-effective technological innovations. By effectively demonstrating the long-term benefits and potential efficiencies of these investments, healthcare facilities can more successfully manage financial challenges and secure the support needed to advance their infrastructure and enhance overall healthcare delivery.

Table 9.1 presents a detailed examination of diverse funding sources for healthcare projects, including grants, partnerships, and loans. It outlines the characteristics of each funding type and provides concrete examples of healthcare projects that have been successfully implemented through these financial avenues. The table is strategically positioned to follow the discussion on external funding, offering a succinct and informative overview that enhances understanding among healthcare professionals and academics of how different funding types propel advancements in medical facilities and technology.

### *9.1.2 Regulatory Hurdles*

Navigating the intricate maze of regulatory requirements poses a formidable barrier to the incorporation of cutting-edge designs and technologies within healthcare facilities. These regulatory frameworks, which encompass building codes, safety standards, and healthcare compliance, are fundamentally designed to safeguard patient safety and uphold the highest standards of care. However, the rigidity and often sluggish pace of regulatory adaptation can impede the seamless integration of innovative solutions that promise to enhance operational efficiency and patient outcomes.

Regulatory bodies set protocols that primarily focus on patient safety, ensuring that all healthcare facilities adhere to stringent quality standards. Despite the essential nature of these regulations, they can inadvertently stifle innovation by failing to keep pace with rapid technological advancements and novel architectural methodologies. This discord between existing regulations and modern innovations can place significant constraints on the ability of healthcare facilities to adopt and benefit from new technologies and improved designs.

The challenge of regulatory hurdles can be navigated through several strategic approaches. Engaging with regulatory bodies at the early stages of design and planning is crucial. Such proactive engagement facilitates a comprehensive understanding

**Table 9.1** Comprehensive analysis of funding mechanisms and their impact on healthcare infrastructure development

| Funding source | Description | Examples of enabled projects |
|---|---|---|
| Grants | Non-repayable funds are typically provided by governmental bodies, non-profit organizations, or private foundations to promote healthcare initiatives | 1. Grant from the Health Resources and Services Administration (HRSA) for developing rural health clinics<br>2. Funding from the Bill & Melinda Gates Foundation for malaria eradication programs in Sub-Saharan Africa |
| Partnerships | Agreements between healthcare entities and external organizations such as universities, technology firms, or other hospitals, facilitate resource sharing and co-development of technologies | 1. Collaboration between a major hospital and a biotech company to create precision medicine solutions<br>2. A partnership between a university medical center and a healthcare startup to develop wearable health monitoring devices |
| Loans | Financial borrowing is characterized by specific repayment terms, often used for large-scale infrastructure projects or technology upgrades that promise long-term returns | 1. A substantial loan secured by a hospital for the construction of a state-of-the-art cancer research center<br>2. Funding through a healthcare finance company for the overhaul of electronic health record systems across a network of clinics |

of the regulatory landscape, enabling healthcare facilities to anticipate compliance challenges and address them before they escalate. Moreover, early discussions with regulators allow for the negotiation of the integration of new designs and technologies, which might not strictly align with existing codes.

Another pivotal strategy is advocacy for regulatory flexibility. Healthcare facilities, often in collaboration with professional associations and other stakeholders, can champion the development of more adaptable regulatory frameworks. These revised frameworks should possess the flexibility to embrace new technologies and innovative designs without compromising safety and quality standards. Advocacy efforts can also focus on urging the revision of obsolete regulations that no longer fulfill their intended purposes or unnecessarily restrict innovation.

The implementation of pilot projects can serve as a practical approach to overcoming regulatory barriers. By demonstrating the safety and effectiveness of new technologies or design concepts through pilots, healthcare facilities can gather concrete evidence to support the case for regulatory adjustments. Similarly, referencing case studies from other regions or nations where similar innovations have been successfully adopted and regulated can bolster the argument for updating local standards.

Consultations with experts in healthcare regulation are also invaluable. These specialists can provide critical insights into aligning innovative projects with regulatory demands and recommend effective risk mitigation strategies associated with non-compliance. Legal and regulatory consultants play a pivotal role in preparing the necessary documentation and presentations for regulatory bodies, helping to articulate a persuasive case for the adoption of novel solutions.

Lastly, continuous education and training for healthcare administrators and facility managers are essential to remain informed about regulatory changes and prepare for potential modifications. Educating staff about the rationale behind regulatory updates and the advantages of new technologies can facilitate smoother implementation and enhance compliance.

In conclusion, while regulatory hurdles in healthcare design and technology present significant challenges, they are not insurmountable. Through early engagement with regulators, advocacy for flexible regulations, the strategic use of pilot projects, expert consultations, and an emphasis on continuous education, healthcare facilities can effectively navigate these obstacles. This multifaceted approach ensures compliance and cultivates an environment conducive to innovation, which is crucial for enhancing the quality of patient care and the efficiency of healthcare operations.

Equation 9.1 outlines the projected compliance cost formula, which is crucial for estimating the financial implications of regulatory changes when introducing new healthcare technologies. Compliance costs are a significant factor for healthcare institutions as they adapt to new regulations, ensuring that their technologies and practices meet updated standards. This equation assists in preparing for the financial impact and strategic planning necessary to implement these technologies effectively within the regulatory framework.

The projected compliance cost ($C$) is calculated based on multiple variables that account for the nature and scope of the regulatory changes, the complexity of the technology being introduced, and the specific requirements that must be met. These include the initial assessment cost ($A$), which covers the expenses involved in evaluating how the new regulations affect existing systems; the technology modification cost ($M$), representing the financial outlay required to modify or replace current technologies to meet new standards; the training cost ($T$), which is the investment in training staff to use and maintain the new technologies proficiently; and the ongoing compliance monitoring cost ($O$), which involves continuous oversight and regular audits to ensure sustained compliance. The cumulative compliance cost ensures that healthcare providers can effectively manage the transition without compromising on safety, effectiveness, or efficiency.

By incorporating these diverse costs, healthcare administrators can comprehensively understand the total financial investment needed to align with regulatory standards. This foresight aids in budget allocation and resource planning, ultimately contributing to a smoother transition to safer and more efficient healthcare practices. The equation is formulated as follows to provide a clear and practical approach to calculating projected compliance costs:

$$C = A + M + T + O. \tag{9.1}$$

This formulation is essential for healthcare decision-makers, providing them with a tool to forecast and manage the costs associated with regulatory compliance effectively. It ensures that healthcare institutions are not only compliant but also financially prepared for the introduction of innovative technologies, thus supporting the overarching goal of enhancing patient care through advanced medical practices.

### *9.1.3 Resistance to Change*

Resistance to change is an inherent challenge in well-established healthcare systems, particularly when it involves adopting new designs and technologies. This resistance originates from a variety of sources, such as skepticism about the benefits of new technologies, fear of the unknown, and concerns over potential disruptions during implementation. Several factors contribute to this resistance: comfort with the current routines that make new systems seem disruptive, fears that new technologies might fail to deliver promised outcomes or might introduce worse results than existing ones, financial anxieties concerning the cost of new systems and the specter of unforeseen expenses, and a lack of training that leaves staff feeling unprepared to effectively utilize new technologies or navigate newly designed spaces.

To effectively overcome this resistance, comprehensive strategies are necessary. First, it is crucial to implement extensive education and training programs. These programs should not only instruct staff on how to use new systems but also help them understand the benefits and reasons behind the changes. This approach helps mitigate fears and builds confidence in the new technologies. Continuous education efforts are also vital to keep all stakeholders regularly updated on the progress and developments related to new designs and technologies.

Developing an effective communication strategy is another critical step. This strategy should aim to keep all stakeholders well-informed throughout the technological adoption or infrastructural redesign process. Emphasizing transparency in communications builds trust and reduces resistance. It is important to use multiple communication channels, such as staff meetings, newsletters, and digital platforms, to ensure that information is accessible to everyone involved.

Engagement and involvement of staff, patients, and community members in the planning and implementation phases are also paramount. By involving these stakeholders, the planning process incorporates diverse perspectives and fosters a sense of ownership, which can lead to better outcomes and reduced resistance. Establishing mechanisms for feedback allows individuals to express their concerns and suggestions, which should be taken seriously and used to adjust plans as necessary.

Pilot projects and demonstrations provide practical examples of how new technologies or designs function in a real-world setting. Successful pilots can act as proof of concept, visibly demonstrating the benefits and helping to dispel skepticism. Additionally, sharing case studies and success stories from other healthcare facilities that have successfully integrated new technologies can be very persuasive in overcoming resistance.

Finally, it is essential to clearly articulate the long-term benefits of the new systems, such as cost savings, improved patient outcomes, and enhanced staff satisfaction. Providing solid data and evidence that support these benefits can make a compelling and rational case for the changes.

In conclusion, addressing resistance to change in healthcare settings is a complex but manageable challenge that requires a nuanced approach involving education, effective communication, stakeholder engagement, and clear demonstrations of benefits. By tackling the underlying reasons for resistance and actively involving stakeholders in the process of change, healthcare facilities can facilitate smoother transitions and ensure the successful integration of innovative designs and technologies. This strategic approach not only improves the immediate healthcare environment but also contributes to the long-term success and sustainability of healthcare operations.

Table 9.2 presents summarized results from a survey conducted among healthcare staff, comparing attitudes towards innovative practices before and after exposure to new technologies and systems. This table encapsulates shifts in perception and acceptance, illustrating the effective transformation in attitudes facilitated by strategic engagement and involvement. Positioned to complement discussions on resistance to change, it provides empirical evidence that strategic interventions can significantly reduce skepticism and enhance receptivity to innovation within healthcare environments.

**Table 9.2** Empirical analysis of healthcare staff attitudes toward innovation before and after exposure

| Respondent group | Attitude before innovation exposure | Attitude after innovation exposure | Change in attitude |
|---|---|---|---|
| Physicians | Generally skeptical, concerned about the impact on clinical routines | More positive, recognizing the potential for improved patient care and efficiency | Positive shift, increased acceptance |
| Nurses | Wary of increased workload and complexity | Appreciative of ease and efficiency gains after practical demonstrations | Positive shift decreased resistance |
| Administrative staff | Indifferent or unaware of potential impacts | Supportive after seeing improvements in workflow and data management | Significant positive shift |
| Technical staff | Excited but cautious about reliability and integration issues | Highly supportive after successful pilot tests and technical training | Positive shift, solidified support |

### *9.1.4 Technical Challenges*

Technical challenges represent a formidable barrier when integrating new technologies and complex systems into established healthcare facilities. These challenges encompass a range of issues from compatibility problems with existing systems to the intricacies of maintenance and the necessity for specialized skills among staff. Understanding these technical challenges is crucial for successful integration. New technologies may not always seamlessly integrate with existing systems, leading to interoperability issues that hinder data sharing and operational efficiency. Additionally, advanced technologies typically require regular and specialized maintenance to function effectively, presenting a significant challenge if the necessary skills or resources are scarce. Furthermore, the adoption of new technologies may demand that staff acquire or develop specialized skills, posing a substantial hurdle in environments where training resources are limited or inadequate.

To effectively overcome these technical challenges, several strategic measures must be implemented. Continuous training and training for all relevant staff—including facility managers, healthcare providers, and IT personnel—are essential. These programs should cover both the operation and maintenance of new technologies and be viewed as an ongoing process, with refresher courses and updates provided regularly to ensure that staff skills and knowledge keep pace with technological advancements.

The careful selection of technology is also crucial. It is advisable to choose technologies that are known for their compatibility with a broad range of existing systems. This may involve opting for platforms and tools that adhere to widely accepted standards or that include robust integration support. Technologies that offer scalable and modular capabilities allow for customization to meet the specific needs and constraints of the healthcare facility, making them particularly valuable.

Collaborating with reliable vendors is another key strategy. Establishing strong partnerships with technology vendors that provide not only high-quality products but also comprehensive support services is vital. Reliable vendors should offer training, technical support, and regular updates to their systems, and it is beneficial to select vendors with a proven track record of successful implementations in healthcare settings like their own, ensuring they understand the unique challenges and requirements of the industry.

A phased implementation approach can further facilitate the integration of new technologies. Introducing new technologies gradually, starting with pilot areas before extending implementation hospital-wide, allows for the identification and resolution of any technical issues on a smaller scale, thereby reducing the risk of widespread disruption. Feedback from the pilot phase should be used to adjust and provide additional training, ensuring that staff are fully prepared, and systems are optimized before full-scale deployment.

Lastly, developing a proactive IT support strategy is crucial. This strategy should include regular check-ups and maintenance schedules to preemptively address potential technical issues. Investing in a robust IT infrastructure that supports new technologies and allows for easy updates and maintenance is also imperative.

In conclusion, overcoming the technical challenges associated with integrating new technologies in healthcare facilities demands a comprehensive approach that includes ongoing training, careful technology selection, collaboration with reliable vendors, a phased implementation approach, and proactive IT support. By addressing these critical elements, healthcare facilities can enhance their technological capabilities efficiently and effectively, thereby improving patient care and operational efficiency.

Equation 9.2 delineates the return on investment (ROI) calculation for technological upgrades within healthcare settings, an essential measure for evaluating the economic value derived from investing in advanced medical technologies. This calculation is critical for healthcare organizations to determine the feasibility and value of upgrading their equipment or systems, ensuring that the benefits, in terms of enhanced efficiency and patient care quality, justify the costs involved.

The return on investment for technological upgrades is quantified by considering several key financial and operational parameters. These include the initial cost of the upgrade ($I$), which encompasses the purchase price and installation expenses; the maintenance cost ($M$), accounting for ongoing expenses to keep the technology operating optimally; the expected lifespan of the technology ($L$), which impacts the depreciation and the time over which the investment is amortized; and the efficiency gains ($E$), reflecting the increase in operational efficiency and the potential for cost savings or increased revenue due to the upgrade. These factors are crucial as they collectively determine the net financial impact of the investment over its useful life.

The ROI is then calculated by dividing the sum of the efficiency gains over the expected lifespan of the technology by the total cost of the upgrade (including both initial and maintenance costs), subtracted from 1, to represent the percentage return. This is expressed as follows:

$$\mathrm{ROI} = \left( \frac{E \times L - (L + M)}{I + M} \right) \times 100. \tag{9.2}$$

This equation serves as a pivotal tool for healthcare administrators and financial officers, providing a clear and concise metric to assess the cost-effectiveness of technological enhancements. By calculating ROI, healthcare facilities can make informed decisions about where to allocate their resources, prioritizing investments that offer substantial returns through improved service delivery and operational efficiency. This strategic financial planning is instrumental in maintaining a high standard of patient care while ensuring the sustainability and growth of the healthcare institution.

### 9.1.5 Cultural Barriers

Cultural barriers introduce distinct challenges when implementing advanced architectural designs, particularly in the global context where diverse cultural norms and expectations significantly shape the perception and utilization of healthcare facilities. These barriers can profoundly influence both the accessibility and acceptance of these facilities.

Understanding the nuances of cultural preferences and practices is crucial. In many societies, the design of healthcare facilities can directly impact their accessibility. For instance, certain cultures may require more private spaces due to gender norms, or they may prefer architectural styles that echo local aesthetics and historical significance. Additionally, the engagement level of local communities during the planning and design phases can make or break the success of a healthcare facility. Without sufficient community involvement, there is a risk that the final designs will not align with the community's needs or values, potentially leading to resistance or underutilization. Another significant aspect is ensuring that facilities are accessible and welcoming to all community segments. This includes accommodating linguistic diversity, religious practices, and various social norms within the facility's design and operational policies.

To effectively navigate these cultural barriers, several strategic approaches can be employed. Engaging with local communities through thorough consultations is imperative. These consultations should occur early in the planning process and continue throughout the lifecycle of the project, ensuring that community feedback is consistently incorporated. Involving community leaders and representatives can help tailor the project to meet cultural expectations and enhance local acceptance and support.

Incorporating culturally relevant design elements into the architecture of healthcare facilities is also essential. This may include the use of local materials, incorporation of local art, and adherence to building styles that resonate with the community's identity. Furthermore, designing flexible spaces that can adapt to various cultural practices—such as prayer rooms, gender-separated areas, or family consultation spaces—can significantly enhance the facility's cultural compatibility.

Training and sensitization for healthcare staff on cultural competence are critical. Staff should be well-versed in the cultural backgrounds of the communities they serve to ensure respectful and effective interaction. Additionally, healthcare facilities should develop multilingual resources and signage to address the linguistic diversity of the patient population they serve.

Building physical accessibility and inclusivity into the design of healthcare facilities is another crucial strategy. Facilities should be accessible to individuals of all abilities and ages and consider socio-economic factors that might affect access to healthcare. Designing services and spaces that are welcoming and non-intimidating to underserved or marginalized groups is vital for promoting equitable healthcare access.

**Table 9.3** Impact of culturally inclusive design elements on patient satisfaction metrics

| Facility | Patient satisfaction before | Patient satisfaction after | Notable changes in design elements |
|---|---|---|---|
| Hospital A | 68% satisfied | 85% satisfied | Introduction of multi-faith prayer rooms and signage in multiple languages |
| Clinic B | 73% satisfied | 92% satisfied | Design adjustments to include private consultation areas respecting gender norms |
| Health center C | 80% satisfied | 95% satisfied | Use of local art and materials reflecting the community's cultural heritage |
| Community clinic D | 65% satisfied | 89% satisfied | Establishment of family consultation spaces and culturally specific dietary accommodations in the cafeteria |

In conclusion, addressing cultural barriers in the design and operation of healthcare facilities requires a deliberate and inclusive approach. By integrating cultural preferences into architectural designs, actively engaging with local communities, and ensuring facility inclusiveness, healthcare facilities can effectively overcome these barriers. Such proactive measures not only enhance the acceptance and operational success of these facilities but also ensure they effectively meet the diverse needs of the populations they serve, thereby contributing to better health outcomes across communities. As the field of healthcare design evolves, these cultural considerations will remain integral to the strategic recommendations for integrating advanced concepts into practical and impactful healthcare solutions.

Table 9.3 provides a comparative analysis of patient satisfaction metrics before and after the integration of culturally inclusive design elements in healthcare facilities. This table illustrates the tangible benefits of such designs, underscoring their effectiveness in enhancing patient satisfaction by accommodating diverse cultural needs and preferences. Situated after discussions on cultural barriers, it serves as empirical evidence of the positive outcomes that arise from culturally sensitive architectural planning, reinforcing the argument for inclusive design practices in global healthcare settings.

## 9.2 Strategic Recommendations for Future Developments

To ensure that healthcare facilities are not only responsive to the current needs but also adaptable to future demands and innovations, strategic planning and forward-thinking approaches are essential. This section outlines key recommendations that can guide future developments in healthcare design, aiming to enhance functionality, patient care, and sustainability.

### 9.2.1 Embrace Technology Integration

As the healthcare landscape continues to evolve at a rapid pace, the integration of technology into facility design has transitioned from being merely beneficial to essential. The future of healthcare facilities hinges on their capability to support and leverage advancements in artificial intelligence (AI), the Internet of Things (IoT), telemedicine, and other digital innovations. This requires a foundational shift towards building robust technological infrastructures.

Embracing technology integration involves several key components. Firstly, the infrastructure for emerging technologies must be meticulously planned. Designing healthcare facilities that can support big data analytics is crucial, which necessitates robust data centers, high-speed network capabilities, and secure cybersecurity systems. These facilities must efficiently and safely handle large volumes of sensitive patient data. Additionally, the infrastructure should support IoT devices that monitor patient health in real-time, manage hospital resources effectively, and ensure the optimal functioning of critical medical equipment.

The incorporation of dedicated spaces for telemedicine is increasingly becoming a standard necessity as more patients and healthcare providers opt for remote consultations. These spaces require high-quality video conferencing equipment, effective soundproofing, and reliable internet connectivity to facilitate clear and uninterrupted communication between providers and patients.

Furthermore, with the rise in the use of mobile health applications, healthcare facilities must integrate these technologies into their daily operations. This integration can significantly enhance patient engagement, empower patients to take more control over their health, and streamline communication between patients and healthcare providers. The seamless integration of these technologies should aim to enhance rather than disrupt existing clinical workflows. This involves a strategic approach that includes comprehensive training for medical staff and ongoing support to ensure that the technologies are used effectively and integrated smoothly into daily operations.

Healthcare facilities must also consider how new technologies interact with existing electronic health records (EHR) systems and other clinical management tools. Proper integration can substantially enhance diagnostic and treatment capabilities, making a significant impact on the efficiency and effectiveness of healthcare delivery.

Lastly, the design of healthcare facilities should be future-proof, allowing for flexibility to accommodate upcoming technological advancements. This might involve creating modular spaces that can be easily reconfigured or updated as new technologies and treatments emerge. Planning for technological scalability is essential, as it ensures that facilities can adapt without requiring extensive overhauls, thus saving time and resources in the long term.

In conclusion, the design of future healthcare facilities must pivot significantly toward integrating advanced technologies. By preparing for the broad adoption of AI, IoT, telemedicine, and mobile health applications, healthcare facilities can ensure they remain at the forefront of delivering high-quality, efficient, and patient-centered

**Table 9.4** Comparative analysis of healthcare delivery metrics before and after telemedicine integration

| Metric | Before telemedicine | After telemedicine | Description of impact |
|---|---|---|---|
| Patient satisfaction | 70% satisfied | 88% satisfied | Improved access and convenience increased patient satisfaction scores |
| Cost savings | Minimal | Substantial | Reduction in travel and physical infrastructure costs led to significant savings |
| Reach | Localized | Extended nationally and internationally | Telemedicine enabled services to remote and underserved areas, greatly expanding reach |

care. This proactive approach not only enhances healthcare delivery but also positions facilities to rapidly adapt to future innovations, ensuring their long-term relevance and sustainability.

Table 9.4 provides a structured comparison of key healthcare delivery metrics including patient satisfaction, cost savings, and reach, before and after the implementation of telemedicine facilities. Positioned immediately after the discussion on dedicated telemedicine areas, this table quantifies the enhancements brought by telemedicine in terms of operational efficiency and patient care. It underscores the expanded accessibility and cost-effectiveness, serving as a compelling evidence base for the adoption of digital health strategies in modern healthcare settings.

### 9.2.2 *Focus on Flexibility and Modularity*

In the dynamic and rapidly evolving landscape of healthcare, the concepts of flexibility and modularity in facility design are paramount. These design principles are critical for accommodating changing healthcare needs and the seamless integration of emerging technologies. By emphasizing flexibility and modularity, healthcare facilities can significantly enhance their ability to adapt to new treatments, patient care models, and technological advancements, ensuring they remain equipped to respond to the evolving demands of healthcare delivery.

The advantages of embracing flexible and modular designs are manifold. Designing healthcare spaces with adaptable room layouts is essential; this can be achieved with movable walls and modular design elements that allow for the expansion, reduction, or repurpose of rooms according to current demands. Such adaptability proves invaluable in situations such as emergencies or during sudden spikes in patient volume, like those seen in epidemics.

Investing in modular furniture and equipment is another crucial aspect of this design philosophy. Such items can be easily moved, adjusted, or repurposed, aiding in customizing patient care environments to meet specific needs and facilitating the integration of new technologies and equipment without necessitating significant renovations. Furthermore, facilities should incorporate scalable infrastructure capable of adjusting or upgrading as new technological advancements emerge. This includes designing easily accessible utility conduits and ceiling systems that accommodate various medical technologies, alongside versatile floor plans that support a range of activities without requiring structural changes.

The modular design also minimizes disruptions when changes are necessary. By constructing healthcare spaces to be modular, facilities can implement modifications more quickly and at lower costs, with minimal impact on patient care and facility operations. This approach preserves the functionality of the facility, even during renovations or technology upgrades, thus maintaining continuous patient care and operational efficiency.

To implement flexible and modular designs effectively, collaboration with skilled design professionals is crucial. Engaging architects and designers who specialize in flexibility ensures that new or renovated healthcare facilities are genuinely adaptable. These professionals can leverage the latest trends and technologies in modular construction to provide innovative solutions that meet the evolving demands of healthcare. Additionally, incorporating input from stakeholders, including clinicians, nurses, and other healthcare staff during the design process is vital. Their insights ensure that the designed spaces meet practical needs and enhance workflow efficiency. Patient feedback is also invaluable in guiding design choices, ensuring that the spaces fulfill their needs and expectations.

Regular reviews and updates of space utilization and adaptability are also essential. Facilities should conduct ongoing assessments to determine if the designs continue to meet current needs and anticipate future requirements. This continual evaluation facilitates timely adjustments to the infrastructure, ensuring the facility remains responsive to changes.

In conclusion, focusing on flexibility and modularity in healthcare facility design transcends aesthetic or temporary convenience—it embodies a strategic approach that guarantees long-term relevance, efficiency, and responsiveness to the dynamic nature of healthcare. By adopting these principles, healthcare facilities can enhance their capability to deliver high-quality care cost-effectively, adapting seamlessly to the demands and opportunities the future of healthcare may present.

Equation 9.3 provides a mathematical framework for conducting a cost-benefit analysis of investing in modular designs in healthcare facilities. Modular furniture and equipment are increasingly favored in medical settings for their flexibility and adaptability, allowing spaces to be reconfigured with minimal disruption and cost. This adaptability is particularly valuable in healthcare, where the ability to respond to changing treatment protocols and technological advancements without extensive renovations can lead to significant cost savings over time.

The cost-benefit analysis for modular designs quantifies the savings realized from reduced renovation needs against the initial investment costs. The equation considers the initial cost of acquiring modular furniture and equipment ($I$), the average annual savings from avoided renovations ($S$), and the expected lifespan of the modular designs ($L$). These savings accrue annually as the need for frequent renovations is diminished, extending the useful life of the facilities without requiring costly and disruptive updates.

The total cost savings (TCS) generated by the investment in modular designs is calculated by multiplying the annual savings by the lifespan of the equipment and then subtracting the initial investment cost. This provides a net value that reflects the financial advantage of choosing modular options:

$$\mathrm{TCS} = (S \times L) - I. \tag{9.3}$$

This equation is instrumental for healthcare administrators and facility managers when assessing the long-term financial implications of their furnishing choices. By understanding the cost–benefit dynamics of modular designs, healthcare facilities can make strategic decisions that reduce long-term capital expenditures and enhance operational flexibility. These decisions support a healthcare environment that can adapt quickly to evolving medical practices and patient needs, thereby improving overall healthcare delivery and patient outcomes.

### 9.2.3 Prioritize Patient-Centered Design

Patient-centered design is a cornerstone of contemporary healthcare architecture, emphasizing the creation of spaces that foster healing by addressing the holistic needs of patients. This design philosophy prioritizes environments that support physical, psychological, and emotional well-being, thereby improving patient outcomes and enhancing the overall healthcare experience.

Key aspects of patient-centered design include the integration of therapeutic elements that contribute to healing. One crucial element is the strategic use of natural light, which has been shown to boost mood and accelerate recovery. Incorporating large windows and skylights can bring healing natural light into patient areas and common spaces, significantly enhancing the ambiance and therapeutic potential of these environments.

Another fundamental component is noise reduction. Effective noise control is essential as excessive noise can increase stress and impede recovery. Utilizing sound-absorbing materials in walls, ceilings, and floors can drastically reduce noise levels. Additionally, careful planning of the layout to minimize noise transmission from high-traffic areas to patient rooms is vital for maintaining a tranquil environment conducive to healing.

Balancing privacy with communal spaces within healthcare facilities is also crucial. Private rooms enhance patient dignity and confidentiality, key elements of psychological comfort. Simultaneously, communal areas provide necessary spaces for social interaction, which is particularly beneficial for mental health, especially among long-term patients. Furthermore, the integration of green spaces, such as gardens, courtyards, and rooftop gardens, can significantly enhance the healing environment. These areas not only improve aesthetics but also offer psychological benefits by reducing stress and promoting relaxation.

Patient-centered design also necessitates the seamless integration of technology to enhance patient care without being obtrusive. This includes incorporating telemedicine capabilities within patient rooms, ensuring easy access to charging stations for personal devices, and integrating advanced medical equipment that blends smoothly into the care environment. Additionally, respecting the cultural values and norms of the patient population is imperative. Design elements that accommodate dietary preferences, language needs, and religious practices can make patients feel valued and understood, thereby improving their overall satisfaction and comfort.

Implementing patient-centered design requires active engagement with patients and their families. Involving these primary stakeholders in the design process can yield invaluable insights into what makes a space truly healing and comfortable. Furthermore, interdisciplinary collaboration between architects, designers, healthcare professionals, and even art therapists can foster innovative solutions that enhance patient well-being. Each participant brings a unique perspective, enriching the design process with diverse ideas and knowledge.

Continuous evaluation through post-occupancy evaluations is crucial to assess how effectively the spaces serve patients and staff. Regular feedback and ongoing assessments enable facilities to continually refine and enhance their environments to better meet the needs of their users.

In conclusion, patient-centered design transcends mere aesthetics; it is a vital strategy that directly impacts health outcomes and the overall satisfaction of healthcare facility users. By prioritizing design elements that promote healing and comfort, healthcare facilities can offer care that genuinely supports the well-being of their patients, ensuring both immediate benefits and long-term success.

Table 9.5 showcases the differential impact of patient-centered design features on recovery times and patient satisfaction scores. Located after the comprehensive discussion on patient-centered design elements, this table quantitatively demonstrates the benefits of environments crafted around patient needs. It compares environments equipped with design features that support physical, psychological, and emotional well-being to those without such features, underlining the effectiveness of patient-centered design in enhancing health outcomes and patient experience.

### *9.2.4 Enhance Disaster Preparedness*

Enhancing disaster preparedness in the design of healthcare facilities is crucial for ensuring that hospitals and medical centers can respond effectively to a range of emergencies, from pandemics to natural calamities. This proactive approach not only protects the facility's infrastructure but also guarantees that patient care continues uninterrupted under various adverse conditions.

**Table 9.5** Impact of patient-centered design on recovery times and satisfaction

| Design feature | Average recovery time (with feature) (days) | Average recovery time (without feature) (days) | Patient satisfaction score (with feature) (%) | Patient satisfaction score (without feature) (%) |
|---|---|---|---|---|
| Natural light | 4.5 | 5.8 | 92 | 75 |
| Noise reduction | 4.7 | 6.1 | 89 | 68 |
| Access to nature | 4.3 | 6.5 | 95 | 70 |
| Privacy and communal spaces | 4.9 | 6.2 | 91 | 72 |

Key strategies for enhancing disaster preparedness begin with the design of flexible and convertible spaces within healthcare facilities. These spaces can quickly transform to meet the demands of emergencies; for instance, conference rooms, outpatient areas, and even parking lots can be adapted to serve as treatment areas or triage centers during crises. Implementing modular units that can be easily installed or removed as needed provides additional capacity and versatility during emergencies.

Building robustness is another critical strategy. Healthcare facilities should be constructed to withstand extreme weather events such as hurricanes, earthquakes, and floods, utilizing materials and construction techniques that are suited to the local environmental risks. Critical areas of the hospital, such as emergency departments and intensive care units, should be designed to be particularly resilient, ensuring they remain operational regardless of external conditions.

Redundancy in critical systems is essential for disaster preparedness. Facilities should install redundant power supplies, including generators and solar panels, to ensure continuous operation during power outages. These systems must be regularly tested and maintained. Additionally, redundant water supply and telecommunications systems are vital for maintaining essential services during a disaster.

Emergency and evacuation protocols are fundamental to disaster preparedness. Facilities should design clear and accessible evacuation routes that accommodate all patients, including those with mobility issues, ensuring these paths are well-marked and free from obstructions. Regular training in emergency procedures and conducting drills are necessary to ensure that staff are prepared and know their roles during a disaster.

Healthcare continuity planning is also crucial. Facilities should develop and regularly update a comprehensive disaster response plan that includes strategies for scaling operations up or down based on the severity of the event. Establishing partnerships with local emergency services, other hospitals, and community organizations can help coordinate responses and share resources during widespread emergencies.

Considering pandemics like COVID-19, incorporating design features that help control infections is imperative. This includes improved ventilation systems, surfaces that are easily sanitized, and isolation rooms that can be activated during outbreaks. Additionally, leveraging technology to enhance disaster response capabilities, such

**Table 9.6** Impact of facility redesign on emergency response times and patient outcomes

| Emergency type | Response time before redesign (min) | Response time after redesign (min) | Patient outcomes before | Patient outcomes after |
|---|---|---|---|---|
| Natural disaster | 15 | 9 | 60% positive outcomes | 85% positive outcomes |
| Fire | 20 | 12 | 50% positive outcomes | 78% positive outcomes |
| Mass casualty | 30 | 18 | 45% positive outcomes | 80% positive outcomes |
| Pandemic response | 25 | 15 | 55% positive outcomes | 90% positive outcomes |

as employing telemedicine to provide care remotely when access to the facility is compromised, is increasingly important.

Securing critical data with robust cybersecurity measures and backing up data offsite ensures that it is preserved and accessible during and after a disaster. This is crucial for maintaining continuity of care and protecting sensitive patient information.

In conclusion, the imperative for healthcare facilities to be well-prepared for a variety of disasters cannot be overstated. By incorporating flexible design, robust construction, redundant systems, comprehensive emergency protocols, and advanced technology, healthcare facilities can ensure they remain safe, functional, and responsive in the face of disasters. This comprehensive preparedness not only safeguards the infrastructure and staff but most importantly, ensures that patient care continues without interruption, ultimately saving lives during critical times.

Table 9.6 illustrates the improvements in emergency response times and subsequent patient outcomes following a strategic redesign focused on enhanced disaster preparedness. Placed after the detailed implementation strategies for disaster preparedness, this table offers a clear before-and-after comparison, quantifying the efficacy of infrastructure and operational enhancements in reducing response times during emergencies. It serves as compelling evidence of the critical importance of proactive facility design in improving overall emergency responsiveness and patient care during disasters.

### *9.2.5 Invest in Sustainable Practices*

Sustainability in healthcare facility design and operation is crucial not only for environmental conservation but also for promoting healthier environments for patients, staff, and the broader community. Investing in sustainable practices enhances a facility's efficiency and can significantly reduce long-term operational costs, making it a strategic imperative for modern healthcare environments.

Key strategies for investing in sustainable practices begin with the use of energy-efficient materials and systems. Utilizing high-performance building materials improves insulation and enhances energy efficiency, which reduces the need for extensive heating and cooling—major sources of energy expenditure in healthcare facilities. Additionally, installing energy-efficient lighting systems, such as LED lighting coupled with smart sensors that adjust lighting based on occupancy and natural light levels, can substantially decrease energy usage.

The integration of renewable energy sources is another vital strategy. Facilities can significantly reduce their reliance on fossil fuels and lower greenhouse gas emissions by incorporating solar panels, wind turbines, or geothermal systems into their energy infrastructure. Innovative options like building-integrated photovoltaics (BIPV) not only provide power but also add to the aesthetic appeal of the facility.

Sustainable water management practices are essential for reducing water usage. This includes installing water-saving fixtures and appliances, such as low-flow faucets, toilets, and showerheads. Additionally, implementing rainwater harvesting systems to collect and reuse water for non-potable purposes like gardening, flushing toilets, and cooling systems can lead to significant water conservation.

Green building design optimizes the natural climate conditions to minimize energy usage. Strategic placement of windows and skylights maximizes natural light and harnesses heat during colder months while reducing heat ingress during warmer months. Incorporating green roofs and living walls not only improves insulation and reduces energy costs but also enhances air quality and adds natural beauty to the facility.

The incorporation of green spaces within healthcare facilities is also a strategic sustainable practice. Creating therapeutic gardens and accessible green spaces benefits the environment and promotes patient recovery and staff well-being. Exposure to nature has been proven to reduce stress and improve health outcomes, making these spaces vital to the holistic healing environment.

Promoting active transport is another facet of sustainable healthcare design. Providing facilities such as bike racks, showers, and locker rooms encourages staff and visitors to bike or walk to the hospital, reducing carbon emissions and promoting healthier lifestyles. Enhancing the connectivity of public transport and offering incentives for using public or shared transportation modes can further decrease the facility's carbon footprint.

Finally, holistic environmental impact reduction involves comprehensive waste management practices, including recycling and composting programs. Proper segregation of waste facilitates recycling processes and reduces the volume of waste sent to landfills. Using environmentally friendly and non-toxic cleaning and maintenance products decreases the chemical load within the facility, contributing to a healthier indoor environment.

In conclusion, investing in sustainable practices is not just an environmental responsibility but a strategic approach that positions healthcare facilities as leaders in environmental stewardship and public health promotion. By embracing sustainability, healthcare facilities can create healthier, more pleasant spaces for patients and staff, demonstrating a commitment to the well-being of both people and the planet, and setting a positive example for the community.

Equation 9.4 is a crucial mathematical model used to quantify the environmental impact reduction attributable to the adoption of sustainable practices in healthcare facilities. This equation helps measure the effectiveness of initiatives aimed at reducing carbon emissions and water usage, critical components of environmental sustainability in the medical sector. Implementing these practices not only supports global environmental goals but also enhances the reputation of healthcare facilities as responsible community stakeholders.

The equation assesses the reduction in environmental impacts by calculating the decrease in carbon footprint and water usage before and after the implementation of sustainable practices. It includes the initial carbon footprint ($C_i$) and water usage ($W_i$) before the implementation of sustainability measures, and the reduced carbon footprint ($C_r$) and water usage ($W_r$) after these measures have been put into place. The reductions are expressed as differences between the initial and reduced values, reflecting the direct environmental benefits of the sustainability initiatives.

To provide a comprehensive view, the environmental impact reduction (EIR) is calculated by summing the reductions in both carbon footprint and water usage, weighted by their respective environmental significance factors ($f_c$ for carbon and $f_w$ for water). These factors reflect the relative importance or impact of carbon and water savings in the specific environmental context of the healthcare facility, allowing for a tailored assessment of sustainability performance:

$$\text{EIR} = (C_i - C_r) \times f_c + (W_i - W_r) \times f_w. \tag{9.4}$$

This equation is indispensable for environmental analysts and sustainability officers in healthcare settings, enabling them to quantify and report on the effectiveness of their green initiatives. By calculating the environmental impact reduction, healthcare facilities can verify the success of their sustainability strategies, adjust practices as necessary, and communicate these achievements in a quantifiable manner to stakeholders, including patients, regulatory bodies, and the community. This transparency in environmental management not only demonstrates a commitment to sustainability but also aligns with broader public health goals by fostering a healthier, more sustainable environment.

### *9.2.6 Foster Community Engagement*

Fostering community engagement is crucial for healthcare facilities that aspire to serve and integrate effectively with the communities they inhabit. By involving residents and stakeholders in the planning, design, and ongoing development of healthcare services, facilities can enhance their relevance and efficacy, ensuring that they meet the specific needs of their community.

Key strategies for fostering community engagement include designing healthcare facilities to be open and inviting all community members. This involves creating easily accessible public areas such as cafes, gardens, and educational spaces that encourage the public to view the facility not just as a clinical space but as a community hub. It is essential to ensure that these spaces are inclusive, accommodating diverse populations with provisions for different languages, cultural practices, and physical abilities.

Implementing health education and prevention programs that address local health concerns is another vital strategy. By providing ongoing programs, healthcare facilities can take a proactive role in improving community health outcomes. Tailoring these programs to the specific demographics and needs of the community—for instance, focusing on chronic disease prevention, maternal health, or nutrition—can make them more relevant and impactful.

Community involvement in the planning and development processes is also crucial. Engaging community members through forums, surveys, and focus groups ensures that the services provided align with the actual needs and priorities of the community. Forming a community advisory board that includes a diverse cross-section of residents can provide ongoing feedback and suggestions for improvement, ensuring that the facility remains responsive to community needs.

Building partnerships with local schools, businesses, non-profits, and faith-based organizations is essential to extend health initiatives beyond the healthcare facility itself. These partnerships can amplify the reach and impact of health programs and involve the facility in community events that promote health and wellness, such as health fairs, screening days, and educational seminars.

Maintaining transparency in facility operations and developments is crucial for building trust with the community. Regular updates shared through newsletters, social media, or public meetings can keep the community informed and engaged. Establishing open lines of communication where community members can express concerns, ask questions, and provide feedback about the healthcare facility and its services is also vital.

In conclusion, by fostering robust community engagement, healthcare facilities not only enhance their capacity to meet local health needs but also build trust and loyalty among community members. This approach transforms healthcare facilities into community assets that support not only medical needs but also the social and educational aspects of health. Implementing these strategic recommendations requires a commitment to continuous improvement and community partnership, ensuring that healthcare facilities can adapt and respond to the evolving challenges and opportunities they face.

Table 9.7 provides a quantitative assessment of the improvements in community health metrics following the implementation of engagement programs focused on education and preventive care. Situated at the end of the section on community engagement, this table highlights the direct correlation between active community

**Table 9.7** Efficacy of community engagement programs in enhancing health outcomes

| Program type | Health metric before program | Health metric after program | Percentage improvement | Description of program impact |
|---|---|---|---|---|
| Diabetes education | 12% of the community at risk | 7% of the community at risk | 42% improvement | Reduced diabetes risk through regular educational workshops and personalized nutrition counseling |
| Hypertension screening | 20% identified with the condition | 15% identified with the condition | 25% improvement | Enhanced early detection and management through mobile screening units and follow-up care programs |
| Maternal health services | 75% prenatal care coverage | 90% prenatal care coverage | 20% improvement | Improved maternal health outcomes and higher prenatal care engagement through community-based health services |
| Nutrition and fitness campaigns | 30% of the community obese | 22% of the community obese | 27% improvement | Lower obesity rates are achieved by integrating fitness classes and nutritional guidance into community centers |

involvement in healthcare initiatives and tangible public health benefits. It illustrates how diverse engagement strategies, from educational workshops to prevention campaigns, contribute significantly to better health outcomes across community demographics.

Next, we will critically assess current design trends in healthcare to identify areas where improvements are necessary to meet future challenges more effectively, ensuring that healthcare environments continue to evolve in ways that best serve both patients and their communities.

## 9.3 Limitations of Current Design Trends

While contemporary healthcare design has made significant strides in addressing the needs of patients and healthcare providers, certain limitations persist that can hinder the effectiveness and adaptability of healthcare facilities. Acknowledging these limitations is crucial for guiding future improvements and ensuring that design trends evolve to meet the complex demands of modern healthcare.

### *9.3.1 Overemphasis on Aesthetics*

In the realm of healthcare facility design, the prioritization of aesthetics often aims to create environments that are welcoming and soothing for both patients and staff. However, this emphasis on visual appeal must be carefully balanced with the essential requirements of functionality, efficiency, and adaptability, which are crucial for the smooth operation of healthcare services. The challenge arises when aesthetic considerations begin to overshadow these critical operational aspects, potentially compromising the primary mission of effective healthcare delivery.

Facilities designed with a primary focus on aesthetics can suffer from compromised functionality. For example, such designs might not allocate sufficient space for essential medical equipment or fail to accommodate the complex workflows typical in healthcare settings, leading to operational inefficiencies that could negatively impact patient care. Furthermore, design choices that favor visual appeal may utilize materials and layouts that are challenging to maintain and clean, thereby posing significant hygiene and maintenance challenges. These issues highlight the need for a design philosophy that does not sacrifice practical utility for aesthetic allure.

Another significant challenge posed by an overemphasis on aesthetics is inadequate traffic flow within healthcare facilities. Designs driven by aesthetic considerations can lead to floor plans that obstruct efficient movement, creating congestion and potentially increasing the risk of errors and accident issues that are particularly critical in emergencies where every second counts. Moreover, visually appealing but poorly configured corridors and public spaces can hinder the smooth movement of patients, staff, and critical emergency equipment, further complicating the operational effectiveness of the facility.

The issue of insufficient flexibility in design is another concern. Healthcare facilities must be adaptable, and able to evolve with advancements in medical treatments, technologies, and changing patient needs. A design approach overly focused on aesthetics can lack the necessary flexibility to integrate new equipment or reconfigure spaces to meet the evolving demands of healthcare practices. Static and inflexible design elements can make necessary renovations or upgrades disruptive and costly, which is far from ideal in dynamic healthcare environments.

To address these challenges, adopting an integrated design approach is crucial. This approach should consider aesthetics alongside functionality right from the design's inception. Involving healthcare professionals in the planning process can ensure that the facility's layout and features support both clinical operations and patient care effectively. Employing design simulations and mock-ups can also be beneficial, as these methods allow for the practical implications of design choices to be evaluated and adjusted based on feedback from healthcare staff and patient care simulations.

Incorporating adaptable and modular design elements that can be easily reconfigured as needs change is another strategic approach. This includes the use of movable walls, adjustable lighting, and modular furniture, which support multiple uses and facilitate updates with minimal disruption. Selecting materials and finishes that are both aesthetically pleasing and functional—such as durable, easy-to-clean surfaces—can significantly enhance the facility's maintenance and hygiene.

Finally, while aesthetic standards are maintained, it is paramount to prioritize elements that directly enhance the patient and staff experience. This encompasses considerations such as natural light, soundproofing, and privacy, which significantly contribute to comfort and well-being. Designing public and communal areas to be welcoming yet functional ensures these spaces meet the social and emotional needs of patients and visitors without compromising their utility for healthcare delivery.

In conclusion, while aesthetics plays a vital role in creating healthcare environments that promote healing and comfort, they must not detract from the functionality and adaptability essential to a facility's primary mission of care delivery. By adopting a balanced and integrated approach to design, healthcare facilities can ensure that their environments are not only visually appealing but also highly functional and responsive to the dynamic nature of healthcare.

Table 9.8 categorizes common design features in healthcare settings according to their aesthetic appeal and functional impact, providing examples of how these features have either compromised or enhanced healthcare delivery. Positioned after the discussion of the challenges posed by an overemphasis on aesthetics, this table serves as a practical reference, illustrating concrete instances where aesthetic decisions have affected the functionality of healthcare environments. It aims to support a balanced approach in healthcare design by emphasizing the integration of both aesthetic and functional elements.

### *9.3.2 Inadequate Integration of Technology*

The rapid evolution of medical technology presents significant challenges for healthcare facility design, especially when these advances are not adequately anticipated in the architectural layout. If not effectively integrated, these technologies can lead to operational inefficiencies and diminish the potential benefits for patient care, highlighting the necessity for thoughtful integration strategies during the design phase.

One of the primary challenges in integrating technology into healthcare facilities is the difficulty associated with retrofitting. Facilities initially designed without consideration for future technological advancements often encounter significant hurdles when updating or integrating new systems. Retrofitting modern technologies into existing structures can be not only costly but also disruptive and technically complex. The absence of proper conduits for wiring and designated spaces for equipment means

**Table 9.8** Comparative analysis of aesthetic and functional design features in healthcare facilities

| Design feature | Aesthetic appeal | Functional impact | Example of impact on healthcare delivery |
|---|---|---|---|
| Large water features | High—enhances visual appeal and patient relaxation | Low—may interfere with traffic flow and space utility | Installed in a hospital atrium, the feature obstructed pathways, causing navigational difficulties for patients and emergency services |
| Open-plan layouts | High—promotes a modern and spacious environment | Mixed—can improve or impede operational efficiency depending on layout specifics | Used in a clinic, it facilitated better natural light distribution but increased noise levels, disrupting patient consultations |
| High-end finishes | High—contributes to a luxurious atmosphere | Low—often requires high maintenance and may not withstand heavy usage | Chosen for patient rooms, such materials led to higher maintenance costs and were less durable under frequent cleaning protocols |
| Modular walls | Medium—can be aesthetically customized | High—offers flexibility for space reconfiguration according to needs | Implemented in an outpatient center, allowed for quick adaptation to changing treatment requirements, enhancing functionality |

that adding new technology could necessitate extensive construction. Such disruptions can significantly impact patient care and inflate costs, emphasizing the need for foresight in initial designs to accommodate technological evolutions.

Moreover, inadequate integration of technology can severely disrupt healthcare delivery. For instance, if digital health records are not seamlessly accessible due to insufficient IT infrastructure, it can slow down operations, extend wait times, and negatively affect the overall patient experience. Additionally, essential medical devices that are not well integrated into patient care areas can hinder medical procedures, affecting the quality of care and patient outcomes. This underlines the critical role that technology plays in modern healthcare and the adverse impacts of its poor integration.

Furthermore, failing to effectively incorporate advanced technologies can result in missed opportunities for enhancing patient care. Technology has the potential to significantly improve patient outcomes, increase operational efficiency, and reduce errors. Advanced tools like real-time health monitoring, telemedicine, and robotic surgery require sophisticated integration into healthcare settings. Facilities that are ill-prepared for such integration may not fully leverage the capabilities of these advanced technologies, thus missing out on potential improvements in patient care and facility operations.

To effectively address these challenges, it is imperative to anticipate future technological needs during the design phase. This involves considering current and emerging technologies and planning for their potential impacts on healthcare delivery. Design strategies might include creating modular spaces that can be easily adapted, ensuring sufficient electrical and data capacity, and designing flexible room layouts that can accommodate changes. Engaging with technology experts to predict future trends and requirements ensures that designs are forward-thinking and inclusive of upcoming innovations.

Incorporating a flexible infrastructure is also crucial. Designing healthcare facilities with adaptable infrastructures that can evolve with technological advances is essential. This includes installing easily accessible conduits for wiring, creating adaptable spaces for new equipment, and establishing robust IT networks that can manage increased data loads. Utilizing standard interfaces and open platforms can enhance compatibility with various technologies, facilitating easier updates and integration.

Ongoing collaboration with technologists is another strategic approach. By fostering a collaborative environment where architects, designers, and technologists work together throughout the design and construction process, it ensures that technological needs are considered at every stage and integrated seamlessly into the facility. Regular updates to design and infrastructure plans based on technological advancements and feedback from clinical staff can help address operational needs and challenges effectively.

Finally, implementing pilot tests and feedback mechanisms before finalizing the design can identify potential issues and allow for necessary adjustments. Establishing continuous feedback mechanisms with healthcare providers who use these spaces helps in perpetually improving technology integration and resolving issues that arise post-construction.

In conclusion, effective integration of technology into healthcare facilities is crucial for maximizing the benefits of modern medical advancements. By anticipating future technological needs, incorporating flexible infrastructures, and fostering ongoing collaboration with technologists, healthcare facilities can ensure that they remain at the forefront of delivering high-quality, efficient, and technologically advanced patient care. This proactive approach not only enhances patient outcomes but also positions the facility to swiftly and effectively adapt to future innovations, ensuring sustained operational excellence and improved healthcare delivery.

Equation 9.5 addresses the economic analysis between retrofitting technology in existing healthcare facilities versus integrating such technologies during the initial design phase. This equation is fundamental for healthcare facility planners and financial managers to evaluate the cost-effectiveness of proactive versus reactive approaches to technology integration. By quantifying these costs, healthcare institutions can make informed decisions that optimize financial resources and operational efficiency.

The equation compares the total cost of retrofitting technology ($C_r$) with the cost of initial integration ($C_i$). The cost of retrofitting includes not only the price of the technology itself but also additional expenses such as structural modifications, extended downtime, and potential disruptions to healthcare services. Conversely, the cost of initial integration encompasses the technology cost within the original construction budget, usually resulting in lower overall expenditure due to planned integration and fewer operational disruptions.

To provide a clear financial comparison, the equation calculates the cost difference and relates it to the cost of initial integration to express it as a percentage, highlighting the potential savings from proactive technology planning:

$$\text{Cost Difference Percentage} = \left(\frac{C_r - C_i}{C_i}\right) \times 100. \tag{9.5}$$

This formulation helps healthcare administrators understand the financial benefits of integrating advanced technologies during the design phase rather than retrofitting them later. By presenting the cost difference as a percentage, the equation clearly illustrates the potential cost savings and supports strategic financial planning. This proactive approach not only saves money but also enhances the adaptability and future readiness of healthcare facilities, ensuring that they remain at the forefront of medical technology without incurring unnecessary costs or disruptions. This strategic foresight is crucial in maintaining high standards of patient care and operational efficiency in a rapidly evolving healthcare landscape.

### *9.3.3 Insufficient Disaster Preparedness*

The COVID-19 pandemic exposed significant vulnerabilities in the design and preparedness of healthcare facilities worldwide, highlighting the urgent need for improved disaster readiness to handle large-scale emergencies. This reality has led to a reevaluation of architectural practices in healthcare, with a newfound emphasis on incorporating robust disaster preparedness into facility design.

A primary challenge highlighted by the pandemic was the limited capacity of many healthcare facilities to handle surge events. During the peak periods of the pandemic, numerous facilities were overwhelmed due to their inability to scale space quickly to accommodate a sudden influx of patients. This lack of capacity severely hindered effective patient flow and the maintenance of standard care levels. Additionally, the insufficient number of isolation areas further complicated response efforts, as there was inadequate infrastructure to segregate infectious patients effectively, thereby increasing the risk of cross-contamination and further spreading the disease.

Another critical area of concern is inadequate infection control in traditional healthcare designs, which were often ill-equipped to handle the control of infectious diseases on a large scale. Many existing facilities lacked advanced air filtration systems and were designed with materials and patient flow considerations that did not minimize the spread of pathogens adequately. This was particularly evident with airborne pathogens, such as the novel coronavirus, where many facilities found themselves without the specialized ventilation systems necessary to control the spread effectively.

In response to these challenges, it is imperative to design healthcare facilities with adaptable and scalable spaces that can be swiftly reconfigured to serve various purposes, such as expanding patient capacity or creating isolation zones. These designs might include multipurpose rooms equipped with movable walls and convertible furnishings and consider overflow areas outside traditional hospital settings, such as modular units or repurposed buildings that can be rapidly deployed during emergencies.

Furthermore, implementing advanced infection control designs is crucial. This involves integrating state-of-the-art ventilation and filtration systems specifically designed to manage airborne diseases more effectively. Facilities should also be capable of modifying airflow patterns and maintaining negative pressure in designated areas to prevent the spread of infections. Additionally, the choice of surfaces and materials should prioritize ease of cleaning and resistance to pathogens, with innovations such as antimicrobial surfaces and hands-free technologies to further reduce disease transmission risks.

Equally important is the establishment of designated emergency operation centers (EOCs) within the facility, equipped with the necessary technology and resources to coordinate response efforts effectively during a disaster. These EOCs should be designed to function autonomously, with independent power and communication systems, to ensure operability in the event of broader infrastructure failures.

Developing robust emergency protocols is also essential. These protocols should include detailed plans for a variety of disaster scenarios—pandemics, natural disasters, and other emergencies—with regular training and drills conducted to keep staff prepared and ensure the effectiveness of these protocols. Moreover, it is crucial to address the psychological impacts of disasters on both staff and patients by providing support systems and resources to manage stress and trauma effectively.

Lastly, healthcare facilities should be designed with community integration in mind, enabling a coordinated response with local health providers, emergency services, and government agencies. This collaborative approach ensures a unified strategy for disaster response and resource sharing. Engaging with the community to educate and prepare the public for emergency responses and recovery processes is also vital.

In conclusion, enhancing disaster preparedness in healthcare facilities is critical to ensuring they can respond effectively to emergencies and pandemics. By incorporating flexible designs, advanced infection control measures, robust emergency operation capabilities, and comprehensive community collaboration, facilities can better protect patients, staff, and the broader community during crises. This strategic approach not only addresses the immediate needs during disasters but also strengthens the overall resilience of healthcare systems against future challenges.

Table 9.9 provides a detailed checklist of essential design and operational features aimed at enhancing disaster preparedness in healthcare facilities. Positioned after the section on enhancing disaster preparedness, this checklist serves as a practical tool for healthcare administrators and facility managers to assess and improve their current infrastructure. It encompasses a broad range of criteria from structural adaptations to procedural enhancements, ensuring facilities are equipped to handle emergencies efficiently and effectively.

**Table 9.9** Comprehensive disaster readiness checklist for healthcare facilities

| Category | Checklist item | Description of feature or action |
|---|---|---|
| Structural adaptability | Movable walls and convertible spaces | Facilities should incorporate flexible designs that can quickly transform to accommodate surge capacity during emergencies |
| | Emergency operation centers (EOCs) | Designate and equip specific areas to serve as command centers during a crisis, with independent power and communication systems |
| Infection control | Advanced ventilation systems | Install systems capable of altering airflow patterns and maintaining negative pressure in isolation zones |
| | Antimicrobial surfaces | Use materials that are easy to clean and reduce the transmission of pathogens |
| Emergency protocols | Comprehensive emergency action plans | Develop and regularly update plans for various types of emergencies, including drills and staff training |
| | Psychological support mechanisms | Establish protocols to support the mental health of staff and patients during and after disasters |
| Resource management | Stockpile essential supplies | Maintain a reserve of critical medical supplies and personal protective equipment to last through extended emergency periods |
| | Backup systems for power and water | Ensure the availability of independent backup systems for power and water to maintain operational continuity |
| Community integration | Collaboration with local emergency services | Form partnerships with local health agencies, emergency responders, and community organizations for coordinated disaster response |
| | Public education and engagement | Engage with the community to educate and prepare the public for emergency responses and recovery processes |

### *9.3.4 Neglect of Staff Workspaces*

The emphasis on patient-centric designs in healthcare facilities is undeniably important, yet it should not lead to the neglect of well-designed staff workspaces. Providing optimal environments for healthcare staff is essential as inadequately considered staff areas can contribute to inefficiencies, elevated stress levels, and, ultimately, adversely affect patient care.

One of the primary challenges of neglecting staff workspace is the increased risk of staff stress and burnout. Inadequately designed staff areas can lead to physical discomfort and mental fatigue, which are significant contributors to staff burnout. This not only deteriorates morale but can also result in higher turnover rates and a decline in the quality of patient care. For instance, if the staff areas are cramped, poorly lit, or lack necessary facilities, it can make daily tasks more strenuous and stressful, compounding fatigue and dissatisfaction.

Inefficiencies in workflow also arise from poorly designed workspaces. Workspaces that fail to support efficient workflows can lead to delays and errors in patient care. For example, if nurse stations are too far from patient rooms or lack necessary resources, it leads to unnecessary travel and time wastage. Such inefficiencies not only strain the staff but also impact on the timeliness and quality of care provided to patients.

Additionally, the lack of adequate rest areas for staff can severely impact their ability to recharge during breaks. Without proper spaces to relax and decompress, staff may not be able to regain their alertness and productivity, directly influencing patient safety and the quality of care. These areas are essential for staff to mentally and physically detach from their clinical responsibilities, even if momentarily, which is crucial for maintaining long-term effectiveness and well-being.

To address these challenges, it is vital to adopt an inclusive design process. Involving healthcare staff in the design process ensures that their needs are considered, enabling the creation of workspaces that are functional, comfortable, and supportive of their duties. Regular feedback sessions can also be instrumental in identifying issues with current designs and guiding necessary improvements, ensuring the spaces evolve with the changing needs of the staff.

Ergonomic and efficient workspace designs are crucial. Workspaces should be tailored to the tasks that staff need to perform, featuring adjustable furniture, adequate lighting, and easy access to necessary tools and technology. Strategic placement of workstations, medication dispensers, and other frequently used equipment can minimize unnecessary movement and save time, thereby enhancing workflow efficiency.

Moreover, creating dedicated rest and recharge zones for staff is essential. These areas should be designed to offer a genuine break from the clinical environment, equipped with amenities such as comfortable seating, relaxation tools, and facilities for eating and personal care. Such spaces support the staff's physical and mental well-being, allowing them to return to their duties refreshed and focused.

The implementation of supportive technology can also significantly reduce the workload on staff by streamlining tasks. Automated systems for administrative tasks, patient monitoring, and communication can free up staff time for direct patient care. Ensuring that technology interfaces are intuitive and providing adequate training can maximize the benefits of these tools.

Focusing on the mental health and well-being of healthcare staff is equally important. Designing programs and spaces that support mental health, offering access to resources like counseling and stress management programs, and promoting a culture of wellness that encourages staff to take necessary breaks and seek help when needed, are all critical for sustaining a healthy work environment.

In conclusion, balancing the design of healthcare facilities to meet both patient and staff needs is crucial for the effective delivery of care. By creating workspaces that enhance productivity, reduce fatigue, and promote overall well-being, healthcare facilities can ensure that their staff are well-supported. This not only improves the working conditions but also has a direct impact on the quality of care provided to patients. Incorporating these considerations into future healthcare designs will foster more resilient and effective healthcare environments.

**Table 9.10** Effect of staff workspace redesign on turnover rates in healthcare facilities

| Facility type | Turnover rate before redesign (%) | Turnover rate after redesign (%) | Description of redesign |
|---|---|---|---|
| General hospital | 22 | 15 | Introduction of ergonomic workstations and improved rest areas |
| Pediatric clinic | 18 | 11 | Redesign of nurse stations for better accessibility and workflow efficiency |
| Community health center | 25 | 17 | Enhancement of staff break rooms and addition of private consultation areas for staff |
| Specialist surgical unit | 20 | 12 | Implementation of advanced technology solutions to reduce administrative burdens |

Table 9.10 contrasts staff turnover rates before and after specific design improvements targeted at enhancing staff workspaces. Positioned following the discussion on the importance of well-designed staff areas, this table quantitatively demonstrates the tangible benefits derived from investing in ergonomic and supportive work environments for healthcare staff. It underscores the link between thoughtful workspace design and reduced staff turnover, highlighting the positive outcomes on staff satisfaction and retention.

### *9.3.5 Environmental Sustainability Shortfalls*

While the healthcare sector is increasingly recognizing the importance of environmental sustainability, many facilities still face significant shortfalls in effectively integrating these practices. Sustainable design is crucial not only for reducing the environmental footprint of healthcare facilities but also for promoting healthier environments for patients and staff.

One of the primary challenges in this area is the lack of energy-efficient systems. Many healthcare facilities operate with outdated systems that consume high levels of energy, leading to increased operational costs and a larger environmental impact. Common areas needing improvement include inefficient heating, ventilation, air conditioning, and lighting systems. These outdated systems are not only costly to operate but also contribute significantly to the carbon footprint of the facility, underscoring the need for modern, energy-efficient replacements.

Another significant issue is the insufficient use of sustainable materials in construction and renovation projects. Often, materials used are not sourced sustainably or possess a high carbon footprint. Additionally, there is a frequent lack of recycling or reuse of materials, contributing further to environmental degradation. This practice not only affects the sustainability of the healthcare facility but also misses an opportunity to set a standard for responsible material usage in the community.

Furthermore, many healthcare designs lack features that could significantly reduce the facility's environmental impact. For example, the absence of green roofs, rainwater harvesting systems, and facilities designed to minimize waste are missing opportunities for reducing the environmental footprint and promoting sustainability.

To address these challenges, it is imperative to integrate energy-efficient systems as part of the healthcare facility's infrastructure. Upgrading to energy-efficient HVAC, lighting, and water heating systems and investing in technologies such as LED lighting, energy-efficient appliances, and smart energy management systems can drastically reduce energy consumption. Additionally, healthcare facilities could consider installing solar panels or integrating geothermal energy solutions to supplement their energy needs with renewable sources.

The use of sustainable building materials should also be prioritized. By opting for sustainable, non-toxic, and recycled materials in construction and renovations, healthcare facilities can significantly reduce their environmental impact. Choosing materials such as bamboo, recycled metal, and low-VOC paints not only supports sustainability but also enhances the healthiness of the environment for patients and staff. Designing with longevity in mind—selecting materials and constructions that require less maintenance and replacement over the facility's lifetime—can further enhance sustainability.

Implementing water conservation measures is another crucial strategy. Installing fixtures that reduce water usage and incorporating systems for rainwater harvesting and greywater recycling can greatly decrease a facility's demand for local water resources and reduce treatment costs.

Developing comprehensive waste management strategies is essential for reducing the environmental impact of healthcare operations. These strategies should include robust recycling programs, proper segregation of medical and non-medical waste, and initiatives aimed at reducing the use of single-use materials. Educating staff and patients on waste reduction practices can also play a significant role in enhancing sustainability efforts.

Additionally, incorporating green spaces within and around healthcare facilities can improve air quality, reduce heat islands, and provide therapeutic benefits to patients and staff. Designing landscapes to support local biodiversity, using native plants, and creating habitats that encourage local wildlife can further bolster environmental efforts.

Finally, aiming for sustainability certifications such as LEED, WELL, or Green Globes can provide a framework for implementing sustainable practices and enhance the facility's reputation as a leader in environmental stewardship.

In conclusion, sustainability must be a foundational principle in the design and operation of healthcare facilities. By embracing sustainable practices, healthcare providers can demonstrate leadership in environmental stewardship and significantly contribute to global efforts in combating climate change. Sustainable healthcare facilities often see benefits in reduced operational costs and improved patient and staff well-being, making a compelling case for their widespread adoption.

Equation 9.6 is designed to estimate the reduction in carbon footprint resulting from specific sustainable design choices in healthcare facilities. This formula is essential for healthcare administrators and sustainability officers committed to reducing environmental impacts through targeted design innovations and practices. By quantifying the reductions in carbon emissions, healthcare facilities can align more closely with environmental sustainability goals, which is increasingly important in the context of global climate initiatives and regulatory demands.

The equation calculates the reduction in the carbon footprint by considering the annual carbon emissions before the implementation of sustainable design choices ($C_b$) and the annual carbon emissions after these choices have been implemented ($C_a$). The difference between these two values gives the absolute reduction in carbon emissions attributable to sustainable design choices. To further refine the calculation, this absolute reduction can be adjusted by an effectiveness factor ($E$), which accounts for the overall efficiency and impact of the sustainable designs over time:

$$\text{Carbon Footprint Reduction (CFR)} = (C_b - C_a) \times E. \tag{9.6}$$

This effectiveness factor ($E$) is a critical component of the equation, as it provides a way to measure the true impact of the sustainable designs' relative to their intended environmental performance outcomes. It considers various factors such as the longevity and durability of the designs, the maintenance required to sustain their efficiency, and their adaptability to future technological advancements or environmental regulations.

By applying this equation, healthcare facilities can not only assess the immediate benefits of their sustainable design choices but also plan for long-term environmental and financial sustainability. The calculated carbon footprint reduction serves as a key indicator of the success of their environmental strategies, supporting continuous improvement and helping to justify investments in sustainable technologies and designs. This proactive approach enhances the facility's reputation, supports compliance with environmental standards, and contributes to broader public health goals by promoting a healthier environment.

### 9.3.6 Cultural and Social Sensitivity Gaps

Addressing the gaps in cultural and social sensitivity within healthcare facility design is crucial. Addressing the gaps in cultural and social sensitivity within healthcare facility design is increasingly crucial as globalization and demographic changes lead to more diverse populations. Ensuring that facilities are welcoming and accessible to everyone is fundamental to providing equitable healthcare, as these elements directly influence patient comfort and satisfaction.

One significant challenge is the lack of cultural representation within many healthcare facilities. Often, these environments fail to reflect the cultural diversity of the communities they serve, which can manifest in various ways. For instance, signage may only be available in one language, or the artwork and decor might not reflect the local cultures. Additionally, the absence of spaces designed for specific cultural practices can make these facilities feel less welcoming or accommodating to individuals who observe customs.

Another issue is the inadequate accommodation of social needs. Healthcare facilities frequently overlook varied family structures, gender identities, and social norms, potentially creating environments that feel exclusionary. Examples of this include the lack of family-friendly spaces or gender-neutral restrooms, which can alienate segments of the population and deter them from seeking necessary medical care.

Accessibility issues extend beyond the physical; they also encompass linguistic accessibility and accommodations for sensory impairments, which are often neglected in traditional healthcare designs. Ensuring broad accessibility is crucial for inclusivity, allowing individuals with diverse needs to navigate and utilize healthcare facilities without undue hardship.

To enhance cultural and social sensitivity in healthcare design, inclusive design consultations are vital. Involving a diverse group of stakeholders, including community leaders, cultural advisors, and representatives from various demographic groups—in the design process can yield more inclusive choices that resonate with the broader community. Their insights can ensure that the facility caters to the cultural and social nuances of its users.

Implementing multilingual resources and signage within the facility is another key strategy. This approach accommodates non-native speakers and reduces linguistic barriers, making healthcare more accessible and reducing the stress associated with navigating medical environments.

Incorporating culturally relevant design elements can significantly impact the perceived inclusivity of a facility. Art, decor, and architectural elements that reflect the cultural diversity of the local community can create a more welcoming environment and demonstrate respect for various cultural identities.

Designing spaces that accommodate a range of cultural practices, such as prayer rooms, communal gathering spaces, or private family rooms, ensures that all groups feel supported in maintaining their cultural traditions within the healthcare setting. This consideration can greatly enhance the comfort and acceptance of diverse populations.

Providing ongoing cultural competence training for staff is essential. Such training improves interactions, and the quality of services provided to diverse populations, covering aspects of cultural awareness, communication styles, and specific health concerns prevalent among different community groups.

Finally, establishing mechanisms to regularly assess the effectiveness of cultural and social sensitivity initiatives is crucial. Feedback from patients and staff can provide valuable insights into areas needing improvement and help tailor services to better meet the needs of the community.

In conclusion, incorporating a broader range of cultural and social considerations into the design and operation of healthcare facilities is essential for creating truly inclusive and accessible environments. By embracing these principles, healthcare facilities can improve patient satisfaction, enhance the quality of care, and ensure that all community members feel valued and supported. This approach not only benefits patients but also contributes to a more equitable healthcare system.

Table 9.11 displays the outcomes of a cultural sensitivity audit conducted in healthcare facilities, comparing pre-audit and post-audit results to gauge progress in cultural inclusivity. Situated at the end of the section on enhancing cultural and social sensitivity, this table highlights the areas where significant improvements have been made and identifies persistent challenges. It serves as a diagnostic tool for healthcare administrators to understand better and address the gaps in cultural representation and accommodation within their facilities.

**Table 9.11** Assessment of cultural sensitivity in healthcare facilities: pre- and post-audit results

| Facility type | Cultural sensitivity score pre-audit | Cultural sensitivity score post-audit | Areas of improvement | Ongoing challenges |
|---|---|---|---|---|
| Community clinic | 60/100 | 75/100 | Increased use of multilingual signage; introduction of cultural decor | Limited spaces for diverse cultural practices; ongoing need for staff training |
| Regional hospital | 55/100 | 80/100 | Implementation of culturally diverse patient advisory boards; expanded training programs | Integrating more diverse dietary options in the cafeteria |
| Urban health center | 70/100 | 85/100 | Enhanced accessibility features for non-English speakers; addition of private family rooms | Adapting physical spaces to accommodate larger family groups |
| Specialized care facility | 65/100 | 78/100 | Added spaces for cultural practices like prayer rooms; increased cultural artifacts in inpatient areas | Struggles with fully integrating culturally relevant medical practices |

Next, we will explore how advocating for interdisciplinary collaboration can further enhance healthcare design. This approach brings together diverse expertise—from medical professionals to architects and cultural consultants—to create facilities that are not only functional and efficient but also responsive to the complex needs of modern healthcare.

## 9.4 Advocacy for Interdisciplinary Collaboration

In the realm of healthcare architecture, embracing interdisciplinary collaboration is crucial for developing facilities that are not only innovative and efficient but also deeply attuned to the multifaceted needs of patients, staff, and the broader community. Advocating for a collaborative approach among various professionals—ranging from architects and engineers to healthcare providers, patients, and technology experts—can drive significant improvements in the design and functionality of healthcare environments.

### *9.4.1 Integrating Diverse Expertise*

The importance of integrating diverse expertise in the design of healthcare facilities cannot be overstated. This approach brings together professionals from various fields including architecture, engineering, and healthcare, to create environments that are not only efficient and state-of-the-art but also attuned to the nuanced needs of those they serve. A multidisciplinary strategy is essential in addressing the complex demands of contemporary healthcare settings. Architects and engineers contribute their knowledge of sustainable and efficient building practices, which is crucial for creating spaces that promote health and well-being. Their innovative designs can enhance patient recovery by facilitating natural light, minimizing noise, and using materials that are conducive to a healing environment.

Furthermore, healthcare professionals provide invaluable insights into the operational needs of medical facilities. Their expertise ensures that the layouts and designs enhance workflow efficiency, patient care processes, and overall medical service delivery. The practical knowledge they bring ensures that facilities are not only beautifully designed but are also practical and functional, meeting the real-world demands of healthcare delivery. Technology specialists also play a crucial role by introducing advanced solutions that streamline operations, improve patient outcomes, and secure sensitive data. Their involvement is critical in effectively integrating digital health technologies, such as telemedicine and automated systems, within healthcare environments.

The integration of user-centered design principles marks another significant aspect. Including patients and community members in the planning process ensures that the facilities truly reflect the needs and preferences of those who will use them daily. This inclusive approach is key to creating environments that are not just functional but also comforting and conducive to recovery. Moreover, the collaboration of diverse expertise helps in designing healthcare facilities that are adaptable and resilient. This foresight is essential for ensuring that facilities can evolve with the changing demands of healthcare services and technological advancements, which is crucial for their long-term sustainability and relevance.

To effectively integrate diverse expertise, it is recommended to establish interdisciplinary design teams that include professionals from the fields. These teams should collaborate throughout the project, ensuring a holistic approach to healthcare facility design. Engaging community members and patients through workshops, focus groups, and surveys is also vital. This direct involvement helps gather essential feedback that can be integrated into the design process, making the facilities more attuned to the needs of the community they serve. Furthermore, promoting continuous learning and adaptation within the design team can foster innovative solutions and ensure the facility remains at the forefront of healthcare delivery. Regular updates based on new technologies and changes in healthcare practices are necessary to maintain the relevance and effectiveness of healthcare facilities.

Prototyping and pilot testing of design concepts should also be conducted to evaluate their practicality in real-world scenarios. This could include creating mock-ups of patient rooms or testing new technological systems to assess their functionality under everyday conditions. Additionally, establishing feedback mechanisms during and after construction can ensure that the facility continuously aligns with user needs and industry advancements. By fostering an environment of continuous feedback and adaptation, healthcare facilities can remain responsive and effective in meeting the evolving needs of patients and staff.

In conclusion, the integration of diverse expertise is fundamental in the design and operation of modern healthcare facilities. This collaborative approach not only leads to the creation of spaces that are technologically advanced and efficient but also ensures that these environments are welcoming and supportive of both patients and healthcare staff. This strategy is indispensable for fostering innovation and ensuring that healthcare facilities are equipped to face the challenges of contemporary healthcare delivery. Through such interdisciplinary collaboration, healthcare facilities can achieve a balance of form and function that not only meets today's standards but also anticipates future needs.

Table 9.12 provides a detailed breakdown of the various professional roles involved in the design of healthcare facilities, outlining their unique contributions and insights. Positioned after the introduction to diverse expertise, this table highlights how architects, engineers, healthcare professionals, technology experts, and patients each play a crucial role in creating efficient, user-centered, and adaptable healthcare environments. It underscores the value of integrating diverse perspectives to achieve a holistic design approach that meets modern healthcare demands.

**Table 9.12** Contributions of interdisciplinary teams in healthcare facility design

| Professional role | Contributions | Specific insights or actions |
|---|---|---|
| Architects | Design optimization for health and efficiency | Innovate in the use of sustainable materials and spaces promoting well-being; focus on patient and staff flow within facilities |
| Engineers | Technical solutions for building functionality | Develop robust systems for energy efficiency, structural integrity, and environmental sustainability |
| Healthcare professionals | Practical insights into operational needs | Provide essential details on patient care processes, equipment needs, and safety protocols to ensure functionality and care quality |
| Technology experts | Integration of cutting-edge technological solutions | Implement telemedicine systems, patient data management, and security solutions to streamline operations and enhance care delivery |
| Patients and community members | User-centered design feedback | Offer perspectives on facility accessibility, comfort, and supportive features that enhance the healing environment |

### 9.4.2 *Enhancing Design Solutions*

Interdisciplinary collaboration is essential for designing healthcare facilities that are not only visually appealing but also functionally effective in addressing the diverse and complex requirements of modern medical care. This approach ensures that design solutions significantly enhance every aspect of healthcare delivery, including patient comfort, safety, technological integration, and operational efficiency. By combining the unique perspectives and expertise of various professionals—such as medical staff, designers, IT specialists, and sustainability experts, the design process results in holistic healthcare environments tailored to the needs of patients and healthcare providers alike.

One of the key benefits of this interdisciplinary approach is the creation of optimal patient rooms. Medical staff, by sharing their insights on the layout necessary for effective medical procedures and patient care, collaborate with designers who ensure these spaces are not only functional but also comforting and welcoming. Such rooms enhance the patient's healing process by incorporating elements that reduce stress and promote well-being. Furthermore, IT specialists are instrumental in integrating advanced medical technologies into these environments. They ensure that the healthcare facilities are equipped with the necessary infrastructure to support sophisticated medical equipment and robust data management systems, facilitating seamless integration of technology in patient care, which is essential for effective diagnostics and treatment.

Safety and accessibility are also paramount in healthcare facility design. Architects and engineers collaborate closely with medical professionals to ensure that all aspects of the facility's design adhere to the highest standards of safety and are accessible to all users. This includes careful planning of non-slip floors, adequate lighting, handrails in patient areas, and accessible entrances and pathways, which are critical for ensuring patient and staff safety and mobility. Additionally, sustainability experts contribute to the design of energy-efficient systems and eco-friendly building practices. Their expertise helps in reducing the environmental impact of healthcare facilities while also curbing operational costs through the implementation of sustainable technologies such as high-efficiency HVAC systems and solar panels.

To enhance these design solutions, several strategies are employed. Regular interdisciplinary meetings are crucial during the planning and design phases to ensure a unified approach to the healthcare facility's development. These meetings facilitate open dialogue and idea sharing, allowing every team member to contribute their expertise toward achieving the best outcomes. Establishing shared goals and objectives at the project's outset aligns all team members with the healthcare facility's mission, fostering a collaborative effort focused on enhancing patient care, improving efficiency, and promoting sustainability.

Moreover, the use of prototyping and user feedback mechanisms, such as virtual reality simulations or full-scale mock-ups, is vital for testing design solutions in real-world scenarios. This approach allows for adjustments to be made based on direct input from end-users, ensuring the facility meets actual needs and functions effectively. Continuous education and cross-training among team members are also encouraged to deepen understanding across disciplines, enhancing appreciation for the collaborative nature of projects and leading to more integrated and innovative solutions.

Documenting and sharing the learnings from these collaborative processes is another critical aspect. By making this information available not only within the organization but also to the broader community, healthcare facility design can continuously evolve, setting new standards in the field and fostering environments that support healing, efficiency, and innovation. Enhanced design solutions achieved through interdisciplinary collaboration result in healthcare facilities that are not merely structures but comprehensive environments that effectively support the diverse needs of their users and are well-equipped to adapt to future challenges in healthcare delivery.

Equation 9.7, the Efficiency Improvement Index, quantifies the enhancement in healthcare operations achieved through interdisciplinary design solutions. This index is vital for healthcare managers and architects who integrate varied disciplinary insights to optimize facility layout, technology usage, and workflow processes. The efficiency gained through such interdisciplinary approaches is critical in improving patient care quality, reducing operational costs, and increasing the capacity of healthcare services.

The index measures efficiency improvement by considering variables that directly reflect operational performance enhancements. These include time savings ($T$), which represents the reduction in the time required to perform healthcare operations; error reduction ($E$), which quantifies the decrease in mistakes or inaccuracies in medical procedures or administrative tasks; and patient throughput ($P$), the increase in the number of patients that can be effectively managed within a given timeframe. Each variable is weighted ($w_T$, $w_E$, $w_P$) according to its relative importance to the overall efficiency of the healthcare facility, allowing for a tailored assessment based on specific operational goals.

The Efficiency Improvement Index (*EII*) is calculated by aggregating the weighted improvements across these variables, providing a comprehensive measure of the operational efficiency gains:

$$\mathrm{EII} = (T \times w_T) + (E \times w_E) + (P \times w_P). \tag{9.7}$$

This equation enables healthcare facilities to objectively evaluate the effectiveness of interdisciplinary design solutions in enhancing operational efficiency. By calculating the EII, facilities can prioritize areas for improvement, allocate resources more effectively, and ultimately deliver better healthcare outcomes. This metric not only underscores the value of integrating diverse expertise in facility design and management but also aligns with broader healthcare objectives of improving patient satisfaction, safety, and care delivery efficiency.

### 9.4.3 Streamlining Implementation

Streamlining the implementation of new healthcare facilities or renovations is pivotal to minimizing conflicts, delays, and cost overruns. By emphasizing collaborative planning from the outset, healthcare projects can transition smoothly and efficiently, reducing the typical barriers that arise when projects operate in segmented or isolated ways. This collaborative approach fosters effective communication among all stakeholders involved, ensuring that every phase of the project is aligned and cohesively executed.

The benefits of a streamlined implementation through collaborative planning are manifold. Early involvement of all key stakeholders enables a comprehensive review of the project plans, allowing for the identification and resolution of potential issues before construction begins. This proactive stance prevents costly and time-consuming corrections during the later stages, thus maintaining project momentum and integrity. Collaborative planning also fosters the creation of integrated schedules that account for the inputs and constraints of all disciplines involved. Such integration prevents delays that can occur when one team's progress depends on the completion of tasks by another, ensuring a smooth flow of project activities.

Moreover, involving all stakeholders early helps in more accurately forecasting budgets. Each team can provide insights into their specific costs, which helps in identifying and mitigating potential financial oversights early in the planning process. This is crucial for maintaining financial control and preventing budget overruns, which are common in large-scale projects like healthcare facilities.

To effectively streamline implementation, several strategies are employed. Comprehensive stakeholder engagement from the conceptual phase is crucial; involving architects, engineers, healthcare professionals, IT specialists, and even patient representatives ensures that the project benefits from diverse insights, leading to more holistic planning and design. Regular coordination meetings throughout the project lifecycle help maintain open lines of communication, ensuring all teams are aware of upcoming tasks and potential bottlenecks that could affect the schedule.

Utilization of project management tools is another vital strategy. These tools provide real-time updates and visibility into the project's progress for all stakeholders, facilitating task tracking, timeline management, and prompt alerts to potential delays. Establishing clear protocols for conflict resolution, involving mediation and negotiation tailored to the project's needs, ensures that any disputes are resolved swiftly, maintaining project momentum.

Quality control and assurance processes agreed upon by all parties involved are essential. Regular quality checks are necessary to ensure that all work adheres to the agreed standards and specifications, thereby reducing the likelihood of rework. Additionally, incorporating feedback mechanisms to continually assess the efficiency of collaboration and make necessary adjustments is key. Valuing feedback from all team members fosters a cooperative and productive work environment.

In conclusion, the streamlining of healthcare facility projects through collaborative planning is crucial for ensuring that these complex undertakings are completed on time, within budget, and to the required quality standards. This approach not only benefits the construction phase but also enhances the overall functionality and effectiveness of the finished healthcare facility. By fostering early and ongoing communication among a diverse group of stakeholders, potential problems can be anticipated and addressed promptly, leading to smoother project execution and superior outcomes. This strategy not only facilitates the practical aspects of construction but also enriches the eventual delivery of healthcare services, making the facilities more responsive to the needs of their users.

Table 9.13 documents instances of common conflicts encountered during healthcare facility projects and outlines how interdisciplinary collaboration facilitated effective resolutions. Positioned within the section on strategies for streamlining implementation; this table provides practical examples detailing the nature of each conflict, the collaborative solution employed, and the positive outcomes achieved. It serves as a guide for project teams on how to leverage diverse expertise to overcome challenges, ensuring smoother project execution and enhanced outcomes.

**Table 9.13** Conflict resolution in healthcare facility projects through interdisciplinary collaboration

| Conflict scenario | Collaborative solution | Outcome of resolution |
|---|---|---|
| Scheduling overlaps | Architects, engineers, and construction managers used integrated project management tools to realign schedules | Delays were minimized, and the project remained on track for timely completion |
| Budget overruns in equipment installation | Healthcare professionals and financial planners reviewed the procurement process to identify cost-saving alternatives without compromising quality | The budget stabilized, and essential equipment was installed without reducing the project scope |
| Design misalignments | Regular coordination meetings were held to ensure all design modifications were communicated across disciplines, including feedback from healthcare staff | Design realigned to operational needs, preventing costly post-construction modifications |
| Compliance issues | Compliance specialists and legal advisors worked together to navigate regulatory requirements, updating the project team regularly | The project met all regulatory standards, avoiding legal and safety issues post-completion |
| Material shortages | A cross-functional team including logisticians and suppliers implemented a real-time inventory management system | Material shortages were addressed promptly, reducing downtime and keeping the project on schedule |

### *9.4.4 Innovating for Future Needs*

The interdisciplinary approach in healthcare design serves as a crucial platform for innovation, especially in anticipating and adapting to the evolving demands of healthcare delivery. By harnessing the collective expertise of various fields—medical, architectural, technological, and more—this approach enables teams to foresee future challenges and seamlessly integrate solutions that render facilities more adaptable and future-ready.

One significant advantage of this method is the ability to anticipate demographic changes. As population dynamics evolve, including aging and shifts in population size, an interdisciplinary team equipped with diverse insights can design facilities that are flexible and adaptable to meet changing healthcare needs. This foresight allows for the creation of healthcare environments that can adjust to varying patient populations and their specific health requirements without the need for extensive modifications.

Additionally, the rapid pace of technological innovation in healthcare necessitates a design approach that can accommodate these changes efficiently. An interdisciplinary team, including IT specialists and medical technologists, is better positioned to predict technological advancements and incorporate them into facility designs in a way that is both smooth and cost-effective. This ensures that healthcare facilities remain at the forefront of technology, enhancing patient care and operational efficiency.

The continual evolution of healthcare protocols, often in response to new medical research or changes in healthcare policy, also underscores the need for adaptable design. Having a diverse team ensures that new healthcare facilities are built with flexibility to accommodate these changes, facilitating updates to infrastructure and operations without significant overhauls, thus maintaining relevance and effectiveness over time.

To advocate for and bolster interdisciplinary collaboration, several actions can be undertaken. Hosting workshops and seminars that bring together professionals from healthcare, architecture, technology, and other relevant fields is crucial. These events act as platforms for knowledge exchange, sparking innovation and fostering relationships that underpin continuous collaboration. Establishing ongoing forums or working groups focused on healthcare facility design is another effective strategy. These groups provide a structured environment for regular reviews and discussions on the latest developments in healthcare needs and technologies, ensuring that projects remain cutting-edge and aligned with current and future requirements.

Developing cross-disciplinary education and training programs is also vital. These programs should encourage healthcare professionals to learn design principles and, conversely, train architects, engineers, and designers in the basics of healthcare operations and patient care. This educational exchange enhances mutual understanding and collaboration, leading to more informed and holistic healthcare facility designs.

In conclusion, by advocating for and implementing interdisciplinary collaboration, healthcare facility design can achieve new levels of functionality and innovation. This approach not only enriches the creative process but also ensures that healthcare facilities are well-equipped to address both present and future challenges effectively. Through structured workshops, continuous forums, and cross-disciplinary education, stakeholders can cultivate a cohesive community dedicated to innovation. This leads to the development of healthcare environments that are not only efficient and adaptable but also focused on improving patient outcomes and building more resilient healthcare systems.

Equation 9.8, termed the Adaptability Quotient, captures the flexibility and future readiness of healthcare facilities by quantifying their architectural flexibility, technological readiness, and responsiveness to evolving healthcare trends. This quotient is particularly relevant for healthcare facility managers and designers who need to ensure that their buildings and systems can adapt effectively to rapid changes in medical technology, patient care standards, and regulatory environments. The Adaptability Quotient (AQ) is calculated by considering three key dimensions:

- Architectural Flexibility ($A$): This measures the ability of the facility's physical structure to accommodate changes such as expansions, modifications, or reconfigurations without significant disruptions or excessive costs. This flexibility is vital for extending the useful life of the facility and for supporting evolving medical practices and technologies.
- Technological Readiness ($T$): This assesses the extent to which the facility is prepared to integrate new and emerging technologies. This includes the infrastructure to support digital health technologies, telemedicine, and advanced diagnostic and treatment equipment, which are crucial for modernizing patient care and enhancing facility efficiency.
- Responsiveness to Healthcare Trends ($R$): This dimension evaluates how quickly and effectively a facility can respond to changes in healthcare practices, patient demographics, and care protocols. This responsiveness ensures that the facility remains relevant and capable of meeting the shifting demands of healthcare delivery.

Each of these dimensions is weighted ($w_A$, $w_T$, $w_R$) according to their relative importance to the overall adaptability of the facility. The weights allow for customization based on specific strategic priorities and operational contexts. The Adaptability Quotient is then computed as follows:

$$\text{AQ} = (A \times w_A) + (T \times w_T) + (R \times w_R). \tag{9.8}$$

By quantifying the Adaptability Quotient, healthcare administrators can assess and benchmark their facility's capacity to adapt to future challenges and opportunities. This metric not only guides strategic investments in infrastructure and technology but also supports planning for long-term sustainability and competitiveness in the healthcare sector. This approach underscores the importance of adaptability in healthcare facility management, promoting a proactive rather than reactive approach to changes in the healthcare landscape.

## 9.5 Economic Impact of Healthcare Facilities

The economic impact of healthcare facilities extends far beyond their primary function of providing medical services. This subsection explores how well-designed healthcare facilities can influence the local and broader economy, addressing both direct and indirect economic impacts and offering insights into maximizing these benefits through strategic design and operation.

### 9.5.1 Direct Economic Contributions

The economic impact of healthcare facilities stretches well beyond their fundamental mission of delivering medical services. These institutions serve as pivotal economic engines, contributing significantly to both local and broader economies. This influence is expressed not only through their roles in health service provision but also through their substantial economic contributions that support community development and economic stability.

Healthcare facilities are integral economic catalysts within their communities. They generate considerable economic contributions through direct employment and significant capital investments, making them crucial components of community infrastructure. As one of the largest employers in any region, healthcare facilities offer a wide array of job opportunities. These range from highly specialized medical roles such as doctors and nurses to essential support positions including administrative, maintenance, and security staff. The diversity of employment opportunities within a healthcare facility contributes to a robust local job market and helps stabilize the local economy, especially during economic downturns when healthcare services remain in constant demand.

The construction, renovation, and expansion of healthcare facilities represent significant capital investments that extend their economic impact. These projects not only aim to enhance healthcare service delivery but also stimulate local economies by providing substantial business to local construction firms, architects, and engineering consultants. The continual need for modern and upgraded facilities ensures a steady influx of capital into the community, fostering growth in related sectors such as the manufacturing of medical and building materials.

In addition to direct economic impacts, healthcare facilities also support a broad spectrum of local businesses indirectly. The everyday operations of these facilities generate substantial economic activity as employees and visitors spend on local services such as restaurants, parking, hotels, and retail outlets. Moreover, many healthcare facilities prioritize sourcing from local suppliers to reduce costs and support the community, further amplifying local business revenues.

Investments in new medical technologies not only enhance the quality of healthcare but also bolster the high-tech industry. Healthcare facilities frequently invest in advanced medical equipment, IT infrastructure, and software solutions, which contributes to the growth of these sectors. Furthermore, facilities often allocate resources for training and development, which enhances the skill levels within the community and promotes broader economic development.

Healthcare facilities can also act as magnets for related industries, such as medical research, biotechnology firms, and other healthcare-related enterprises. The presence of a major healthcare institution can attract businesses that seek to collaborate with medical experts or leverage hospital facilities, enhancing the region's appeal as a hub for healthcare and research. This clustering effect can transform regions into vibrant centers of healthcare and research, stimulating further economic and developmental growth.

**Table 9.14** Economic contributions of healthcare facilities through employment and investment

| Aspect | Metric | Description | Data point |
|---|---|---|---|
| Direct employment | Number of jobs created | Jobs directly associated with healthcare services | 1200 jobs |
| Indirect employment | Number of jobs supported | Jobs in local businesses supported by facility operations | 800 jobs |
| Capital investments | Total investment | Total capital spent on construction, technology, and facility upgrades | $500 million |
| Investment period | Time frame | Duration over which the investments were made | 2015–2020 |
| Local business support | Economic impact on local businesses | Contribution to local economy through operational spending | $200 million annually |

In conclusion, healthcare facilities are more than just centers for medical care; they are vital to the economic fabric of their communities. By driving employment, supporting local businesses, and making substantial investments in technology and infrastructure, they play a significant role in fostering community development and economic stability. Understanding and supporting the growth and sustainability of these institutions are essential for ensuring continued economic vitality and an improved quality of life for residents. Recognizing these facilities' broad economic impacts allows communities to appreciate the full value of investing in healthcare infrastructure, which in turn supports a thriving, resilient local economy.

Table 9.14 details the economic impact of healthcare facilities by showcasing the number of jobs created and the scale of capital investments over a specified period. Situated after discussions on employment opportunities and capital investments in healthcare infrastructure, this table quantifies these contributions, highlighting their significant role as economic catalysts within communities. It illustrates not only the direct and indirect employment benefits but also the substantial capital inflow associated with the development and expansion of healthcare facilities.

### 9.5.2 *Inducing Local Business Growth*

The establishment of a healthcare facility within a community significantly influences local economic development, catalyzing growth in both healthcare-related and broader commercial sectors. This dynamic relationship extends the impact of such facilities beyond their primary role in health service delivery, fostering a vibrant economic ecosystem. The presence of a hospital or clinic heightens the demand for nearby pharmacies and medical supply stores, which are strategically positioned to provide easy access for patients requiring prescriptions and medical supplies following treatments. The proximity of these businesses to healthcare facilities ensures a consistent flow of customers, contributing to their financial stability and longevity.

Moreover, specialist medical providers such as imaging centers, laboratories, and therapeutic services tend to cluster around major healthcare institutions. This congregation facilitates better coordination of patient care and streamlines the referral processes, which is crucial for integrated health services. The close association with larger facilities not only enhances the convenience for patients but also increases business volumes for these specialized providers through direct referrals, thereby boosting their operational viability.

In addition to healthcare-related businesses, the local non-medical business sector also benefits from the presence of healthcare facilities. Restaurants, cafés, and catering services in the vicinity often experience an increase in customer base consisting of facility staff, patients, and visitors. These businesses may adapt their offerings or operational hours to cater to the health-conscious or accommodate the schedules of shift workers, thereby embedding themselves further into the community linked to the healthcare facility. Similarly, retail outlets like gift shops, florists, and convenience stores enjoy increased sales due to frequent visits by those buying gifts or personal items for patients.

Furthermore, the service industries such as parking, hotels, and transportation services expand to meet the needs of the healthcare facility's visitors, particularly in urban settings or where the facility serves a larger regional population. Hotels might experience longer stays from families of patients undergoing extended treatments, which supports not only the service sector but also the local tourism industry.

To maximize the potential for local business growth, healthcare facilities can engage in strategic partnerships and collaborative marketing with local enterprises. By negotiating discounts with local services for patient families or jointly promoting service offerings, hospitals can enhance their community integration and mutual economic benefit. Additionally, healthcare facilities can drive community engagement by participating in local business associations, hosting health-related events, and advocating for community-centric developments.

In conclusion, the economic influence of healthcare facilities profoundly drives growth across a multitude of sectors, directly and indirectly related to medical services. By fostering a cooperative environment where healthcare and community interests intersect, these facilities can significantly enhance their role as key economic drivers, thereby benefiting not only patients but the entire local economy in measurable and sustainable ways. This complex interplay underscores the importance of healthcare facilities as pivotal elements in both healthcare delivery and economic development within their communities.

Equation 9.9 focuses on quantifying the economic multiplier effect of healthcare facilities on local business growth. This effect is crucial for understanding the broader economic impact of healthcare investments, especially in how such facilities contribute to local economies by generating employment, supporting local sourcing, and stimulating indirect business activities. This calculation is pivotal for healthcare administrators and policymakers to justify and optimize investments in healthcare infrastructure from an economic development perspective. The economic multiplier effect is calculated by integrating variables that capture the direct and indirect economic contributions of healthcare facilities:

- Employment Generation ($E$): This variable measures the number of jobs directly created by the healthcare facility, including medical staff, administrative personnel, and support services. Employment generation is a primary driver of economic activity, providing income and stimulating spending within the community.
- Local Sourcing ($L$): This reflects the facility's procurement of goods and services from local businesses, which supports local industries and promotes regional economic stability and growth. It includes everything from medical supplies to facility maintenance services.
- Indirect Business Stimulation ($I$): This variable captures the additional economic activity generated because of the healthcare facility's operations, such as increased demand for local retail and food services by employees, patients, and visitors.

Each variable is weighed according to its impact on the local economy ($w_E$, $w_L$, $w_I$), allowing the equation to reflect different regional economic conditions and strategic priorities. The Economic Multiplier Effect (*EME*) is then expressed as follows:

$$\text{EME} = (E \times w_E) + (L \times w_E) + (I \times w_I). \tag{9.9}$$

This equation provides a comprehensive measure of how healthcare facilities act as economic engines in their communities, beyond their primary role of delivering health services. By calculating the Economic Multiplier Effect, healthcare leaders can demonstrate the value of healthcare facilities in economic terms, supporting efforts to attract further investment, enhance community relations, and foster sustainable local economic development. This insight is invaluable for making informed decisions about facility expansion, service diversification, and community engagement strategies, ensuring that healthcare investments contribute positively to the broader economic ecosystem.

### 9.5.3 Innovation and Research Development

Healthcare facilities, especially those affiliated with academic institutions, are foundational in propelling advancements in medical innovation and research. These institutions stand as beacons not only for exceptional patient care but also as vital centers for the development of new medical technologies, pharmaceuticals, and innovative treatment methodologies. Their active engagement in research and development

(R&D) significantly elevates the medical field while simultaneously stimulating local economies through the creation of a robust ecosystem of innovation.

Such facilities often collaborate with research institutions and the private sector to pioneer medical technologies. These partnerships frequently yield groundbreaking medical devices and software solutions that enhance patient outcomes and streamline operations within healthcare settings. Moreover, healthcare facilities with dedicated research departments are instrumental in pharmaceutical advancements. By conducting clinical trials and research studies, these facilities contribute to the discovery of new drugs and therapies. These activities not only advance medical science but also attract substantial investment into the region from pharmaceutical companies and research grants, further underscoring their pivotal role in healthcare innovation.

Additionally, the research undertaken within these facilities significantly enhances treatment protocols and healthcare practices, often culminating in publications in prestigious medical journals. This dissemination of new knowledge not only sets global healthcare standards but also cements the reputation of these institutions as leaders in the medical community.

From an economic perspective, facilities known for their innovative and research capabilities attract investments from large corporations, venture capitalists, and governmental bodies. This influx of funds not only facilitates further research and development activities but also creates a self-sustaining cycle of investment and innovation, which in turn propels local economic growth. The focus on R&D in these institutions generates high-value employment opportunities in the research, science, and technology sectors. These positions are typically well-compensated and come with substantial benefits, contributing significantly to the economic prosperity of the area.

Moreover, regions that house innovative healthcare facilities become magnets for highly skilled professionals, including researchers, clinicians, and technologists. The influx of top-tier talent encourages more businesses and institutions to establish themselves in the area, enhancing the local economy's dynamism and making it an attractive place to live and work. Additionally, innovations originating from these healthcare facilities often lead to the creation of spin-off companies and startups. These new enterprises focus on commercializing research findings, thereby stimulating the local startup ecosystem, fostering entrepreneurship, and further diversifying the economy.

To amplify their impact, healthcare facilities should intensify their partnerships with universities, industry, and government agencies. Strengthening these collaborations can enhance research capabilities and resources, providing additional funding, access to specialized equipment, and a broader pool of expertise. Moreover, engaging with local communities and educational institutions can broaden public understanding of the significance of their research activities. Such engagement might include public lectures, workshops, and school programs aimed at inspiring future medical professionals and researchers.

Furthermore, healthcare facilities can collaborate with local governments and other stakeholders to develop innovation clusters. These clusters, which co-locate

**Table 9.15** Investment in medical research and its returns in healthcare innovation

| Investment area | Financial investment (million) | Technological advancements | Pharmaceutical developments | Clinical improvements |
|---|---|---|---|---|
| Medical devices | $150 | Development of a new non-invasive glucose monitoring device | N/A | Enhanced patient compliance and monitoring |
| Drug research | $200 | N/A | Introduction of a new class of antiviral drugs | Reduced hospitalization rates for viral infections |
| Clinical procedure innovations | $50 | Implementation of AI-driven diagnostic algorithms | N/A | Increased accuracy of diagnoses; reduced diagnostic time |
| Collaborative research projects | $100 | Development of integrated care platforms | Improvement in personalized medicine treatments | Better patient outcomes through tailored treatment plans |

startups, research labs, and academic institutions, facilitate collaboration and knowledge exchange, thereby accelerating innovation and economic growth.

In conclusion, healthcare facilities engaged in research and development serve as crucial catalysts for both medical innovation and economic development. By fostering an environment supportive of innovation and research, these institutions not only push medical science forward but also enhance the economic vitality and attractiveness of their regions, proving indispensable to both healthcare and local economic landscapes.

Table 9.15 provides a detailed breakdown of investments in medical research by healthcare facilities and the returns in terms of technological advancements, pharmaceutical developments, and clinical improvements. Positioned near the discussion on pharmaceutical advancements and new medical technologies, this table effectively links financial inputs with innovative outputs, demonstrating how substantial investments in research translate into significant advancements in medical science and healthcare delivery.

### *9.5.4 Healthcare as a Public Good*

Healthcare facilities are indispensable as stewards of public health, playing a pivotal role in the economic and social dynamics of communities. By enhancing the health status of populations, these institutions contribute significantly to the broader economic stability and productivity, underscoring the necessity of perceiving healthcare not merely as a service but as an essential public good. The efficient management and accessibility of healthcare facilities can have profound effects, making them central to community well-being and economic resilience.

The impact of healthcare on economic stability is multifaceted. Healthy individuals are inherently more productive and capable of contributing effectively to the workforce. Facilities that provide timely and effective treatment ensure the workforce remains robust, directly influencing economic output and efficiency. Furthermore, preventive care and early intervention in treating illnesses can mitigate the severity of diseases, enabling quicker recovery times and reducing absences from work, thus maintaining a more consistent and productive workforce. Effective healthcare reduces absenteeism rates, which benefits employers and the economy at large. Chronic conditions and acute illnesses that are well-managed mean fewer days lost to sickness.

Additionally, investing in preventive care and efficient disease management can diminish the long-term costs associated with chronic diseases, which are among the most expensive to treat. By reducing the incidence and severity of these conditions, healthcare facilities not only conserve resources but also alleviate the financial strain on healthcare systems. Efficient delivery of healthcare services that prevent complications and manage chronic conditions effectively can curtail the necessity for more expensive emergency care and specialized treatments, leading to substantial savings in healthcare expenditures.

The broader social benefits of efficient and accessible healthcare are equally significant. Enhanced healthcare services improve the quality of life for individuals, fostering greater community satisfaction and well-being. Improvements in healthcare can extend life expectancy, reduce rates of disability, and foster a healthier population that enjoys more years in good health. Moreover, accessible healthcare plays a crucial role in educational outcomes. Healthy children are more likely to attend school and perform better academically, meaning that healthcare facilities that offer pediatric care and preventive services contribute to more promising educational achievements.

Reducing inequality is another critical advantage of accessible healthcare. Quality healthcare is a fundamental element in addressing social disparities. Facilities that provide equitable services to all community segments help bridge gaps in health equity, promoting more inclusive and equitable societies.

To further enhance healthcare's role as a public good, facilities can implement several strategies. Developing community outreach programs focused on preventive care, health education, and screenings can preemptively address health issues, reducing the need for extensive hospital visits and lowering healthcare costs. Collaborations with government entities, non-profits, and private sectors can expand healthcare services, particularly in underserved areas, fostering innovative solutions that benefit the entire community. Additionally, advocating for policies that improve healthcare access and affordability, such as supporting public health funding and insurance coverage, can enhance the overall healthcare landscape.

In conclusion, healthcare facilities are critical to the economic and social vitality of communities by bolstering public health. Viewing healthcare as a public good emphasizes the comprehensive benefits of investing in efficient and accessible healthcare systems. These investments pay dividends not only through enhanced economic productivity and reduced healthcare costs but also in fostering a healthier, more equitable society. This holistic approach highlights the transformative power of healthcare in shaping prosperous and resilient communities.

Equation 9.10 is devised to perform a comprehensive cost-benefit analysis of public health investments made by healthcare facilities. This analysis is essential in quantifying the economic returns from investments in public health programs, such as preventive care, wellness programs, and community health initiatives. By assessing these investments, healthcare facilities can demonstrate their value in economic terms, justifying the allocation of resources towards public health measures that yield significant community benefits. The cost-benefit analysis considers several crucial factors:

- Reduced Absenteeism ($A$): This measures the decrease in workplace absenteeism due to improved public health, reflecting lower sickness rates and higher attendance levels. Reduced absenteeism is directly linked to higher productivity and stability in the workforce, providing a clear economic benefit to the community.
- Increased Productivity ($P$): This factor quantifies the improvement in workforce productivity that results from better overall health. Healthier individuals can perform their tasks more efficiently, contributing more significantly to their workplaces and the economy.
- Lower Long-Term Healthcare Costs ($L$): Investments in public health can lead to reduced demand for medical treatment by preventing illnesses and promoting healthier lifestyles. This reduction in healthcare utilization decreases long-term healthcare costs for communities and healthcare systems alike.

Each factor is weighted ($w_A$, $w_P$, $w_L$) according to its economic impact, allowing the equation to be tailored to specific healthcare settings and regional economic conditions. The Public Health Cost-Benefit Ratio (PHC-BR) is calculated as follows:

$$\text{PHC-BR} = \frac{(A \times w_A) + (P \times w_P) + (L \times w_L)}{\text{Cost of Public Health Investment } (C)}. \tag{9.10}$$

This ratio provides a clear indication of the economic return on investment (ROI) for public health initiatives, helping healthcare facilities to evaluate the effectiveness of their programs in terms of both health outcomes and economic benefits. By using this equation, healthcare administrators can make informed decisions about where to allocate resources to maximize the benefits of public health investments, thus supporting not only the health of the population but also the economic vitality of the community.

### 9.5.5 Challenges in Balancing Cost and Quality

Balancing the cost of healthcare with the delivery of high-quality care presents a perennial challenge for healthcare facilities. This balance is crucial not only for the sustainability of healthcare providers but also for ensuring that patients receive the best possible care without encountering undue financial burdens. The complexities involved in this balance often stem from the high costs associated with advanced technologies, operational inefficiencies, and the pressures exerted by regulatory and market dynamics.

Investments in the latest medical technologies and state-of-the-art facilities are essential for enhancing the quality of healthcare. These advancements can provide more precise diagnostics, more effective treatments, and ultimately, better patient outcomes. However, the high acquisition and maintenance costs associated with cutting-edge technology can significantly inflate healthcare expenses, creating a substantial financial challenge for healthcare providers. Additionally, inefficiencies in healthcare operations, such as redundant processes, underutilized resources, or outdated practices, can further escalate costs without correspondingly enhancing the quality of care. These inefficiencies must be addressed to manage costs effectively.

Healthcare providers also contend with various pressures from insurance companies, government regulations, and the dynamics of the healthcare market. These pressures often compel providers to minimize costs, which can sometimes result in the reluctance to adopt innovative but costly treatments or technologies, potentially stifling advancements in patient care.

To navigate these challenges, healthcare facilities can employ several strategies aimed at balancing cost and quality effectively. Strategic investments in technology should focus on tools that provide clear, measurable improvements in patient outcomes and operational efficiencies. Employing cost-benefit analyses can help in determining which technologies will yield the best returns in terms of patient care enhancements and long-term cost reductions.

Implementing lean healthcare principles can also play a critical role in reducing costs while maintaining or improving the quality of care. These principles involve eliminating waste, simplifying procedures, and optimizing workflows, thereby ensuring that resources are concentrated on aspects that directly benefit patient outcomes. Additionally, adopting integrated care models that coordinate various aspects of patient care through a streamlined system can reduce duplicative services and enhance patient experiences, which can lead to better health outcomes and cost savings.

Shifting towards value-based care models, where providers are reimbursed based on patient health outcomes rather than the volume of services provided, is another effective strategy. This approach encourages healthcare facilities to focus on delivering high-quality care efficiently. Furthermore, utilizing data analytics to monitor and improve care quality and operational efficiency can enable healthcare facilities to make informed decisions that balance quality and cost-effectiveness effectively.

Investing in continuous training and education for healthcare providers ensures that they are up to date with the latest, most efficient practices and technologies. Well-trained staff are crucial for efficient resource use, which can reduce waste and enhance patient care. Additionally, focusing on patient-centered care that tailors treatments to individual patient needs can lead to more effective and efficient use of healthcare resources. Engaging patients in their care decisions not only improves satisfaction and health outcomes but can also reduce the need for costly follow-up treatments and readmissions.

In conclusion, while balancing cost and quality in healthcare is complex, it is achievable through strategic planning, investment in efficiency, and a focus on value-based care. By aligning operational strategies to improve patient outcomes at sustainable costs, healthcare facilities can ensure the provision of high-quality care while also managing their financial health effectively. This balanced approach is essential for the long-term viability of healthcare providers and the overall well-being of the communities they serve.

Table 9.16 lists common cost-cutting measures implemented in healthcare facilities alongside their potential impacts on healthcare quality, including case studies that illustrate successful balance achievements. Situated within the section on operational inefficiencies, this table provides valuable insights into how various healthcare organizations manage the delicate balance between reducing costs and maintaining high-quality care. It serves as a practical reference for healthcare administrators seeking to optimize budget allocations without compromising patient outcomes.

### *9.5.6 Strategies for Maximizing Economic Impact*

Healthcare facilities have the potential to be significant economic hubs within their Healthcare facilities can serve as significant economic hubs within their communities, catalyzing growth not only through direct healthcare services but also by exerting a substantial influence on local economies in various beneficial ways. To fully harness this potential, these institutions can implement a series of strategic actions designed to extend their impact beyond traditional healthcare delivery, thereby maximizing their economic contributions.

One effective strategy is the adoption of sustainable practices. By designing and retrofitting healthcare facilities with energy-efficient technologies—such as LED lighting, high-efficiency HVAC systems, and solar panels can significantly reduce long-term operating costs and contribute to environmental sustainability. Implementing comprehensive waste management strategies that include recycling, composting, and proper segregation of medical waste not only reduces operational costs but also enhances the facility's reputation as an environmentally responsible community member.

Another pivotal strategy is the development of centers of excellence in specialized medical fields such as cardiology, oncology, or orthopedics. Establishing these centers enhances the facility's reputation, attracts top-tier medical professionals,

**Table 9.16** Balancing cost and quality in healthcare: comparative analysis of cost-cutting measures and their impact on care

| Cost-cutting measure | Potential impact on quality | Case study example | Outcome of balance |
|---|---|---|---|
| Reduction in non-critical staff | May increase the workload for the remaining staff, affecting care quality | General hospital A streamlined administrative roles | Achieved savings with minimal impact on patient care by enhancing operational efficiency |
| Generic medication substitution | Lower cost but potential variability in patient response | Clinic B implemented a generic-first prescription policy | Maintained treatment effectiveness with significant cost reductions |
| Outsourcing non-medical services | Can reduce operational costs, risk of service quality variability | Health center C outsourced facility management | Reduced operational expenses while maintaining high cleanliness and safety standards |
| Utilization of telemedicine | Reduces physical infrastructure costs, but requires technological adaptation | Specialized unit D expanded telehealth services | Expanded access to care at reduced cost, with high patient satisfaction |
| Centralized purchasing | Economies of scale can lower costs, risk of over-dependence on specific suppliers | Regional hospital E centralized medical supplies procurement | Lowered supply costs with improved stock management practices |

and draws patients from broader geographical areas, all of which contribute to local and regional economic development. Investing in research and development to pioneer new treatments and medical technologies not only elevates patient care but also boosts the facility's standing in the medical community, potentially creating additional revenue streams through patents and innovations.

Promoting medical tourism is yet another strategic avenue. Obtaining international healthcare accreditations can enhance a facility's global credibility and attract international patients seeking high-quality medical care. Offering packages that integrate medical treatment with local hospitality services, including accommodations and cultural experiences, can appeal to medical tourists and stimulate local tourism.

Community partnerships and economic development also play a crucial role in maximizing economic impact. Collaborating with local businesses to meet various needs of the healthcare facility—from catering and laundry services to IT support—helps keep economic benefits within the community and supports local enterprises. Partnering with educational institutions to offer training programs can fill employment needs within the facility and enhance the community's overall employment qualifications by equipping residents with the necessary skills in healthcare professions.

**Table 9.17** Strategies for maximizing economic impact in healthcare facilities

| Strategy | Action example | Outcome example |
|---|---|---|
| Adopting sustainable practices | Installation of solar panels and high-efficiency HVAC systems | Reduced energy costs by 30%, improved environmental sustainability ratings |
| Developing centers of excellence | Establishment of a specialized cardiology center | Attracted top cardiologists, increased patient referrals by 40% |
| Promoting medical tourism | Gained international accreditation and partnered with local tourism businesses | Drew patients globally, increased local tourism spending by 20% |
| Community partnerships and economic development | Partnered with local technical colleges to train healthcare workers | Enhanced local employment provided a steady pipeline of qualified staff |
| Leveraging technology for efficiency and outreach | Expanded telemedicine services across state lines | Increased patient base by 25%, reduced operational costs through enhanced efficiency |

Lastly, leveraging technology for efficiency and outreach is essential. Expanding services beyond physical boundaries through telemedicine not only makes healthcare more accessible but also allows the facility to serve a larger population, thus enhancing its economic impact. Utilizing advanced health information technology can streamline operations, improve patient outcomes, and lead to more efficient resource use, thereby increasing the facility's capacity for patient care.

In conclusion, by integrating these strategic actions, healthcare facilities can transcend their traditional roles and significantly enhance their economic impact. Embracing sustainable practices, specialization in medical fields, promoting medical tourism, fostering community partnerships, and leveraging technology are key strategies that can transform these institutions into cornerstone entities for economic stability and growth in their regions. This comprehensive approach ensures that healthcare facilities contribute robustly not only to the health and wellness of their communities but also to their economic vitality and sustainability.

Table 9.17 summarizes strategic actions healthcare facilities can undertake to maximize their economic impact, featuring examples of each strategy and the outcomes achieved. Positioned after the detailed discussion of these strategies, this table serves as a quick reference guide for healthcare administrators and planners looking to enhance the economic contributions of their facilities. It illustrates how targeted actions can extend a facility's influence beyond traditional healthcare delivery, fostering substantial economic growth within their communities.

# Further Reading


Aceto, G., Persico, V., & Pescapé, A. (2018). The role of Information and Communication Technologies in healthcare: taxonomies, perspectives, and challenges. *Journal of Network and Computer Applications*, 107, 125–154. https://doi.org/10.1016/j.jnca.2018.02.008

Ahmed, S. (2019). Integrating DMAIC approach of Lean Six Sigma and theory of constraints toward quality improvement in healthcare. *Reviews on Environmental Health*, 34(4), 427–434. https://doi.org/10.1515/reveh-2019-0003

Alam, M., Zou, P. X., Stewart, R. A., Bertone, E., Sahin, O., Buntine, C., & Marshall, C. (2019). Government championed strategies to overcome the barriers to public building energy efficiency retrofit projects. *Sustainable Cities and Society*, 44, 56–69. https://doi.org/10.1016/j.scs.2018.09.022

Al Shamsi, H., Almutairi, A. G., Al Mashrafi, S., & Al Kalbani, T. (2020). Implications of language barriers for healthcare: a systematic review. *Oman Medical Journal*, 35(2), e122. https://doi.org/10.5001/omj.2020.40

Ali, P. A., & Watson, R. (2018). Language barriers and their impact on provision of care to patients with limited English proficiency: Nurses' perspectives. *Journal of Clinical Nursing*, 27(5–6), e1152–e1160. https://doi.org/10.1111/jocn.14204

Al-Dmour, H., Salman, A., Abuhashesh, M., & Al-Dmour, R. (2020). Influence of social media platforms on public health protection against the COVID-19 pandemic via the mediating effects of public health awareness and behavioral changes: integrated model. *Journal of Medical Internet Research*, 22(8), e19996. https://doi.org/10.2196/19996

Attaran, M. (2022). Blockchain technology in healthcare: Challenges and opportunities. International *Journal of Healthcare Management*, 15(1), 70–83. https://doi.org/10.1080/20479700.2020.1843887

Auschra, C. (2018). Barriers to the integration of care in inter-organisational settings: a literature review. *International Journal of Integrated Care*, 18(1). https://doi.org/10.5334/ijic.3068

Bantham, A., Ross, S. E. T., Sebastião, E., & Hall, G. (2021). Overcoming barriers to physical activity in underserved populations. *Progress in Cardiovascular Diseases*, 64, 64–71. https://doi.org/10.1016/j.pcad.2020.11.002

Berlinger, N., Wynia, M., Powell, T., Hester, D. M., Milliken, A., Fabi, R., & Jenks, N. P. (2020). Ethical framework for health care institutions responding to novel Coronavirus SARS-CoV-2 (COVID-19) guidelines for institutional ethics services responding to COVID-19. *The Hastings Center*, 12(3), 1–12.

Bertone, E., Stewart, R. A., Sahin, O., Alam, M., Zou, P. X., Buntine, C., & Marshall, C. (2018). Guidelines, barriers and strategies for energy and water retrofits of public buildings. *Journal of Cleaner Production*, 174, 1064–1078. https://doi.org/10.1016/j.jclepro.2017.11.065

Bestsennyy, O., Chmielewski, M., Koffel, A., & Shah, A. (2022). From facility to home: how healthcare could shift by 2025. McKinsey & Company.

Bonawitz, K., Wetmore, M., Heisler, M., Dalton, V. K., Damschroder, L. J., Forman, J., … & Moniz, M. H. (2020). Champions in context: which attributes matter for change efforts in healthcare?. *Implementation Science*, 15, 1–10. https://doi.org/10.1186/s13012-020-01024-9

Bombard, Y., Baker, G. R., Orlando, E., Fancott, C., Bhatia, P., Casalino, S., … & Pomey, M. P. (2018). Engaging patients to improve quality of care: a systematic review. *Implementation Science*, 13, 1–22. https://doi.org/10.1186/s13012-018-0784-z

Borkowski, N., & Meese, K. A. (2020). Organizational behavior in health care. Jones & Bartlett Learning.

Brottman, M. R., Char, D. M., Hattori, R. A., Heeb, R., & Taff, S. D. (2020). Toward cultural competency in health care: a scoping review of the diversity and inclusion education literature. *Academic Medicine*, 95(5), 803–813. https://doi.org/10.1097/ACM.0000000000002995

Cantor, M. N., & Thorpe, L. (2018). Integrating data on social determinants of health into electronic health records. *Health Affairs*, 37(4), 585–590. https://doi.org/10.1377/hlthaff.2017.1252

Challa, S. S. S., Tilala, M., Chawda, A. D., & Benke, A. P. (2022). Quality management systems in regulatory affairs: Implementation challenges and solutions. *Journal for Research in Applied Sciences and Biotechnology*, 1(3), 278–284. https://doi.org/10.55544/jrasb.1.3.36

Chen, P. T., Lin, C. L., & Wu, W. N. (2020). Big data management in healthcare: Adoption challenges and implications. *International Journal of Information Management*, 53, 102078. https://doi.org/10.1016/j.ijinfomgt.2020.102078

Christensen, H. B., Floyd, E., & Maffett, M. (2020). The only prescription is transparency: The effect of charge-price-transparency regulation on healthcare prices. *Management Science*, 66(7), 2861–2882. https://doi.org/10.1287/mnsc.2019.3330

Clark, L. T., Watkins, L., Piña, I. L., Elmer, M., Akinboboye, O., Gorham, M., … & Regnante, J. M. (2019). Increasing diversity in clinical trials: overcoming critical barriers. *Current Problems in Cardiology*, 44(5), 148–172. https://doi.org/10.1016/j.cpcardiol.2018.11.002

Crown, W., Buyukkaramikli, N., Sir, M. Y., Thokala, P., Morton, A., Marshall, D. A., … & Pasupathy, K. S. (2018). Application of constrained optimization methods in health services research: report 2 of the ISPOR optimization methods emerging good practices task force. *Value in Health*, 21(9), 1019–1028. https://doi.org/10.1016/j.jval.2018.05.003

Delaney, L. J. (2018). Patient-centred care as an approach to improving health care in Australia. Collegian, 25(1), 119–123. https://doi.org/10.1016/j.colegn.2017.02.005

Deline, M. B. (2019). Framing resistance: identifying frames that guide resistance interpretations at work. *Management Communication Quarterly*, 33(1), 39–67. https://doi.org/10.1177/0893318918793731

Dinh-Le, C., Chuang, R., Chokshi, S., & Mann, D. (2019). Wearable health technology and electronic health record integration: scoping review and future directions. *JMIR mHealth and uHealth*, 7(9), e12861. https://doi.org/10.2196/12861

Dixit, M. K., Venkatraj, V., Ostadalimakhmalbaf, M., Pariafsai, F., & Lavy, S. (2019). Integration of facility management and building information modeling (BIM) A review of key issues and challenges. *Facilities*, 37(7/8), 455–483. https://doi.org/10.1108/F-03-2018-0043

Duggan, C., Dvaladze, A., Rositch, A. F., Ginsburg, O., Yip, C. H., Horton, S., … & Anderson, B. O. (2020). The breast health global initiative 2018 global summit on improving breast healthcare through resource-stratified phased implementation: methods and overview. *Cancer*, 126, 2339–2352. https://doi.org/10.1002/cncr.32891

Errida, A., & Lotfi, B. (2021). The determinants of organizational change management success: Literature review and case study. *International Journal of Engineering Business Management*, 13, 18479790211016273. https://doi.org/10.1177/18479790211016273

Figueroa, C. A., Harrison, R., Chauhan, A., & Meyer, L. (2019). Priorities and challenges for health leadership and workforce management globally: a rapid review. *BMC Health Services Research*, 19, 1–11. https://doi.org/10.1186/s12913-019-4080-7

Frederix, I., Caiani, E. G., Dendale, P., Anker, S., Bax, J., Böhm, A., … & van der Velde, E. (2019). ESC e-Cardiology Working Group Position Paper: Overcoming challenges in digital health implementation in cardiovascular medicine. *European Journal of Preventive Cardiology*, 26(11), 1166–1177. https://doi.org/10.1177/2047487319832394

Fung-Kee-Fung, M., Maziak, D. E., Pantarotto, J. R., Smylie, J., Taylor, L., Timlin, T., … & Stewart, D. J. (2018). Regional process redesign of lung cancer care: a learning health system pilot project. *Current Oncology*, 25(1), 59. https://doi.org/10.3747/co.25.3719

Ginsburg, O., Yip, C. H., Brooks, A., Cabanes, A., Caleffi, M., Dunstan Yataco, J. A., … & Anderson, B. O. (2020). Breast cancer early detection: A phased approach to implementation. *Cancer*, 126, 2379–2393. https://doi.org/10.1002/cncr.32887

Gonzalez, M. E. (2019). Improving customer satisfaction of a healthcare facility: reading the customers' needs. *Benchmarking: An International Journal*, 26(3), 854–870. https://doi.org/10.1108/BIJ-01-2017-0007

He, W., Zhang, Z. J., & Li, W. (2021). Information technology solutions, challenges, and suggestions for tackling the COVID-19 pandemic. *International Journal of Information Management*, 57, 102287. https://doi.org/10.1016/j.ijinfomgt.2020.102287

Henderson, S., Horne, M., Hills, R., & Kendall, E. (2018). Cultural competence in healthcare in the community: A concept analysis. *Health & Social Care in the Community*, 26(4), 590–603. https://doi.org/10.1111/hsc.12556

Holmes, L., Cresswell, K., Williams, S., Parsons, S., Keane, A., Wilson, C., ... & Starling, B. (2019). Innovating public engagement and patient involvement through strategic collaboration and practice. *Research Involvement and Engagement*, 5, 1–12. https://doi.org/10.1186/s40900-019-0160-4

Jaeger, F. N., Pellaud, N., Laville, B., & Klauser, P. (2019). Barriers to and solutions for addressing insufficient professional interpreter use in primary healthcare. *BMC Health Services Research*, 19, 1–11. https://doi.org/10.1186/s12913-019-4628-6

Jongen, C., McCalman, J., & Bainbridge, R. (2018). Health workforce cultural competency interventions: a systematic scoping review. *BMC Health Services Research*, 18, 1–15. https://doi.org/10.1186/s12913-018-3001-5

Kaimahranen, A., Hietapakka, L., & Heponiemi, T. (2019). Increasing cultural awareness: qualitative study of nurses' perceptions about cultural competence training. *BMC Nursing*, 18, 1–9. https://doi.org/10.1186/s12912-019-0363-x

Karamat, J., Shurong, T., Ahmad, N., Waheed, A., & Khan, S. (2018). Barriers to knowledge management in the health sector of Pakistan. *Sustainability*, 10(11), 4155. https://doi.org/10.3390/su10114155

Karatas, M., Eriskin, L., Deveci, M., Pamucar, D., & Garg, H. (2022). Big Data for Healthcare Industry 4.0: Applications, challenges and future perspectives. *Expert Systems with Applications*, 200, 116912. https://doi.org/10.1016/j.eswa.2022.116912

Kaye, A. D., Okeagu, C. N., Pham, A. D., Silva, R. A., Hurley, J. J., Arron, B. L., ... & Cornett, E. M. (2021). Economic impact of COVID-19 pandemic on healthcare facilities and systems: International perspectives. *Best Practice & Research Clinical Anaesthesiology*, 35(3), 293–306. https://doi.org/10.1016/j.bpa.2020.11.009

Khullar, D., Bond, A. M., & Schpero, W. L. (2020). COVID-19 and the financial health of US hospitals. *Jama*, 323(21), 2127–2128. https://doi.org/10.1001/jama.2020.6269

Kluge, H., Martín-Moreno, J. M., Emiroglu, N., Rodier, G., Kelley, E., Vujnovic, M., & Permanand, G. (2018). Strengthening global health security by embedding the International Health Regulations requirements into national health systems. *BMJ Global Health*, 3(Suppl 1), e000656. https://doi.org/10.1136/bmjgh-2017-000656

Lee, D., & Yoon, S. N. (2021). Application of artificial intelligence-based technologies in the healthcare industry: Opportunities and challenges. *International Journal of Environmental Research and Public Health*, 18(1), 271. https://doi.org/10.3390/ijerph18010271

Lee, S. M., & Lee, D. (2021). Opportunities and challenges for contactless healthcare services in the post-COVID-19 Era. *Technological Forecasting and Social Change*, 167, 120712. https://doi.org/10.1016/j.techfore.2021.120712

Mackey, T. K., Kuo, T. T., Gummadi, B., Clauson, K. A., Church, G., Grishin, D., ... & Palombini, M. (2019). 'Fit-for-purpose?'–challenges and opportunities for applications of blockchain technology in the future of healthcare. *BMC Medicine*, 17, 1–17. https://doi.org/10.1186/s12916-019-1296-7

Maki, O., Alshaikhli, M., Gunduz, M., Naji, K. K., & Abdulwahed, M. (2022). Development of digitalization road map for healthcare facility management. *IEEE Access*, 10, 14450–14462. https://doi.org/10.1109/ACCESS.2022.3146341

McGregor, B., Belton, A., Henry, T. L., Wrenn, G., & Holden, K. B. (2019). Improving behavioral health equity through cultural competence training of health care providers. *Ethnicity & Disease*, 29(Suppl 2), 359. https://doi.org/10.18865/ed.29.S2.359

McIntyre, D., Obse, A. G., Barasa, E. W., & Ataguba, J. E. (2018). Challenges in financing universal health coverage in sub-Saharan Africa. In *Oxford Research Encyclopedia of Economics and Finance*. https://doi.org/10.1093/acrefore/9780190625979.013.28

Melles, M., Albayrak, A., & Goossens, R. (2021). Innovating health care: key characteristics of human-centered design. *International Journal for Quality in Health Care*, 33(Supplement_1), 37–44. https://doi.org/10.1093/intqhc/mzaa127

Menear, M., Blanchette, M. A., Demers-Payette, O., & Roy, D. (2019). A framework for value-creating learning health systems. *Health Research Policy and Systems*, 17, 1–13. https://doi.org/10.1186/s12961-019-0477-3

Miller, C. J., Barnett, M. L., Baumann, A. A., Gutner, C. A., & Wiltsey-Stirman, S. (2021). The FRAME-IS: a framework for documenting modifications to implementation strategies in healthcare. *Implementation Science*, 16, 1–12. https://doi.org/10.1186/s13012-021-01105-3

Morris, A. M., Bai, A., Burry, L., Dresser, L. D., Ferguson, N. D., Lapinsky, S. E., … & Bell, C. M. (2019). Long-term effects of phased implementation of antimicrobial stewardship in academic ICUs: 2007–2015. *Critical Care Medicine*, 47(2), 159–166. https://doi.org/10.1097/CCM.0000000000003514

MULUKUNTLA, S. (2022). Generative AI-Benefits, Limitations, Potential risks and Challenges in Healthcare Industry. *EPH-International Journal of Medical and Health Science*, 8(4), 1–9. https://doi.org/10.53555/eijmhs.v8i4.211

Mühlebach, S. (2018). Regulatory challenges of nanomedicines and their follow-on versions: a generic or similar approach?. *Advanced Drug Delivery Reviews*, 131, 122–131. https://doi.org/10.1016/j.addr.2018.06.024

Murrar, S., & Brauer, M. (2019). Overcoming resistance to change: Using narratives to create more positive intergroup attitudes. *Current Directions in Psychological Science*, 28(2), 164–169. https://doi.org/10.1177/0963721418818552

Nair, L., & Adetayo, O. A. (2019). Cultural competence and ethnic diversity in healthcare. *Plastic and Reconstructive Surgery–Global Open*, 7(5), e2219. https://doi.org/10.1097/GOX.0000000000002219

Nilsen, P., Seing, I., Ericsson, C., Birken, S. A., & Schildmeijer, K. (2020). Characteristics of successful changes in health care organizations: an interview study with physicians, registered nurses and assistant nurses. *BMC Health Services Research*, 20, 1–8. https://doi.org/10.1186/s12913-020-4999-8

Nyblade, L., Stockton, M. A., Giger, K., Bond, V., Ekstrand, M. L., Lean, R. M., … & Wouters, E. (2019). Stigma in health facilities: why it matters and how we can change it. *BMC Medicine*, 17, 1–15. https://doi.org/10.1186/s12916-019-1256-2

Olatomiwa, L., Blanchard, R., Mekhilef, S., & Akinyele, D. (2018). Hybrid renewable energy supply for rural healthcare facilities: An approach to quality healthcare delivery. *Sustainable Energy Technologies and Assessments*, 30, 121–138. https://doi.org/10.1016/j.seta.2018.09.007

Origlia Ikhilor, P., Hasenberg, G., Kurth, E., Asefaw, F., Pehlke-Milde, J., & Cignacco, E. (2019). Communication barriers in maternity care of allophone migrants: Experiences of women, healthcare professionals, and intercultural interpreters. *Journal of Advanced Nursing*, 75(10), 2200–2210. https://doi.org/10.1111/jan.14093

Oleribe, O. O., Momoh, J., Uzochukwu, B. S., Mbofana, F., Adebiyi, A., Barbera, T., … & Taylor-Robinson, S. D. (2019). Identifying key challenges facing healthcare systems in Africa and potential solutions. *International Journal of General Medicine*, 395–403.

Pacifico Silva, H., Lehoux, P., Miller, F. A., & Denis, J. L. (2018). Introducing responsible innovation in health: a policy-oriented framework. *Health Research Policy and Systems*, 16, 1–13. https://doi.org/10.1186/s12961-018-0362-5

Rowe, A. K., Rowe, S. Y., Peters, D. H., Holloway, K. A., Chalker, J., & Ross-Degnan, D. (2018). Effectiveness of strategies to improve health-care provider practices in low-income and middle-income countries: a systematic review. The Lancet Global Health, 6(11), e1163–e1175.

Safi, S., Thiessen, T., & Schmailzl, K. J. (2018). Acceptance and resistance of new digital technologies in medicine: qualitative study. *JMIR Research Protocols*, 7(12), e11072. https://doi.org/10.2196/11072

Santana, M. J., Manalili, K., Jolley, R. J., Zelinsky, S., Quan, H., & Lu, M. (2018). How to practice person-centred care: A conceptual framework. *Health Expectations*, 21(2), 429–440. https://doi.org/10.1111/hex.12640

Seyhan, A. A. (2019). Lost in translation: the valley of death across preclinical and clinical divide–identification of problems and overcoming obstacles. *Translational Medicine Communications*, 4(1), 1–19. https://doi.org/10.1186/s41231-019-0050-7

Shahbaz, M., Gao, C., Zhai, L., Shahzad, F., & Hu, Y. (2019). Investigating the adoption of big data analytics in healthcare: the moderating role of resistance to change. *Journal of Big Data*, 6(1), 1–20. https://doi.org/10.1186/s40537-019-0170-y

Sharifi, N., Adib-Hajbaghery, M., & Najafi, M. (2019). Cultural competence in nursing: A concept analysis. *International Journal of Nursing Studies*, 99, 103386. https://doi.org/10.1016/j.ijnurstu.2019.103386

Sheikh, A., Anderson, M., Albala, S., Casadei, B., Franklin, B. D., Richards, M., … & Mossialos, E. (2021). Health information technology and digital innovation for national learning health and care systems. *The Lancet Digital Health*, 3(6), e383–e396.

Shrank, W. H., Keyser, D. J., & Lovelace, J. G. (2018). Redistributing investment in health and social services—the evolving role of managed care. *Jama*, 320(21), 2197–2198. https://doi.org/10.1001/jama.2018.14987

Smith, R. L., Soeters, M. R., Wüst, R. C., & Houtkooper, R. H. (2018). Metabolic flexibility as an adaptation to energy resources and requirements in health and disease. *Endocrine Reviews*, 39(4), 489–517. https://doi.org/10.1210/er.2017-00211

Stouten, J., Rousseau, D. M., & De Cremer, D. (2018). Successful organizational change: Integrating the management practice and scholarly literatures. *Academy of Management Annals*, 12(2), 752–788. https://doi.org/10.5465/annals.2016.0095

Stubbe, D. E. (2020). Practicing cultural competence and cultural humility in the care of diverse patients. *Focus*, 18(1), 49–51. https://doi.org/10.1176/appi.focus.20190041

Teisberg, E., Wallace, S., & O'Hara, S. (2020). Defining and implementing value-based health care: a strategic framework. *Academic Medicine*, 95(5), 682–685. https://doi.org/10.1097/ACM.0000000000003122

Uyarra, E., Zabala-Iturriagagoitia, J. M., Flanagan, K., & Magro, E. (2020). Public procurement, innovation and industrial policy: Rationales, roles, capabilities and implementation. *Research Policy*, 49(1), 103844. https://doi.org/10.1016/j.respol.2019.103844

Vaccari, M., Tudor, T., & Perteghella, A. (2018). Costs associated with the management of waste from healthcare facilities: an analysis at national and site level. *Waste Management & Research*, 36(1), 39–47. https://doi.org/10.1177/0734242X17739968

Vaishnavi, V., Suresh, M., & Dutta, P. (2019). A study on the influence of factors associated with organizational readiness for change in healthcare organizations using TISM. *Benchmarking: An International Journal*, 26(4), 1290–1313. https://doi.org/10.1108/BIJ-06-2018-0161

van der Merwe, E., Grobbelaar, S., & Bam, W. (2020). Exploring the functional dynamics of innovation for inclusive development innovation systems: a case study of a large scale maternal mHealth project in South Africa. *Innovation and Development*. https://doi.org/10.1080/2157930X.2019.1567884

Vayena, E., Haeusermann, T., Adjekum, A., & Blasimme, A. (2018). Digital health: meeting the ethical and policy challenges. *Swiss Medical Weekly*, 148, w14571. https://doi.org/10.4414/smw.2018.14571

World Health Organization. (2020). Operational framework for primary health care: transforming vision into action.

Yazdani, A., & Wells, R. (2018). Barriers for implementation of successful change to prevent musculoskeletal disorders and how to systematically address them. *Applied Ergonomics*, 73, 122–140. https://doi.org/10.1016/j.apergo.2018.05.004

# Chapter 10
# Architecting the Future of Healthcare: Integrative Design Strategies for Enhanced Care and Community Well-Being

This chapter synthesizes insights from the comprehensive exploration of healthcare facility design, underscoring the profound impact that architectural strategies have on patient care, staff well-being, and community health. This chapter delves into several crucial aspects: the role of patient-centric designs in enhancing healthcare experiences, the integration of advanced technology to streamline care, the importance of sustainability and resilience in construction, and the adaptability of facilities to future healthcare challenges. It also highlights the significance of inclusivity and accessibility in fostering environments that cater to diverse populations. Furthermore, this chapter explores visionary perspectives on the future of hospital architecture, emphasizing the integration of smart technologies, personalized healthcare environments, and advanced modular designs. The discussion extends to the pivotal role of mental and emotional well-being in healthcare settings, advocating for environments that support holistic healing. Additionally, this chapter projects future trends, including the increasing integration of community-centric spaces within healthcare facilities, aiming to transform them into hubs of health education and preventive care. Conclusively, it calls for a collaborative approach among architects, healthcare professionals, and policymakers to foster spaces that are not only healing but also innovative and reflective of advanced medical practices. This forward-looking outlook emphasizes the need for adaptable, patient-focused, and technologically integrated healthcare environments that resonate with global health dynamics and community needs.

## 10.1 Synthesis of Key Insights

Throughout this exploration of global healthcare design, several key insights have emerged that underscore the critical role architecture plays in shaping the efficacy, efficiency, and humanitarian impact of healthcare facilities. These insights not only highlight the complexities involved in designing modern healthcare environments but

 
A. A. Firoozi and A. A. Firoozi, *Transformative Design in Healthcare*,
https://doi.org/10.1007/978-3-032-08019-6_10

also demonstrate how thoughtful design can profoundly influence patient outcomes, staff well-being, and community health.

### 10.1.1 Patient-Centric Design

In the realm of healthcare architecture and facility planning, a patient-centric design philosophy is instrumental. This approach transcends traditional architectural aesthetics, focusing intensely on the needs and experiences of patients to facilitate an environment conducive to healing and comfort. The core premise of patient-centric design is to create healthcare settings that are not only visually appealing but are also functionally tailored to support the health and recovery of patients. This involves a meticulous consideration of various elements, such as the physical layout, which should promote easy navigation and minimize stress for both patients and their families. Navigation in healthcare facilities is often a source of significant stress, particularly in large hospitals where the complexity of the layout can be overwhelming. Effective patient-centric designs address these challenges by incorporating clear, intuitive pathways, ample signage, and a layout that logically connects different departments, thereby easing the cognitive load on visitors and patients alike.

Furthermore, accessibility is a critical component, ensuring that all patients, including those with limited mobility, can move freely and independently within the facility. This not only involves adherence to legal standards for accessibility but also extends to thoughtful touches like the placement of handrails, wheelchair-accessible entrances and facilities, and room designs that accommodate various medical needs and equipment. In addition to these functional aspects, the integration of natural elements plays a pivotal role in patient-centric design. Research has consistently shown that views of nature, natural light, and spaces that offer a connection to the outdoors can significantly expedite recovery times and enhance the overall well-being of patients. This biophilic design approach harnesses the restorative powers of nature to create spaces that are not merely functional but truly healing environments.

The implementation of patient-centric design principles is a complex, multidisciplinary effort that requires collaboration among architects, healthcare professionals, and even patients themselves. It demands a deep understanding of how the environment can affect health outcomes, as well as a commitment to creating spaces that respect and respond to the needs of those who use them. In doing so, healthcare facilities have become more than buildings; they are dynamic environments that contribute actively to the process of healing and comfort, embodying the ethos of compassion and care that is at the heart of the medical profession.

Table 10.1 presents a detailed checklist essential for integrating patient-centric design within healthcare facilities, emphasizing accessibility, navigation, and biophilic elements. It outlines critical design features such as wheelchair-accessible entrances with no steps, handrails at optimal heights, and restroom facilities that accommodate all mobility levels. For navigation, the checklist specifies clear, multilingual signage and a logical layout with color-coded paths to simplify wayfinding.

**Table 10.1** Comprehensive checklist for implementing patient-centric design in healthcare facilities

| Design element category | Checklist item | Details/specifications |
|---|---|---|
| Accessibility | Wheelchair-accessible entrances | Entrances should have no steps and doors should be automated |
| | Handrails in corridors | Handrails are to be installed on both sides at a height of 34–38 inches |
| | Accessible restroom facilities | Restrooms to accommodate wheelchairs; include support bars and emergency call buttons |
| Navigation | Clear signage in multiple languages | Signs should be visually clear, using universal symbols where possible |
| | Logical layout and departmental signage | Pathways to departments should be direct and marked by color-coded lines or shapes |
| Biophilic design | Incorporation of natural light | Utilize large windows and skylights where structural design permits |
| | Indoor plant installations | Non-allergenic plants in common areas and patient rooms to enhance air quality and aesthetics |
| | Outdoor patient-accessible gardens | Gardens should be easily accessible, including pathways suitable for all mobility levels |

The biophilic design criteria include large windows to maximize natural light, indoor plant installations to enhance air quality, and accessible outdoor gardens, creating a healing environment that improves patient recovery and staff well-being. This table serves as a practical guide for healthcare administrators aiming to create environments that are not only functional but also supportive of holistic health.

### *10.1.2 Technological Integration*

The seamless integration of technology within healthcare environments represents a cornerstone in the evolution of modern medical facilities. As the healthcare sector continues to embrace digital transformation, the architectural design of these facilities must not only accommodate but also enhance the functionality and accessibility of advanced medical technologies. This integration is vital, ranging from telemedicine platforms to sophisticated diagnostic and treatment technologies that have the potential to revolutionize patient care.

At the forefront of this technological adoption is telemedicine, which has rapidly expanded access to healthcare services, particularly in remote or underserved regions. Architectural designs must therefore include dedicated spaces that are optimized for virtual consultations, ensuring privacy and quiet. Furthermore, these designs should incorporate robust data connectivity to support real-time, high-quality video communications. Beyond telemedicine, healthcare facilities increasingly rely on advanced diagnostic tools, such as MRI machines and CT scanners, which require specialized spaces designed to accommodate their size, technical specifications, and safety requirements, including radiation shielding.

Moreover, the integration of technology extends to patient management systems that streamline the workflow of medical personnel and enhance patient experience. Architectural designs must include areas that facilitate quick and easy access to these systems for healthcare providers. For instance, nursing stations can be strategically placed and designed to provide optimum visibility to patient rooms and central monitoring systems, allowing for efficient care coordination and continuous patient monitoring.

The challenge lies in ensuring that these technological integrations enhance rather than complicate the delivery of care. This requires a design philosophy that foresees the needs of future technologies while remaining adaptable. For example, infrastructural flexibility could be built into the design, allowing for easy upgrades and modifications as new technologies emerge. Additionally, care must be taken to ensure that the introduction of technology into healthcare settings does not introduce barriers between patients and medical staff but rather serves as a bridge to facilitate communication, streamline processes, and ultimately enhance the overall quality of care.

In conclusion, the architectural design of healthcare facilities must be visionary, anticipating the rapid advancements in technology and preparing environments that support and advance these innovations. This approach not only future-proofs medical facilities but also ensures that they remain at the cutting edge of patient care, effectively blending high-tech with high-touch service in the pursuit of optimal health outcomes.

Equation 10.1 describes the method for calculating the return on investment ($ROI$) for technological integrations within healthcare settings. The equation provides a comprehensive framework to quantify the financial benefits derived from implementing advanced medical technologies, considering several critical factors. These factors include cost savings from operational efficiencies, enhanced patient outcomes due to improved diagnostic and treatment capabilities, and the reduction in procedure times which contributes to higher throughput in patient care services. The ROI is a vital metric for healthcare administrators to assess the effectiveness of investments in technology and to justify future expenditures in upgrading or expanding technological infrastructure. The formula integrates the initial investment cost ($I$), the annual savings or gains from the technology ($S$), and the expected lifespan of the technology ($L$), to calculate the overall return on investment. This calculation not only helps in making informed decisions about where to allocate resources but also in evaluating the impact of technology on improving care quality and operational efficiency. The return on investment for technological implementation in healthcare is therefore given by:

$$\text{ROI} = \left( \frac{S \times L - I}{I} \right) \times 100. \tag{10.1}$$

This approach ensures that the financial assessments are aligned with the strategic goals of healthcare facilities and that investments are made in technologies that offer significant value in terms of cost, efficiency, and patient care outcomes.

### *10.1.3 Sustainability and Resilience*

The imperative for sustainability and resilience in the design of healthcare facilities has become increasingly urgent as the environmental impact of these structures gains prominence. This shift towards eco-friendly architecture is not just about mitigating harm to the environment but also about harnessing design as a tool to promote better health outcomes and create more supportive work environments for staff. Sustainable healthcare facilities are designed with a holistic approach that considers energy efficiency, environmentally friendly materials, and the incorporation of green spaces, all of which contribute significantly to the well-being of patients and the efficiency of medical operations.

Energy efficiency is a critical component of sustainable healthcare design, involving sophisticated heating, ventilation, and air conditioning (HVAC) systems that reduce energy consumption while maintaining optimal indoor air quality, essential for patient recovery and staff performance. The use of high-performance insulation and glazing, energy-efficient lighting systems, and renewable energy sources such as solar panels also play a vital role in reducing the carbon footprint of healthcare facilities. Furthermore, the choice of sustainable materials that are non-toxic, recycled, or sustainably sourced—reduces environmental impact and enhances the building's overall healthfulness.

Beyond these technical specifications, the integration of green spaces within and around healthcare facilities offers substantial health benefits. These areas provide tranquil settings for relaxation and recovery, reducing stress and promoting mental well-being for patients, visitors, and staff alike. Moreover, natural landscaping around buildings can improve air quality and reduce heat island effects, contributing to the facility's resilience against climate change-related impacts.

Additionally, designing resilience involves preparing facilities to withstand natural disasters and other emergencies, ensuring continuity of care during critical times. This may include structural reinforcements that allow buildings to resist seismic activities, flood-proofing measures, and the strategic placement of critical infrastructure to ensure that power and water supplies are maintained during disruptions.

The challenge for architects and planners lies in balancing these sustainability goals with the complex needs of a healthcare setting, ensuring that facilities are not only environmentally responsible but also robust and adaptable to changing conditions. This requires innovative thinking and a commitment to long-term sustainability goals, positioning healthcare facilities as leaders in the global movement towards more sustainable and resilient urban development. By adopting these principles, healthcare facilities can set a benchmark for performance that aligns with the broader societal shifts towards sustainability and resilience, ultimately creating environments that heal both patients and the planet.

Table 10.2 outlines the utilization of sustainable materials in healthcare facilities, detailing their properties, environmental impacts, and effects on patient health. This table highlights materials such as recycled glass, which is non-toxic and durable, helping reduce landfill waste and minimizing patient exposure to toxins. Bamboo is noted for its rapid renewability and strength, which not only curbs deforestation but also enhances the aesthetic quality of patient environments, contributing to overall well-being. Low-VOC paints are showcased for their role in improving indoor air quality by emitting fewer pollutants, crucial for patients with respiratory conditions. Cork flooring is appreciated for its sustainable harvesting and biodegradable nature, offering a comfortable, antimicrobial surface that diminishes noise, thus creating a quieter, more peaceful setting conducive to healing. Recycled steel is emphasized for its strength and recyclability, reducing the need for new material extraction and ensuring easy maintenance. Collectively, these materials exemplify how healthcare facilities can integrate eco-friendly practices that promote environmental sustainability while simultaneously improving patient care environments, aligning with broader health and safety standards.

### *10.1.4 Flexibility and Adaptability*

In the dynamic landscape of healthcare, the ability of facilities to adapt to changing conditions represents a critical component of design strategy. Flexibility and adaptability are not merely advantageous attributes; they are essential qualities that ensure healthcare facilities can respond effectively to emergent health crises, such as pandemics, and adapt to the rapidly evolving landscape of medical practices and technologies. This foresight in design equips facilities to maintain high standards of care, regardless of the circumstances, reflecting a deep understanding of the fluid nature of healthcare needs.

Designing flexibility means creating spaces that can be easily converted or expanded based on immediate needs. For instance, patient rooms might be designed with modular features that allow them to be transformed into intensive care units during a health crisis or reconfigured to accommodate different types of medical equipment as treatment technologies evolve. Likewise, the integration of movable partitions, retractable walls, and modular furniture can quickly change the layout of

**Table 10.2** Utilization of sustainable materials in healthcare facilities: Properties and impacts

| Sustainable material | Properties | Environmental impact | Impact on patient health |
|---|---|---|---|
| Recycled glass | Non-toxic, highly durable, easy to clean | Reduces landfill waste, low energy in production | Reduces exposure to toxins, easy maintenance enhances sanitary conditions |
| Bamboo | Rapidly renewable, strong, lightweight | Lowers deforestation rates, biodegradable | Non-toxic, promotes a natural aesthetic that can enhance patient well-being |
| Low-VOC paints | Low volatile organic compound levels | Reduces air pollution, minimal environmental toxicity | Improves indoor air quality, essential for patient respiratory health |
| Cork flooring | Renewable, excellent insulation, sound dampening | Sustainable harvest practices, biodegradable | Antimicrobial, comfortable underfoot, reduces noise stress in the environment |
| Recycled steel | High-strength, recyclable | Reduces energy consumption and mining impacts | Durable and low maintenance, improving environmental cleanliness |

spaces to increase capacity or modify their function without significant disruptions to the facility's operations.

Adaptability extends to the infrastructure of the building as well. This involves installing advanced mechanical, electrical, and plumbing systems that can be easily upgraded as technology advances or as environmental regulations change. It also includes the capability to enhance digital infrastructure swiftly, essential for incorporating new health information technologies or expanding telehealth services.

Moreover, the concept of adaptability in healthcare design also considers the future scalability of facilities. This foresight involves planning for physical expansion opportunities, such as adding floors or new wings and doing so in a way that minimizes impact on ongoing operations. Zoning plans might designate areas for future development, and foundational structures could be built to support potential vertical expansions.

The importance of flexibility and adaptability was particularly underscored during the COVID-19 pandemic when many healthcare facilities had to quickly repurpose and reorganize spaces to accommodate a surge in patients while simultaneously adapting areas for testing and vaccination. This ability to pivot rapidly not only ensured the continued delivery of medical services but also safeguarded public health and safety.

In conclusion, incorporating flexibility and adaptability into the design of healthcare facilities is crucial for future proofing against the unpredictable nature of global health trends and technological advancements. By embracing these principles, healthcare facilities can remain resilient and responsive, ensuring they continue to meet the highest standards of care and adapt seamlessly to whatever challenges the future may hold.

Equation 10.2 outlines the method for conducting a cost–benefit analysis of flexible design features in healthcare facility planning and construction. This equation allows healthcare administrators and facility planners to quantitatively assess the economic viability of incorporating adaptable and modifiable design elements into their infrastructure. Flexible design features are increasingly recognized for their ability to accommodate evolving medical technologies and patient care needs without necessitating extensive renovations, thereby providing substantial long-term savings.

The equation considers the initial costs ($C_{\text{initial}}$) associated with integrating these flexible designs, which might include specialized construction materials and architectural designs that facilitate future modifications. It also factors in the anticipated savings ($S_{\text{savings}}$) from reduced renovation needs over the facility's operational life. These savings arise from the decreased necessity for disruptive and costly restructuring as medical practices evolve or expand. Furthermore, the equation accounts for potential revenue ($R_{\text{potential}}$) generated from the ability to repurpose and adapt spaces for various uses, which could include anything from expanding patient care areas to integrating new diagnostic services that respond to growing healthcare demands.

By providing a financial metric, the cost–benefit ratio (*CBR*), this analysis supports strategic decision-making concerning facility investments, ensuring that expenditures on flexible designs are justified by the projected economic returns. This is crucial in managing healthcare budgets effectively while maintaining the adaptability of healthcare environments to future changes. The cost–benefit analysis for flexible design features is thus expressed as follows:

$$\text{CBR} = \frac{S_{\text{savings}} + R_{\text{potential}}}{C_{\text{initial}}}. \tag{10.2}$$

This formula helps decision-makers to understand the break-even points and profitability of investing in flexible designs, thereby optimizing resource allocation in healthcare infrastructure development.

### *10.1.5 Inclusivity and Accessibility*

Inclusivity and accessibility stand as fundamental principles in the design and operation of contemporary healthcare facilities. Ensuring that these environments are welcoming and accessible to everyone, irrespective of physical ability, age, or cultural background, is crucial not only from an ethical standpoint but also for the practical efficacy of delivering healthcare services to diverse populations. The imperative to embody these principles in both the architectural design and the service delivery models of healthcare facilities underscores a broader commitment to universal design that is usable by all people to the greatest extent possible without the need for adaptation or specialized design.

The physical design of healthcare facilities must meticulously address a range of needs to ensure that all individuals can access and utilize the services effortlessly. This includes barrier-free access features such as ramps, wide doorways, and elevators capable of accommodating wheelchairs and other mobility aids. Inside the facility, corridors, waiting areas, and restrooms must also be designed to be easily navigable by people with various types of disabilities. Attention to sensory inclusivity is also paramount, incorporating features such as tactile flooring for the visually impaired and hearing loop systems for the hearing impaired.

Moreover, inclusivity transcends physical accessibility, extending into the cultural and linguistic realms. Healthcare facilities must cater to a culturally diverse clientele, which involves providing signage, informational materials, and digital content in multiple languages. Furthermore, staff training is essential to equip healthcare providers with the skills to deliver culturally sensitive care, ensuring they can effectively communicate and empathize with patients from various backgrounds. This cultural competence helps in building trust and improving health outcomes by ensuring that all patients feel understood, respected, and valued during their healthcare experiences.

In terms of delivery service, inclusivity means offering a range of services that meet the diverse needs of the community the facility serves. This might include offering specialized services for different age groups, such as pediatric and geriatric care, and ensuring that services are sensitive to the cultural practices and needs of diverse ethnic groups. The availability of gender-sensitive services and consideration for the LGBTQ+ community are also critical aspects of inclusive healthcare provision.

These insights into inclusivity and accessibility form the bedrock of effective, responsive, and humane healthcare architecture. As the global population continues to grow and diversify, the lessons drawn from these foundational principles will increasingly inform the development and evolution of healthcare facilities worldwide. The ongoing challenge for healthcare architects and providers is to adapt and innovate continually, ensuring that facilities not only meet current standards of inclusivity and accessibility but also anticipate and adapt to future needs, thereby contributing to a more equitable and effective healthcare system.

Table 10.3 serves as an exhaustive guide to implementing accessibility features in healthcare facilities, aligned with compliance standards and best practices to enhance inclusivity. This table categorizes essential features into mobility access, navigational aids, elevators, restroom accessibility, accommodations for hearing impairments, visual elements, and cultural and linguistic inclusivity. Each category details required features such as ramps with gentle slopes and anti-slip surfaces for mobility, tactile floor indicators for the visually impaired, and elevators equipped with audible signals and Braille buttons. Compliance standards are mapped against the Americans with Disabilities Act (ADA) and local disability acts to ensure facilities meet legal requirements. Best practices go beyond compliance, suggesting wider doorways, multilingual signage, and staff training in cultural competence to create an environment that is accessible and welcoming to all patients, regardless of physical ability or cultural background. This table is a crucial resource for healthcare administrators aiming to develop facilities that prioritize patient care through comprehensive accessibility.

**Table 10.3** Comprehensive guide to accessibility features and compliance in healthcare facilities

| Feature type | Required feature | Compliance standard | Best practice |
|---|---|---|---|
| Mobility access | Ramps and zero-threshold entrances | ADA (Americans with Disabilities Act) compliance | Ramps with a gentle slope, anti-slip surfaces |
| | Wide doorways and corridors | ADA standards for accessible design | Doorways at least 36 inches wide for easy passage |
| Navigational aid | Tactile floor indicators for visually impaired | ADA guidelines for the visually impaired | Use of different textures on floors leading to critical areas like emergency rooms |
| Elevators | Elevators with sufficient space for wheelchairs | ADA minimum elevator dimensions | Elevators with audible signals and Braille buttons |
| Restroom accessibility | Accessible toilet stalls, sinks, and grab bars | ADA restroom requirements | Stalls wide enough for a wheelchair to turn, sinks with space for wheelchairs |
| Hearing impairments | Hearing loop systems in reception and common areas | Compliance with local disability acts | Installation in all service points to assist hearing-impaired patients |
| Visual elements | Signage with high contrast and large print | ADA requirements for signage | Signage in multiple languages, including Braille where necessary |
| Cultural and linguistic inclusivity | Multilingual staff and signage | Local health service standards | Training programs for staff on cultural competence, signage in predominant community languages |

## 10.2 Future Visions in Hospital Architecture

Looking to the future, hospital architecture is poised to undergo transformative changes that will further enhance the delivery of healthcare services and the overall patient experience. These future visions encompass several progressive trends and innovations that are expected to redefine how healthcare facilities operate and interact with their communities. Here is what we can anticipate:

### *10.2.1 Integration of Smart Technologies*

The integration of smart technologies into hospital architecture heralds a transformative era in healthcare delivery, characterized by increased efficiency, precision, and personalized care. The future of hospital design is intrinsically linked with the advancements in Internet of Things (IoT) devices, artificial intelligence (AI), and robotics. These technologies promise not only to enhance operational efficiency but also to significantly improve patient care outcomes. Smart hospitals leverage these innovations to create a highly interconnected environment where real-time data collection and analysis inform clinical decisions and streamline hospital operations.

IoT devices play a pivotal role in this new era, enabling continuous monitoring of patient health through wearable sensors and embedded devices in hospital beds and other medical equipment. These devices collect a vast array of data points, from vital signs to movement patterns, which can be analyzed to tailor patient care more effectively. For instance, wearable devices that monitor heart rate and blood pressure can provide critical data to healthcare providers almost instantaneously, allowing for timely interventions.

Artificial intelligence extends these capabilities by integrating complex algorithms that can analyze the massive volumes of data generated by IoT devices to optimize treatment plans and predict patient outcomes. AI systems can identify patterns and anomalies in patient data that might elude human observers, enabling earlier diagnosis of potential complications and more precise treatment regimens. Furthermore, AI can assist in administrative tasks, from scheduling appointments to managing supply chains, thereby reducing the administrative burden on healthcare professionals and allowing them to focus more on patient care.

Robotics also plays a crucial role in smart hospitals, performing tasks ranging from surgical assistance to routine logistical tasks like delivering supplies and medications. Robots can operate with precision and efficiency, reducing human error and increasing safety in procedures. Additionally, robotics can help in disinfecting facilities, a critical aspect in managing hospital-acquired infections.

To support these technologies, hospital facilities must be designed with robust data networks that can handle the high volume and velocity of data transmission required by smart technologies. This includes not only high-speed internet connections but also secure data storage solutions to protect patient privacy. Moreover, the physical spaces within hospitals need to be adaptable, and able to accommodate new technologies as they are developed. This might involve modular designs that can be reconfigured as needed or infrastructure that can support the heavy power and cooling requirements of high-tech equipment.

In conclusion, the future of hospital design is an exciting intersection of architecture and technology. By integrating smart technologies like IoT, AI, and robotics, hospitals can become more responsive and effective, providing care that is both cutting-edge and deeply human-centered. The challenge for architects and planners will be to create spaces that are both flexible enough to adapt to rapid technological advancements and robust enough to support the complex networks these technologies require. This proactive approach to hospital design will not only revolutionize patient care but also set a new standard for the healthcare facilities of the future.

Table 10.4 compares traditional and smart hospitals across several dimensions including operational efficiency, patient monitoring, patient care, cost management, safety, administrative tasks, and adaptability. Traditional hospitals often rely on manual processes that can lead to inefficiencies and higher operational costs, while smart hospitals utilize automated systems and AI to streamline operations, enhance patient care, and reduce costs. This table highlights how smart hospitals achieve greater efficiency and accuracy through continuous real-time monitoring and AI-driven personalized care plans, offering significant advantages in safety, cost management, and the ability to quickly adapt to new technologies.

## *10.2.2 Personalized Healthcare Environments*

The evolution of personalized medicine is set to revolutionize the architectural design of healthcare facilities, shifting towards environments that adapt to the unique needs of each patient. As medical treatment becomes more tailored to individual genetic profiles and lifestyle factors, so too must the spaces in which healthcare is delivered. This shift promises to enhance patient comfort, increase treatment efficacy, and improve overall patient outcomes by creating a more intimate and responsive healthcare experience.

One of the key aspects of personalized healthcare environments is the ability for patients to control and adjust their immediate surroundings. Advances in smart building technologies allow for the customization of room conditions, including lighting, temperature, and even wall colors, which can be adjusted to suit personal preferences and medical needs. For instance, lighting in patient rooms can be designed to mimic natural circadian rhythms, aiding in better sleep patterns and faster recovery. Temperature and air quality controls can be finely tuned to individual preferences, providing comfort and reducing stress.

Furthermore, the personalization of healthcare environments extends to how treatments are tailored to patients. With the integration of genetic information and lifestyle data, treatments can be specifically designed to work most effectively for everyone. This level of customization would require healthcare facilities to integrate advanced

**Table 10.4** Operational and clinical impact: Traditional hospitals versus smart hospitals

| Aspect | Traditional hospital | Smart hospital | Comparative advantage |
|---|---|---|---|
| Operational efficiency | Manual data entry, paper-based tracking | Automated systems for real-time data capture and analysis | Smart hospitals streamline operations, reduce human error, and enhance data accuracy |
| Patient monitoring | Periodic monitoring based on routine checks | Continuous real-time monitoring using wearable and embedded devices | Smart hospitals provide dynamic patient monitoring, leading to timely interventions |
| Patient care | Standardized care protocols are based on generalized patient data | Personalized care plans driven by AI analytics of real-time data | Enhanced precision in treatment, tailored to individual patient needs |
| Cost management | Higher operational costs due to inefficiencies and manual processes | Lower long-term costs through automation and optimized resource management | Cost-effective operations with better resource utilization |
| Safety and error reduction | Higher potential for human error in medication administration and data recording | Robotics and automated systems minimize human error and enhance procedural safety | Increase safety with technology-driven checks and balances |
| Administrative tasks | Time-consuming manual scheduling, billing, and record-keeping | AI-driven automated scheduling, billing, and electronic health records | Reduced administrative burden, freeing up staff for patient care |
| Adaptability | Limited adaptability to technological advancements without significant renovations | Designed for flexibility, it can integrate new technologies with minimal structural changes | Smart hospitals remain technologically agile, adapting quickly to new healthcare innovations |

diagnostic tools directly within the patient rooms or at least within the architectural layout, facilitating immediate and ongoing adjustments to treatment plans as more data becomes available.

Architectural spaces themselves can also be designed to be more flexible, accommodating different medical equipment and setups tailored to specific treatments. This could mean modular furniture and fixtures that can be easily moved or reconfigured, or rooms designed to quickly convert from a standard patient room to a more specialized treatment area. The use of adaptive technologies, such as augmented reality, can further enhance this personalization, allowing patients to visualize and control their environments and treatment plans more directly.

The concept of personalized healthcare environments also implies a deeper integration of patient care teams, who need space that enables them to provide responsive, tailored care. This might include more collaborative workspaces close to patient rooms, enhanced communication systems that keep all team members updated on patient status, and technology-enabled data points that provide real-time feedback on patient responses to treatments.

In conclusion, as personalized medicine continues to advance, the design of healthcare facilities will increasingly focus on creating environments that adapt to and cater to individual patient needs. This approach not only promises to enhance patient comfort and autonomy but also aligns with the broader goals of medical treatment, increasing effectiveness and improving health outcomes. The challenge for healthcare architects will be to anticipate these needs and create spaces that are not only functional and flexible but also deeply human-centered, ultimately transforming the patient's experience in profound and lasting ways.

Equation 10.3 provides a formula for evaluating the efficiency and effectiveness of personalized care in healthcare settings. Personalized care, tailored to the individual needs of patients, has become a cornerstone of modern medical practice. It enhances patient outcomes by addressing specific health conditions with tailored treatments and interventions. This approach not only improves clinical outcomes but also boosts patient satisfaction by involving patients actively in their care processes.

The equation integrates key metrics that reflect the success of personalized care strategies. These include patient recovery times ($R$), which measure the speed of patient improvement post-treatment; patient satisfaction scores ($S$), which gauge the personal experiences and contentment of patients with their care; and resource utilization rates ($U$), which assess the efficiency with which healthcare resources are deployed to achieve desired outcomes. Together, these variables provide a comprehensive view of the operational effectiveness and patient-centric performance of personalized care initiatives.

Efficiency in personalized care is quantified by weighing the benefits in terms of patient outcomes and satisfaction against the resources expended. This metric is vital for healthcare providers to optimize their care delivery models, ensuring that personalized interventions are not only effective but also economically viable. The efficiency metric for personalized care is calculated as follows:

$$\text{Efficiency} = \frac{(R \times S)}{U}. \tag{10.3}$$

This formula underscores the balance between providing highly tailored care that meets individual patient needs and managing the overall resource demands. It aids healthcare organizations in refining their approaches to care, ensuring that personalization enhances both clinical results and patient experiences without undue strain on healthcare resources.

### 10.2.3 Advanced Modular and Flexible Designs

As healthcare continues to evolve at a rapid pace, driven by technological advancements and shifting patient needs, the architectural design of hospitals must also adapt. The future of hospital design emphasizes advanced modular and flexible construction techniques that allow facilities to rapidly adjust to changing healthcare demands. This approach is not merely a response to immediate crises, such as pandemics or natural disasters, but also a proactive strategy for accommodating long-term shifts in healthcare delivery, such as the expansion of outpatient services or community health programs.

Modular construction stands at the forefront of this flexible design paradigm. By using prefabricated modules that can be quickly assembled, disassembled, and reconfigured, hospitals can expand or modify their layouts with minimal disruption to ongoing operations. This flexibility is crucial not only for responding to emergencies, requiring rapid scaling of capacity (such as adding isolation wards during a disease outbreak), but also for adapting to the evolving landscape of healthcare, which increasingly favors less invasive treatments and shorter hospital stays.

Moreover, the design of hospital spaces to serve multiple purposes significantly enhances the utility and efficiency of healthcare facilities. Rooms designed with modular features can be easily converted from patient rooms to consultation spaces or even to community meeting areas, depending on current needs. This adaptability extends to the infrastructure within these spaces, such as utility hookups and technological interfaces, which can support various medical and communication equipment configurations.

The use of flexible design also facilitates the integration of community health initiatives into the hospital environment. For example, spaces can be repurposed for community health education programs, vaccination drives, or as wellness centers focusing on preventive care. This not only maximizes the use of hospital space but also strengthens the hospital's role as a community health resource, thereby enhancing public health outcomes.

The strategic incorporation of adaptable, multipurpose spaces within hospitals necessitates a sophisticated level of planning and design. It requires architects and planners to envision multiple future scenarios and incorporate elements that allow for easy transitions between different functions. This might involve modular walls, flexible power and data supply systems, and multi-use furniture that can be easily moved or reconfigured.

In conclusion, the adoption of advanced modular and flexible designs in hospital architecture is crucial for creating healthcare facilities that are resilient, responsive, and capable of meeting both current and future healthcare challenges. By embracing these innovative construction techniques, hospitals can ensure they remain at the cutting edge of healthcare delivery, providing high-quality care that adapts swiftly to the needs of their patients and communities. This approach not only benefits patient care but also contributes to the sustainability and efficiency of healthcare infrastructure in the face of ever-changing global health dynamics.

**Table 10.5** Analyzing the benefits and challenges of modular construction in healthcare facilities

| Aspect | Benefits | Challenges or limitations |
|---|---|---|
| Speed of construction | Faster construction times due to off-site fabrication. Reduces overall project timeline | Potential delays in module delivery can impact schedules |
| Flexibility | Allows for easy expansion, reduction, or reconfiguration of spaces based on healthcare needs | Design limitations may arise due to the need for standardization in modules |
| Cost-effectiveness | Reduced on-site labor costs and minimal resource wastage. Economies of scale in manufacturing | Initial investment in design and fabrication setup can be high |
| Quality control | Construction in a controlled environment improves quality and consistency | Modifications on-site can be restricted; any changes require factory adjustments |
| Sustainability | Less waste production and reduced site disturbance during construction | Transportation of modules can offset environmental benefits if not managed properly |
| Scalability | Prefabricated elements can be quickly added to adapt to increasing patient demands or technological upgrades | Structural constraints might limit the addition of multiple floors or extensive horizontal expansions |
| Community impact | Reduced construction time minimizes disruption to hospital operations and nearby areas | Large-scale modular projects may require significant logistical coordination, impacting local traffic and services temporarily |

Table 10.5 outlines the benefits and challenges associated with implementing modular construction techniques in healthcare facilities. Key advantages include faster construction due to off-site fabrication, enhanced flexibility for space reconfiguration, cost-effectiveness through reduced labor and waste, and superior quality control in a controlled manufacturing environment. However, challenges include potential delays in module delivery, design constraints due to standardization needs, higher initial investments, and possible negative environmental impacts from module transportation. This table also discusses modular construction's scalability and its impact on the community, noting the balance between quick scalability and the logistical demands on local infrastructure during large-scale projects.

### *10.2.4 Emphasis on Mental and Emotional Health*

In the evolving landscape of healthcare, there is an increasing recognition of the integral role that mental and emotional well-being plays in overall health outcomes. Future hospital designs are pivoting towards creating environments that nurture not only the physical but also the mental and emotional health of patients. This holistic approach to healthcare architecture recognizes that healing is a multi-faceted process, where the surroundings can significantly influence patient recovery and staff well-being.

The emphasis on mental and emotional health in hospital design can be seen in the incorporation of therapeutic elements that engage the senses and foster a serene, healing atmosphere. Healing gardens are a prime example of such an initiative, offering a natural sanctuary where patients and staff can find solace away from the clinical environment. These gardens are thoughtfully designed with diverse plant life, soothing water features, and gentle pathways that encourage leisurely walks and reflective moments. The benefits of such spaces are supported by research showing that exposure to nature can lower blood pressure, reduce stress hormone levels, and enhance emotional well-being.

Art installations also play a crucial role in enriching the mental and emotional climate of healthcare facilities. Art can transcend cultural and language barriers, providing comfort, inspiration, and sometimes a necessary distraction from health concerns. Hospitals are incorporating artworks that range from calming landscapes to abstract pieces that stimulate contemplation and conversation. These installations can transform sterile hospital corridors and waiting areas into dynamic galleries where art becomes a part of the healing process.

Moreover, the architectural design itself is being reconsidered to support mental and emotional health. This includes the use of natural light, which has been shown to boost mood and energy, and the selection of color schemes that have calming effects. The layout of the hospital is designed to promote privacy and dignity, with features like private rooms and quiet zones where patients can have moments of solitude or intimate conversations with family members.

Future hospital designs also consider the acoustic environment, as excessive noise can be a significant source of stress and anxiety for both patients and staff. Sound-absorbing materials and design strategies that minimize disruptive noises help create a quieter, more peaceful environment conducive to recovery and concentrated work.

In conclusion, the greater emphasis on mental and emotional health in hospital design reflects a comprehensive understanding of healthcare, where the healing process is supported by a nurturing environment. By integrating natural elements, art, thoughtful use of light and color, and noise control, hospitals can become places of refuge and healing, not only treating the body but also attending to the mind and spirit. This approach not only enhances the patient's experience but also supports the well-being of the healthcare professionals who work within these spaces, ultimately leading to improved health outcomes and a more humane healthcare system.

Equation 10.4 delineates a method for conducting a cost–benefit analysis specifically focused on investments in therapeutic design elements within healthcare environments. Therapeutic design in healthcare involves integrating architectural and environmental elements that contribute to the well-being and recovery of patients. Such designs can include access to natural light, soundproofing, green spaces, and other stress-reducing features that have been shown to improve patient outcomes by decreasing stress levels, reducing dependency on medications, and potentially shortening hospital stays.

This equation allows healthcare administrators and facility managers to quantitatively assess the economic impacts of these design elements. It integrates the savings from decreased medication use ($M$), which represents the reduction in pharmaceutical costs; the benefits derived from shortened hospital stays ($H$), which not only improve patient throughput but also reduce operational costs; and the value assigned to reduced patient stress levels ($S$), which might be quantified in terms of better patient satisfaction scores and lower readmission rates.

By calculating the total benefits ($B$) obtained from these therapeutic elements and comparing them to the initial and ongoing costs ($C$) associated with their implementation, the equation provides a clear financial metric. This metric helps decision-makers understand the return on investment ($ROI$) for therapeutic design strategies, guiding investments that promote patient health while ensuring financial sustainability. The cost–benefit analysis for therapeutic design investments is therefore formulated as follows:

$$\text{ROI} = \frac{M + H + S}{C}. \tag{10.4}$$

This equation underscores the importance of investing in healthcare environments that support therapeutic outcomes, ensuring that such investments are justified by tangible improvements in patient care and cost savings. It is an essential tool for aligning financial objectives with the overarching goal of enhancing patient well-being in healthcare settings.

### *10.2.5 Eco-Friendly and Sustainable Practices*

As the awareness of environmental impact deepens, sustainability has emerged as a critical guiding principle in the design and operation of healthcare facilities. Future healthcare architecture is increasingly focused on minimizing the ecological footprint of hospitals, recognizing that environmental health is inextricably linked to human health. This commitment to sustainability involves a comprehensive approach, including the use of sustainable materials, the integration of energy-efficient systems, and a thoughtful consideration of the local climate and ecological context.

The selection of sustainable materials is a foundational aspect of eco-friendly hospital design. These materials are chosen for their low environmental impact, which may include factors such as reduced energy consumption during production, lower emissions, and high recyclability. Sustainable building materials like bamboo, recycled steel, and low-volatile organic compounds (VOC) paints not only contribute to the structural and aesthetic quality of hospital buildings but also promote a healthier indoor environment, free from harmful chemicals and allergens.

Energy efficiency is another cornerstone of sustainable healthcare architecture. Hospitals are intensive energy users, with around-the-clock operations that require reliable power for lighting, heating, cooling, and medical equipment. To address this, modern hospital designs incorporate advanced heating, ventilation, and air conditioning (HVAC) systems, LED lighting, and energy management systems that optimize energy use. Additionally, many hospitals are now incorporating renewable energy sources, such as solar panels and wind turbines, to reduce reliance on fossil fuels and decrease greenhouse gas emissions.

The design of a hospital is also increasingly responsive to the local climate and ecology, an approach known as bioclimatic design. This strategy involves designing buildings based on the local environment to naturally regulate temperatures and lighting. For example, in hot climates, hospitals might feature reflective roofs, shaded façades, and extensive greenery to reduce heat absorption and enhance energy efficiency. In colder regions, maximizing solar gain through strategic window placement and using thermal mass to store heat can reduce heating requirements.

Moreover, hospitals are extending their commitment to sustainability through the implementation of green spaces not only for patient and staff use but also as part of their stormwater management strategies. Green roofs and rain gardens can help manage runoff, reduce flooding risks, and increase biodiversity, contributing to local ecological health.

Finally, the role of hospitals extends beyond healing the sick to encompass broader stewardship of environmental health. This is reflected in practices such as comprehensive recycling programs, the reduction of medical waste, and the use of non-toxic cleaning agents that are effective without being harmful to the environment. Hospitals that embrace these eco-friendly and sustainable practices set a precedent for other institutions, demonstrating that healthcare facilities can be both healers of people and protectors of the planet.

In conclusion, the shift towards eco-friendly and sustainable practices in healthcare architecture not only reflects a commitment to environmental stewardship but also aligns with the broader goals of public health. By designing hospitals that minimize environmental impact and promote ecological health, the healthcare sector can lead by example, showing how large institutions can operate sustainably and responsibly in an increasingly ecological-conscious society.

Table 10.6 presents a comprehensive guide to sustainable materials used in hospital construction, highlighting the environmental benefits, durability, and cost-effectiveness of each option. It showcases materials such as bamboo, known for its rapid renewability and carbon capture efficiency, and recycled steel, which minimizes landfill use and reduces energy consumption. The table also evaluates low-VOC paints that lessen indoor air pollutants, cork flooring that offers natural resistance to mold and pests, recycled glass suitable for non-load-bearing applications, and rammed earth, celebrated for its low environmental impact and excellent thermal mass. Each material is assessed on its durability and cost implications, providing vital information for making informed choices in sustainable hospital design.

**Table 10.6** Guide to sustainable materials used in hospital construction

| Sustainable material | Environmental benefits | Durability | Cost-effectiveness |
|---|---|---|---|
| Bamboo | Rapidly renewable, captures more carbon dioxide than hardwood | Highly durable, comparable to traditional hardwood | Generally cost-effective, depending on local availability |
| Recycled steel | Reduces landfill use, less energy consumption in production than new steel | Extremely durable, does not lose strength when recycled | Costs vary, often cheaper than new steel depending on market conditions |
| Low-VOC paints | Minimizes indoor air pollutants, better for environmental health | Comparable durability to conventional paints | Slightly more expensive, but costs offset by health benefits |
| Cork flooring | Renewable, biodegradable, and easy to recycle | Naturally resistant to mold, fungi, and pests; good for indoor air quality | Mid-range cost; durability makes it a cost-effective option long-term |
| Recycled glass | Reduces glass waste in landfills, low energy consumption in production | Less durable than traditional glass, use in non-load-bearing areas | Cost-effective, especially if sourced locally |
| Rammed earth | Low environmental impact, utilizes natural materials from the site | Extremely durable, excellent thermal mass | High initial costs, but savings in energy consumption can offset this over time |

### *10.2.6 Community-Centric Spaces*

The future trajectory of hospital architecture envisions a more profound integration with the communities they serve, transforming hospitals from mere venues for acute medical treatment into vibrant hubs of health education and preventive care. This paradigm shift in healthcare facility design reflects an expansive view of health that includes not just the treatment of illness but the promotion of overall wellness within the community. To realize this vision, hospital designs are increasingly incorporating community-centric spaces that foster interaction, education, and accessibility.

Architectural innovations are making way for the inclusion of public spaces that draw the community into the hospital's daily life. Open-air markets on hospital grounds, for example, can promote nutritional education and provide access to fresh, healthy foods, which are fundamental to preventive healthcare. Similarly, integrating public parks within or around hospital campuses offers a dual benefit: they provide serene environments that promote mental and physical well-being for patients and staff, and they serve as recreational spaces for the community, encouraging physical activity and social interaction.

Furthermore, the integration of educational facilities such as community health classrooms and interactive workshops within hospitals can enhance public health literacy. These spaces allow for ongoing health education on a wide range of topics—from chronic disease prevention to maternal health—which can have profound implications for community health outcomes. Hospitals can host regular seminars and workshops, led by healthcare professionals, that empower community members with the knowledge to manage their health proactively.

By blurring the lines between healthcare facilities and community centers, hospitals can strengthen their role in the community, becoming centers of learning, wellness, and social connection. This approach not only enhances the hospital's capacity to impact public health positively but also redefines the spatial and functional design of healthcare facilities, making them more inclusive and responsive to the needs of the communities they serve.

These visions for the future of hospital architecture are a testament to a holistic approach to health that acknowledges the interconnectedness of physical, mental, and environmental well-being. The integration of community-centric spaces within hospital designs is a strategic move towards creating healthcare environments that are not only focused on healing but are also proactive in health promotion, prevention, and community engagement.

In the concluding remarks of this discussion, we will reflect on the overarching themes that have emerged from this exploration of future hospital designs. We will summarize the key innovations and strategic directions that promise to shape resilient, inclusive, and patient-centric healthcare environments. This synthesis will not only highlight the transformative potential of these design principles but also underscore the importance of integrating community and environmental considerations into the fabric of healthcare infrastructure, paving the way for a more holistic and sustainable approach to health and wellness.

Equation 10.5 describes the calculation of the Community Engagement Index (CEI), a metric designed to quantify the level of community engagement fostered through hospital-led initiatives. Community engagement in healthcare is crucial for extending the impact of health services beyond the confines of hospital walls and into the broader community, aiming to improve public health outcomes, enhance health literacy, and encourage active participation in health-promoting activities.

This equation considers several key variables: community attendance rates ($A$), which reflect the participation level in hospital-initiated programs and events; health literacy improvements ($L$), measuring the increase in community members' understanding of health practices and medical information as a result of engagement efforts; and community health outcomes ($O$), which assess the tangible health benefits realized by the community, such as reduced incidence of chronic diseases or improved overall health markers.

The Community Engagement Index is thus a comprehensive measure, capturing the effectiveness of hospitals in engaging with their local communities and the direct benefits of such engagement. The index not only serves as a benchmark for the success of current initiatives but also helps in planning and improving future community outreach activities. The *CEI* is calculated by integrating these factors into a single formula, thereby providing a robust indicator of engagement success:

$$\text{CEI} = \frac{(A \times L \times O)}{100}. \tag{10.5}$$

This formula ensures that the contributions of various engagement aspects are collectively considered, providing a holistic view of how well a healthcare institution interacts with and benefits its community. The index supports healthcare providers in gauging the efficacy of their community programs and in making data-driven decisions to enhance their outreach and engagement strategies.

## 10.3 Final Remarks: The Road Ahead for Healthcare Facilities

As we look ahead, the future of healthcare facilities is marked by a commitment to innovation, adaptability, and a deeper understanding of the holistic needs of patients. The evolving landscape of healthcare, driven by technological advancements and changing societal needs, demands that healthcare architecture not only adapts but also anticipates future challenges and opportunities.

### *10.3.1 Holistic Approach to Health*

The future direction of healthcare facilities is increasingly steering towards a holistic approach to health, integrating the physical, mental, and social aspects of well-being into every facet of healthcare delivery. This comprehensive perspective on health is transforming healthcare facilities from places primarily focused on treating illness to vibrant centers of wellness and prevention. Such an approach reflects a deep understanding that health is a multifaceted experience influenced by a myriad of factors extending beyond medical treatments.

A holistic approach in healthcare facilities involves an expansion of services to include programs that address mental health, nutritional advice, exercise regimes, and social well-being. These facilities are becoming equipped to provide not only acute medical care but also comprehensive wellness programs that support the overall health of individuals and communities. For example, hospitals are increasingly incorporating integrative medicine centers that offer services such as acupuncture, meditation, and massage therapy, which are recognized for their benefits in stress reduction and overall health enhancement.

Moreover, this approach advocates for stronger partnerships between various healthcare providers, social workers, and community organizations to deliver care that addresses the broad spectrum of factors affecting health. By doing so, healthcare facilities can better support patients' needs throughout their lives, from preventive care and wellness to acute care and chronic disease management. This interconnectedness ensures that care is seamless, patient-centered, and responsive to the evolving health dynamics of the population.

The design of healthcare facilities themselves must reflect this holistic approach by creating environments that promote healing and well-being. This includes designing spaces that are not only functional but also comforting and welcoming. Features like natural lighting, patient-controlled environmental settings, and the inclusion of green spaces contribute to a therapeutic atmosphere that can enhance the healing process. The layout of these facilities is also adapting to promote more interaction, providing spaces that encourage social connections among patients, families, and caregivers, which is a vital component of emotional and psychological health.

In embracing a holistic approach to health, healthcare facilities are poised to become true wellness hubs within their communities, offering a range of services and programs that emphasize health maintenance and disease prevention. This shift not only aligns with emerging health paradigms but also sets a new standard for healthcare delivery, ensuring that facilities are better equipped to meet the diverse and changing health needs of the population they serve. This strategic direction not only promises to enhance individual health outcomes but also contributes to healthier, more resilient communities.

Table 10.7 outlines the diverse range of services provided by holistic healthcare facilities, each designed to enhance patient health comprehensively. The table details integrative medicine practices that blend traditional and alternative therapies to improve overall well-being and accelerate recovery. It highlights the benefits of physical therapy in enhancing mobility and alleviating pain, and mental health services that support emotional resilience. Nutritional counseling and fitness programs are shown to play critical roles in disease prevention and overall health enhancement. Additionally, the table covers social work services that aid community integration, chiropractic care to address musculoskeletal issues, and therapeutic approaches like massage and art and music therapies, which help reduce stress and promote healing. Each service is described alongside its specific health benefits, illustrating the holistic approach to patient care that addresses physical, emotional, and social health needs.

### *10.3.2 Patient-Centric Innovation*

The future of healthcare architecture is fundamentally tied to the principle of patient-centric innovation. This approach prioritizes the patient experience in every aspect of facility design and service delivery, crafting environments that are not only aesthetically pleasing but also functionally supportive of healing and comfort. The emphasis is on creating spaces that embody care and consideration, reflecting a deep understanding of the needs and experiences of patients as well as the staff who care for them.

**Table 10.7** Comprehensive services offered in holistic healthcare facilities: Enhancing patient health

| Service type | Description | Benefits to patient health |
|---|---|---|
| Integrative medicine | Combines traditional medicine with alternative therapies such as acupuncture, yoga, and herbal treatments | Enhances overall well-being, reduces stress, and improves recovery rates |
| Physical therapy | Offers rehabilitation services for post-operative patients and those recovering from injuries | Aids faster recovery, improves mobility, and reduces pain |
| Mental health services | Provides counseling, therapy, and support groups for mental health issues | Supports emotional and psychological well-being, essential for holistic recovery |
| Nutritional counseling | Tailors' dietary plans to individual health needs and supports chronic disease management | Promotes better dietary habits, essential for disease prevention and health management |
| Fitness programs | Includes guided exercise regimes tailored to patient capabilities and health goals | Enhances physical fitness, crucial for overall health and preventing illness |
| Social work services | Assists with community integration and access to social support networks | Facilitates coping with health conditions within a community context, improving social well-being |
| Chiropractic care | Focuses on the diagnosis and treatment of mechanical disorders of the musculoskeletal system, especially the spine | Reduces pain, improves functional abilities, and enhances quality of life |
| Massage therapy | Offers therapeutic massage to relieve pain, reduce stress, and promote relaxation | Helps in reducing muscle tension, improving circulation, and boosting recovery |
| Art and music therapy | Utilizes the arts to improve and enhance physical, mental, and emotional well-being | Supports expressive therapy methods, and aids in stress relief and emotional balance |

Patient-centric innovation in healthcare architecture involves an intuitive design that simplifies navigation through the healthcare facility. This means minimizing the often complex maze-like layouts that can confuse and stress patients and their families. Wayfinding aids such as clear signage, interactive directories, and straightforward layouts help ease movement within the facility, making it more accessible and less daunting for everyone.

Moreover, the design of patient rooms and public areas focuses on comfort and tranquility, integrating elements such as natural light, soundproofing materials, and views of nature which have been shown to speed up recovery times and reduce stress levels. For example, large windows that offer scenic views or internal courtyards that provide green spaces can significantly enhance the therapeutic environment. Such design choices not only uplift the spirits of patients but also provide respite for healthcare workers, offering them a more pleasant and stress-reducing work environment.

Functionality is also a key component of patient-centric innovation, ensuring that every design element serves the purpose of improving patient outcomes. This includes ergonomic patient beds, efficiently designed nurse stations that allow for optimal patient monitoring, and the integration of technology that supports health management. Telehealth facilities, for example, are becoming an integral part of patient rooms, allowing for seamless communication with specialists who may not be on-site, thereby extending the reach and efficacy of medical consultations.

Additionally, patient-centric designs are inclusive, catering to a diverse range of needs including those of the elderly, children, and individuals with disabilities. This involves adjustable fixtures, wheelchair-accessible facilities, and sensory-friendly environments designed for patients with autism and other sensory sensitivities. The goal is to ensure that all patients feel welcomed and supported, regardless of their age, physical ability, or health condition.

In conclusion, innovation in healthcare architecture will continue to refine and emphasize patient-centric designs, integrating advancements that enhance functionality, comfort, and aesthetic quality. This approach not only improves the care experience for patients and families but also supports the well-being and efficiency of healthcare staff. By focusing on intuitive, comforting, and supportive design elements, healthcare facilities can truly transform into spaces that facilitate healing and foster well-being for everyone involved.

Equation 10.6 presents a model for calculating the Patient Satisfaction Score ($PSS$), which evaluates the impact of patient-centric design elements within healthcare facilities on overall patient satisfaction. These design elements, such as access to natural light and quiet rooms, are increasingly recognized for their significant role in enhancing patient experience by providing a more comfortable and healing environment. The inclusion of these factors in facility design has been linked to improved patient outcomes, reduced stress, and increased satisfaction with healthcare services.

This model quantitatively assesses how specific design features correlate with patient satisfaction levels. It incorporates variables such as the presence of natural light ($N$), which has been shown to boost mood and decrease recovery time, and the availability of quiet rooms ($Q$), which provide patients with a peaceful environment conducive to rest and recovery. These elements are weighted by their perceived importance to patient satisfaction, represented by the coefficients $\alpha$ for natural light and $\beta$ for quiet rooms, which are determined based on empirical research or patient feedback surveys.

The Patient Satisfaction Score is then calculated by combining these weighted factors, which allows healthcare administrators to directly see the contribution of each design element to overall patient satisfaction. This insight is crucial for guiding future investments in facility design and for optimizing the patient environment to enhance satisfaction and well-being. The equation for the Patient Satisfaction Score is given as follows:

$$\mathrm{PSS} = \alpha \cdot N + \beta \cdot Q. \tag{10.6}$$

This equation provides a clear, actionable framework for measuring and enhancing patient satisfaction through targeted improvements in the healthcare environment, ensuring that facilities are not only functional but also supportive of patients' psychological and physical health.

### 10.3.3 Interdisciplinary Collaboration

The future of healthcare facility design and management is increasingly dependent on interdisciplinary collaboration. As healthcare challenges grow in complexity, the necessity for a cohesive and unified approach becomes more evident. This collaborative strategy involves architects, healthcare professionals, technologists, and even patients themselves, each contributing their unique perspectives and expertise to create environments that are truly patient-centered and responsive to the needs of the healthcare community.

This interdisciplinary approach ensures that healthcare facilities are not only designed with clinical efficiency in mind but also with a keen awareness of the technological advancements and patient care innovations that characterize modern medicine. Architects and designers bring their expertise in creating spaces that are both functional and healing, incorporating elements like natural light, patient flow optimization, and environmentally sustainable materials. Healthcare professionals provide insights into the operational needs of medical facilities, emphasizing ease of access, infection control, and the integration of medical technology.

Technologists play a crucial role in this collaborative process, integrating cutting-edge technologies such as telemedicine, electronic health records, and automated systems directly into the facility. Their input is vital in ensuring that these technologies are seamlessly incorporated into the building design, enhancing patient care and staff workflow without becoming obtrusive or burdensome.

Patients, often overlooked in traditional design processes, are increasingly recognized as essential contributors. By involving patients in the design process, healthcare facilities can ensure that their spaces truly meet the needs and preferences of those they serve. This can include considerations for privacy, personal space, and family involvement, which are crucial for patient satisfaction and comfort.

Moreover, this collaborative approach extends to the management of healthcare facilities, where continuous feedback loops between all stakeholders can lead to ongoing improvements in both the design and operation of healthcare environments. For instance, regular consultations with healthcare staff can identify areas where design adjustments may enhance efficiency or reduce stress. Similarly, patient feedback can lead to better customization of care environments to meet diverse patient needs.

In conclusion, interdisciplinary collaboration in healthcare facility design and management is critical for crafting spaces that are not only efficient and sustainable but also adaptable to the rapidly evolving landscape of healthcare. This collaborative model promises to bring about facilities that not only meet the current standards of care but also anticipate future developments, ensuring that healthcare environments continue to evolve in alignment with advances in medicine and technology. By embracing the diverse perspectives of all stakeholders involved, healthcare facilities can become more responsive, innovative, and truly centered on delivering the best possible care to the communities they serve.

Table 10.8 presents the significant advantages of interdisciplinary collaboration in the design and operation of healthcare facilities, emphasizing how diverse professional insights enhance healthcare delivery. The benefits range from sparking enhanced innovation through a melding of ideas from various fields, which directly addresses complex healthcare challenges, to improving operational efficiency by optimizing workflows and reducing redundancies. This collaborative approach also significantly enhances patient care, ensuring facilities are tailored to the actual needs and preferences of users, thus increasing patient satisfaction. Cost-effectiveness is achieved by minimizing waste and optimizing resource use. Additionally, the input from multiple disciplines results in flexible designs that can adapt to future healthcare needs and robust risk management by improving the identification and mitigation of potential issues. This table underscores the tangible outcomes of integrating diverse expert insights into healthcare planning and management, leading to more effective, efficient, and patient-centered healthcare environments.

**Table 10.8** Enhancing healthcare through interdisciplinary collaboration: Key benefits

| Benefit category | Specific benefit | Description |
|---|---|---|
| Enhanced innovation | Broader range of ideas and solutions | Collaboration across disciplines leads to innovative solutions that address complex healthcare needs more effectively |
| Improved efficiency | Streamlined operations and processes | Input from diverse fields helps optimize workflows, reduce redundancies, and improve operational efficiency |
| Enhanced patient care | Tailored and responsive design | Contributions from healthcare professionals and patients ensure facilities are designed with direct input on needs and preferences, improving the care environment |
| Increased patient satisfaction | Environments aligned with patient needs | Facilities designed with interdisciplinary input often result in higher patient satisfaction due to the consideration of comprehensive care needs |
| Cost-effectiveness | Reduced waste and optimized resource use | Efficient design and operation, driven by collaborative planning, lead to cost savings and better resource allocation |
| Adaptability | Flexible designs suited to evolving needs | Interdisciplinary teams are more likely to consider future adaptability in designs, ensuring longevity and relevance of healthcare facilities |
| Risk management | Enhanced risk identification and management | Diverse perspectives contribute to better risk assessment and management, leading to safer patient environments |

### 10.3.4 Sustainability and Resilience

As healthcare systems continue to evolve, the imperatives of environmental sustainability and resilience have become central to the design and operation of healthcare facilities. Recognizing that the health sector significantly impacts the environment, there is a compelling need to ensure that new facilities are designed to be energy-efficient, environmentally friendly, and resilient to the various challenges posed by climate change. This focus on sustainability and resilience is not just about reducing the environmental footprint but also about enhancing the health and well-being of patients and staff by creating healthier spaces.

Energy efficiency is a critical aspect of sustainable healthcare design. This involves the integration of advanced systems that reduce energy consumption without compromising patient care. For example, the use of high-efficiency HVAC systems, LED lighting, and energy management systems are becoming standard practices in new healthcare facilities. Additionally, incorporating renewable energy sources such as solar panels and wind turbines can significantly decrease a hospital's reliance on fossil fuels and reduce its carbon emissions.

Beyond energy efficiency, sustainable design also emphasizes the use of environmentally friendly materials that are non-toxic and sourced sustainably. These materials contribute to better indoor air quality and reduce the overall environmental impact of building construction and maintenance. For instance, using recycled, reclaimed, or rapidly renewable materials can help minimize the ecological footprint of healthcare facilities.

Resilience in healthcare facility design refers to the ability of buildings to withstand and continue functioning in the face of natural disasters, such as floods, hurricanes, or earthquakes. This involves strategic site selection, robust building techniques, and the integration of emergency power systems that ensure the continuity of critical medical services during and after disasters. The resilience of a healthcare facility is crucial not only for patient safety but also for community recovery in the aftermath of a crisis.

Moreover, sustainable healthcare facilities often incorporate green spaces, which not only mitigate the building's impact on the local environment but also provide therapeutic benefits to patients and staff. These green spaces can help manage stormwater, reduce heat island effects, and offer areas for relaxation and physical activity, thereby contributing to the overall well-being of the hospital community.

In conclusion, the incorporation of sustainability and resilience into healthcare facility design represents a forward-thinking approach that aligns the objectives of environmental stewardship with the core mission of healthcare: to promote and maintain health. By prioritizing these principles, healthcare facilities can serve as models of sustainability, demonstrating how large institutions can operate effectively while also safeguarding the planet and enhancing the health of their occupants. These efforts not only respond to immediate environmental concerns but also prepare healthcare facilities to be robust and adaptive to future challenges, ensuring long-term sustainability and resilience.

Equation 10.7 focuses on calculating the reduction in carbon footprint achieved through sustainable practices implemented in healthcare facilities. As healthcare providers increasingly prioritize sustainability, understanding the environmental impact of their operations becomes essential. This equation helps quantify the effectiveness of measures like reducing energy consumption, increasing the use of renewable energy sources, and enhancing waste management practices in lowering the overall carbon emissions associated with healthcare services.

The calculation integrates several key components: energy consumption ($E$), which accounts for the total energy used by the facility, adjusted by the proportion of that energy that is sourced from renewable resources ($R$); and effective waste management ($W$), which includes recycling, composting, and other practices that reduce the amount of waste sent to landfills, thus decreasing greenhouse gas emissions from waste processing. The coefficients $\alpha$ and $\beta$ represent the impact weight of renewable energy usage and waste management, respectively, reflecting their relative importance in reducing the carbon footprint.

By assessing these sustainable practices, healthcare facilities can monitor and improve their environmental strategies, aiming for a significant reduction in their carbon emissions. The formula provides a clear metric by which the cumulative impact of these sustainability efforts can be evaluated, helping facilities meet both regulatory requirements and their own sustainability goals. The equation for calculating the carbon footprint reduction in healthcare facilities is expressed as follows:

$$\text{Carbon Reduction} = (\alpha \cdot E \times R) + (\beta \cdot W). \tag{10.7}$$

This formula underscores the critical roles of energy efficiency, renewable energy adoption, and waste management in achieving environmental sustainability in the healthcare sector. It guides facilities in prioritizing initiatives that have the most substantial impact on reducing their carbon footprint, thereby contributing to broader environmental protection efforts.

### *10.3.5 Embracing Technology*

The rapid advancement of digital technologies has revolutionized the healthcare landscape, making the integration of these technologies into healthcare facilities a critical aspect of modern healthcare delivery. As we move forward, the ability of healthcare facilities to seamlessly embrace and utilize cutting-edge technology will not only determine their effectiveness but also their relevance in an increasingly digital world. This necessitates a forward-thinking approach in architectural and operational planning, ensuring that healthcare environments are equipped to harness the full potential of technological innovations.

Telemedicine has become a staple in expanding access to healthcare, allowing patients to consult with healthcare professionals remotely. Facilities must be designed with dedicated spaces equipped with the necessary technology and privacy to facilitate these virtual interactions. This includes high-quality video conferencing equipment, secure internet connections, and soundproofed rooms that ensure confidential and effective communication between patients and healthcare providers.

Beyond telemedicine, the integration of AI-driven diagnostics and treatment options is transforming patient care. AI algorithms can analyze vast amounts of medical data much faster than human providers, offering diagnostics and recommending treatments with high accuracy. Healthcare facilities must have the infrastructure to support the computational and data storage demands of AI technologies. This includes robust IT systems, high-speed data processing capabilities, and advanced cybersecurity measures to protect sensitive patient data.

Furthermore, the role of robotics in healthcare continues to expand, from robotic surgical assistants to automated systems for logistics and cleaning. These technologies improve precision in surgical procedures and enhance efficiency in hospital operations. Designing spaces that can accommodate robotic technologies while ensuring safety and functionality is crucial. This might involve structural reinforcements for heavy equipment, larger operating rooms to accommodate robotic systems, and integration of control stations where staff can monitor and control automated systems.

The architecture of healthcare facilities must also be adaptable, allowing for easy updates and modifications as technology evolves. Modular design elements, scalable IT infrastructure, and flexible spaces are key to accommodating future technological advancements. Facilities should be planned with expansion in mind, considering the rapid pace of technological innovation and the subsequent need for periodic upgrades to hardware and software systems.

In conclusion, embracing technology in healthcare facility design is not merely about installing new systems but creating an environment that is fundamentally built to adapt to and incorporate technological advancements. This includes careful planning of physical spaces, IT infrastructure, and operational protocols. By integrating state-of-the-art technology, healthcare facilities can enhance their service delivery, improve patient outcomes, and remain at the forefront of the healthcare industry. The strategic embrace of technology ensures that healthcare facilities are not only equipped to meet current needs but are also prepared for the future, ultimately leading to more efficient, effective, and responsive healthcare environments.

Table 10.9 compares traditional and technologically advanced healthcare facilities across various aspects such as patient outcomes, operational efficiency, healthcare delivery, and cost-efficiency. It highlights how advanced technologies like AI, IoT, and data analytics enhance patient care by providing more accurate diagnostics and personalized treatments, improving recovery rates and patient satisfaction. These facilities also increase operational efficiency through automation, reduce operational costs with efficient resource management, and expand healthcare access via telemedicine. Additionally, the integration of digital health records improves data management, ensuring better security and compliance. Overall, technologically advanced facilities offer significant advantages over traditional settings, leading to higher staff satisfaction and better patient care.

**Table 10.9** Differential impacts on healthcare: Traditional versus technologically advanced facilities

| Aspect | Traditional facilities | Technologically advanced facilities | Comparative advantage |
|---|---|---|---|
| Patient outcomes | Standard outcomes with variability due to manual processes and slower diagnostics | Improved accuracy in diagnostics and personalized treatment plans due to AI and data analytics | Technologically advanced facilities offer more precise and tailored healthcare interventions, enhancing patient recovery rates and satisfaction |
| Operational efficiency | Reliant on manual processes leading to potential errors and delays | Automation and AI integration streamline operations, reducing errors and increasing throughput | Higher efficiency in technologically advanced facilities reduces wait times and improves healthcare delivery |
| Healthcare delivery | Conventional in-person consultations and treatments | Enhanced capabilities for remote consultations and telemedicine, reducing the need for physical visits | Technologically advanced facilities provide greater accessibility and convenience to patients, leading to broader healthcare reach |
| Cost-efficiency | Higher long-term operational costs due to inefficiencies and less effective resource management | Reduced operational costs through efficient resource management and predictive maintenance via IoT devices | Technologically advanced facilities are more cost-effective in the long run, despite higher initial investments |
| Patient data management | Paper-based records prone to loss and errors | Digital records with high-level encryption and easy access for authorized personnel | Better data security and accessibility in technologically advanced facilities enhance patient care and compliance with health regulations |
| Innovation capacity | Limited capacity for rapid adaptation to new medical advancements | Built-in flexibility to incorporate the latest medical technology and practices swiftly | Technologically advanced facilities remain at the cutting edge of medical science, continually improving their care capabilities |
| Staff satisfaction | Lower due to operational burdens and administrative overhead | Higher satisfaction from streamlined workflows and reduced administrative tasks, allowing more focus on patient care | Enhanced staff morale and retention in technologically advanced facilities due to better work conditions and fewer operational frustrations |

### 10.3.6 Community Integration

The evolving role of healthcare facilities as integral components of community infrastructure highlights a shift towards more inclusive and accessible healthcare models. This trend towards community integration sees healthcare facilities not only as places for treating illness but also as vibrant hubs for education, prevention, and wellness. By extending their services to include more community-oriented programs, healthcare facilities can play a crucial role in demystifying healthcare processes, making them more transparent and less intimidating to the public.

Community-focused healthcare approaches aim to break down the barriers that often exist between medical institutions and the populations they serve. By offering educational programs on health and wellness, facilities can empower individuals with knowledge about disease prevention, healthy lifestyles, and self-care practices. This proactive sharing of information fosters a well-informed community that is better equipped to manage its health and engage more confidently with healthcare systems.

Moreover, the integration of preventive care services within healthcare facilities can significantly reduce the incidence and impact of chronic diseases. Facilities that provide regular screening programs, vaccination drives, and early intervention services can address health issues before they develop into more serious conditions. These services, readily accessible within the community, ensure that preventive care is not just available but is also conveniently reached by all segments of the population.

Wellness services, including fitness classes, nutritional counseling, and mental health support groups, further extend the role of healthcare facilities beyond traditional medical care. By creating spaces within the healthcare environment that cater to holistic wellness, physical, mental, and emotional facilities can support the overall well-being of the community. This holistic approach not only improves individual health outcomes but also contributes to the vitality of the entire community.

As healthcare facilities continue to evolve into community hubs, the design, and management of these spaces must consider best to serve the diverse needs of the community effectively. This includes creating welcoming, non-clinical spaces within facilities where community members can gather, learn, and participate in wellness activities. It also involves fostering partnerships with local organizations and stakeholders to ensure that the services offered are aligned with the community's needs and priorities.

In conclusion, the future of healthcare facilities is intrinsically linked to their ability to integrate deeply with the communities they serve. By transitioning from isolated care providers to community-centered hubs, healthcare facilities can play a pivotal role in reshaping how healthcare is perceived and delivered. This shift not only promises to improve patient care and staff well-being but also enhances community health on a broader scale. Embracing this community-focused approach will undoubtedly pose challenges, but the potential to transform healthcare into a more inclusive, preventive, and integrated system offers immense opportunities. Moving forward, the commitment to these principles will set new standards for healthcare environments, promising a future that is not only promising but truly transformative.

Equation 10.8 is designed to calculate the Community Health Impact Score (*CHIS*), a comprehensive metric that assesses the overall impact of healthcare facilities on the health of the communities they serve. This score is instrumental in understanding how the presence, accessibility, and activities of a healthcare facility contribute to community health outcomes, thereby providing valuable insights for policymakers, healthcare administrators, and community planners.

The *CHIS* incorporates three primary factors: accessibility ($A$), which measures how easily community members can access the healthcare facility in terms of location, transportation options, and physical or financial barriers; range of services offered ($S$), which evaluates the diversity and comprehensiveness of medical services available that cater to various health needs of the community; and community engagement activities ($E$), which reflect the facility's efforts to reach out, educate, and involve the community in health-promoting activities and programs.

Each of these factors is assigned a coefficient ($\alpha$, $\beta$, $\gamma$), representing their relative importance based on community needs and the facility's strategic health objectives. These coefficients are derived from empirical data or stakeholder consultations, ensuring that the score accurately reflects local health priorities and the facility's capacity to address them. The Community Health Impact Score is thus calculated using the following equation:

$$\text{CHIS} = \alpha \cdot A + \beta \cdot S + \gamma \cdot E. \tag{10.8}$$

This equation not only quantifies the direct and indirect contributions of a healthcare facility to community health but also helps in identifying areas where improvements can be made. By evaluating the *CHIS* regularly, healthcare facilities can strategically adjust their services and community interactions to maximize their positive impact on community health, ultimately leading to a healthier, more engaged public.

## 10.4 Implications for Policy and Practice

The insights and trends explored throughout this book illuminate the evolving landscape of healthcare facility design and underscore the essential interplay between architectural practices, healthcare policy, and clinical outcomes. This subsection discusses how these insights should influence health policy and architectural practice, offering recommendations for Policymakers, healthcare administrators, and architects on integrating and leveraging best practices for future developments.

### 10.4.1 Policymakers

Policymakers play a pivotal role in shaping the landscape of healthcare architecture by creating environments that encourage and facilitate innovation. Through the development and implementation of thoughtful policies, they can significantly influence the adoption of advanced design practices and technologies in healthcare facilities. These policies can help ensure that healthcare environments not only meet current health standards but are also prepared to adapt to future challenges and advancements in medical care.

One effective strategy is the establishment of incentives for the adoption of innovative design practices and technologies. Tax incentives, grants, and targeted funding opportunities are powerful tools that can drive the integration of cutting-edge sustainable designs and technologies in healthcare facilities. These incentives not only make it financially viable for healthcare organizations to invest in advanced technologies and sustainable practices but also encourage them to prioritize these investments over traditional, less efficient options.

As the fields of medical technology and architectural design evolve, it is also crucial that regulations governing facility design and operations are continually updated to keep pace with innovation. Outdated regulations can stymie progress and prevent the adoption of new technologies that could enhance patient care and operational efficiency. Policymakers must ensure that these regulations are flexible and adaptive, fostering an environment where new ideas and technologies can be implemented without undue bureaucratic hurdles.

Additionally, policies should promote interdisciplinary collaboration, which is essential in creating healthcare facilities that are truly responsive to the needs of patients and staff. Supporting initiatives that bring together architects, healthcare professionals, technologists, and community representatives can lead to more holistic and effective design solutions. This could be achieved through funding collaborative research and development projects, or through regulatory frameworks that require interdisciplinary consultation during the planning and design phases of new healthcare facilities.

In conclusion, policymakers have a critical role in ensuring that healthcare facilities are not only equipped to meet the challenges of today but are also prepared for the innovations of tomorrow. By implementing supportive policies that incentivize innovation, update regulations in line with current technologies, and encourage interdisciplinary collaboration, Policymakers can significantly enhance the quality and effectiveness of healthcare environments. These actions will not only benefit patients and healthcare providers but will also contribute to the overall health and well-being of the community.

Table 10.10 outlines various policy incentives and their direct impacts on healthcare innovation and implementation. It examines the efficacy of tax breaks for sustainable design, grants for technological advancements, funding for research and development, incentives for interdisciplinary collaboration, and regulatory flexibility. Each category details the types of incentives offered, specific healthcare innovations

**Table 10.10** Policy incentives driving healthcare innovations: Examples and outcomes

| Policy incentive | Description | Healthcare innovation | Outcome and impact |
|---|---|---|---|
| Tax breaks for sustainable design | Reductions in tax liabilities for incorporating eco-friendly designs and materials | Implementation of energy-efficient systems and green building certifications like LEED | Enhanced environmental sustainability, reduced operational costs, and improved patient environments |
| Grants for technological advancement | Financial grants are provided to facilities adopting new medical technologies | Adoption of telehealth platforms and AI diagnostic tools | Expanded access to care, improved diagnostic accuracy, and patient outcomes |
| Funding for research and development | Public and private funding is allocated for healthcare R&D projects | Development of new biomedical devices and procedures | Accelerated introduction of innovative treatment options, leading to better health outcomes |
| Incentives for interdisciplinary collaboration | Support for projects involving multiple fields of expertise | Collaborative projects between architects, engineers, and medical researchers | Creation of multifunctional spaces that adapt to the changing needs of healthcare delivery |
| Regulatory flexibility | Policies allow quicker adaptation and implementation of new technologies | Faster approval and integration of cutting-edge treatments and devices in clinical settings | Shortened innovation cycles, rapid responsiveness to health crises, and continuous improvement in care quality |

they promote, and the outcomes achieved. For example, tax breaks encourage the adoption of energy-efficient systems resulting in more sustainable facilities with reduced operational costs. Grants facilitate the integration of advanced technologies like telemedicine and AI, enhancing diagnostic precision and expanding access to care. This table demonstrates how policy incentives effectively foster healthcare advancements that lead to improved patient care environments, more efficient care delivery, and rapid advancements in medical technology.

### *10.4.2 Healthcare Administrators*

Healthcare administrators are crucial in the effective implementation of advanced design and operational strategies within healthcare facilities. Their leadership and strategic vision play a pivotal role in ensuring that healthcare environments are not only efficient and state-of-the-art but also responsive to the evolving needs of patients and communities. To achieve this, administrators must prioritize flexibility in both design and operation, enabling quick adaptation to changing healthcare demands and technological advancements.

Investing in modular and adaptable infrastructure is a key strategy for maintaining this flexibility. Modular designs allow for spaces within healthcare facilities to be reconfigured based on current needs, whether accommodating an influx of patients due to a seasonal illness or adapting to long-term changes in healthcare delivery, such as the increase in telemedicine services. This adaptability extends to the integration of new patient care technologies, ensuring that facilities can quickly update or enhance technological capabilities without extensive renovations.

Training and development of staff are equally critical in maximizing the benefits of innovative facility designs and technologies. Healthcare administrators should ensure that all staff members are proficient in utilizing the full capabilities of the facility's technological assets. Regular training programs are essential, not only to familiarize staff with new systems and procedures but also to continually enhance their skills as technology evolves. This investment in staff training ensures that the human resources within the facility can work synergistically with technological innovations to provide optimal care.

Moreover, healthcare administrators should focus on strengthening community ties through thoughtful facility design. By creating spaces that serve both public and clinical functions, healthcare facilities can become integrated parts of the community fabric. Examples might include public seminar rooms where health education workshops are held, open green spaces for community use, or café areas where patients, families, and community members can interact in a non-clinical setting. These spaces help demystify healthcare settings, making them less intimidating and more accessible to the public.

In conclusion, healthcare administrators must be visionary in their approach to integrating operational and design strategies. By fostering flexible, adaptable infrastructures, ensuring comprehensive staff training, and enhancing community integration, they can transform healthcare facilities into dynamic environments that meet clinical needs and promote public health. These strategies not only improve patient care and operational efficiency but also position healthcare facilities as integral and approachable parts of the communities they serve.

Equation 10.9 addresses the return on investment (*ROI*) for modular infrastructure in healthcare settings, an essential calculation for determining the financial viability of such systems. Modular infrastructure, which includes elements like prefabricated hospital rooms and mobile health units, offers the potential for rapid deployment and flexibility in managing healthcare facility needs. These systems are particularly valuable for adapting to fluctuating patient volumes or expanding healthcare services without the need for permanent construction.

This *ROI* calculation incorporates the total cost of the modular infrastructure ($C$), which includes both the initial purchase and any installation costs; the savings in implementation time ($T$), which often translates to earlier operational capabilities and thus faster revenue generation; and the impact on patient care ($P$), which could reflect improvements in service delivery, patient throughput, and overall treatment quality.

The *ROI* for modular infrastructure is calculated by evaluating the economic benefits derived from the reduced time to deploy and the enhanced capacity for patient care, against the initial and ongoing costs. The equation helps healthcare administrators and financial planners to assess whether the benefits of modular infrastructure justify the investment, crucial for making informed decisions about upgrading or expanding healthcare facilities. The formula is as follows:

$$\text{ROI} = \frac{(T + P)}{C} \times 100. \tag{10.9}$$

This metric is instrumental in guiding strategic investments in healthcare infrastructure, ensuring that resources are allocated efficiently to improve patient care delivery and facility responsiveness. It also highlights the broader economic and operational advantages of adopting modular systems in the healthcare sector.

### *10.4.3 Architects*

Architects play a critical role in shaping the future of healthcare facilities, tasked with translating emerging medical practices and technological advancements into functional, sustainable, and healing environments. As custodians of healthcare architecture, architects should continue to champion and implement evidence-based design principles, integrating insights from research to create spaces that enhance patient care, staff efficiency, and overall health outcomes.

A key aspect of this approach involves maximizing natural light and integrating nature into healthcare settings. Studies have consistently shown that exposure to natural light and views of nature can speed up patient recovery, reduce stress, and improve overall well-being. Architects should focus on designs that incorporate large windows, atriums, and indoor gardens, not just as aesthetic features but as integral elements of the healing environment. Additionally, the layout of healthcare facilities must be meticulously planned to promote efficient workflows. This includes optimizing the placement of nursing stations, minimizing distances between frequently used spaces, and designing intuitive pathways that facilitate easy navigation for both staff and patients.

Another vital practice for architects is the development of mechanisms for ongoing feedback from those who use healthcare spaces daily—patients and staff. Engaging with these key stakeholders through surveys, interviews, and observational studies can provide invaluable insights into the effectiveness of design choices and areas needing improvement. Post-occupancy evaluations should become a standard practice, not only to assess the success of architectural innovations but also to identify opportunities for refinement and optimization in real-time. These evaluations help ensure that the designs not only meet theoretical expectations but also function effectively in real-world healthcare settings.

As leaders in the design process, architects also have a unique opportunity to advocate for and implement sustainable practices. This commitment to sustainability involves adhering to rigorous environmental standards and seeking innovative design solutions that reduce energy consumption, minimize waste, and utilize eco-friendly materials. By pushing for these innovations, architects can set new benchmarks for sustainable design in healthcare, influencing the broader architectural community and ensuring that healthcare facilities contribute positively to their environmental context.

In conclusion, by embracing evidence-based design principles, actively seeking feedback to refine their designs, and championing sustainability, architects can significantly influence the development of healthcare facilities. These practices not only enhance the functionality and efficiency of healthcare environments but also ensure that these spaces are responsive to the needs of patients, staff, and the community. Through thoughtful design and continuous innovation, architects can help create healthcare facilities that are not only effective and efficient but also nurturing and sustainable, truly serving as models for future developments in the field.

Table 10.11 presents a series of impactful case studies where direct feedback from patients and staff led to significant design and operational improvements in healthcare facilities. Each case study lists the specific feedback received, the subsequent design changes implemented, and the measurable impacts on the facility. For instance, Community Hospital A responded to complaints about overcrowded outpatient areas by expanding waiting spaces and adding private rooms, leading to enhanced privacy and reduced waiting times. Similarly, Regional Medical Center B improved emergency response efficiency by redesigning the ER layout to optimize patient flow and equipment access. This table shows how integrating feedback can directly enhance patient care, staff efficiency, and overall facility effectiveness, underlining the value of patient and staff input in shaping healthcare environments.

## 10.5 Integrating Global Health Trends

As global health security becomes a paramount concern, healthcare facilities worldwide must adapt to effectively manage and mitigate international health threats. This section explores how the integration of design features that enhance surveillance, response, and resilience can prepare healthcare facilities to handle global health emergencies more effectively, ensuring they are not only reactive but also proactive in their approaches.

The design of healthcare facilities plays a critical role in their ability to respond to health emergencies that transcend national borders, such as pandemics or widespread environmental disasters. Facilities must be equipped with the infrastructure necessary to handle a surge in patient numbers, including expandable spaces, isolation units, and flexible areas that can be quickly converted for emergency use. The strategic placement of these facilities to serve large populations and regions prone to specific health risks is also crucial for ensuring timely and effective responses.

**Table 10.11** Case studies: Enhancing healthcare facilities through patient and staff feedback

| Case study | Feedback provided | Design improvements made | Impact on facility |
|---|---|---|---|
| Community Hospital A | Patients found the outpatient area too crowded | Redesign to expand waiting areas and add more private consultation rooms | Improved patient privacy and reduced waiting times |
| Regional Medical Center B | Staff reported inefficiencies in the emergency room layout | Restructured layout to streamline patient flow and optimize equipment placement | Enhanced emergency response times and staff workflow efficiency |
| City Hospital C | Feedback highlighted the lack of natural light in patient rooms | Incorporation of larger windows and skylights in inpatient and communal areas | Increased patient satisfaction and reduced recovery times |
| Suburban Clinic D | Comments on the difficulty in navigating the facility | Improved signage and the introduction of an intuitive wayfinding system | Better patient and visitor experience with reduced confusion and stress |
| Specialty Hospital E | Staff desired more restful spaces for breaks | Creation of dedicated staff lounges with access to outdoor areas | Enhanced staff well-being and job satisfaction, reducing turnover |

Enhanced surveillance capabilities are integral to the early detection and management of health crises. This involves the architectural integration of technology that supports data collection and analysis, such as sensors to monitor environmental conditions and systems that track patient health data across networks. These technologies enable healthcare providers to detect anomalies early, potentially preventing the spread of disease within communities and across borders.

Resilience in healthcare facility design extends beyond structural durability to include systems that ensure continuity of operations during disruptions. This includes redundant power supplies, water systems, and communication networks that can operate independently in the event of a failure. Designing with resilience in mind also means facilities should have the capacity to coordinate with international health organizations and comply with global health regulations to ensure a unified response to health threats.

Moreover, the integration of telehealth and remote care technologies can greatly enhance a facility's responsiveness by allowing it to extend its reach without compromising the safety of patients and healthcare workers. Telehealth stations, designed into the structure of healthcare facilities, can facilitate remote diagnostics and care, crucial during times when traditional healthcare delivery might be disrupted.

Educational and training spaces within healthcare facilities are also essential. These spaces can be used to disseminate information about global health trends and emergency protocols not only to healthcare professionals but also to the community. Regular workshops and simulations based on current global health insights can prepare both staff and the public for potential health threats, fostering a community well-informed about the best practices in health emergencies.

Incorporating these elements into the design and operation of healthcare facilities reflects a forward-thinking approach that recognizes the interconnectedness of global health. By preparing facilities to handle, respond to, and recover from international health threats, we can build a more resilient global healthcare infrastructure. This proactive approach ensures that healthcare facilities are not just care providers but also integral components of global health security.

Equation 10.10 formulates the Health Surveillance Efficiency Model, a crucial metric for evaluating the effectiveness of health surveillance systems that are integrated into healthcare facility designs. Such systems are key in the early detection and management of infectious disease outbreaks, thereby playing a significant role in preventing their spread within and beyond the healthcare environment.

The efficiency of these health surveillance systems is measured by considering two main variables: the number of early detections ($D$), which quantifies how many potential health threats (like infectious diseases) are identified at an early stage by the system; and the effectiveness in containment of these outbreaks ($C$), which assesses how successfully these threats are managed and controlled following detection.

The model uses these variables to provide a comprehensive view of the surveillance system's operational success. By quantifying the capability of these systems to not only detect but also contain health threats, healthcare administrators can assess the direct impact of their surveillance infrastructure on overall facility safety and public health outcomes. The equation for the Health Surveillance Efficiency Model is as follows:

$$\text{Efficiency} = \frac{D \times C}{100}. \tag{10.10}$$

This efficiency score is pivotal for healthcare facilities to evaluate and enhance their surveillance practices. It assists in determining whether current systems are adequate or if further investments are needed to improve their efficacy, ultimately supporting a proactive approach to healthcare management and disease prevention.

## 10.6 The Role of Community and Patient Feedback

The integration of community and patient feedback into the design and continuous improvement of healthcare facilities is not just a beneficial practice but a necessary one for creating environments that truly meet the needs of those they serve. This subsection explores the importance of such involvement, discussing effective methods for community engagement and detailing how feedback can be systematically integrated into ongoing facility operations and upgrades.

Engaging community members and patients in the planning and design process helps ensure that healthcare facilities cater to the specific needs and preferences of the populations they serve. This participatory approach can lead to improved patient satisfaction, better health outcomes, and stronger community support for healthcare initiatives. To effectively involve the community and patients, healthcare administrators and planners can employ a variety of strategies:

### *10.6.1 Community Workshops and Public Consultations*

Engaging with the community through workshops and public consultations is a fundamental approach in the planning stages of new healthcare facilities or major renovations. These interactions offer a platform for community members to express their opinions, share their experiences, and contribute ideas, ensuring that the services provided align with local needs and expectations.

The process of conducting these workshops and consultations should be structured to be inclusive and accessible, encouraging participation from a broad cross-section of the community. This includes scheduling meetings at convenient times and locations, providing necessary accommodation for disabilities, and potentially offering virtual participation options to increase accessibility. Efforts should also be made to reach out to underrepresented groups to ensure diverse input that reflects the entire community.

During these sessions, it is crucial to communicate the scope of the project, the objectives, and how community input will be integrated into the planning process. This transparency helps build trust between healthcare providers and the community, fostering a sense of partnership and mutual respect. Facilitators should use these opportunities to gather specific feedback on various aspects of healthcare facility design and operation, such as accessibility, services offered, and patient care priorities.

The insights gained from community workshops and public consultations are invaluable. They can reveal local health concerns that might not be apparent from statistical data alone and provide innovative ideas from those who will directly benefit from the facilities. Moreover, involving the community early in the planning process helps to build a sense of ownership and pride in the local healthcare services, which can lead to greater community support and smoother implementation of projects.

In summary, community workshops and public consultations are essential for creating healthcare facilities that truly meet the needs of the people they serve. By actively involving community members in the planning process, healthcare administrators and architects can ensure that the resulting facilities are not only functional and efficient but also cherished and supported by the community they serve. This approach not only improves the relevance and quality of healthcare services but also strengthens the overall relationship between healthcare facilities and the communities they are part of.

Table 10.12 illustrates how healthcare facilities have successfully translated community feedback into tangible design and operational enhancements, ultimately

**Table 10.12** Transforming feedback into functionality: Outcomes of community engagement in healthcare design

| Community feedback example | Design/operational change implemented | Impact on facility |
|---|---|---|
| Need for a more elderly-friendly design | Introduction of wheelchair-accessible facilities and non-slip floors | Improved accessibility and safety for elderly patients, leading to higher satisfaction |
| Request for childcare services | Creation of an on-site childcare center for patients and staff | Increased convenience for staff and visitors, supporting better work-life balance and patient visitation |
| Demand for mental health services | Integration of a dedicated mental health wing with private, serene environments | Enhanced mental health services provision, meeting a critical community need |
| Concerns over emergency service accessibility | Expansion of the emergency department and improved signage for emergency access | Faster and more efficient emergency response, leading to better patient outcomes |
| Feedback on lack of community space | Development of community rooms available for public use and health workshops | Strengthened community ties and improved health education opportunities |

improving facility functionality and patient care. The table highlights diverse community feedback examples, the specific changes implemented in response, and the positive impacts these changes have had on the facilities. For example, responding to the need for more elderly-friendly designs, one facility introduced wheelchair-accessible features and non-slip floors, significantly enhancing safety and accessibility for elderly patients. Another facility addressed the demand for mental health services by integrating a dedicated mental health wing, thereby meeting a crucial community need and enhancing service provision. These case studies underscore the importance of community involvement in healthcare design, demonstrating how such engagement directly contributes to more effective and responsive healthcare environments.

### *10.6.2 Surveys and Feedback Forms*

Regular collection of feedback through surveys and forms is a critical component in the continuous improvement of healthcare facilities. By systematically distributing these instruments to patients and community members, healthcare administrators can gather valuable data on the effectiveness of facility designs, services, and overall patient satisfaction. This ongoing feedback loop enables healthcare providers to stay attuned to the needs and preferences of those they serve and adapt accordingly.

Surveys and feedback forms should be designed to be comprehensive yet concise, covering various aspects of the healthcare experience, from the physical environment and accessibility to the quality of care and staff interactions. To encourage higher response rates, these surveys should be made easily accessible and user-friendly. Options might include digital formats that can be completed on smartphones or tablets, as well as paper forms for those who prefer or require them. Facilities might also consider offering surveys in multiple languages to accommodate non-English speakers and ensure that all community members can provide feedback.

The process of analyzing feedback should be structured and systematic. Data collected should be regularly reviewed to identify trends and common issues that may not be immediately obvious. This analysis can highlight areas of success where the facility excels, as well as pinpoint areas needing improvement. For instance, consistent comments about the difficulty in navigating the facility could prompt a review of signage and wayfinding aids.

Implementing changes based on survey feedback demonstrates to patients and community members that their opinions are valued, and that the healthcare facility is committed to making tangible improvements. This can enhance patient trust and loyalty, which are crucial for the long-term success of any healthcare provider.

Moreover, surveys and feedback forms can also serve as tools to validate the impact of newly implemented designs or service upgrades. By comparing data collected before and after these changes, administrators can quantitatively assess their effectiveness. This evidence-based approach to facility management not only supports better strategic planning but also helps justify investments in certain areas, whether it is upgrading technology, redesigning patient rooms, or enhancing common areas.

In conclusion, regular surveys and feedback forms are indispensable tools for healthcare facilities aiming to provide high-quality care and environments. By leveraging the insights gained from these tools, healthcare facilities can evolve and adapt in ways that continually improve patient experiences and outcomes, thereby fulfilling their mission to serve their communities effectively.

Equation 10.11, the Feedback Impact Assessment Formula, is designed to quantitatively evaluate the effectiveness of changes implemented in healthcare settings based on patient and community feedback. This assessment is crucial for healthcare facilities committed to continuous improvement and patient-centered care. It provides measurable outcomes that reflect how well the facility responds to feedback and adapts its practices to meet the needs and expectations of those it serves.

This formula incorporates two significant variables: patient satisfaction scores ($S$), which gauge the responses from patients regarding their perceptions of care quality and overall experience post-implementation of changes; and health outcome improvements ($H$), which measure the tangible health benefits observed following the changes, such as reduced readmission rates, better recovery statistics, or improved chronic disease management.

These variables are weighted by their importance, represented by coefficients $\alpha$ and $\beta$, which are determined based on the facility's strategic objectives and the expected impact of the feedback-driven changes. The Feedback Impact Assessment Formula calculates a score that reflects the overall success of these changes in enhancing patient satisfaction and health outcomes, thereby providing a comprehensive view of the return on investment from engaging with and acting on patient and community insights. The equation is expressed as follows:

$$\text{Impact Score} = \alpha \cdot S + \beta \cdot H. \tag{10.11}$$

This formula helps healthcare providers quantify the benefits of patient and community involvement in shaping healthcare services, ensuring that the facility not only listens to but effectively acts on feedback to improve care delivery and patient well-being.

### *10.6.3 Focus Groups*

Focus groups serve as a valuable tool for gaining deeper and more nuanced insights into the healthcare needs and preferences of diverse communities. By organizing focus groups that include a range of community representatives, healthcare administrators can tap into a rich source of qualitative data that complements the broader quantitative data gathered through surveys and feedback forms.

The structured yet open-ended nature of focus group discussions allows participants to express their thoughts and feelings in a way that is often more detailed and complex than what is possible in a survey. These discussions can uncover subtle but critical insights into how different community members perceive healthcare services, the physical layout of facilities, and the overall accessibility of healthcare. For healthcare facilities planning renovations or new constructions, focus groups can review preliminary plans and provide feedback on how these plans might meet or fail to meet community expectations and needs.

To maximize the effectiveness of focus groups, it is important to ensure that the groups are diverse and representative of the community's demographic and socioeconomic makeup. This diversity helps ensure that the feedback received is comprehensive and inclusive, capturing a wide range of experiences and perspectives. For example, including members from various age groups, ethnic backgrounds, and different health needs can provide a rounded view of how healthcare services are perceived and what improvements are necessary.

During focus group sessions, participants can discuss the potential impacts of facility designs and service models on the community. They can evaluate how friendly the facilities are to elderly patients, how well they accommodate patients with disabilities, or how effectively they serve families with young children. These sessions can also delve into specific issues such as privacy concerns, the efficiency of patient flow, or the adequacy of waiting areas.

The insights gained from these discussions can be incredibly valuable, often highlighting opportunities for improvements that might not be evident through other forms of consultation. Participants might suggest small but significant changes, such as the placement of seating in hallways for patients who may need to rest while navigating a large facility, or the need for more culturally sensitive artwork and decor.

To ensure that focus groups lead to actionable outcomes, it is crucial to follow up these discussions with clear plans for how the insights will be integrated into facility planning and operations. This follow-through demonstrates to participants that their contributions are valued and taken seriously, fostering a stronger relationship between the healthcare facility and the community it serves.

In conclusion, focus groups are a crucial strategy for engaging with community members in a meaningful way, providing them with a platform to voice detailed feedback and recommendations. These groups not only enrich the decision-making process but also enhance community trust and support, crucial components for the long-term success and integration of healthcare facilities within their communities.

Table 10.13 encapsulates the actionable outcomes derived from focus group discussions within healthcare settings, highlighting how community input has directly influenced design and operational enhancements in facilities. This table outlines various focus group themes, the key insights gathered, the specific changes implemented based on these insights, and the resulting outcomes for the facilities. For instance, in response to feedback on elder accessibility, one facility improved safety and navigation by installing additional handrails and non-slip flooring, markedly boosting satisfaction rates among elderly patients. Another facility enhanced family support within the healthcare environment by creating dedicated family waiting areas and private consultation rooms, fostering a more supportive care environment. These examples demonstrate the effectiveness of incorporating focused community feedback into healthcare design, significantly enriching patient and family experiences and broadening the scope of care provided.

### *10.6.4 Patient Advisory Councils*

Establishing patient advisory councils is a strategic approach that healthcare facilities can use to foster ongoing engagement with patients and ensure that services remain patient-centered and responsive to community needs. These councils create a formal mechanism for continuous dialogue between healthcare providers and the patient population, allowing for a structured and effective exchange of insights and recommendations.

Patient advisory councils typically comprise patients, family members, and sometimes community representatives who have firsthand experience with the healthcare system. By involving these stakeholders in discussions about facility operations, policies, and service design, healthcare administrators can gain a deeper understanding of patient needs and the effectiveness of existing services. This continuous input is invaluable in guiding improvements and adaptations that enhance the quality of care and patient satisfaction.

The role of a patient advisory council extends beyond simple feedback. Members can serve as advocates for patient rights and interests, ensuring that the facility's decisions and policies consider the patient's perspective. For example, council members can provide insights into how hospital environments can be made more comfortable

**Table 10.13** Translating community voices: Key insights from focus groups in healthcare design

| Focus group theme | Key insight | Influence on design/operation | Outcome |
|---|---|---|---|
| Elder accessibility | Need for more navigable spaces for the elderly | Installation of additional handrails and non-slip flooring in corridors and bathrooms | Improved safety and accessibility for elderly patients, resulting in higher satisfaction rates |
| Family-friendly facilities | Request for family-oriented spaces | Creation of family waiting areas and private patient-family consultation rooms | Enhanced comfort for patients and families, promoting a supportive care environment |
| Cultural sensitivity | Diversity in artwork and decor was lacking | Incorporation of culturally diverse and locally sourced art in public and patient areas | Increased community connection and respect for local culture, enhancing the facility's community integration |
| Mental health considerations | Discussion on the need for quiet zones | Design of dedicated quiet spaces for mental health, which are located away from busy clinical areas | Better mental health support within the facility, leading to broader service satisfaction |
| Technology integration | Desire for more interactive health technology | Implementation of digital information kiosks and enhanced patient room tech amenities | Improved patient engagement and autonomy in health management, streamlining operations |

for long-term patients, suggest enhancements to patient communication practices, or advocate for better integration of mental health services within the hospital setting.

Furthermore, the council can help the facility stay attuned to the changing dynamics of patient care and community health needs. As healthcare evolves with advancements in medicine and shifts in public health trends, the feedback from council members can help healthcare providers anticipate and respond to new challenges. For instance, during the COVID-19 pandemic, patient advisory councils could have played a critical role in advising on the adjustments necessary to manage patient flow, visitor policies, and telehealth services.

For the councils to be effective, they must be given a clear mandate, adequate resources, and genuine support from healthcare leadership. Meetings should be held regularly, and the council's recommendations should be taken seriously, with transparent follow-up on how their input is being integrated into facility operations and decision-making processes.

In conclusion, patient advisory councils are a vital component of a modern, responsive healthcare facility. By formalizing the involvement of patients in the governance and improvement of healthcare services, facilities can ensure that they not only meet the current needs of their patients but are also well-prepared to adapt to future demands. This ongoing dialogue enhances the facility's accountability, improves service quality, and strengthens the relationship between healthcare providers and the communities they serve.

Table 10.14 showcases a collection of impactful changes initiated by patient advisory councils across various healthcare facilities, demonstrating the tangible benefits of integrating patient and family insights into healthcare management and design. It presents specific issues that were addressed at each institution, such as the need for more private rooms at City General Hospital and the improvement of wayfinding systems at Community Hospital. The table details the councils' innovative recommendations, including the adoption of digital scheduling systems and the enhancement of multilingual support, which led to significant enhancements in patient care processes and satisfaction. These case studies exemplify the vital role that patient advisory councils play in bridging the gap between healthcare providers and the communities they serve, leading to a healthcare environment that is not only more responsive to the needs of patients but also more efficient and inclusive. The successes highlighted here underscore the profound impact of active patient and family participation in healthcare decision-making and emphasize the importance of such councils in fostering continuous improvement and adaptability in healthcare settings.

**Table 10.14** Empowering change: Success stories from patient advisory councils

| Hospital name | Issue addressed | Council recommendation | Outcome and impact |
|---|---|---|---|
| City general hospital | Lack of privacy in shared rooms | The suggested redesign is to include more private rooms and partition improvements | Increased patient satisfaction and privacy, leading to higher ratings in patient care surveys |
| Regional health center | Inefficient outpatient scheduling | Recommended the implementation of a new digital scheduling system | Enhanced appointment efficiency, reduced waiting times, and improved patient flow |
| Community hospital | Poor wayfinding and signage | Advised on a comprehensive overhaul of signage throughout the facility | Improved navigation within the hospital, reducing patient and visitor frustration |
| Downtown medical clinic | Insufficient support for non-English speakers | Proposed the hiring of multilingual staff and the use of translation services | Better communication with a diverse patient base, enhancing overall care quality and accessibility |
| Suburban healthcare facility | Inadequate family support amenities | Suggested the creation of family rest areas and on-site childcare services | Improved family support during patient visits, contributing to a more comforting healthcare experience |

### 10.6.5 Digital Platforms for Engagement

In today's digital age, leveraging online tools and platforms is a crucial strategy for enhancing community engagement in healthcare. By utilizing digital platforms such as online forums, social media, and dedicated mobile apps, healthcare facilities can facilitate broader dialogues with the communities they serve. These digital channels are especially effective in connecting with younger demographics and providing access to those who may face barriers to attending in-person meetings due to physical limitations, geographical distances, or time constraints.

Online Forums: Establishing online forums can provide a space for continuous, asynchronous discussions that allow community members to engage at their convenience. These forums can be moderated by healthcare professionals and serve as a platform for patients and community members to ask questions, share experiences, and provide feedback about the healthcare services they receive. This ongoing interaction not only fosters a sense of community but also provides healthcare administrators with real-time insights into patient concerns and suggestions.

Social Media Platforms: Social media has become a powerful tool for community engagement and information dissemination. Healthcare providers can use these platforms to share updates about services, changes in policies, and health education resources. Additionally, social media allows for real-time engagement, where patients and community members can interact directly with healthcare providers through comments, direct messages, and live sessions. This level of interaction enhances transparency and builds trust between healthcare facilities and the communities they serve.

Dedicated Mobile Apps: Developing dedicated mobile apps for healthcare facilities can centralize many aspects of patient engagement. These apps can offer functionalities like appointment scheduling, access to medical records, telehealth services, and real-time notifications about health alerts or new services. Furthermore, apps can be designed to collect feedback from users, providing another channel for patients to express their opinions and satisfaction with the services they receive.

The use of digital platforms for community engagement also has the potential to reach demographics that are traditionally less engaged in healthcare settings, such as young adults or working-age populations who might prefer digital communication over traditional methods. By tailoring the digital content and functionalities to meet the needs and preferences of these groups, healthcare providers can significantly enhance their outreach and engagement efforts.

In conclusion, digital platforms offer a versatile and effective means for engaging with the community in the context of healthcare. By integrating these technologies into their engagement strategies, healthcare facilities can ensure that they are accessible and responsive to a broader spectrum of the community, thereby enhancing the overall effectiveness and responsiveness of their services. As digital technology continues to evolve, so too should the methods by which healthcare providers connect with and serve their communities, ensuring that everyone has a voice in shaping the healthcare they receive.

Table 10.15 offers a detailed analysis of the utilization and impact of various digital platforms in enhancing community involvement within healthcare settings. The table outlines the specific features of each platform, such as the asynchronous discussion capabilities of online forums and the real-time interaction facilitated by social media platforms. It provides user statistics that highlight engagement levels, such as the 25,000 monthly active users participating in online forums, and the tangible benefits these platforms have brought to healthcare delivery and patient engagement. Notable improvements include better chronic disease self-management through forum discussions, increased public health awareness via social media, and more efficient patient service experiences through mobile apps. The table underscores the effectiveness of these digital tools in fostering a more informed and engaged patient community, demonstrating their pivotal role in modern healthcare ecosystems where technology and patient interaction converge to improve health outcomes and care accessibility.

**Table 10.15** Enhancing community involvement: A comparative overview of digital platforms in healthcare

| Digital platform | Features | User statistics | Tangible improvements in healthcare |
|---|---|---|---|
| Online forums | Asynchronous discussions, and topic threads, moderated by healthcare professionals | 25,000 active monthly users, with high engagement in chronic illness discussions | Increased patient knowledge and self-management of chronic conditions |
| Social media platforms | Real-time updates, multimedia content, and direct patient engagement through comments and DMs | 50,000+ followers on major platforms, high share, and interaction rates on health tips | Improved health awareness, faster dissemination of health alerts, and increased patient recruitment for health programs |
| Dedicated mobile apps | Appointment scheduling, telehealth, access to medical records, and feedback features | Over 10,000 downloads with a 4.5-star average rating, and 30% monthly active users | Enhanced patient convenience, increased appointment adherence, and direct feedback loop leading to service improvements |
| Email newsletters | Regular updates, personalized health information, and preventive care reminders | 15,000 subscribers with a 40% open rate, and high click-through rate for preventive care content | Higher engagement in preventive care programs, and improved patient compliance with treatment protocols |
| Interactive webinars | Live sessions with healthcare experts, Q&A segments, and downloadable resources | Average of 500 attendees per session, high engagement in post-session surveys | Greater patient understanding of health issues increased trust and transparency between patients and providers |

### 10.6.6 Integrating Feedback into Operations

To effectively integrate community and patient feedback into healthcare operations, healthcare facilities must establish robust and well-defined processes that facilitate the collection, analysis, and incorporation of this feedback into everyday practices. This systemic approach ensures that the voices of patients and community members are not only heard but are also instrumental in shaping the operational dynamics of healthcare institutions.

Initially, healthcare facilities must prioritize the establishment of mechanisms that enable continuous and systematic collection of feedback. This involves deploying diverse methodologies such as digital surveys, suggestion boxes, patient forums, and direct interviews, which collectively ensure a comprehensive aggregation of perspectives. Such methodologies must be accessible and promoted to encourage active participation from a broad demographic, reflecting the true diversity of the community.

Once collected, the feedback needs to be meticulously analyzed to discern patterns, identify pressing issues, and understand the expectations and experiences of the community. This analysis should be conducted regularly to capture the evolving needs and concerns of patients and community members. Advanced analytical tools and interdisciplinary teams, including data analysts, healthcare professionals, and patient advocacy representatives, can be pivotal in interpreting this data accurately and impartially.

Integrating this feedback into healthcare operations is the next critical step. This can be achieved by linking feedback outcomes to performance metrics for both individual healthcare providers and the facility. Such metrics not only incentivize improvements in patient care but also help in setting benchmarks for service quality. Moreover, operational policies and practices should be periodically revised to reflect the insights gained from patient and community feedback, ensuring that the facility adapts to meet emerging needs and expectations.

Transparent communication about how feedback is being utilized is equally important. Healthcare facilities should regularly inform patients and the community about the changes being implemented based on their suggestions. This transparency fosters trust and reinforces the community's role in co-creating a healthcare environment that truly resonates with their needs and aspirations.

By valuing and systematically incorporating community and patient feedback, healthcare facilities transform into dynamic, patient-centered environments that not only provide medical treatment but also advance public health and well-being. This continuous loop of feedback and improvement ensures that healthcare facilities remain aligned with the community's needs, fostering an atmosphere of trust and collaboration. Ultimately, this approach propels healthcare facilities towards excellence in service delivery, making them the true pillars of the community they serve.

Equation 10.12 details the Operational Improvement Index (*OII*), which is designed to quantify the effectiveness of operational changes made in healthcare settings based on feedback. This index is crucial for evaluating how modifications in operations—stemming from internal assessments, patient feedback, or staff suggestions—impact the overall efficiency and effectiveness of healthcare delivery.

The *OII* considers several key factors: cost savings ($C$), which measures the reduction in operational expenses because of the implemented changes; improved patient outcomes ($P$), which reflect enhancements in health results, such as reduced complications or improved recovery rates; and enhanced staff satisfaction ($S$), indicating improvements in workplace conditions and morale, which can directly affect service quality and patient care.

Each factor is weighted by a coefficient that represents its relative importance to the healthcare facility's goals. These coefficients ($\alpha$, $\beta$, $\gamma$) are determined based on strategic priorities and the specific objectives of the operational changes. The Operational Improvement Index thus provides a comprehensive measure of how well the changes achieve desired outcomes in cost-efficiency, patient care, and staff welfare. The index is calculated using the following formula:

$$\mathrm{OII} = \alpha \cdot C + \beta \cdot P + \gamma \cdot S. \tag{10.12}$$

By using this index, healthcare managers can effectively gauge the return on investment from operational changes and adjust strategies accordingly to maximize benefits across financial, clinical, and human resource dimensions. This systematic approach ensures that operational improvements are not only implemented but are also continuously refined to support high-quality healthcare delivery.

## 10.7 Sustainability and Ethical Considerations

Sustainability and ethics are increasingly recognized as foundational elements in the design and operation of healthcare facilities. This subsection summarizes the ethical imperatives for sustainable design and underscores the necessity for environmentally friendly and socially responsible practices in healthcare architecture. These practices not only contribute to the well-being of the environment but also enhance the health outcomes of patients and the community at large.

The integration of sustainability into healthcare design involves more than just the adoption of green building technologies or materials; it represents a holistic approach that considers the long-term impacts of healthcare operations on the environment and society. Sustainable practices in healthcare settings can significantly reduce waste, lower energy consumption, and minimize the carbon footprint of these essential services.

### 10.7.1 Ethical Imperatives for Sustainability

The ethical imperatives for embedding sustainability within healthcare design are deeply rooted in the foundational principle of medical ethics, "do not harm." This principle, traditionally focused on direct patient care, extends its relevance to encompass the environmental and social ramifications of healthcare operations. In the broader context, sustainability in healthcare is not merely an operational or strategic choice but a moral imperative that aligns with the core values of the medical profession.

Sustainable healthcare design goes beyond energy-efficient buildings and recycled materials; it embodies a holistic approach that considers the health of the planet as integral to human health. By adopting sustainable practices, healthcare providers take a proactive stance in preventing potential harm caused by environmental degradation. This degradation can manifest in various forms, such as air and water pollution, which are known to contribute to a wide range of health issues from respiratory problems to infectious diseases. Therefore, the commitment to sustainable design is a commitment to prevent these indirect, yet profound, harms.

Furthermore, the integration of sustainable practices in healthcare infrastructure also contributes to the resilience of these facilities. Hospitals and clinics designed with sustainability in mind are often better equipped to withstand environmental challenges such as natural disasters, climate change-induced weather events, and resource scarcity. For example, the use of renewable energy sources and sustainable water management systems can ensure that facilities remain operational during power outages or water shortages. Thus, sustainable design directly contributes to the continuity of care, a critical component during adverse conditions when the community needs medical services most.

Incorporating sustainability into healthcare design also reflects a commitment to social responsibility. It sends a strong message about the healthcare sector's role in leading by example, demonstrating that health and well-being extend beyond the clinic or hospital and into the community and the environment at large. This approach encourages a shift in societal norms towards more sustainable lifestyles, promoting long-term health benefits that are shared by the entire community.

Thus, the ethical imperatives for sustainability in healthcare are multifaceted and compelling. They advocate for a comprehensive approach to healthcare that views environmental stewardship as essential to medical ethics and public health. By designing and operating sustainably, healthcare providers not only "not harm" but actively contribute to the healing and health of the planet and its inhabitants. This paradigm shift is fundamental to the future of healthcare and the well-being of generations to come.

Table 10.16 delineates a variety of sustainability initiatives and their multifaceted impacts on healthcare delivery and community well-being, underscoring the ethical dimensions of adopting eco-friendly practices within healthcare settings. This table highlights initiatives such as the implementation of energy-saving technologies like LED lighting and solar panels, which not only lower operational costs—freeing up resources for enhanced patient care—but also contribute significantly to reducing the environmental footprint of healthcare facilities. It also details waste reduction strategies that improve patient and staff safety by minimizing exposure to harmful substances, while simultaneously benefiting the local ecosystem by reducing landfill waste. Sustainable water management practices are shown to safeguard critical water supplies, essential for both hospital operations and community health, especially in times of scarcity. The table further explores the adoption of green building designs and the sourcing of local materials, emphasizing their roles in promoting a healthier recovery environment and supporting the local economy. Each entry in the table is linked to direct ethical impacts on patient care, such as improved recovery conditions, and broader community health benefits, including economic support and environmental protection, illustrating the profound influence of sustainable practices in shaping a more ethically responsible approach to healthcare.

**Table 10.16** Advancing ethical healthcare: The impact of sustainability initiatives

| Sustainability initiative | Description | Direct ethical impact on patient care | Indirect ethical impact on community well-being |
|---|---|---|---|
| Energy savings | Implementation of LED lighting and solar panels | Reduces operational costs, allowing more funds to be allocated to patient services | Reduces carbon footprint, contributing to cleaner air and a healthier community environment |
| Waste reduction programs | Enhanced recycling protocols and use of biodegradable materials | Minimizes exposure to harmful waste and chemicals for patients and staff | Reduces landfill use and environmental toxins, promoting a healthier local ecosystem |
| Sustainable water management | Systems to reduce water use and increase water recycling | Ensures availability of water resources for hygiene and care, crucial in preventing hospital-acquired infections | Preserves local water supplies, crucial for community health during drought conditions |
| Green building design | Use of non-toxic, sustainable building materials and incorporation of green spaces | Promotes a healthier, more therapeutic environment for patients' recovery | Enhances local biodiversity and provides community green spaces for mental and physical well-being |
| Local sourcing of materials | Purchasing from local, sustainable suppliers | Reduces supply chain risks and enhances the reliability of medical supplies | Supports the local economy and reduces transportation emissions, leading to improved community health outcomes |

### *10.7.2 Environmental Stewardship*

Healthcare facilities hold a distinctive position of influence and responsibility when it comes to environmental stewardship. As large consumers of energy and resources, these institutions have the potential to set significant precedents in sustainable practices. By leading through example, healthcare facilities can play a pivotal role in promoting environmental health, which is intrinsically linked to public health.

The integration of energy-efficient systems within healthcare facilities is a foundational step towards environmental stewardship. This includes the adoption of LED lighting, high-efficiency heating, ventilation, and air conditioning (HVAC) systems, and other technologies that reduce energy consumption while maintaining optimal care conditions. Energy efficiency not only lowers operational costs but also minimizes the carbon footprint of healthcare operations, contributing to the mitigation of climate change.

Further, the use of renewable energy sources such as solar, wind, and geothermal energy represents a forward-thinking approach to sustainable healthcare. By transitioning to renewables, healthcare facilities can decrease their reliance on fossil fuels, reduce greenhouse gas emissions, and demonstrate the feasibility and benefits of clean energy integration in high-demand settings. This shift not only supports global efforts towards a more sustainable energy future but also aligns with health sector goals of promoting a healthier environment as a fundamental component of public health.

Responsible waste management is another critical aspect of environmental stewardship for healthcare facilities. Proper segregation, recycling, and disposal of medical and non-medical waste are essential to prevent environmental contamination and reduce the spread of infectious diseases. Moreover, innovative approaches to waste reduction, such as reprocessing medical devices and composting organic waste, can further enhance the sustainability efforts of healthcare facilities.

Additionally, the design of hospital landscapes plays a vital role in environmental stewardship. Creating green spaces within healthcare premises can enhance local biodiversity and provide serene settings that support the recovery and well-being of patients. Studies have shown that access to natural views and green environments can speed up recovery times, reduce the need for pain medications, and improve overall mental health. These spaces also offer community members a place for relaxation and interaction, fostering a stronger connection between the healthcare facility and the community it serves.

Through these practices, healthcare facilities not only adhere to their mission of promoting health and preventing illness but also contribute actively to environmental conservation. This comprehensive approach to environmental stewardship not only enhances the resilience of healthcare institutions but also sets a benchmark for sustainability that can inspire other sectors and communities to adopt similar practices.

Table 10.17 presents an enlightening overview of renewable energy adoption across healthcare facilities worldwide, detailing the types of renewable energy

**Table 10.17** Global trends in renewable energy adoption by healthcare facilities

| Country/ region | Type of renewable energy | Adoption rate (%) | Performance outcomes |
|---|---|---|---|
| United States | Solar | 45 | Reduced energy costs by 30%, and cut $CO_2$ emissions by 25% annually |
| Germany | Wind | 60 | Increased energy self-sufficiency, and reduced operational costs by 40% |
| Japan | Geothermal | 35 | Enhanced energy stability, and lowered external energy dependence by 50% |
| Australia | Solar and wind | 50 | Cut greenhouse gas emissions by 30%, and energy costs down by 35% |
| Brazil | Hydroelectric | 25 | Reduced reliance on non-renewable energy by 40%, increased cost-efficiency |
| India | Solar | 40 | Achieving 60% of the energy requirement from solar, reduced energy expenditures significantly |
| South Africa | Solar | 30 | Improved energy availability during power shortages, operational savings of 25% |

sources utilized, their adoption rates by country or region, and the consequent performance outcomes. This table illustrates the significant strides being made in diverse geographical contexts towards sustainable healthcare operations. For instance, in the United States, 45% of healthcare facilities harness solar energy, achieving a reduction in energy costs by 30% and a notable decrease in $CO_2$ emissions. In contrast, Germany shows a higher engagement with wind energy, with a 60% adoption rate leading to a 40% reduction in operational costs and enhanced energy self-sufficiency. The table also highlights Japan's adoption of geothermal energy, which has decreased its dependence on external energy sources by 50%, showcasing the country's unique geographic advantage. Meanwhile, Australia's combined approach using solar and wind energy has notably reduced greenhouse emissions and energy costs. This comparative data not only underscores the varied renewable energy strategies employed based on regional strengths and needs but also reflects the broad, positive impacts of these initiatives on reducing healthcare facilities' operational costs and environmental footprints, thereby contributing to more sustainable and economically efficient healthcare systems globally.

### *10.7.3 Social Responsibility*

The construction and operation of healthcare facilities are intrinsically linked to social responsibility, given their profound impact on local communities. Sustainable design in healthcare goes beyond environmental considerations, embedding social equity and community well-being at the core of its principles. These facilities must be designed and operated in ways that not only meet medical needs but also promote social justice and economic inclusivity.

A cornerstone of social responsibility in healthcare infrastructure is ensuring that these facilities are accessible to everyone, irrespective of socioeconomic status. This means that healthcare services must be physically accessible to all community members, including those with disabilities, and financially accessible, ensuring that cost does not become a barrier to care. Design considerations might include the strategic location of facilities to serve underrepresented areas, the provision of transportation services for remote populations, or even mobile healthcare units that bring services directly to the people.

Moreover, the construction and operation of healthcare facilities should be seen as opportunities to contribute positively to the local economy. This can be achieved by prioritizing local labor and materials, which helps to stimulate the local economy and reduce the environmental impact associated with transporting resources over long distances. Job creation, particularly in economically disadvantaged areas, not only provides financial benefits but also boosts community morale and cohesion.

Healthcare facilities should also consider the broader socioeconomic context by integrating services that address the specific health challenges of the local community. This might involve offering specialized services that target prevalent local health issues, providing educational programs that address specific community health needs, or collaborating with local organizations to ensure a holistic approach to health and well-being.

Furthermore, the benefits of new and expanded healthcare services must be equitably distributed to ensure that all segments of the community, especially underserved populations, can access them. This involves thoughtful community engagement to understand the diverse needs and challenges faced by different groups, followed by the implementation of tailored services to meet these needs. Effective community engagement can be facilitated through regular consultations, partnerships with local leaders and organizations, and transparent communication strategies.

In sum, the social responsibility of healthcare facilities extends far beyond the provision of medical services. It encompasses a commitment to improving the overall quality of life for local communities through thoughtful, inclusive, and equitable practices in healthcare design and operation. By adopting these principles, healthcare facilities can truly fulfill their role as stewards of community health and well-being, ensuring that their impact is both positive and enduring.

Equation 10.13 provides a comprehensive formula for assessing the social impact of healthcare facility operations. It is designed to quantify the contributions a facility makes towards community service outreach, local employment opportunities, and socioeconomic enhancements. The equation considers various metrics that reflect the facility's engagement and its effects on the surrounding community.

The social impact ($SI$) is measured to ensure that healthcare facilities are not only centers for medical care but also pivotal community assets that contribute positively to the local economy and societal well-being. This assessment involves parameters such as the extent of community service outreach ($CS$), which includes healthcare programs, education, and other services provided to the community; the level of local employment (*LE*) generated by the facility, which supports economic growth and reduces unemployment; and the overall socioeconomic enhancements (*SE*) that describe improvements in quality of life and economic stability attributed to the facility's operations. These factors are weighted by their respective impacts on the community, with coefficients $\alpha$, $\beta$, and $\gamma$ representing the relative importance of each factor in the overall social impact assessment. These coefficients are adjusted based on regional social priorities and the specific objectives of the healthcare facility.

The formula is crucial for administrators and policymakers to evaluate how well a facility fulfills its role as a community resource, beyond its primary function of providing healthcare. It also helps in strategic planning and in securing funding or support from governmental and non-governmental organizations by quantitatively demonstrating the facility's community impact. The formula for calculating the social impact is as follows:

$$\mathrm{SI} = (\alpha \times \mathrm{CS}) + (\beta \times \mathrm{LE}) + (\gamma \times \mathrm{SE}). \tag{10.13}$$

This equation underscores the facility's role in fostering a thriving community by integrating healthcare services with social and economic contributions, ensuring a holistic approach to health and wellness.

### *10.7.4 Promoting Health Beyond Care*

Sustainable healthcare facilities have a profound role in promoting health that extends well beyond the confines of direct medical care. By thoughtfully designing environments that inherently contribute to well-being, these facilities can positively influence the health outcomes of individuals and the broader community. This dual focus on healing and health promotion requires a holistic approach, integrating both indoor and outdoor environments to support overall health and wellness.

Indoor environments play a crucial role in patient recovery and staff well-being. Ensuring high indoor air quality is paramount, as poor air quality has been linked to various health issues, including respiratory infections, heart disease, and worsened allergy and asthma conditions. Healthcare facilities can improve air quality by incorporating advanced ventilation systems that filter and circulate air more effectively, as well as by choosing materials and finishes that minimize or eliminate the emission of volatile organic compounds (VOCs). Furthermore, maximizing natural lighting through architectural design not only reduces the reliance on artificial lighting, which is energy-intensive but also provides significant psychological and physiological benefits. Exposure to natural light has been shown to improve mood, enhance sleep quality, and increase the efficiency of staff and the comfort of patients.

Externally, the design of healthcare facilities should include spaces that encourage physical activity and interaction with nature. Landscaped gardens, walking paths, and therapeutic gardens are examples of outdoor features that can significantly enhance mental and physical health. These spaces provide opportunities for physical exercise, which is vital for maintaining physical health, and offer serene environments that aid in mental recovery, reduce stress, and increase patient and staff satisfaction. The accessibility of these spaces to all users, including those with mobility issues, is essential to ensure that everyone can benefit from these health-promoting features.

Moreover, the integration of community spaces within healthcare facilities can foster greater community interaction and support networks. Community gardens, public plazas, and open spaces not only serve as additional areas for physical activity and relaxation but also promote social connections among community members. Social interactions are crucial for mental health and can provide emotional support that complements traditional medical treatments.

In promoting health beyond care, sustainable healthcare facilities act as community hubs that support a comprehensive approach to health. This approach recognizes that health is influenced by a variety of factors, including the physical environment, and seeks to create spaces that nurture well-being at every level. By doing so, healthcare facilities not only enhance their direct care capabilities but also contribute to the building of healthier communities. This proactive stance on health promotion through environmental design is a testament to the evolving role of healthcare facilities in modern society, where the focus is on preventing illness as much as it is on treating it.

Table 10.18 meticulously catalogues the multifaceted benefits of incorporating natural elements into healthcare environments, offering a detailed comparison of their impacts on both patient and staff well-being. The table delineates several types of natural elements, such as indoor plants, water features, natural lighting, therapeutic gardens, and open green spaces, each associated with specific health benefits. For instance, the presence of indoor plants is noted not only for improving air quality but also for their significant role in reducing stress and enhancing mood stability among patients, while similarly bolstering job satisfaction and reducing fatigue among staff. Water features like indoor fountains contribute to creating a serene ambiance that aids mental relaxation, crucial for both patients in recovery and staff in high-stress roles. The strategic use of natural lighting through skylights and large windows is highlighted for its ability to boost vitamin D levels and improve sleep quality for patients, alongside increasing alertness and mood for healthcare providers. Therapeutic gardens and open green spaces extend these benefits further by facilitating physical rehabilitation and promoting mental health recovery for patients, while also providing staff with valuable spaces for physical activity and stress relief. This comprehensive overview underscores the intrinsic value of integrating natural elements into healthcare settings, illustrating that such design choices can profoundly enhance the therapeutic and work environments, contributing to overall health and well-being for all facility users.

**Table 10.18** Enhancing health with nature: Benefits of natural elements in healthcare environments

| Natural element | Description | Health benefits for patients | Health benefits for staff |
|---|---|---|---|
| Indoor plants | Potted plants and living walls | Improves air quality, reduces stress, enhances mood stability | Reduces fatigue, improves job satisfaction, lowers stress levels |
| Water features | Indoor fountains and water walls | Creating a calming ambiance, aids in stress reduction, promotes mental relaxation | Enhances work environment, reduces burnout rates, fosters a calming work atmosphere |
| Natural lighting | Skylights and large windows | Boosts vitamin D levels, enhances mood, improves sleep quality | Increases alertness, improves mood, reduces eye strain |
| Therapeutic gardens | Outdoor gardens are accessible to patients and staff | Facilitates physical rehabilitation, promotes mental health recovery, encourages relaxation | Provides stress relief, promotes physical activity, offers a peaceful retreat |
| Open green spaces | Expanses of grassy areas and walkable paths | Encourages physical activity, reduces feelings of confinement, supports cognitive function | Encourages exercise, reduces mental stress, enhances connectivity with nature |

### *10.7.5 Accountability and Transparency*

To genuinely anchor sustainability and ethical considerations in healthcare design, robust mechanisms for accountability and transparency are essential. Healthcare facilities must not only adopt sustainable practices but also demonstrate their commitment through clear and consistent reporting on their environmental impacts and progress toward sustainability goals. This level of transparency is crucial for building trust among patients, staff, and the wider community, and it underscores the facility's dedication to ethical standards.

Regular reporting on environmental and sustainability metrics enables healthcare facilities to maintain an open dialogue with stakeholders, including patients, local communities, regulatory bodies, and environmental groups. Such reporting should be comprehensive, covering aspects like energy consumption, waste management, water usage, and greenhouse gas emissions. By making this information accessible, healthcare facilities can foster a sense of collective responsibility and encourage community involvement in their sustainability initiatives.

Involving stakeholders is not merely about sharing information but also about engaging these groups in meaningful discussions about the facility's practices and policies. This could be facilitated through regular stakeholder meetings, sustainability forums, or even digital platforms that allow for continuous feedback and interaction.

Stakeholder engagement helps to ensure that the facility's sustainability goals align with community expectations and health needs, thereby enhancing the relevance and effectiveness of their initiatives.

Furthermore, embedding sustainability into the organizational culture of healthcare facilities is critical for ensuring that these practices are not peripheral but are central to every decision and operation. This requires not only policies and procedures that prioritize sustainability but also training and education for all staff members. By educating healthcare professionals and administrative staff about the importance of sustainability, facilities can cultivate an environment where sustainable practices are the norm, not the exception.

Reinforcing these commitments through accountability and transparency ensures that healthcare facilities can truly fulfill their promise of promoting and protecting health in its broadest sense. Through thoughtful design and responsible operations, healthcare architecture not only creates environments that are effective in treating illness but also actively contributes to a sustainable and equitable future. This approach allows healthcare facilities to lead by example in the global shift towards more sustainable and ethically grounded practices, setting a standard for others to follow in the quest for health and well-being for all.

Equation 10.14 outlines the methodology for calculating the Environmental Performance Index (*EPI*) of healthcare facilities. This index is essential for assessing how effectively healthcare institutions manage their environmental footprint through waste reduction, energy efficiency, and resource conservation. The EPI serves as a critical tool in guiding facilities towards more sustainable practices, which is crucial in minimizing the impact of healthcare operations on the environment.

The *EPI* calculation incorporates several environmental metrics that are relevant to healthcare operations. These include waste reduction (WR), which measures the facility's success in minimizing waste through strategies such as recycling, composting, and proper waste segregation; energy efficiency (EE), which assesses the facility's use of energy resources and its implementation of energy-saving measures like LED lighting, energy-efficient appliances, and smart building technologies; and resource conservation (*RC*), which evaluates the efficiency in using water and other essential resources, encouraging practices like rainwater harvesting and the use of low-flow fixtures.

Each of these factors is assigned a weight based on its relative importance and impact, denoted by coefficients a, b, and c. These weights allow the *EPI* to reflect the specific environmental goals and regulatory requirements of the region where the healthcare facility operates. By quantifying these practices, the *EPI* helps facilities identify areas for improvement and align their operations with environmental sustainability goals. The formula for calculating the Environmental Performance Index is as follows:

$$\mathrm{EPI} = (a \times \mathrm{WR}) + (b \times \mathrm{EE}) + (c \times \mathrm{RC}). \tag{10.14}$$

This equation is pivotal for healthcare administrators aiming to enhance their facility's sustainability credentials. It not only aids in regulatory compliance but also promotes a culture of environmental responsibility, which is increasingly recognized as a component of quality healthcare delivery.

## Further Reading


Aceto, G., Persico, V., & Pescapé, A. (2018). The role of Information and Communication Technologies in healthcare: taxonomies, perspectives, and challenges. *Journal of Network and Computer Applications*, 107, 125–154. https://doi.org/10.1016/j.jnca.2018.02.008

Aghapour, A. H., Yazdani, M., Jolai, F., & Mojtahedi, M. (2019). Capacity planning and reconfiguration for disaster-resilient health infrastructure. *Journal of Building Engineering*, 26, 100853. https://doi.org/10.1016/j.jobe.2019.100853

Ahad, A., Tahir, M., & Yau, K. L. A. (2019). 5G-based smart healthcare network: architecture, taxonomy, challenges and future research directions. *IEEE Access*, 7, 100747–100762. https://doi.org/10.1109/ACCESS.2019.2930628

Alilyyani, B., Wong, C. A., & Cummings, G. (2018). Antecedents, mediators, and outcomes of authentic leadership in healthcare: A systematic review. *International Journal of Nursing Studies*, 83, 34–64. https://doi.org/10.1016/j.ijnurstu.2018.04.001

Al-Jaroodi, J., Mohamed, N., & Abukhousa, E. (2020). Health 4.0: on the way to realizing the healthcare of the future. *IEEE Access*, 8, 211189–211210. https://doi.org/10.1109/ACCESS.2020.3038858

Altomonte, S., Allen, J., Bluyssen, P. M., Brager, G., Heschong, L., Loder, A., ... & Wargocki, P. (2020). Ten questions concerning well-being in the built environment. *Building and Environment*, 180, 106949. https://doi.org/10.1016/j.buildenv.2020.106949

Bennett, C. L., & Rosner, D. K. (2019, May). The promise of empathy: Design, disability, and knowing the "other". In *Proceedings of the 2019 CHI Conference on Human Factors in Computing Systems* (pp. 1–13). https://doi.org/10.1145/3290605.3300528

Bird, M., McGillion, M., Chambers, E. M., Dix, J., Fajardo, C. J., Gilmour, M., ... & Carter, N. (2021). A generative co-design framework for healthcare innovation: development and application of an end-user engagement framework. *Research Involvement and Engagement*, 7, 1–12. https://doi.org/10.1186/s40900-021-00252-7

Bower, I., Tucker, R., & Enticott, P. G. (2019). Impact of built environment design on emotion measured via neurophysiological correlates and subjective indicators: A systematic review. *Journal of Environmental Psychology*, 66, 101344. https://doi.org/10.1016/j.jenvp.2019.101344

Brambilla, A., Sun, T. Z., Elshazly, W., Ghazy, A., Barach, P., Lindahl, G., & Capolongo, S. (2021). Flexibility during the COVID-19 pandemic response: Healthcare facility assessment tools for resilient evaluation. *International Journal of Environmental Research and Public Health*, 18(21), 11478. https://doi.org/10.3390/ijerph182111478

Brooks, S. K., Dunn, R., Amlôt, R., Rubin, G. J., & Greenberg, N. (2018). A systematic, thematic review of social and occupational factors associated with psychological outcomes in healthcare employees during an infectious disease outbreak. *Journal of Occupational and Environmental Medicine*, 60(3), 248–257. https://doi.org/10.1097/JOM.0000000000001235

Brown, K. E., Fohner, A. E., & Woodahl, E. L. (2023). Beyond the individual: community-centric approaches to increase diversity in biomedical research. *Clinical Pharmacology & Therapeutics*, 113(3), 509–517. https://doi.org/10.1002/cpt.2808

Brunet, J., Billaquois, C., Viellard, H., & Courari, F. (2024). Eco-friendly hospital architecture. Journal of Visceral Surgery. https://doi.org/10.1016/j.jviscsurg.2023.11.008

Cenci, M. P., Scarazzato, T., Munchen, D. D., Dartora, P. C., Veit, H. M., Bernardes, A. M., & Dias, P. R. (2022). Eco-friendly electronics—a comprehensive review. *Advanced Materials Technologies*, 7(2), 2001263. https://doi.org/10.1002/admt.202001263

Chibuike, M. C., Sara, G. S., & Adele, B. (2024). Overcoming challenges for improved patient-centric care: a scoping review of platform ecosystems in healthcare. *IEEE Access*. https://doi.org/10.1109/ACCESS.2024.3356860

Chukwu, E., & Garg, L. (2020). A systematic review of blockchain in healthcare: frameworks, prototypes, and implementations. *IEEE Access*, 8, 21196–21214. https://doi.org/10.1109/ACCESS.2020.2969881

Cummings, G. G., Tate, K., Lee, S., Wong, C. A., Paananen, T., Micaroni, S. P., & Chatterjee, G. E. (2018). Leadership styles and outcome patterns for the nursing workforce and work environment: A systematic review. *International Journal of Nursing Studies*, 85, 19–60. https://doi.org/10.1016/j.ijnurstu.2018.04.016

Daú, G., Scavarda, A., Scavarda, L. F., & Portugal, V. J. T. (2019). The healthcare sustainable supply chain 4.0: The circular economy transition conceptual framework with the corporate social responsibility mirror. *Sustainability*, 11(12), 3259. https://doi.org/10.3390/su11123259

Dutta, P. K., Singh, B., Towfeek, A. S. K., Adamopoulou, J. P., Bardavouras, A. N., Bamwerinde, W., ... & Ayiga, N. (2024). IoT Revolutionizes Humidity Measurement and Management in Smart Cities to Enhance Health and Wellness. *Mesopotamian Journal of Artificial Intelligence in Healthcare*, 2024, 110–117. https://doi.org/10.58496/MJAIH/2024/013

Duque-Uribe, V., Sarache, W., & Gutiérrez, E. V. (2019). Sustainable supply chain management practices and sustainable performance in hospitals: a systematic review and integrative framework. *Sustainability*, 11(21), 5949. https://doi.org/10.3390/su11215949

Fallah-Aliabadi, S., Ostadtaghizadeh, A., Ardalan, A., Fatemi, F., Khazai, B., & Mirjalili, M. R. (2020). Towards developing a model for the evaluation of hospital disaster resilience: a systematic review. *BMC Health Services Research*, 20, 1–11. https://doi.org/10.1186/s12913-020-4915-2

Frank, L. D., Iroz-Elardo, N., MacLeod, K. E., & Hong, A. (2019). Pathways from built environment to health: A conceptual framework linking behavior and exposure-based impacts. *Journal of Transport & Health*, 12, 319–335. https://doi.org/10.1016/j.jth.2018.11.008

Gallotti, R., Sacco, P., & De Domenico, M. (2021). Complex urban systems: challenges and integrated solutions for the sustainability and resilience of cities. *Complexity*, 2021(1), 1782354. https://doi.org/10.1155/2021/1782354

Ghazal, T. M., Hasan, M. K., Alshurideh, M. T., Alzoubi, H. M., Ahmad, M., Akbar, S. S., ... & Akour, I. A. (2021). IoT for smart cities: Machine learning approaches in smart healthcare—A review. *Future Internet*, 13(8), 218. https://doi.org/10.3390/fi13080218

Giambattista, A., Di Lucchio, L., & Zolotova, M. (2021). Shaping the hospital of the future: improve the user experience in the public healthcare sector through service design education. *Design Management Journal*, 16(1), 31–47. https://doi.org/10.1111/dmj.12067

Hafez, F. S., Sa'di, B., Safa-Gamal, M., Taufiq-Yap, Y. H., Alrifaey, M., Seyedmahmoudian, M., ... & Mekhilef, S. (2023). Energy efficiency in sustainable buildings: a systematic review with taxonomy, challenges, motivations, methodological aspects, recommendations, and pathways for future research. *Energy Strategy Reviews*, 45, 101013. https://doi.org/10.1016/j.esr.2022.101013

Habibi Rad, M., Mojtahedi, M., & Ostwald, M. J. (2021). The integration of lean and resilience paradigms: A systematic review identifying current and future research directions. *Sustainability*, 13(16), 8893. https://doi.org/10.3390/su13168893

Humphreys, H., Vos, M., Presterl, E., & Hell, M. (2023). Greater attention to flexible hospital designs and ventilated clinical facilities are a pre-requisite for coping with the next airborne pandemic. *Clinical Microbiology and Infection*, 29(10), 1229–1231.

Intrevado, P., Verter, V., & Tremblay, L. (2019). Patient-centric design of long-term care networks. *Health Care Management Science*, 22, 376–390. https://doi.org/10.1007/s10729-018-9445-7

Iflaifel, M., Lim, R. H., Ryan, K., & Crowley, C. (2020). Resilient health care: a systematic review of conceptualisations, study methods and factors that develop resilience. *BMC Health Services Research*, 20, 1–21. https://doi.org/10.1186/s12913-020-05208-3

Jaušovec, M., & Gabrovec, B. (2023). Architectural Evaluation of Healthcare Facilities: A Comprehensive Review and Implications for Building Design. *Buildings*, 13(12), 2926. https://doi.org/10.3390/buildings13122926

Javed, A. R., Shahzad, F., ur Rehman, S., Zikria, Y. B., Razzak, I., Jalil, Z., & Xu, G. (2022). Future smart cities: Requirements, emerging technologies, applications, challenges, and future aspects. *Cities*, 129, 103794. https://doi.org/10.1016/j.cities.2022.103794

Johnson, J., Hall, L. H., Berzins, K., Baker, J., Melling, K., & Thompson, C. (2018). Mental healthcare staff well-being and burnout: A narrative review of trends, causes, implications, and recommendations for future interventions. *International Journal of Mental Health Nursing*, 27(1), 20–32. https://doi.org/10.1111/inm.12416

Karahoca, A., Karahoca, D., & Aksöz, M. (2018). Examining intention to adopt to internet of things in healthcare technology products. *Kybernetes*, 47(4), 742–770. https://doi.org/10.1108/K-02-2017-0045

Kaur, M., Singh, D., Kumar, V., Gupta, B. B., & Abd El-Latif, A. A. (2021). Secure and energy efficient-based E-health care framework for green internet of things. *IEEE Transactions on Green Communications and Networking*, 5(3), 1223–1231. https://doi.org/10.1109/TGCN.2021.3081616

Kim, S., Baek, S., Sluyter, R., Konstantinov, K., Kim, J. H., Kim, S., & Kim, Y. H. (2023). Wearable and implantable bioelectronics as eco-friendly and patient-friendly integrated nanoarchitectonics for next-generation smart healthcare technology. *EcoMat*, 5(8), e12356. https://doi.org/10.1002/eom2.12356

Kumar, A., Krishnamurthi, R., Nayyar, A., Sharma, K., Grover, V., & Hossain, E. (2020). A novel smart healthcare design, simulation, and implementation using healthcare 4.0 processes. *IEEE Access*, 8, 118433–118471. https://doi.org/10.1109/ACCESS.2020.3004790

Lake, E. T., Sanders, J., Duan, R., Riman, K. A., Schoenauer, K. M., & Chen, Y. (2019). A meta-analysis of the associations between the nurse work environment in hospitals and 4 sets of outcomes. *Medical Care*, 57(5), 353–361. https://doi.org/10.1097/MLR.0000000000001109

Luke, J., Franklin, R., Aitken, P., & Dyson, J. (2021). Safer hospital infrastructure assessments for socio-natural disaster–a scoping review. *Prehospital and Disaster Medicine*, 36(5), 627–635. https://doi.org/10.1017/S1049023X21000650

Lyng, H. B., Macrae, C., Guise, V., Haraldseid-Driftland, C., Fagerdal, B., Schibevaag, L., & Wiig, S. (2022). Capacities for resilience in healthcare; a qualitative study across different healthcare contexts. *BMC Health Services Research*, 22(1), 474. https://doi.org/10.1186/s12913-022-07887-6

Majid, U., & Wasim, A. (2020). Patient-centric culture and implications for patient engagement during the COVID-19 pandemic. *Patient Experience Journal*, 7(3), 5–16. http://bit.ly/ExperienceFramework

Maleki Minbashrazgah, M., & Shabani, A. (2019). Eco-capability role in healthcare facility's performance: Natural-resource-based view and dynamic capabilities paradigm. *Management of Environmental Quality: An International Journal*, 30(1), 137–156. https://doi.org/10.1108/MEQ-07-2017-0073

Marques, G., Pitarma, R., M. Garcia, N., & Pombo, N. (2019). Internet of things architectures, technologies, applications, challenges, and future directions for enhanced living environments and healthcare systems: a review. *Electronics*, 8(10), 1081. https://doi.org/10.3390/electronics8101081

Mathew, P., Chandy, S. J., Sivaraman, S., Ranjalkar, J., Ali, H. M., & Thomas, S. A. (2024). Formulating a community-centric indicator framework to quantify one health drivers of antibiotic resistance: a preliminary step towards fostering 'antibiotic-smart communities'. *Antibiotics*, 13(1), 63. https://doi.org/10.3390/antibiotics13010063

Mbunge, E., Muchemwa, B., & Batani, J. (2021). Sensors and healthcare 5.0: transformative shift in virtual care through emerging digital health technologies. *Global Health Journal*, 5(4), 169–177. https://doi.org/10.1016/j.glohj.2021.11.008

Moldovan, F., Moldovan, L., & Bataga, T. (2023). The environmental Sustainability assessment of an Orthopedics emergency hospital supported by a new innovative framework. *Sustainability*, 15(18), 13402. https://doi.org/10.3390/su151813402

Morley, J., Machado, C. C., Burr, C., Cowls, J., Joshi, I., Taddeo, M., & Floridi, L. (2020). The ethics of AI in health care: a mapping review. *Social Science & Medicine*, 260, 113172. https://doi.org/10.1016/j.socscimed.2020.113172

Nuzzo, J. B., Meyer, D., Snyder, M., Ravi, S. J., Lapascu, A., Souleles, J., ... & Bishai, D. (2019). What makes health systems resilient against infectious disease outbreaks and natural hazards? Results from a scoping review. *BMC Public Health*, 19, 1–9. https://doi.org/10.1186/s12889-019-7707-z

Onasanya, A., & Elshakankiri, M. (2021). Smart integrated IoT healthcare system for cancer care. *Wireless Networks*, 27(6), 4297–4312. https://doi.org/10.1007/s11276-018-01932-1

Pace, P., Aloi, G., Gravina, R., Caliciuri, G., Fortino, G., & Liotta, A. (2018). An edge-based architecture to support efficient applications for healthcare industry 4.0. *IEEE Transactions on Industrial Informatics*, 15(1), 481–489. https://doi.org/10.1109/TII.2018.2843169

Patrício, L., Sangiorgi, D., Mahr, D., Čaić, M., Kalantari, S., & Sundar, S. (2020). Leveraging service design for healthcare transformation: Toward people-centered, integrated, and technology-enabled healthcare systems. *Journal of Service Management*, 31(5), 889–909. https://doi.org/10.1109/JOSM-11-2019-0332

Prada, M., Prada, I. F., Cristea, M., Popescu, D. E., Bungău, C., Aleya, L., & Bungău, C. C. (2020). New solutions to reduce greenhouse gas emissions through energy efficiency of buildings of special importance–Hospitals. *Science of the Total Environment*, 718, 137446. https://doi.org/10.1016/j.scitotenv.2020.137446

Prakash, G., & Srivastava, S. (2020). Exploring value-dense environment in the healthcare service delivery: a patient-centric perspective. *The TQM Journal*, 32(2), 331–347. https://doi.org/10.1108/TQM-04-2019-0093

Qadri, Y. A., Nauman, A., Zikria, Y. B., Vasilakos, A. V., & Kim, S. W. (2020). The future of healthcare internet of things: a survey of emerging technologies. *IEEE Communications Surveys & Tutorials*, 22(2), 1121–1167. https://doi.org/10.1109/COMST.2020.2973314

Ranjan, R., & Ch, B. (2024). A comprehensive roadmap for transforming healthcare from hospital-centric to patient-centric through healthcare internet of things (IoT). *Engineered Science*, 30, 1175. https://doi.org/10.30919/es1175

Ranjbari, M., Esfandabadi, Z. S., Zanetti, M. C., Scagnelli, S. D., Siebers, P. O., Aghbashlo, M., ... & Tabatabaei, M. (2021). Three pillars of sustainability in the wake of COVID-19: A systematic review and future research agenda for sustainable development. *Journal of Cleaner Production*, 297, 126660. https://doi.org/10.1016/j.jclepro.2021.126660

Roy, S., Prasanna Venkatesan, S., & Goh, M. (2021). Healthcare services: A systematic review of patient-centric logistics issues using simulation. *Journal of the Operational Research Society*, 72(10), 2342–2364. https://doi.org/10.1080/01605682.2020.1790306

Schuh, C., de Bruin, J. S., & Seeling, W. (2018). Clinical decision support systems at the Vienna General Hospital using Arden Syntax: Design, implementation, and integration. *Artificial Intelligence in Medicine*, 92, 24–33. https://doi.org/10.1016/j.artmed.2015.11.002

Sepe, M. (2021). Covid-19 pandemic and public spaces: improving quality and flexibility for healthier places. *Urban Design International*, 26(2), 159. https://doi.org/10.1057/s41289-021-00153-x

Setola, N., Naldi, E., Arnetoli, M. V., Marzi, L., & Bologna, R. (2022). Hospital responses to COVID-19: evidence from case studies to support future healthcare design research. *Facilities*, 40(1/2), 131–145. https://doi.org/10.1108/F-03-2021-0023

Shanafelt, T., Trockel, M., Ripp, J., Murphy, M. L., Sandborg, C., & Bohman, B. (2019). Building a program on well-being: key design considerations to meet the unique needs of each organization. *Academic Medicine*, 94(2), 156–161. https://doi.org/10.1097/ACM.0000000000002415

Stergiopoulos, S., Michaels, D. L., Kunz, B. L., & Getz, K. A. (2020). Measuring the impact of patient engagement and patient centricity in clinical research and development. *Therapeutic Innovation & Regulatory Science*, 54, 103–116. https://doi.org/10.1007/s43441-019-00034-0

Søvold, L. E., Naslund, J. A., Kousoulis, A. A., Saxena, S., Qoronfleh, M. W., Grobler, C., & Münter, L. (2021). Prioritizing the mental health and well-being of healthcare workers: an urgent global public health priority. *Frontiers in Public Health*, 9, 679397. https://doi.org/10.3389/fpubh.2021.679397

Tariverdi, M., Fotouhi, H., Moryadee, S., & Miller-Hooks, E. (2019). Health care system disaster-resilience optimization given its reliance on interdependent critical lifelines. *Journal of Infrastructure Systems*, 25(1), 04018044. https://doi.org/10.1061/(ASCE)IS.1943-555X.0000465

Taylor, R. R., O'Dell, B., & Murphy, J. W. (2023). Human-centric AI: philosophical and community-centric considerations. *AI & SOCIETY*, 1–8. https://doi.org/10.1007/s00146-023-01694-1

Tian, S., Yang, W., Le Grange, J. M., Wang, P., Huang, W., & Ye, Z. (2019). Smart healthcare: making medical care more intelligent. *Global Health Journal*, 3(3), 62-65. https://doi.org/10.1016/j.glohj.2019.07.001

Tortorella, G. L., Fogliatto, F. S., Mac Cawley Vergara, A., Vassolo, R., & Sawhney, R. (2020). Healthcare 4.0: trends, challenges and research directions. *Production Planning & Control*, 31(15), 1245–1260. https://doi.org/10.1080/09537287.2019.1702226

Upadhyay, A., Mukhuty, S., Kumar, V., & Kazancoglu, Y. (2021). Blockchain technology and the circular economy: Implications for sustainability and social responsibility. *Journal of Cleaner Production*, 293, 126130. https://doi.org/10.1016/j.jclepro.2021.126130

Vetterli, C., & Scherrer, C. (2019). Service innovation by patient-centric innovation processes. In *Service Design and Service Thinking in Healthcare and Hospital Management: Theory, Concepts, Practice* (pp. 55–67). https://doi.org/10.1007/978-3-030-00749-2_4

Voulvoulis, N., Giakoumis, T., Hunt, C., Kioupi, V., Petrou, N., Souliotis, I., & Vaghela, C. J. G. E. C. (2022). Systems thinking as a paradigm shift for sustainability transformation. *Global Environmental Change*, 75, 102544. https://doi.org/10.1016/j.gloenvcha.2022.102544

Wei, H., Sewell, K. A., Woody, G., & Rose, M. A. (2018). The state of the science of nurse work environments in the United States: A systematic review. *International Journal of Nursing Sciences*, 5(3), 287–300. https://doi.org/10.1016/j.ijnss.2018.04.010

Winter, A., Stäubert, S., Ammon, D., Aiche, S., Beyan, O., Bischoff, V., ... & Löffler, M. (2018). Smart medical information technology for healthcare (SMITH). *Methods of Information in Medicine*, 57(S 01), e92–e105. https://doi.org/10.3414/ME18-02-0004

Zavratnik, V., Podjed, D., Trilar, J., Hlebec, N., Kos, A., & Stojmenova Duh, E. (2020). Sustainable and community-centred development of smart cities and villages. *Sustainability*, 12(10), 3961. https://doi.org/10.3390/su12103961

MIX
Papier aus verantwortungsvollen Quellen
Paper from responsible sources
FSC® C105338

If you have any concerns about our products,
you can contact us on
**ProductSafety@springernature.com**

In case Publisher is established outside the EU,
the EU authorized representative is:
**Springer Nature Customer Service Center GmbH**
**Europaplatz 3, 69115 Heidelberg, Germany**

Printed by Libri Plureos GmbH
in Hamburg, Germany